The Correspondence of John Tyndall

VOLUME 9

THE CORRESPONDENCE OF JOHN TYNDALL

VOLUME 9

The Correspondence, February 1865–December 1866

Edited by
Michael D. Barton, Iwan Rhys Morus,
and James C. Ungureanu

General Editors
Roland Jackson, Bernard Lightman, and Michael S. Reidy

Associate Editors
Michael D. Barton, Erin Grosjean, and Michael Laurentius

UNIVERSITY OF PITTSBURGH PRESS

Published by the University of Pittsburgh Press, Pittsburgh, Pa., 15260

Manufactured in the United States of America
Printed on acid-free paper
10 9 8 7 6 5 4 3 2 1

Cataloging-in-Publication data is available from the Library of Congress

ISBN 13: 978-0-82290-4608-3
ISBN 10: 0-82290-4608-4

CONTENTS

ACKNOWLEDGMENTS

This volume of the Tyndall correspondence has been a long time in the making, and its completion delayed by the onset of the pandemic. As volume editors we would like first of all to thank the general editors of the correspondence, Roland Jackson, Bernie Lightman, and Michael Reidy, for their help and forbearance throughout the process—particular thanks are due to Bernie for the unfailing courtesy of his requests that we get on with it! We are also grateful to James Elwick, Erin Grosjean, and Michael Laurentius for help along the way with the editorial process. Before other duties pulled him away, Geoff Belknap made vital contributions to the early preparations for the volume. For help in the archives, we would like to thank Jennie De Protani at the Athenaeum; Zoe Stanstell, manuscript reference specialist at the British Museum; Katrina Dean, keeper of manuscripts at Cambridge University Library; and Jane Harrison and Charlotte New at the Royal Institution of Great Britain. We are immensely grateful for the numerous people who have been willing to lend us their expertise and respond to our queries on specific issues. In particular, we would like to thank Melinda Baldwin, Ruth Barton, Richard Bellon, Ken Corbett, Skye Doney, Diarmid Finnegan, Ian Hesketh, Rebekah Higgitt, David Hutchings, Roland Jackson, Frank James, Max Meulendijks, Elizabeth Neswald, Simon Schaffer, and Efram Sera-Shriar. We would also like to thank the wider community of #twitterstorians for enthusiastic assistance in the reading of difficult words.

LIST OF ABBREVIATIONS

ADB	*Allgemeine Deutsche Biographie*
Alum. Cantab.	J. A. Venn, *Alumni Cantabrigienses, 1752–1900*, 6 vols. (Cambridge: Cambridge University Press, 1940–54)
ANB	*American National Biography*
Ascent of John Tyndall	Roland Jackson, *The Ascent of John Tyndall: Victorian Scientist, Mountaineer, and Public Intellectual* (Oxford: University of Oxford Press, 2018)
BAAS	British Association for the Advancement of Science
BL	British Library
Brit. Assoc. Rep.	*Reports of the British Association for the Advancement of Science*
Brock & MacLeod, *Hirst Journals*	William H. Brock and Roy M. MacLeod, eds., *Natural Knowledge in Social Context: The Journals of Thomas Archer Hirst FRS* (London: Mansell microfiche, 1980)
CDSB	*Complete Dictionary of Scientific Biography*, 27 vols. (Detroit: Charles Scribner's Sons, 2008)
CUL SC	Cambridge University Library, Stokes Correspondence
Faraday Correspondence	F.A.J.L. James, ed., *The Correspondence of Michael Faraday*, vol. 6, 1861–1867 (London: Institution of Engineering and Technology, 2011)
FRS	Fellow of the Royal Society
Glaciers of the Alps	John Tyndall, *The Glaciers of the Alps* (London: John Murray, 1860)
Heat (1863)	John Tyndall, *Heat Considered as a Mode of Motion: Being a Course of Twelve Lectures Delivered at the Royal Institution of Great Britain in the Season of 1862* (London: Longman, Green, Longman, Roberts, & Green, 1863)
Heat (1865)	John Tyndall, *Heat as a Mode of Motion*, 2nd ed. (London: Longman, Green, Longman, Roberts, & Green, 1865)
HLS	*Historisches Lexikon der Schweiz*
Hours of Exercise in the Alps	John Tyndall, *Hours of Exercise in the Alps*, 1st ed. (London: Longmans, Green, and Co., 1871)

IC HP	Imperial College of Science, Technology and Medicine, College Archives, Huxley Papers
Journal	John Tyndall's journal, 1855–72, typescript (RI MS JT/2/13c)
LT	Louisa Tyndall, John Tyndall's widow
Mountaineering in 1861	John Tyndall, *Mountaineering in 1861. A Vacation Tour* (London: Longman, Green, Longman, and Roberts, 1862)
NDB	*Neue Deutsche Biographie*
ODNB	*Oxford Dictionary of National Biography*
OED	*Oxford English Dictionary*
On Radiation	John Tyndall, *On Radiation. The "Rede" Lecture Delivered in the Senate-House before the University of Cambridge on Tuesday, May 16, 1865* (London: Longman, Green, Longman, Roberts & Green, 1865)
Phil. Mag.	*Philosophical Magazine*
Phil. Trans.	*Philosophical Transactions of the Royal Society of London*
PMG	*Pall Mall Gazette*
Poggendorff's Annal.	*Poggendorff's Annalen der Physik und Chemie*
Remembrances	William Frederick Pollock, *Personal Remembrances of Sir Frederick Pollock: Second Baronet, Sometime Queen's Remembrancer*, 2 vols. (London: MacMillan and Co., 1887)
RI	Royal Institution of Great Britain
RI MS JT	Royal Institution, London, John Tyndall Papers
Roy. Inst. Proc.	*Proceedings of the Royal Institution of Great Britain*
Roy. Soc. Proc.	*Proceedings of the Royal Society of London*
RGO MS.RGO 6	Royal Greenwich Observatory Archives, Papers of George Airy
RS	Royal Society of London
RS HS	Royal Society, Herschel Papers
Sound (1867)	John Tyndall, *Sound: A Course of Eight Lectures Delivered at the Royal Institution of Great Britain* (London: Longmans, Green, and Co., 1867)
Tyndall Correspondence, vol. 6	Michael D. Barton, Janet Browne, Ken Corbett, and Norman McMillan, eds., *The Correspondence of John Tyndall, Volume 6: The Correspondence, November 1856–February 1859* (Pittsburgh: University of Pittsburgh Press, 2019)
Tyndall Correspondence, vol. 8	Piers J. Hale, Elizabeth Neswald, Nathan N. Kapoor, and Michael D. Barton, eds., *The Correspondence of John Tyndall, Volume 8: The Correspondence, June 1862–January 1865* (Pittsburgh: University of Pittsburgh Press, 2020)

Tyndall Correspondence, vol. 10	Michael D. Barton, Ken Corbett, and Roland Jackson, eds., *The Correspondence of John Tyndall, Volume 10: The Correspondence, January 1867–December 1868* (Pittsburgh: University of Pittsburgh Press, 2021)
X Club	Ruth Barton, *The X Club: Power and Authority in Victorian Science* (Chicago: University of Chicago Press, 2018)

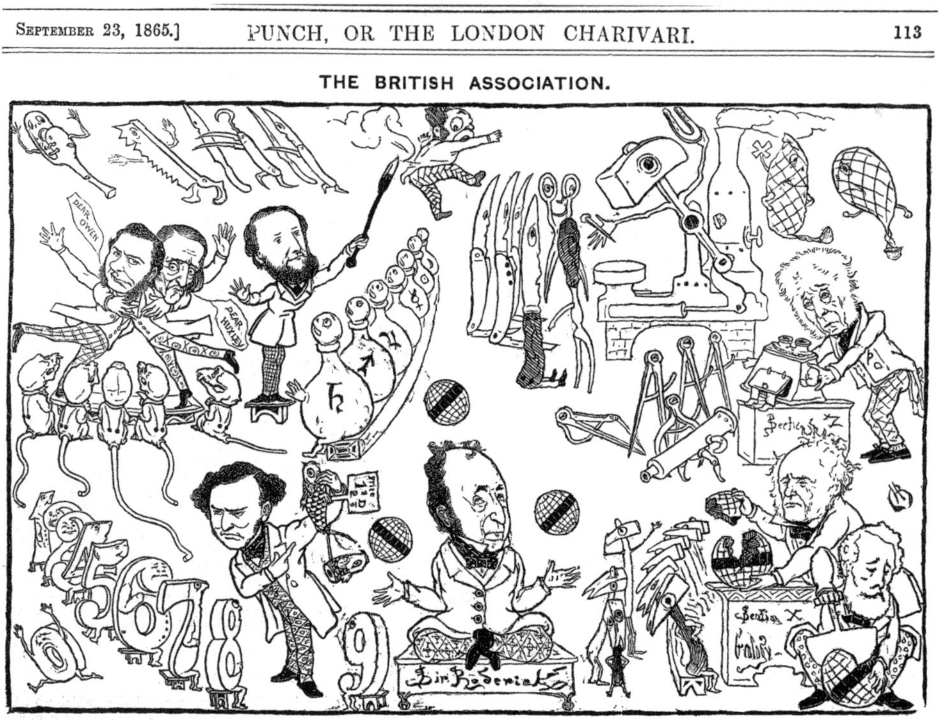

John Tyndall is included in this caricature in the 23 September 1865 issue of the satirical magazine *Punch*, which depicts members of the of the British Association for the Advancement of Science at its annual meeting in Birmingham held earlier that month. Demonstrating their scientific practice in humorous fashion, the caricature includes Thomas Huxley, Richard Owen, Charles Babbage, Charles Lyell, George Scrope, David Brewster, and Roderick Impey Murchison. Tyndall is shown pricking the geologist Joseph Beete Jukes with a hot poker, a reference to their reported dispute over coal measures. Juliet Pollock referred to this caricature in a letter to Tyndall on 24 September 1865, asking, "Who is the man with the poker in Punch?" (letter 2307). This was probably one of Tyndall's earliest appearances in the pages of *Punch*. Credit: Paul D. Stewart/Science Photo Library.

INTRODUCTION TO VOLUME 9

This ninth volume of the Tyndall correspondence contains 314 letters and covers the period from February 1865 to November 1866. Tyndall was by this time in his mid-forties (impossible to be more specific given the uncertainties surrounding the actual year of his birth) and in the prime of life. His career as a man of science was firmly established and flourishing. He had been professor of natural philosophy at the Royal Institution of Great Britain (RI) for more than a dozen years. He had established himself as Michael Faraday's successor and during the intervening years had taken on an increasing number of Faraday's wide range of duties. This volume also covers the period of Faraday's increasing illness and withdrawal from public life, which had a significant impact on Tyndall both personally and in terms of his standing in the scientific world. Since Tyndall's appointment at the RI in 1853 Faraday had been a mentor and a close personal friend. Despite their differences in age, politics, religion, and science, the two men clearly held each other in great affection.[1] Correspondence between them is few in number in this volume, but letters from Tyndall to Faraday's niece Jane Barnard make the affection clear.[2]

The letters in this volume demonstrate the increasing scope of Tyndall's activities. There are exchanges with Henry Bence Jones discussing RI affairs, for example.[3] There are letters between Tyndall and other European men of science such as Hermann Helmholtz and Rudolf Clausius.[4] He discusses translations of his books with publishers in France.[5] Correspondence with Jane and Thomas Carlyle, and Juliet and William Frederick Pollock, provide glimpses of the high cultural and literary circles in which the boy from rural Ireland now moved.[6] Several of Tyndall's letters over the summer of 1866 describe the year's Alpine mountaineering excursion that had by this time become an annual pilgrimage.[7] In that same year William Ginty, one of his closest friends from his days as a surveyor, died suddenly, and Tyndall found himself dealing with the financial affairs of his friend's widow.[8] There are hints in the correspondence of controversies to come, as Tyndall speculates about the nature of prayer and miracles.[9] A darker side to Tyndall is also revealed, as he joined in the blatantly racist defense of Edward Eyre, the disgraced governor of Jamaica, following the murderous suppression of an uprising on the island and the execution of a prominent local politician.[10]

Gentleman of Science

Two brief letters near the beginning of this volume offer a preliminary insight into just where Tyndall stood in the scientific world by this point. Written on 7 and 8 February 1865, respectively, they are addressed to English polymath John Herschel and mathematician and astronomer George Biddell Airy.[11] Their content is relatively trivial, but they reveal that Tyndall was moving in some fairly rarefied scientific circles. Throughout this volume we see Tyndall in correspondence with leading figures in the previous generation of men of science—those whom Jack Morrell and Arnold Thackray characterized as the "gentlemen of science."[12] This is testimony to the extent to which he was by this period in his life recognized by the metropolitan philosophical elite as being one of them. He was also corresponding, of course, with men such as Joseph Dalton Hooker and Thomas H. Huxley, who were close friends as well as being closer to Tyndall both in age and outlook. Revealing as the letters are concerning Tyndall's own development as a public man of science, they are also fascinating for the insight they offer into the annual routines and rhythms of leading metropolitan men of science. They tell us not just what Tyndall did but what the scientific lives of men like him looked like as well.

The main focus of Tyndall's public life in science was, of course, the RI, where he had been a professor and Faraday's successor since 1853. Correspondence with Henry Bence Jones, the RI's secretary, is particularly revealing concerning Tyndall's involvement in the day-to-day running of the place.[13] As letters show, for example, Bence Jones was seriously ill in the early autumn of 1866. He clearly feared that he was dying, and that even if he survived he would no longer be able to act as the RI's secretary.[14] Bence Jones was keen to consult with Tyndall about possible successors and clear that his successor should be someone with whom Tyndall could get along. Faraday was by now entirely absent from the RI's governance, and the role that he had played was now, both formally and informally, carried out by Tyndall. He was also taking on duties that had previously been carried out by Faraday outside the RI. Tyndall had in effect replaced Faraday as adviser to the Elder Brethren of Trinity House (the Commissioners of Lighthouses). Even though Faraday retained the role, in name at least, until his death, Tyndall was already doing the work.[15]

Correspondence throughout the volume underlines the extent to which Tyndall was by this time an important node in national and international networks. People wrote to him to ask for scientific information of various kinds, or to solicit his views on matters of controversy. Scottish geologist

Charles Lyell, for example, wrote to him asking his opinion of how much of the Earth's heat might be radiated from the poles under different conditions.[16] The economist William Stanley Jevons solicited his views on the fate of the British economy once the coal fields were exhausted.[17] Belgian natural philosopher Joseph Plateau sent him copies of his latest publication, and worried when Tyndall told him they had never arrived, asking Tyndall to check that copies sent to other correspondents had arrived safely.[18] Tyndall's most frequent international correspondents for the period covered by this volume seem to have been the German physicists Rudolf Clausius and Hermann Helmholtz. That his main international contacts were German is unsurprising in view of Tyndall's own history—and his support for Julius Robert Mayer's claims to be the discoverer of the conservation of energy against those of Englishman James Prescott Joule. A throwaway remark in a letter to Clausius about William Thomson's knighthood in the aftermath of the transatlantic telegraph cable project is a telling instance of Tyndall's own animosity toward the North British faction in energy physics.[19]

The annual meetings of the British Association for the Advancement of Science (BAAS) were key events in the annual routines of British men of science, and it is clear that they played an important role in Tyndall's scientific life as well. Tyndall attended both the 1865 meeting in Birmingham and the 1866 meeting in Nottingham. While he had no official role at the Birmingham meeting, he was vice-president of Section A at Nottingham. The annual meetings were also occasions for catching up with old scientific acquaintances and the latest scientific gossip. The Birmingham meeting was relatively uncontentious, but the Nottingham meeting was notable for Welsh man of science William Robert Grove's defense of Darwin's theory of natural selection, delivered in his presidential address.[20] Grove's intervention had been carefully prepared. There is nothing in Tyndall's extant correspondence to indicate that he had been privy to the behind-the-scenes planning that had preceded the address, but his close friend Joseph Dalton Hooker certainly had been. It is tempting to speculate to what extent Tyndall, a supporter of Darwin, was aware beforehand of the explosive contents of Grove's address. It does seem from one reference in a letter to his friend Thomas Hirst that Tyndall was staying with Edward Strutt, the prominent Liberal politician and lord lieutenant of Nottinghamshire, who proposed the vote of thanks to Grove at the end of his address.[21]

That these were years of success in Tyndall's career as a man of science is clear from the honors coming his way. Late in 1865, following the death in September of Robert Walker, the Oxford professor of natural philosophy, he was approached as a possible successor.[22] It was a significant honor, and a measure of the reputation that he had built up over the past decade or so.

Tyndall was deeply tempted, but eventually decided to remain at the RI. As he wrote later to Rudolf Clausius, "[W]hen I thought of my friends in London I could not bring myself to desert them."[23] Already that year, Tyndall had been invited to deliver the annual Rede Lecture at Cambridge. He delivered his lecture, "On Radiation," to a packed Senate House on 13 May.[24] He was back in Cambridge and the Senate House at the end of the month to be awarded the degree of LLD by the university.[25] A year or so later, Tyndall was awarded the same honor by Edinburgh, along with his friends Thomas Huxley and Andrew Ramsay.[26] Tyndall was in Edinburgh in any case. Thomas Carlyle had been installed as rector a few days earlier and Tyndall had escorted the old man there for the occasion.[27]

Private Life

That journey with Carlyle to Edinburgh underscores the degree to which Tyndall's public life in science and his private life intersected. Tyndall had been an admirer of Carlyle's writing since he was young, and he was by the 1860s a close friend of Thomas Carlyle and his wife, Jane. Tyndall sometimes even described himself as Carlyle's "lieutenant."[28] Letters in this volume indicate the lengths to which he went to secure the great man's comfort. The journey to Edinburgh had been made primarily to support his friend as much as to receive his own academic honor, and he wrote to Jane Carlyle immediately after arriving there to assure her that they had survived the journey.[29] "A perfect triumph," he told her in a telegram following her husband's installation, and followed it up with a letter assuring her that the "audience was breathlessly attentive, except at intervals when it burst into irrepressive applause."[30] He was deeply upset a few weeks later to receive news that Jane Carlyle had died suddenly, though his correspondent told Tyndall that his "telegram about the Inaugural Lecture did her so much good."[31] In December he took Thomas Carlyle to Mentone to visit Louisa Baring (Lady Ashburton),[32] and wrote to Scottish geologist Roderick Impey Murchison with a list of his needs for the journey.[33]

Jane Carlyle's was not the only tragic death of a close friend that year. "I have had the terrible news of Ginty's death," he wrote to his friend Thomas Hirst just as he was leaving for the Nottingham meeting of the BAAS.[34] William Ginty was one of Tyndall's oldest friends; they had known each other since Tyndall's days on the Ordnance Survey of Ireland. Ginty had since emigrated to Brazil with this family, but his wife, Margaret, and one of their daughters, Kathleen, were on a visit to England when he suddenly died. Margaret Ginty sent Kathleen to London to ask Tyndall for help, but he was on his annual excursion to the Alps and did not receive her message until his return. He was quick to reassure her, and offered some typically Tyndallian exhortations. "I hardly ever knew a man of energy and general intellectual power greater than those of your husband," he told her. Her duty now was to do the best she could for her son, so that he could take responsibility for her and her daughters. "From all I have heard of your eldest boy he is likely to do credit to the memory of his father and to you," he wrote to her.[35]

Tyndall was now moving in some quite rarefied circles. It is clear from the correspondence in this volume that he was on intimate terms with a number of socially important families—the Pollocks and the Carlyles in particular. Juliet and William Frederick Pollock were among Tyndall's most prolific correspondents, and he was a frequent visitor to their various residences. In fact, he dined with them and their eldest son at Trinity College the night before delivering his Rede Lecture.[36] And while there are no letters in this volume to or from Louisa Baring, it was to her residence in Mentone that Tyndall took the bereaved Carlyle at the end of 1866.[37] Tyndall was clearly a social asset worth cultivating, but there is human warmth in the correspondence as well, with Tyndall anxiously inquiring after sick children, for example.

Tyndall's annual Alpine excursions were another point of contact with his well-heeled acquaintances. William Frederick Pollock was a keen mountain climber, too. As it happens, neither the 1865 nor the 1866 expedition was particularly successful. A little before Tyndall arrived, a party of mountaineers had fallen to their deaths on the Matterhorn and searches for the bodies were still taking place, so Tyndall was unable to climb as he had planned.[38] Back in London he complained to Juliet Pollock that since he "had not [his] due amount of exercise in Switzerland," he did "not find [his] vigour up to high water mark."[39] The following year, the expedition was entirely spoiled by appalling weather. Tyndall wrote to Juliet that "the weather is abominable & defeats all plans."[40] The Pollocks' eldest son, Frederick, was in the Alps that year as well, and wrote to Tyndall complaining that "the rain it raineth every day, and the alpenstock must go into winter quarters."[41] Tyndall himself sent Clausius a long letter describing his efforts to continue the expedition and complaining about the "exceedingly bad weather." He had tried to climb the Aletschhorn, the second-highest peak in the Bernese Alps, but was "defeated in the only attempt which I was able to make by a heavy snowstorm."[42]

The letter to Clausius underlines the extent to which Tyndall's network of personal friendships, as well as professional acquaintances, now stretched beyond the British Isles. Clausius was in fact one of his more prolific correspondents during the period covered by this volume, and again the letters' contents illustrate the slippage between personal and professional life.[43] The correspondence moves easily from discussion of scientific work to scientific gossip, and to questions about family and friends. In the same letter in which he complained to him about the weather, and apologized for being unable to visit him in Zurich, Tyndall even wrote to Clausius asking him to "buy a good strong watch for me."[44] He visited the popular science writer and editor François Moigno in Paris on his way back from the failed Alpine excursion, and Moigno wrote to assure him, "Immediately after your visit, I began the translation of your marvellous Rede lecture."[45] Letters with other correspondents elsewhere in Europe and in North America show Tyndall's scientific reputation to have been international. He exchanged letters and information with American oceanographer Matthew Fontaine Maury when he visited London in 1866.[46] The terms in which the letters were written demonstrate the degree to which these men were friends as well as being scientific acquaintances.

Heat and Sound

Not all of Tyndall's international relationships were as congenial, however. In 1865 he found himself embroiled in a furious controversy with the Hungarian physicist C. K. Akin, who accused Tyndall of attempting to take credit for ideas and discoveries that really belonged to him.[47] According to Akin, Tyndall had stolen his original ideas in the experimental work that he carried out on negative fluorescence during the first half of the 1860s. Knowing of Akin's work, Tyndall had actually undertaken to delay publishing his own experiments until Akin had made his own work public. In October 1864, however, Akin wrote angrily to Tyndall (the letter is no longer extant) accusing him of a serious breach of faith. Tyndall replied with considerable restraint, but in 1865 the debate exploded into the public realm, as Akin accused Tyndall in a letter to the *Philosophical Magazine* (*Phil. Mag.*) and Tyndall responded in kind.[48] He sent a flurry of letters to friends and colleagues seeking their support and explaining his position. George Griffiths, who had collaborated with Akin in some of his researches, funded by a grant from the BAAS, reassured Tyndall that Akin was "one of those unfortunate men who cannot help making themselves disagreeable."[49]

Tyndall's work on heat was the cause of other troubles during this period as well. He had recently sent Johann Conrad Bohn, the professor of mathematics and physics at the University of Giessen, a copy of his recently published *Heat* (1863). Bohn told Tyndall that "a translation of your book into German would be a pleasant thing to many." He suggested that if Tyndall "had the confidence that I could be a truthful interpreter of your thoughts, I would gladly undertake the task of translation and ask you for your kind permission and approbation of my plan."[50] Clearly, though possibly with some trepidation (only the letters from Bohn to Tyndall are extant), Tyndall agreed and Bohn reached an agreement with the Vieweg publishing house in Berlin to publish his translation. However, it was not long before Tyndall started to develop significant reservations about Bohn's activities.[51] He was concerned, for example, that Bohn had entered the publication agreement with Vieweg without any further consultation with him. He also had significant concerns, based on rumors passed on to him by Clausius and Helmholtz, among others, regarding the quality of Bohn's translation. Further concerns emerged when Bohn abruptly changed publishers. "I do not quite understand the movements of Prof. Bohn," he told his friend Heinrich Debus.[52]

When Tyndall persuaded Bohn to allow Gustav Wiedemann to read the proofs it became apparent that Tyndall's concerns regarding the quality of the translation were justified. Tyndall was in the awkward position of having to withdraw his permission to Bohn to translate his book, and find an alternative translator. Tyndall's eventual solution to the problem of finding a trustworthy translator is revealing of the extent to which Tyndall was willing to go to ensure his own reputation was protected. Hermann Helmholtz and Gustav Wiedemann suggested to Tyndall that their wives, Anna and Clara, should carry out the translation, but that to ensure propriety, Hermann's and Gustav's names should appear as translators on the title page when the work was published. Both Anna Helmholtz and Clara Wiedemann were, as their husbands acknowledged, experienced translators with reputations of their own. In fact, it was their acknowledged experience that made them appropriate translators of Tyndall's book. Nevertheless it was not considered appropriate that, as women, their labor in translating a scientific work be made public. Tyndall agreed to his friends' suggestion and with some difficulty succeeded in extricating himself from his agreement with Bohn, who complained bitterly about the damage the episode would inflict on his own reputation as a man of science, since his translation work was already widely known. In due course, the German translation appeared in 1867 as *Die Wärme, betrachtet als eine Art der Bewegung* with Hermann Helmholtz and Gustav Wiedemann's names on the title page as translators.

"I am writing a little book on Sound which I trust will please you," wrote Tyndall at the end of 1865.[53] He was doing a little more than that, in fact, as he explained a few weeks later in a letter to Clausius. "I had this year to give a course of lectures to the young people on Sound," he wrote. "It was Faraday's custom to give those juvenile lectures as they were called, but Faraday will never lecture more, and so the Christmas lectures fall on Frankland and myself."[54] The lectures did indeed result in a book: *Sound: A Course of Eight Lectures Delivered at the Royal Institution of Great Britain* was published in July 1867. On the same day that he wrote to Clausius, he wrote a letter to the American chemist John William Draper, asking him to "kindly mention to Professor Youmans that I am working hard at my book on Sound, and that when it is complete I entertain hopes that he will be pleased with it:—it is a beautiful subject."[55] The three men would later develop both personal and professional relationships with each other. For instance, Draper's 1862 book *The History of the Intellectual Development of Europe* guided Tyndall's own historical narrative in his notorious 1874 "Belfast Address." Edward Livingston Youmans, an American science popularizer and founder and editor of the extremely successful *Popular Science Monthly*, would publish all of Tyndall's works in America in his editorial role for D. Appleton and Company. In March Tyndall delivered another series of lectures on sound at the Royal School of Mines, where he had held the chair of natural philosophy since 1859. By October he told Clausius that "the proof sheets are now coming in, and I may ask you, if you have the time, to glance at them for me."[56]

In addition to the "little book on Sound," Tyndall's lectures generated another publication in 1866: "Mr. Tyndall on the Vibration of Strings," in the *Phil. Mag.* It was an account of experiments with vibrating strings that Tyndall had performed at a Friday Evening Discourse at the RI on 15 June. The account impressed Helmholtz, who wrote to Tyndall, "I already read in the Philosophical Magazine that you have illustrated the composition of sounds for your listeners at the Royal Institution with electric light and I am grateful to you for making the basis of sound impressions comprehensible to people in this way."[57] Tyndall himself was clearly pleased with his demonstration, and seems to have arranged for a private performance a couple of weeks later. He invited Juliet Pollock, telling her, "On Tuesday next at 3 P.M. I expect a small audience here to see the experiments on vibrating strings. The doors shall be open at that hour for you and your friend."[58] He sent a similar letter to Edward Strutt (with whom he would be staying while attending the annual meeting of the BAAS later that summer), inviting him and his wife to the performance.[59] The episode offers a nice example of the interplay between experiment and spectacle in Tyndall's work, and his pleasure in making sure that he put on a good show.

The Eyre Affair

But as Tyndall was entertaining fashionable London with his vibrating strings, he was also becoming embroiled in the most unsavory episode of his career. His involvement in the Eyre affair as it developed during the latter half of 1866 casts a light on attitudes that were deeply racist, even by mid-Victorian standards.[60] The Eyre affair, as it soon became known, took place in the aftermath of the Morant Bay rebellion in Jamaica that erupted during October 1865. The rebellion started when a group of probably several hundred people, led by a local Jamaican Baptist deacon, Paul Bogle, marched on the local courthouse, in the parish of Saint Thomas in the East, to protest poor economic conditions and the arrest of local men for protesting their treatment by the authorities. They were met by local militia and fighting broke out. After several days' insurrection, the governor of Jamaica, Edward Eyre, succeeded in suppressing the rebellion by military force, declared martial law, and summarily arrested, tried, and executed mixed-race Jamaican businessman and member of the Jamaican Assembly George William Gordon for treason, despite the fact that he had no apparent involvement in the insurrection. Bogle and other alleged leaders of the insurrection were also executed, and several hundred of the protesters shot by soldiers.

As an elected member of Jamaica's House of Assembly, Gordon had been a constant critic of Eyre's governorship. He had links with evangelical movements in Britain and was in a position to cause Eyre considerable political embarrassment. As a magistrate in Morant Bay he had complained to Eyre about prison conditions there—and Eyre had responded by removing him from his magistracy. In the immediate aftermath of the rebellion, Eyre had Gordon arrested and then transported from Kingston, which was not under martial law, to Morant Bay, which was. He was then summarily tried for treason in a court martial presided over by Lieutenant Herbert Brand, and executed on 23 October. Eyre was apparently convinced that the poor black peasantry could not have plotted such an insurrection on their own and that Gordon must have been involved in its organization. The execution of Gordon, in particular, caused outrage in Britain. It appeared clear to many that Eyre had deliberately taken advantage of the imposition of martial law to dispose of a political enemy. There were calls for Eyre's removal as governor and his arrest, as well as the arrest of some of the military officers involved, for the unlawful killing of Gordon.

Following the uproar, the British government established a royal commission, headed by the governor of Malta, Sir Henry Storks (1811–74), to inquire into the circumstances surrounding the Morant Bay rebellion and Eyre's response to it. The commission's report was highly critical of Eyre's behavior, and as a result he was removed from the governorship. When Eyre returned to England during the summer of 1866 there were calls for his arrest and trial for murder. The Jamaica Committee, formed by leading liberal intellectuals, including John Stuart Mill, Thomas Huxley, and Charles Darwin, was established to spearhead the campaign for his prosecution. Many of the leading figures of the Jamaica Committee were friends and allies of Tyndall's, so it would not have been unexpected to see him take a prominent part in the campaign as well. Tyndall duly did so, but on the opposing side to those friends. He was soon a leading member of the Eyre Defence Committee, organized by another friend, Thomas Carlyle, to defend the embattled governor against his liberal enemies. The episode brings into stark relief the racism that was so deeply embedded in Victorian culture. Many of the racist attitudes visible here were shared across the political spectrum, even by the liberal gentlemen of the Jamaica Committee.[61]

Much of what Tyndall and other correspondents in this volume wrote in defense of Eyre's actions makes for very difficult reading. In October 1866 the Jamaica Committee sent Tyndall a letter drawing his attention to the reports that they had published with regard to Eyre's activities. Tyndall responded with a long and detailed defense of Eyre that was widely circulated. The "Jamaica negroes," he told them, were "as a general rule, lazy and profligate, while many of the authorities I have consulted, and who have lived among these people, add a third and weightier adjective, which I refrain from writing down."[62] Eyre, he claimed, had been responding to the threat of a rebellion that would have led to the wholesale massacre of white men and women. Joseph Hooker had written to him privately, describing "the Negro in Jamaica" as "a most dangerous savage at the best."[63] Tyndall was clearly uncomfortable to be on the opposite side of the debate from men that he admired—Huxley in particular.[64]

Tyndall received many letters of support for his position nonetheless. English clergyman Frederic Farrar wrote to agree with him that "[u]nless all travellers and the great majority of Ethnologists have wilfully lied, the negro is <u>naturally</u> wild, animal, lazy, sanguinary: he does belong distinctly to an inferior race, and a race which cannot be treated as on an equality with civilised white races."[65] His cousin Helen Tyndall wrote thanking him "for your letter to that would-be-philanthropic committee, who would so willingly forget that though the negro may be 'a man and a brother' he is lower than the brute when his nature is left to itself!"[66] Support for his stance came from men such as George Le Grand Jacob, a former East India Company officer who had been involved in suppressing the Indian Mutiny in 1857.[67] Another member of the Eyre Defence Committee, William Erle, wrote to Tyndall after seeing a copy of the letter: "I hope I am indebted to you for it, at all events I am indebted to you for some of the wholesome vigour which flows into every man's mind who reads and understands that letter."[68] The Eyre affair dragged on beyond the period covered in this volume, as the Jamaica Committee mounted successive attempts to bring Eyre to justice for the atrocities he had committed (but never succeeded).

It is difficult not to see Tyndall as morally diminished by his partisan support for Eyre's vicious actions. It certainly cannot be dismissed as simply an expression of the prevailing racism of the period. Huxley and Darwin's positions—and even Hooker's refusal to allow his private views to be made public[69]—are in this respect telling. But the urgency with which Tyndall's support was canvassed by both sides of the debate is also revealing of the kind of public standing he had acquired over the past decade. Tyndall was by the middle of the 1860s a man who could wield considerable influence. Like his hero Carlyle he was a society lion—he moved easily in the best circles; he was invited to dinner parties and to country house weekends. It also seems clear that the country boy from Leighlin Bridge was proud of what he had become. By dint of his own labors—or so he would see it—he had made himself into someone who mattered. His voice was heard; his influence was considerable. The letters in this volume show Tyndall as a man of science—and as a social man—in his prime.

The center of Tyndall's social and scientific world remained the RI—and his position there became stronger as Faraday faded away. Tyndall had by now taken on the considerable burden of Faraday's duties, as well as continuing with his own. As he acknowledged in a letter to Clausius, "Faraday will never lecture more." His old mentor was "destined I fear soon to fall and pass away as far as his physical frame is concerned. His progress downwards has been exceedingly rapid during the last two years. He seemed to become quite suddenly old, and he is now so feeble as to be hardly able to walk without help—a grand old vine surely, but it is sad to see."[70] Just how important the RI was to Tyndall is clear in his rejection of the professorship at Oxford in favor of remaining in London.[71] His attachment to the RI—and the hard work he continued to carry out to make sure that the institution flourished—was testament to a commitment to a particular vision of metropolitan science as well.

Notes

1. See Geoffrey Cantor, *Michael Faraday: Sandemanian and Scientist: A Study of Science and Religion in the Nineteenth Century* (New York: St. Martin's Press, 1991); and John Tyndall, *Faraday as a Discoverer* (London: Longmans, Green, 1868).
2. For Faraday, see letters 2257, 2298, 2299, 2303, and 2423. For Jane Barnard, see letters 2455, 2467, and 2489.
3. See letters 2324, 2347, 2361, 2463, 2485, 2513, and 2514.
4. For Helmholtz, see letters 2281, 2290, 2331, 2440, and 2511. For Clausius, see letters 2222, 2235, 2242, 2270, 2275, 2282, 2288, 2328, 2329, 2351, 2357, 2387, 2392, 2403, 2405, 2411, 2416, 2426, 2464, 2490, and 2494.
5. For the French publisher François-Napoléon-Marie Moigno, see letters 2339, 2300, 2368, and 2452. For Félix Foucou, who wrote to Tyndall representing the French publisher Pierre-Jules Hetzel, see letters 2255, 2427, 2431, and 2441.
6. For the Carlyles, see letters 2335, 2389, 2398, 2399, 2400, 2402, and 2525. For the Pollocks, see letters 2260, 2268, 2269, 2273, 2297, 2301, 2306, 2307, 2308, 2310, 2311, 2312, 2315, 2325, 2449, 2461, 2477, 2486, and 2497.
7. See letters 2460, 2461, and 2464.
8. See letters 2465, 2466, 2469, 2470, 2473, and 2504.
9. See letters 2314, 2318, 2327, 2328, 2352, 2373, 2384, 2460, and 2476.
10. For the Eyre affair, see letters 2278, 2481, 2491, 2498, 2500, 2505, 2506, 2507, 2510, 2516, 2517, 2518, 2520, 2521, 2522, 2523, 2530, and 2532.
11. See letters 2228 and 2230.

12. Jack Morrell and Arnold Thackray, *The Gentlemen of Science: Early Years of the British Association for the Advancement of Science* (Oxford: Clarendon, 1981).
13. Letters between Tyndall and Bence Jones are listed in n. 3, above.
14. See letter 2485.
15. For Tyndall's role with Trinity House, see letter 2423 and a brief mention in letter 2490.
16. See letter 2243.
17. See letter 2435.
18. See letters 2445, 2453, and 2459.
19. See letter 2490. On the North British faction in energy physics, see Crosbie Smith, *The Science of Energy: A Cultural History of Energy Physics in Victorian Britain* (London: Athlone, 1998).
20. See *Brit. Assoc. Rep. 1867*, liii–lxxxii.
21. See letter 2468.
22. See letters 2320 and 2324.
23. See letter 2351.
24. See letters 2231 and 2232.
25. See letter 2274.
26. See letters 2382, 2384, 2384, 2391, and 2393.
27. See letters 2375, 2384, 2389, 2398, 2399, 2400, and 2402.
28. See letter 2389.
29. See letter 2398.
30. See letters 2399 and 2400.
31. See letter 2414.
32. See letters 2476 and 2531.
33. See letter 2528.
34. See letter 2468. For other letters about Ginty's death, see n. 8, above.
35. See letter 2470.
36. See Tyndall's journal entry of 18 May 1865 (RI MS JT/2/13c/1289).
37. See n. 32, above.
38. See letters 2297 and 2298.
39. See letter 2308.
40. See letter 2461.
41. See letter 2484.
42. See letter 2490.
43. For letters between Tyndall and Clausius, see n. 4, above.
44. See letter 2490.
45. See letter 2300.
46. See letters 2482, 2483, and 2488.
47. The feud began in 1864 (see *Tyndall Correspondence*, vol. 8). For Tyndall's feud with Akin in this volume, see letters 2222, 2225, 2226, 2227, 2235, 2236, 2239, 2241, 2242, 2244, 2245, 2247, 2259, 2270, 2275, and 2305.
48. See letter 2245.
49. See letter 2239.
50. See letter 2296.
51. On Tyndall's dispute with Bohn regarding the German translation and publication of *Heat* (1863), see letters 2387, 2392, 2396, 2401, 2403, 2405, 2406, 2407, 2410, 2411, 2412, 2415, 2416, 2417, 2420, 2426, 2437, 2440, 2446, 2472, and 2476.
52. See letter 2401.
53. See letter 2343.
54. See letter 2351.
55. See letter 2352.

56. See letter 2490.
57. See letter 2511.
58. See letter 2449.
59. See letter 2450.
60. For letters regarding the Eyre affair, see n. 10, above.
61. For attitudes to race among Victorian men of science and more generally, see Douglas A. Lorimer, *Science, Race Relations and Resistance: Britain, 1870–1914* (Manchester: Manchester University Press, 2013); Sadiah Qureshi, *Peoples on Parade: Exhibitions, Empire and Anthropology in Nineteenth-Century Britain* (Chicago: University of Chicago Press, 2011); and Efram Sera-Shriar, *The Making of British Anthropology, 1813–1871* (Pittsburgh: University of Pittsburgh Press, 2013).
62. See letter 2498.
63. See letter 2504.
64. See letter 2500.
65. See letter 2517.
66. See letter 2510.
67. See letter 2520.
68. See letter 2522.
69. Hooker, however, allowed Tyndall to quote him for Tyndall's speech at a meeting of the Eyre Defence Committee, held at Willis's Rooms in London on 18 February 1867 (see *The Correspondence of John Tyndall*, vol. 10).
70. See letter 2351.
71. See n. 22, above.

EDITORIAL PRINCIPLES

Our aim is to include all letters to and from John Tyndall that are currently extant. Some of the letters are from published sources, such as letters written by Tyndall to newspapers or journals, but the majority are either the originals or typescripts produced by Louisa Tyndall from originals. One central editorial principle has been to transcribe, wherever possible, from original letters rather than Louisa's typescripts. This is because of Louisa's conscious and unconscious editorial interventions, a point discussed in more detail below. We have also aimed to reproduce, as accurately as possible, the text of the letters as they were written. Spelling mistakes that have appeared in the handwritten letters have therefore been preserved; mistakes appearing where only the typescript letter has survived are silently corrected, as in our judgment they are almost always typographical errors. But, in general, nonstandard spellings (e.g., the use of dialect) has been retained even when in a typescript-only letter (though there is the possibility that the spelling is Louisa's, not her husband's). Where the transcription is based on a typescript transcription by Louisa (or one of her assistants), this will be indicated by "LT Transcript Only."

We have included illustrations that are part of the letter. In cases where it was possible to reprint the original handwritten image, we have done so. In other cases, volume editors have reconstructed drawings because it was not possible to reprint an image from the handwritten letter (due, for example, to deterioration of the original document).

The letters are presented in chronological order of writing. Where more than one letter has the same date, letters from Tyndall (in alphabetical order of addressee) come first, followed by letters to Tyndall (in alphabetical order of writer), with the exception that if the order of writing can be determined, that order takes precedence. Where letters cannot be dated precisely they are placed at the latest likely date of writing. (In some cases if it is possible, but unlikely, that they were written later, notes indicate such possibilities.) To indicate the beginning of a new year, the date of that year has been inserted in a large font before the first letter from that year. Readers should note that although volumes are arranged in chronological order, letters that are discovered from years already covered in published volumes will appear in the final volume of the correspondence.

For non-English-language letters, both the transcription of the original and the English translation are presented. The transcription of the non-English letter appears first, followed by the English translation in a smaller font. Editorial notes are inserted in the English translation. Some foreign letters are in an original hand, while others exist only as transcriptions by Louisa or others who often did not have a full command of the language. In the latter case these letters contain frequent spelling, grammatical, and interpretation errors that do not seem to have been committed by the original author. In these typescript-only cases, and only in these cases, spelling and grammatical errors have been corrected, as have poor interpretations where a better word choice was obvious from the context.

There is a standard format for each letter. The first line lists the author, if it is to Tyndall, or the recipient, if it is from Tyndall; the date; and the letter number. The second line lists any information supplied by the writer in their salutation, such as the place, date, or time of day when the letter was written. (In some cases, though, where the writer put the date at the foot of the letter we have left it there in order to reproduce the text accurately.) The opening salutation, the text of the letter, and closing then follow. Then come the source and editorial notes, the latter indicated in the text by arabic numerals.

Editorial notes are intended to provide the contextualization necessary to understand the contents of each letter. There are basically six types of editorial notes. Many of them are informative notes on persons mentioned in the letters. If a person referred to in a letter is not described in a note, they have been mentioned more than twice in the volume and they will have an entry in the biographical register at the back of the volume. For the rest—those mentioned once or twice—biographical information will appear in the editorial note appended to the letter in which he or she is first mentioned. When there is a second reference to that person a note will refer back to the letter in which he or she was first mentioned. Biographical notes on well-known figures such as Faraday or Huxley stress what that person was doing at the time when the letter was written rather than providing a full overview of the person's life.

A second type of editorial note identifies allusions and quotations. In the case of quotations, where possible, the first edition of the work is cited unless a different edition is mentioned in the letter. If a quote is not in English a translation is given. Glosses on obscure places, abstruse words, and words used in an unfamiliar sense constitute the third type of editorial note. The fourth type provides information about the context of the letter, drawing on such sources as contemporary publications, scholarly sources, Tyndall's personal journals and field notes, and other letters in the *Tyndall Correspondence*. The fifth type of editorial note is cross-references to other letters in the correspondence, referring to the letter number. The last type is information on significant textual changes that Tyndall made to the original that is thought to be relevant to the meaning of the letter.

There are a series of conventions for indicating when editors have given additional information about the text of a letter:

When Louisa Tyndall has made an annotation or marginal insertion that the editors have deemed important enough to retain, we have included her annotation in an editorial footnote.

When editors other than she have made insertions, they are indicated by unitalicized square brackets; for example, [word].

When certain words are ambiguous and we have had to conjecture their meaning, they are indicated by italicized square brackets; for example, *[*words*]*; when we have considered such words illegible the number of illegible words is indicated in full italicization and square brackets; for example, *[3 words illeg]*.

If a word or series of words are not present or have been destroyed—by inkblots, holes, or mold, for example—they are indicated as missing, italicized and in angular brackets; for example, *<3 words missing>*.

In extremely rare cases where it is clear that Louisa herself has intentionally destroyed a few words, they are indicated as excised, italicized and in angular brackets; for example, *<3 words excised>*.

Finally, a brief word about Louisa. She was responsible for collecting together many of Tyndall's letters in order to write a biography of her late husband. This biography was not completed. It is important to recognize that Louisa was not always reliable as a transcriber. She often threw out the manuscript letter when she had made her transcription, but some original manuscripts remain, which indicate her editing processes. Like any transcriber, she was sometimes inaccurate, but corrections show that she proofread her transcriptions. However, she also edited Tyndall's writing to make his letters seem more consistent or grammatically correct. For example, she made punctuation more formal, corrected grammar, and, occasionally, corrected quotations. She consistently corrected Tyndall's usual (deliberate) lowercase for "god" to "God." Her typewriter was unable to insert the superscripts that Tyndall used in abbreviations, or the "&c" symbols that stood for an address at the foot of the letter. (Modern typefaces are unable to deal with all the variant scrawls that the "etc." symbol took and they are all formalized here to "&c.") Louisa also left out details that she considered too private or too salacious for public consumption. She equated postmark and date of writing and overlooked the time taken to write some letters. When she inferred dates from internal evidence, her reasoning was sometimes faulty. Similarly, her identifications of people are sometimes in error. Independent evidence has always been sought for her dates and identifications. Lastly, the patterns of punctuation differ between LT sources and manuscript sources. Tyndall himself used dashes more often than commas, semicolons, and stops, and he used dashes of many different lengths, which are all standardized here to em-dashes. It is often difficult to distinguish between very short dashes and commas or very short dashes and stops. Similarly, capitalization is often ambiguous; distinguishing *S*/*s*, for "science," for example, is a particular problem. Therefore, scholars for whom punctuation and capitalization are significant will need to consult the manuscripts. Users should also be aware that throughout this project the transcribers and editors have generally worked from scans and photographs rather than the original manuscripts.

NOTE ON MONEY

In the nineteenth century the currency used in England was the pound sterling, usually abbreviated to *£* (or, occasionally, to *L*). A pound consisted of 20 shillings (abbreviated *s*) and each shilling contained 12 pennies (*d*). A guinea was 21 shillings; that is, £1 1s. A crown was 5 shillings (5s), and half a crown equaled two shillings and sixpence (2s 6d). The smallest coin was a farthing, one quarter of a penny (¼d).

Several forms of representation were in contemporary use. Sometimes the numbers for pounds, shillings, and pence were separated by dots, forward slashes, or spaces; thus, for example, 3 pounds, 4 shillings, and 5 pennies could be written as £3.4.5, £3/4/5, or £3 4 5. A zero could be represented by a dash; for example, 7 shillings was often written as 7/- or just 7/. A half crown was often written as 2/6.

The Vereinsthaler was a silver coin issued by Prussia from 1857, to replace the Reichsthaler (issued in 1750), that became a common currency throughout much of Germany in the mid-nineteenth century. The Louis d'or was a French gold coin issued from 1640 to 1795 and still in use in the nineteenth century.

TIMELINE OF JOHN TYNDALL'S LIFE

Year	Event
c. 1822	John Tyndall born at Leighlin Bridge, County Carlow, Ireland
c. 1836	Attends John Conwill's National School in Ballinabranna
1839	Begins employment as civil assistant with Ordnance Survey of Ireland in Carlow
1840	Becomes civil assistant in the Ordnance Survey Office at Youghal, Cork County
1841	First appearance in print (poem in *Carlow Sentinel* under pseudonym "W[alter] S[nooks]")
1842	Joins English Ordnance Survey in Preston
1843	Leads written criticism of Ordnance Survey in *Liverpool Mercury*; dismissed November
1844	Home in Ireland, mostly unemployed, until obtaining a position with the firm of Nevins and Lawton, Surveyors, of Manchester
1845–47	Works for Richard Carter, Surveyor, of Halifax
1845	Meets Thomas Archer Hirst in Halifax
1847	Begins teaching mathematics and surveying at Queenwood College, Hampshire
1848–50	At University of Marburg, working for PhD with Robert Bunsen and Friedrich Stegmann
1850	Returns to England; meets Michael Faraday and William Francis and attends his first British Association meeting, before returning to Marburg
1851	Moves from Marburg to Berlin (April–June); then returns to Queenwood
1852	Elected fellow of the Royal Society
1852	Friendship developed with Thomas Henry Huxley
1853	First lecture at Royal Institution (RI)
1853	Meets Herbert Spencer
1853	Appointed professor of natural philosophy at RI
1853	Awarded Royal Medal for magnetic work; declines when controversy arises over award
1856	Meets Thomas Carlyle
1857	Climbs Mont Blanc for first time with Hirst
1859	Demonstrates existence of greenhouse gases and climatic implications
1859	Joins Government School of Mines as professor of natural philosophy
1860	*The Glaciers of the Alps* published
1860	Meets John Lubbock and his wife, Ellen Lubbock
1861	First ascent of the Weisshorn

Year	Event
1861	Delivers his first Christmas Lectures at RI
1862	*Mountaineering in 1861* published
1862	Begins to champion Mayer's priority in discovery of conservation of energy
1863	*Heat Considered as a Mode of Motion* published
1864	Awarded Royal Society's Rumford Medal
1864	First meeting of X Club
1865	*On Radiation* published
1865	Engages in public controversy over the efficacy of prayer
1866	Succeeds Michael Faraday as scientific adviser to Trinity House
1867	Faraday dies; Tyndall becomes superintendent of the RI
1867	*Sound: A Course of Eight Lectures* published
1868	Successfully climbs Matterhorn
1868	Discovers the cause of light scattering, to be known as "Tyndall Effect"
1868	*Faraday as a Discoverer* published
1868	"Scientific Materialism" lecture at the British Association
1868	Resigns from Royal School of Mines
1869	Joins Metaphysical Society
1870	"On the Scientific Use of the Imagination" lecture at British Association
1870	*Three Scientific Addresses* published
1870	*Researches on Diamagnetism and Magne-Crystallic Action* published
1871	Meets Louis Pasteur for first time while in Paris
1871	*Hours of Exercise in the Alps* published
1871	*Light and Electricity* published
1871	*Fragments of Science* published
1872	Debates over how to measure the efficacy of prayer, aka the "Prayer-Gauge Debate"
1872	*Contributions to Molecular Physics in the Domain of Radiant Heat* published
1872	*The Forms of Water in Clouds and Rivers, Ice and Glaciers* published
1872–73	USA Lecture Tour
1873	*Six Lectures on Light* published
1874	As president of the BAAS, delivers the "Belfast Address"
1876	Marries Louisa Charlotte Hamilton
1877	Develops "Tyndallization" (discontinuous heating and cooling process to destroy heat-resistant spores)
1877	Tyndall and Louisa build summer cottage, Alp Lusgen, at Belalp, northern side of Valais, above Brig
1877	*Fermentation and Its Bearings on Phenomena of Disease* published
1879	*Fragments of Science*, which had gradually expanded, first published in two volumes
1881	*Essays on the Floating Matter of the Air* published
1885	Tyndall and Louisa build Hindhead retreat, Surrey Downs
1887	Resigns from RI
1890	Fight with Gladstone in the *Times* over Irish Home Rule
1892	Hirst dies
1892	*New Fragments* published
1893	Dies at Haslemere, Surrey, accidentally poisoned

TIMELINE OF EVENTS IN JOHN TYNDALL'S LIFE SPECIFIC TO VOLUME 9

Month and year	Biographical details	Notable social, political, and cultural events
January 1865	Engages in controversy with C. K. Akin in the pages of *Phil. Mag.* over discovery of "calcescence." The controversy began with Akin's article "Further Statements concerning the History of Calcescence," *Phil. Mag.* 29 (1865): 136–51. Tyndall answered Akin's charges in "On the History of Calorescence," *Phil. Mag.* 29 (1865): 218–31 Delivers a Friday Evening Discourse at the RI, "On Combustion by Invisible Rays" (20th)	The New York Stock Exchange opens its first permanent headquarters near Wall Street, in New York City (4th) American Civil War, Second Battle of Fort Fisher (13th–15th) Confederate general Robert E. Lee (1807–70) becomes general-in-chief (31st)
February 1865		*PMG* newspaper begins publication (7th) Gregor Mendel reads first of two papers titled "Experiments on Plant Hybridization" at meeting of the Natural History Society of Brünn in Moravia (8th) Tennessee adopts a new constitution that abolishes slavery (22nd) Reform League established (23rd)
March 1865	Alick Cumming, nephew of Mary Anne Coxe, and friend of Tyndall, dies	US president Abraham Lincoln (1809–65) begins his second term (4th) Gregor Mendel reads second of two papers titled "Experiments on Plant Hybridization" at meeting of the Natural History Society of Brünn in Moravia (8th)

Month and year	Biographical details	Notable social, political, and cultural events
March 1865		Confederate States of America agrees to use African American troops (13th) The Congress of the Confederate States of America adjourns for the last time (18th)
April 1865	Travels with Thomas Hirst to Dublin, Ireland, on three-week "walking trip" to lift his spirits after the Akin affair Publishes "On the History of Caloresence" in *Phil. Mag.* Publishes "On the History of Negative Fluorescence" in *Phil. Mag.* Publishes "Note on the Invisible Radiation of the Electric Light" in *Roy. Soc. Proc.* Publishes "Professor Helmholtz on Ice and Glaciers" in *Phil. Mag.*	Richmond, Virginia, is captured by Union forces under General Ulysses S. Grant (1822–85) (3rd) General Robert E. Lee surrenders to General Ulysses S. Grant at Appomattox Court House, effectively ending the American Civil War (9th) Actor and Confederate sympathizer John Wilkes Booth (1838–65) shoots and mortally wounds US president Abraham Lincoln during an evening performance at Ford's Theatre in Washington, DC (14th) Vice-President Andrew Johnson (1808–75) becomes the seventeenth president of the United States (15th) Robert Fitzroy, meteorologist and Royal Navy admiral, earlier captain of HMS *Beagle* with Charles Darwin, dies by suicide (30th)
May 1865	Gives statement on scientific education to the House of Lords Committee on the Public School Bill Delivers the annual Rede Lecture in the Senate House at the University of Cambridge, "On Radiation" (16th), later published as *On Radiation* Receives an honorary LLD from the University of Cambridge (30th)	The Victoria Institute, or Philosophical Society of Great Britain, is established by James Reddie (1820–71) and Philip Henry Gosse (1810–88) in response to Darwin's *On the Origin of Species* (1859) and Charles Lyell's *Geological Evidences of the Antiquity of Man* (1863)
June 1865		Francis Galton (1822–1911) begins formulating his theory of eugenics

Month and year	Biographical details	Notable social, political, and cultural events
June 1865		James Clerk Maxwell (1831–79) publishes *A Dynamical Theory of the Electromagnetic Field* Jumbo, a young male African elephant, arrives at London Zoo and becomes a popular attraction (26th)
July 1865	Publishes book *On Radiation* Leaves London for the Alps, with Thomas Hirst (10th; returns by 26 August)	The Christian Mission, later renamed the Salvation Army, is founded in Whitechapel, London, by William (1829–1912) and Catherine Booth (1829–90) (2nd) The US Secret Service is founded (5th) American political magazine the *Nation* begins publication (6th) A party led by Edward Whymper (1840–1911) makes the first ascent of the Matterhorn (14th). Several members of the party die in an accident on the descent General election is won by the Liberal Party led by Lord Palmerston (1784–1865) Lewis Carroll's (1832–98) *Alice's Adventures in Wonderland* is published by Macmillan in London
August 1865	Attends the 1865 meeting of the Société Helvetique des Sciences Naturelles (Swiss Society of Natural Sciences) in Geneva (21st–23rd) Johann Conrad Bohn proposes to Tyndall to translate his book *On Heat* for a German audience, to which Tyndall agrees. He will later decide against this agreement due to Bohn's inferior translation skills	William Jackson Hooker, botanist, director of the Royal Botanic Gardens, Kew, and father of Joseph Dalton Hooker, who succeeded him as director, dies (12th)
September 1865	Attends the BAAS annual meeting in Birmingham	The thirty-fifth annual meeting of the BAAS is held in Birmingham (6th–13th) William Henry Smyth, astronomer, dies (9th)

Month and year	Biographical details	Notable social, political, and cultural events
October 1865	Engages in public controversy over the efficacy of prayer when he writes a letter to the editor of the *PMG* (11th) Offered the position of reader of experimental philosophy at Oxford, which he ultimately declines because he does not wish to leave London nor the RI	Morant Bay rebellion, an unsuccessful uprising against British rule at Morant Bay, Jamaica, is brutally suppressed with four hundred rebels executed (11th) Prime Minister Lord Palmerston dies (18th); Lord John Russell becomes prime minister (29th)
November 1865	Writes a letter to literary figure and leading women's rights campaigner Frances Power Cobbe (1822–1904) about her *Broken Lights* (1864) (11th)	Mark Twain's (1835–1910) story "The Celebrated Jumping Frog of Calaveras County" is published in the New York weekly the *Saturday Press*, bringing him to national attention Michel Chasles is awarded the RS's Copley Medal for "his historical and original researches in pure geometry" Archibald Smith is awarded the RS's Royal Medal for "his papers in the Philosophical Transactions and elsewhere, on the magnetism of ships" Joseph Prestwich is awarded the RS's Royal Medal for "his numerous & valuable contributions to geological science and more especially for his papers published in the Philosophical Transactions on the general question of the excavation of river valleys, and on the superficial deposits in France and England in which the works of man are associated with the remains of extinct animals"
December 1865	Begins writing "a little book on Sound," which he will eventually publish in July 1867 as *Sound: A Course of Eight Lectures Delivered at the Royal Institution of Great Britain*	Edward Eyre, governor of Jamaica, is dismissed and censured for his excessive actions during the suppression of the recent Morant Bay rebellion (16th) Leopold II (1835–1909) becomes king of Belgium (17th)

Month and year	Biographical details	Notable social, political, and cultural events
December 1865		The thirteenth amendment to the US Constitution (which forever abolishes slavery) is declared ratified by three-quarters of the states of the United States (18th) The Ku Klux Klan is formed by six Confederate Army veterans, with support of the Democratic Party, in Pulaski, Tennessee, to resist Reconstruction and intimidate "carpetbaggers" and "scalawags," as well as to repress freed slaves (24th)
January 1866	Delivers a Friday Evening Discourse at the RI, "On Radiation and Absorption with reference to the Colour of Bodies and their State of Aggregation" (19th)	John Robert Seeley's (1834–95) controversial and anonymous work *Ecce Homo: A Survey in the Life and Work of Jesus Christ* is published
February 1866		Charles Lyell receives the Geological Society of London's Wollaston Medal (16th)
March 1866	Publishes "On Calorescence" in *Phil. Trans.* Publishes "Sixth Memoir on Radiation and Absorption" in *Phil. Trans.* Publishes "On the Black-bulb Thermometer" in *Phil. Mag.* Publishes "Remarks on the Paper of Professor Magnus" in *Phil. Mag.*	The US Congress overwhelmingly passes the Civil Rights Act (13th); US president Andrew Johnson vetoes the bill (27th); Congress overrides the veto on 9 April William E. Gladstone introduces Reform Bill (12th) William Whewell, natural philosopher and philosopher and historian of science, dies (6th)
April 1866	Accompanies Thomas Carlyle to the University of Edinburgh for his speech to the students as rector (2nd) Receives LLD from the University of Edinburgh (2nd)	Alexander II (1818–81) of Russia escapes an assassination attempt in the city of Saint Petersburg (4th) The American Society for the Prevention of Cruelty to Animals is founded in New York City by Henry Bergh (1813–88) (10th)
May 1866	Begins to assist an aging Michael Faraday as scientific adviser to Trinity House (a role he will officially take over in July 1867)	Ferdinand Cohen-Blind makes a failed attempt to assassinate minister president of Prussia Otto von Bismarck (1815–98) in Unter den Linden in Berlin (7th)

Month and year	Biographical details	Notable social, political, and cultural events
June 1866	Delivers a Friday Evening Discourse at the RI, "Experiments on the Vibrations of Strings" (15th); repeats experiments for a private audience at the RI (26th)	The Parliament of Canada meets for the first time in Ottawa (8th) The Austro-Prussian War begins (14th; with an armistice on 22 July) King of Italy Victor Emmanuel II (1820–78) declares war on Austria (19th) Lord Russell resigns as prime minister; Conservative Party takes over, led by the Earl of Derby (26th)
July 1866		Hyde Park demonstrations in Trafalgar Square turn violent (23rd–25th) Tennessee becomes the first US state to be readmitted to the Union following the American Civil War (24th) Successful transatlantic telegraph cable follows an earlier attempt in 1858 (27th)
August 1866	William Ginty, one of Tyndall's oldest friends, dies Serves as one of the vice-presidents for Section A—Mathematics and Physics of the BAAS meeting in Nottingham	US president Johnson formally declares American Civil War over (20th) The thirty-sixth annual meeting of the BAAS is held in Nottingham (22nd–30th) The Treaty of Prague officially ends the Austro-Prussian War (23rd); an armistice had occurred 22 July
September 1866		The last cholera epidemic in London causes over five thousand deaths
October 1866		The Treaty of Vienna ends the war between Austria and Italy (3rd)
November 1866	Writes letter to the Jamaica Committee in support of Edward Eyre (7th), which is published in the *Morning Post* (8th)	Swedish chemist Alfred Nobel (1833–96) invents dynamite Julius Plücker is awarded the RS's Copley Medal for "his researches in analytical geometry, magnetism, & spectral analysis"

Month and year	Biographical details	Notable social, political, and cultural events
November 1866		William Huggins is awarded the RS's Royal Medal for "his researches on the spectra of some of the chemical elements and on the spectra of certain of the heavenly bodies; and especially for his researches on the spectra of the nebulae, published in the Philosophical Transactions" William Kitchen Parker is awarded the RS's Royal Medal for "his researches in comparative osteology, and more especially on the anatomy of the skull, as contained in papers published in the Transactions of the Zoological Society and the Philosophical Transactions"
December 1866	Travels to Mentone, France, with Thomas Carlyle to visit Louisa Baring	George Everest, surveyor and geographer for whom Mount Everest will later be named, dies (1st)

The Correspondence of John Tyndall

1865

From James Clerk Maxwell[1] 30 January [1865][2] 2221

'I have been swinging discs, &c., by the torsion of steel wires a good deal; and I find that after a steel wire has been twisted till it takes a slight set, and so has a new position of equilibrium, it gradually untwists itself with' a very slow motion, so that after four hours it is still creeping back towards its position before it was twisted. A wire twisted and allowed to untwist back to its first position of equilibrium is, therefore, not only a reservoir of energy, like a wound-up spring in a watch at rest, but a moving agent, like a watch going, because it gradually uncoils itself. I suppose the strained state of the external parts is gradually overcome by the constant torsion force arising from the inner parts of the wire, which were never overstrained.

'Here is a case of molecular change going on for hours in a piece of cold steel not subject to vibrations or any other visible dynamical cause of change.

'I find the *set* produced by *very* slight torsions is as the square of the angle of torsion; just as in the case of very slight magnetization the resident magnetism is as the square of the maximum.

'I suppose the law of decrease of power with time in a magnet which has been over-magnetized will be like that of the diminution of torsion in a wire which has been overtwisted.'

'As I had always been careful to preserve the elasticity of my wires I never overstrained them very much in my regular experiments, but the one I used on Saturday was not to be used again, so I gave it three turns in one direction, and it has been untwisting ever since more and more slowly. It was observed first at 12.30 on Saturday, and it has sensibly untwisted since 8 this (Monday) morning. I think there are experiments on the effect of time in twisting wires in "Poggendorff"[3] or "Les Mondes."[4] I forget the experimenter's name.*[5]

'He used to twist his wires by suspending magnets from them, and making terrestrial magnetism do the work of twisting.

'I think my plan is equivalent to establishing equilibrium between the strained outer portion and the unstrained inner part of the wire. The former yields with time to the continual twist of the latter.

'Fairbairn's[6] experiments show that time and repetition of force has not much effect when the iron is not overstrained in any part.

'Professor J. Thomson[7] has explained the condition of an overtwisted wire ('Cambridge and Dublin Mathematical Journal,' Nov. 1848),[8] and has shown that it is twice as *strong* to resist increase of twisting as to resist untwisting, though its *stiffness* is the same for either.

'Mr. Frodsham,[9] watchmaker, Strand, says that steel wire increases in transverse elasticity for about a year after it is made, whether the wire be used as balance-spring in motion or at rest. I find no change of elasticity amounting to 1/1000 in a year, but I do not know how old my wires were when I began. He also says that after twenty years' going, the spring is not to be depended on,—Yours truly, | 'J. CLERK MAXWELL. | '8, Palace Gardens, W., Jan. 30.'

J. Tyndall, 'On The Elasticity of Steel Wires', *The Reader*, 5 (4 February 1865), p. 144

---

1. *James Clerk Maxwell*: James Clerk Maxwell (1831–79), a Scottish physicist and mathematician. He graduated in the Cambridge mathematics Tripos as second Wrangler in 1854 after which he held professorships at Marischal College, Aberdeen (1856–60), King's College, London (1860–5), and from 1871 held the Cavendish professorship at the University of Cambridge. Maxwell is best known for his contributions to the study of electromagnetic radiation. He developed Faraday's field theory and demonstrated that electricity, magnetism, and light were instances of the same phenomenon. Maxwell's work culminated in his 1873 publication, *A Treatise on Electricity and Magnetism*.
2. *[1865]*: the year is given since Tyndall forwarded this letter to *The Reader* and it was published in its 4 February 1865 issue. Tyndall's note accompanying Maxwell's letter is letter 2223.
3. *"Poggendorff"*: *Poggendorff's Annal.*, or *Annalen der Physik und Chemie*, was a German scientific journal edited by the physicist Johann Christian Poggendorf (1796–1877) between 1824–76. The reference here is to: F. Kohlrausch, 'Ueber die elastiche Nachwirkung bei der Torsion', *Poggendorff's Annal.*, 195 (1863), pp. 337–68.
4. *Les Mondes*: *Les Mondes: revue hebdomadaire des sciences et leurs applications aux arts et à l'industrie* was a French scientific journal founded by François-Napoleon-Marie Moigno in 1863. The reference here is to a summary of Kohlrausch's work: 'Variations des constantes élastiques pendant la torsion,' *Les Mondes*, 5 (1864), pp. 497–8.
5. This footnote was provided by Tyndall in the published version of this letter: '**Kohlrausch*. "Les Mondes," July 7, 1864, from "Poggendorff," cxix. p. 337, says Weber found the moment of torsion diminish with the time. Kohlrausch twists glass threads by terrestrial magnetism, and finds continued torsion continues to increase the twist for 46h., and that

when left to itself the glass thread does not return to its old position. W. Thomson finds that glass threads after being twisted and let go, creep back again like my steel wire. The moral of Kohlrausch is "Use glass, but take care of error." Steel, however, does not show these peculiarities till it is cruelly overstrained, and I think it is much less plastic than glass within the limits of elasticity. So says Frodsham with respect to glass chronometer-springs'.

6. *Fairbairn's*: William Fairbairn (1789–1874), a Scottish civil engineer who designed and built bridges and ships utilizing wrought iron construction (*ODNB*).
7. *Professor J. Thomson*: James Thomson.
8. *('Cambridge and Dublin Mathematical Journal,' Nov. 1848)*: J. Thompson, 'On the Strength of Materials, as Influenced by the Existence or Nonexistence of Certain Mutual Strains among the Particles Composing them', *The Cambridge and Dublin Mathematical Journal*, 3 (1848), pp. 252–8.
9. *Mr. Frodsham*: Charles Frodsham (1810–71), an English horologist who established Frodsham & Co., a watch and timepiece business located at 84 Strand, London (V. Mercer, *The Frodshams: The Story of a Family of Chronometer Makers* (Ticehurst: Antiquarian Horological Society, 1981)).

To Rudolf Clausius 1 February 1865 2222

1st. Feb 1865

Mein lieber Clausius

Dein Buch habe ich erhalten,[1] aber leider durch schreckliche Beschäftigung war ich gehindert es durchzulesen.[2]

I was however aware of the whole arrangement regarding its translation.[3] I was asked to write a word of preface introducing it to the English reader, and although it is an impertinence on my part to say one word regarding anything that you write: I have agreed to do what the publisher wished me to do.[4]

Dr. Akin[5] is like a wild beast: he stuck his claws into me[6] without the slightest warning—He has published a long article in the present number of the Philosophical Magazine in which he uses language such as I never thought any mortal would apply to me.—

He does not in the least exasperate me now, though at first I was exceedingly annoyed. I shall answer him mildly, but effectually.[7] I am sure his article will make Faraday & my other friends more indignant than it makes me.

Give my love to little Johnny,[8] and kindest regards to your excellent wife.[9]

Yours ever | John Tyndall

RI MS JT/1/T/190

1. *Dein Buch . . . erhalten*: the previous year Clausius published *Abhandlung uber die Mechanische Wärmtheorie* (Braunschweig: Vieweg und Sohn, 1864), which outlines his mechanical theory of heat.
2. *Dein Buch . . . es durchzulesen*: I received your book, but I was unfortunately unable to read it due to a horrible amount of work (German).
3. *its translation*: the English version of Clausius's book was translated by the mathematician and lifelong friend of Tyndall, Thomas Hirst.
4. *I have agreed . . . me to do*: R. Clausius, *The Mechanical Theory of Heat, with its Applications to the Steam-Engine and to the Physical Properties of Bodies*, trans. T. Hirst (London: John Van Voorst, 1867). This included an introduction to the text by Tyndall.
5. *Dr. Akin*: C. K. Akin.
6. *He stuck his claws into me*: Tyndall was referring to the following article: C. K. Akin, 'Further Statements concerning the History of Calcescence', *Phil. Mag.*, 29 (1865), pp. 136–51; this was a response to his own: J. Tyndall, 'On the History of Negative Fluorescence', *Phil. Mag.*, 29 (1865), pp. 44–55. Akin and Tyndall had come into conflict in December 1864 over the precedence for discovering and naming the phenomena by which heat is turned into light. Akin called this 'Calcescence' while Tyndall used the term 'Negative Fluorescence'.
7. *I shall answer him mildly, but effectually*: J. Tyndall, 'On the History of Calorescence', *Phil. Mag.*, 29 (1865), pp. 218–31 (this is letter 2245 in this volume). The whole debate began with C. K. Akin, 'On the History of Force', *Phil. Mag.*, 28 (1864), pp. 470–7.
8. *Johnny*: Rudolf John Clausius.
9. *excellent* wife: Adelheid Clausius.

To *The Reader* 1 February 1865 2223

ON THE ELASTICITY OF STEEL WIRES. |
Royal Institution, Feb. 1.

THE following remarks on the torsion of steel wires came to me in two private notes.[1] Believing them to be of interest to the readers of this journal, with Professor Maxwell's[2] permission I publish them.

John Tyndall.

[. . .][3]

J. Tyndall, 'On The Elasticity of Steel Wires', *The Reader*, 5 (4 February 1865), p. 144

1. *two private notes*: for Maxwell's letters that Tyndall forwarded to *The Reader*, see letter 2221 (the original letters from Maxwell are missing).
2. *Professor Maxwell's*: see letter 2221, n. 1.
3. Tyndall inserted letter 2221 here.

From Clarence King[1] [1 February 1865][2] 2224

Dear Sir,

I beg leave to offer to you a notice,[3] by Professor Whitney[4] of our explorations in the Southern Sierra Nevada.[5]

Your name, so well known in scientific mountaineering, was often spoken by my companion and myself as we climbed the granite crags and cut our stairways in the trackless snows of a new mountain group.

After three days of dangerous climbing we reached the summit of our peak, just as the chronometer marked noon and ringing our hammers on the topmost pinnacle we called the giant, Mt. Tyndall.[6]

Unlike your Alpine work, ours had neither the interest nor peril of existing glaciers; but we constantly traversed the scored and polished surfaces of their ancient beds.

We hope that you may hereafter extend your travels to the Pacific coast, where you will always find a cordial welcome in the camp of the "California Geological Survey"

I have the honor to be | your obedient servant | Clarence King | Irvington New York State | U.S. | Feb 1st[7]

Professor John Tyndall F.R.S.

RI MS JT/2/13c/1286

---

1. *Clarence King*: Clarence King (1842–1901), an American geologist and student of Josiah Whitney. King worked on the California Geological Survey in the early 1860s and would later go on to help establish the U.S. Geological Survey in 1879 (*CDSB*).
2. *[1 February 1865]*: The date 'Feb 1st' is written next to the signature in a different hand. Tyndall mentioned receiving this letter in his journal entry for 7 March 1865: 'Had a pleasant letter from America; a member of the California Geological Survey—Mr King—has scaled a noble mountain in that region and called it Mount Tyndall. The American Journal of Science for January contains a brief account of his achievement' (RI MS JT/2/13c/1285). For the referred to account, see n. 3.
3. *notice*: J. D. Whitney, 'Notice of the Explorations of the Geological Survey of California, in the Sierra Nevada, during the summer of 1864', *The American Journal of Science and Arts*, 39 (1865), pp. 10–13.
4. *Professor Whitney*: Josiah Dwight Whitney (1819–96), an American geologist, mountaineer, and director of the California Geological Survey. At the time of this letter he had just taken up the inaugural position of Stugis-Hooper Professor of Geology at Harvard University where he would establish the School of Mining and Practical Geology (*CDSB*).
5. *our explorations in the Southern Sierra Nevada*: King and Whitney travelled to the Sierra Nevada in eastern California as part of the Geological Survey of California in the early 1860s.

6. *Mt. Tyndall*: part of the California Geological Survey included an ascent and measurement of a peak (on 6 July 1864) that King named Mount Tyndall (it rises to just over 14,000 feet). This was part of the Mount Whitney mountain range, named after Josiah Whitney. King recounts the ascent and descent of Mt. Tyndall in chapters 3 and 4 in his *Mountaineering in the Sierra Nevada* (Boston: James Osgood & Co., 1872), pp. 49–93.
7. *Feb 1st*: this appears to be written in darker ink, perhaps later, likely in a different hand.

To William Francis 2 February 1865 2225

The Athenaeum | 2nd. February 1865

Dear Francis

Would you, as editor of the Philosophical Magazine, kindly procure for me the note referred to by Dr. Akin at page 148 of your last number,[1] and which Dr. Akin states that he holds at my disposal? It was written to him from the Isle of Wight in April 1864.[2]

Yours very truly | John Tyndall

St Bride Library, Authors' letters

1. *at page 148 of your last number*: C. K. Akin, 'Further Statements concerning the History of Calcescence', *Phil. Mag.*, 29 (1865), pp. 136–51.
2. *written to him . . . in April 1864*: letter missing.

From Louis Jules Duboscq[1] 2 February 1865 2226

J. DUBOSCQ, | Eleve & Succ[esseu]r de SOLEIL, | 21, RUE DE L'ODÉON, | au fond de la Cour. | Paris, le 2 Février 1865 | Monsieur Tyndall | à Londres

Monsieur,

J'ai lu avec un grand intérêt votre belle expérience de la chaleur obscure après le passage des rayons lumineux au travers d'une dissolution d'iode dans le sulfure de carbone. J'ai essayé de la répéter sans pouvoir la réussir; Je vous serai très obligé de me donner quelques détails sur la manipulation; quelle épaisseur a la cuve? Son diamètre? Si elle est fermée avec des plaques de verre ou de quartz? Quelle lentille prenez vous, est-ce celle en verre ou en sel gemme ou est-ce une lentille liquide? Son foyer et son diamètre? S'il y a quelques précautions particulières vous m'obligeriez de me les faire connaître car j'ai le désir de populariser cet expérience, car je fais très souvent de la lumière dans les cours publics.

En attendant l'honneur d'une réponse, veuillez Monsieur, agréer l'assurance de mes sentiments distingués.

Votre Serviteur | M Duboscq

J. DUBOSCQ, | Student & Successor of SOLEIL, |
21 ODÉON STREET, | at the end of the courtyard,[2] |
Paris, 2 February 1865 | Mr. Tyndall | in London

Sir,

I read with much interest about your wonderful experiment on obscure radiation after rays of light are passed through a iodine dissolved in carbon disulfide.[3] I tried to repeat it but was unsuccessful in the attempt. I would be much obliged if you gave me some details about the experiment. What was the thickness of the vat? Its diameter? Is it closed with glass or quartz plates? Which lens did you use, glass or rock salt, or is it a liquid lens? Its relative aperture and its diameter? If there are any special preparations required, kindly let me know, because I wish to popularize this experiment, since I often demonstrate light in public courses.

As I await the honour of your reply, please Sir, accept the assurance of my best regards.

Your Servant | Mr Duboscq

RI MS JT/1/D/182

---

1. *Louis Jules Duboscq*: Louis Jules Duboscq (1817–86), a French optical instrument maker and photographer. Duboscq was apprenticed to Jean Baptiste François Soleil (1798–1878), and later took over his instrument workshop (*CDSB*). Duboscq was particularly innovative in adapting new scientific discoveries to photography, particularly the work of Charles Wheatstone's (1802–75) investigations into stereoscopic vision. See S. Herbert, 'Duboscq, Louis Jules (1817–1886)', in S. Hannavy (ed), *Encyclopedia of Nineteenth-Century Photography*, 2 vols (New York: Routledge, 2008), vol. 1, pp. 445–6.
2. *J. Duboscq . . . end of the courtyard*: this is from the letterhead.
3. *your wonderful experiment . . . in carbon disulfide*: J. Tyndall, 'On Luminous and Obscure Radiation', *Phil. Mag.*, 28 (1864), pp. 329–41. Tyndall had published this article during a priority dispute with C. K. Akin (see letter 2222 to Rudolf Clausius).

To William Armstrong[1] 3 February 1865 2227

Royal Institution | 3rd Feb. 1865

My dear Sir William,

I have been working as you know for several years on the subject of radiant heat.

More than three years ago I discovered that a solution of iodine in bisulphide of carbon which intercepted the light of our most intense sources allowed the invisible radiant heat to pass almost completely. (Phil. Trans. 1862, p. 67. note)[2]

I was diverted from experiments which I then intended to make with this substance, in order to clear up the differences between my results and those of Prof. Magnus[3] regarding the action of aqueous vapour on radiant heat.

My experiments in the autumn of 1863, published in the Phil. Trans. for 1864,[4] added the last links to the chain of evidence regarding the action of aqueous vapour. Towards the end of March last year I had devised a series of experiments on the visible rays of the electric light, intending to employ the substance which I had discovered two years previously to detach the luminous from the non-luminous portion of the radiation.

As is frequently the case in science others were occupied at the same time with ideas of a similar character.

At the Newcastle Meeting of the British Association[5] Dr. Akin,[6] a Hungarian gentleman who arrived in this country in 1862, proposed three experiments, and a grant of money was then voted to Mr Griffith and Dr Akin to carry them out.[7]

Up to the present moment not one of the experiments has been made, and two out of three of them are altogether impracticable. In the third it was proposed to concentrate the sun's rays by a large burning mirror, to cut off the light by "proper absorbents" (not named) and then to operate on the invisible rays.

This idea is Dr. Akin's indisputable right. But he evidently considers that the vote of money to him at Newcastle gives him a monopoly of the subject. I forward you an article[8] which has been published by Dr. Akin in the last number of the Philosophical Magazine, and at page 41 have marked a portion in which Dr. Akin sets forth what he considers to be his rights and mine.

Might I beg of you as President of the British Association,[9] and one who has the interests of science at heart to favour me with your opinion whether the grant of money voted to Dr. Akin at Newcastle deprived me of the right of making the experiments with the electric light which I had proposed to myself last March?[10] I may remark that out of deference to Dr. Akin I did not then make the experiments I intended to perform; but what I want to know is whether Dr. Akin is justified in denying me the right of making them had I chosen to do so?

Yours very Sincerely | J. Tyndall

RI MS JT/1/TYP/1/61–2
LT Typescript Only

---

1. *William Armstrong*: William George Armstrong (1810–1900), first Baron Armstrong, an English arms manufacturer who founded W. G. Armstrong & Co. (1847). Armstrong was well-known for his work on hydraulics and electrostatics, service to the BAAS, and involvement in a number of engineering institutions. He was knighted in 1859 for giving the patent of the Armstrong gun to the Crown and was raised to the peerage as Baron Armstrong of Cragside in 1887 (*ODNB*).
2. *(Phil. Trans. 1862, p. 67. note)*: in J. Tyndall, 'On the Absorption and Radiation of Heat by Gaseous Matter.—Second Memoir', *Phil. Trans.*, 152 (1862), pp. 59–98, Tyndall provided the following note on p. 67: 'A layer of liquid bromine, sufficiently opake to intercept the entire luminous rays of a gas-flame, is highly diathermanous to its obscure rays. An opake solution of iodine in bisulphide of carbon behaves similarly. The details of these experiments shall be published in due time: they were publicly shown in my lectures many months ago.—June 13th, 1862'. For the future publication referred to, see letter 2226, n. 3.
3. *Prof. Magnus*: Heinrich Magnus.
4. *Phil. Trans. for 1864*: J. Tyndall, 'On the Absorption and Radiation of Heat by Gaseous and Liquid Matter.—Fourth Memoir', *Phil. Trans.*, 154 (1864), pp. 201–25.
5. *Newcastle Meeting of the British Association*: the BAAS held its thirty-third meeting at Newcastle Upon Tyne from 26 August to 2 September 1863 (*Brit. Assoc. Rep. 1863*; 'Analysis of Attendances at the Annual Meetings, 1831–1906', *Brit. Assoc. Rep. 1907*, p. lxxxviii).
6. *Dr. Akin*: C. K. Akin. See letters 2222 and 2226.
7. *proposed three experiments . . . to carry them out*: C. K. Akin, 'On the Transmutation of Spectral Rays—Part I', *Brit. Assoc. Rep. 1863*, pp. 93–105. For the announcement of the grant, see p. xxxix of the same publication. At the BAAS meeting the following year in Bath (1864), the *Report* states, 'That Professor Griffith and Dr. Akin be requested to continue their Report on the Transmutation of Spectral Rays' (*Brit. Assoc. Rep. 1864*, p. li).
8. *article*: C. K. Akin, 'On Calcescence', *Phil. Mag.*, 29 (1865), pp. 28–43.
9. *President of the British Association*: Armstrong served as President of the BAAS for 1863, giving a well-received address at the annual meeting on engineering history and coal exhaustion (*Brit. Assoc. Report 1863*, pp. li–lxiv).
10. *favour me with your opinion . . . to myself last March?*: while Armstrong's reply to Tyndall's letter is missing, Tyndall did quote from it in an article in the March issue of *Phil. Mag.*, writing that Armstrong stated that the BAAS 'could not, even if they proposed anything so foolish, confer upon Dr. Akin any exclusive privileges in the matter' (J. Tyndall, 'On the History of Calorescence', *Phil. Mag.*, 29 (1865), pp. 218–31, on p. 222 (this is letter 2245 in this volume).

To John Herschel [7 February 1865][1] 2228

Royal Institution of Great Britain | Tuesday | Feb 7/65

My dear Sir John,

I was misinformed when I stated[2] that the refraction and dispersion of mercury Ethyl & Mercury Methyl were not yet determined. They have been determined by Messrs Gladstone[3] and Dale[4] and reported in Phil. Trans. 1863 p. 328[5]

Yours very truly | John <u>Tyndall</u>

Specific refractive energy

Mercuric Methyl ·1707

Ethyl ·2112

[6 words illeg]

Proceedings RS. N° 55 p. 451

The <u>indices</u> and dispersions are given in the Paper in PT.[6]

RS HS 17/399a

RI MS JT/1/TYP/2/536

1. *[7 February 1865]*: the date 'Feb 7/65' is written later in a different hand, likely Herschel's.
2. *when I stated*: letter missing.
3. *Gladstone*: John Hall Gladstone (1827–1902), an English chemist who trained at University College London before completing his PhD at the University of Giessen under Justus Liebig. Gladstone was FRS, and from 1874 was the Fullerian Professor of Chemistry at the RI (*ODNB*).
4. *Dale*: Thomas Pelham Dale (1821–92), an Anglican clergyman who was trained at the University of Cambridge, and rector of St. Vedest church in London. Dale collaborated regularly with Gladstone on mass spectrometry. In 1880 he was arrested and incarcerated for his belief in 'ritualism', which was seen as an attempt to reintroduce catholic practices into the Church of England (*ODNB*).
5. *Phil. Trans. 1863 p. 328*: J. H. Gladstone and T. P. Dale, 'Researches on the Refraction, Dispersion, and Sensitiveness of Liquids', *Phil. Trans.*, 153 (1863), pp. 317–43.
6. *Specific refractive energy . . . the Paper in PT*: these are notes added to the letter by Herschel. The referred to article in the 'Proceedings' is: J. H. Gladstone and T. P. Dale, 'Researches on the Refraction, Dispersion, and Sensitiveness of Liquids', *Roy. Soc. Proc.*, 12 (1863), pp. 448–53; the 'Paper in PT' is that referenced in n. 5.

From John Herschel 8 February 1865 2229

Collingwood[1] | Feb. 8/65

My dear Sir

Many thanks for your Memoir No 3.[2] also for the Mercury Ethyl & Methyl which arrived quite safely[3]—and which I am very pleased to obtain & beg to thank Messrs Frankland and Duppa[4] for them.

Yours very truly | J F W Herschel

RI MS JT/1/H/104
RI MS JT/1/TYP/2/537

1. *Collingwood*: Herschel's home near Hawkhurst in Kent.
2. *your Memoir No 3.*: J. Tyndall, 'On the Relation of Radiant Heat to Aqueous Vapor', *Phil. Trans.*, 153 (1863), pp. 1–12.
3. *Mercury Ethyl & Methyl which arrived quite safely*: see letter 2228.
4. *Duppa*: Baldwin Francis Duppa (1828–73), an English organic chemist and collaborator with Edward Frankland. He studied at the Royal College of Chemistry, and later worked with Frankland at St. Bartholomew's Hospital on the chemical nature of gasses (C. A. Russell, *Edward Frankland: Chemistry, Controversy and Conspiracy in Victorian England* (Cambridge: Cambridge University Press, 1996), pp. 248–9).

To George Airy 8 February 1865 2230

8th Feb. 1865

My dear Sir

I wish from my heart that I could attend these lectures of yours[1] at South Kensington.[2] The syllabus is an admirable idea.

Yours very truly | John Tyndall.

G. B. Airy Esq[uire] | &c. &c. &c.

RGO MS.RGO 6/690

1. *these lectures of yours*: Airy delivered a course of three lectures on 'Magnetical Errors, Compensations, and Corrections, with Special Reference to Iron Ships and their Compasses' in the Old Lecture Theatre at the Royal School of Mines on 9, 16, and 23 March 1865 (the syllabus for this course was published in *The Nautical Magazine and Naval Chronicle*, 34 (1865), pp. 232–47).

2. *South Kensington*: the South Kensington Museum. This British government department, a subdivision of the Board of Trade, was responsible for promoting education, art, science, technology, and design in Britain and Ireland (from 1853–99). See B. Robertson, 'The South Kensington Museum in context: an alternative history', *Museum and Society*, 2 (2004), pp. 1–14. In 1859, Tyndall became an Examiner for Science for the Department of Science and Art (sometimes Science and Art Department) at the South Kensington Museum, where he also gave lectures.

From Henry Wilkinson Cookson 11 February 1865 2231

S[t]. Peters College, | Cambridge. | Feb. 11, 1865.

Sir

I shall be glad if you will allow me to nominate you Sir Rob[er][t] Rede's Lecturer in this Univ[ersi][ty] for the present year.[1] The Astronomer Royal[2] held the office last year.[3]

A simple lecture is given in the Senate house to a mixed audience, including ladies, and the subject chosen is generally broad enough to excite general interest.

I enclose a paper of last year's announcement. I am afraid that there is not much convenience for any experiments.

I am | Yours faithfully | H. W. Cookson | Vice chancellor.

RI MS JT/2/13c/1282

1. *Sir Rob[er][t] Rede's Lecturer in this Univ[ersi][ty] for the present year*: Tyndall delivered the annual Rede Lecture in the Senate House at the University of Cambridge ('On Radiation') on 16 May 1865, later published as *On Radiation*.
2. *Astronomer Royal*: George Biddell Airy.
3. *last year*: Airy delivered the Rede Lecture on 10 May 1864 on 'The Late Observations of Total Eclipses of the Sun, and the Inferences from them' (G. B. Airy, *A Treatise on Magnetism Designed for the use of Students in the University* (London: Macmillan and Co., 1870), p. v). See letter 2233.

From Jane Hogan[1] 12 February 1865 2232

37 Stanley Street | Macclesfield | Cheshire | 12th Feb. 65

Sir,

I hope you are not displeased at my not paying you that money you so kindly lent me some months ago,[2] it is with much regret that I have never since been able to do so. I have not had since any help from my sister Mrs.

Ginty,[3] but am anxiously expecting it, as she promised to send me assistance soon, and as soon as I receive it from her I shall repay your kindness. I have written to her again by the last Mail, and I hope to have an answer before long. Since Xmas I have had a great deal of trouble, I may say a series of misfortunes. I have been ill myself and my youngest little girl, I could not work and consequently I was obliged not only to part with my clothes, but I had to raise money on my work also, I know it was wrong but at the time what else could I do, I could not see my children perish for want, but it has caused me a great deal of trouble and I now bitterly regret it, I would not on any account do so again even to save our lives, I would not have either done so then only I had expected help from Mrs. Ginty, but I was sadly disappointed—I am now, not only in debt but in danger, and I dare not mention it to any one here, lest the lady[4] to whom it belongs may hear of it, neither must I tell any one who would be likely to tell Mr.[5] or Mrs. Ginty. I am very unhappy about it indeed more so than I can tell. I do not know the moment the lady may call for her work. She sent me word a few days ago that she would call very soon and that I must have it ready for her. I cannot tell you how very miserable I am I have not the least prospect of being able to get out of this trouble and if I cannot get over it I do not know what will become of me and my children for it must be a month and it may be two before I get help from Mrs. Ginty. Why I now write to you and telling you of my trouble is to beg of you that you will once more help me out of this great trouble. If you will kindly lend me one pound and five shillings, it will enable me to get out of this danger and great trouble. I am very much afraid the lady will call on Monday or Tuesday and I am so unhappy about it, if you will kindly let me know by return of post I shall feel indebted to you while I live and please God I shall surely pay you the first opportunity I have which I hope will be soon. I would not ask you only I have no other resource at present. I beg of you Sir to let me know by return of post, I shall ever be obliged to you.

I am Sir, Yours most respectfully | and grateful. | Jane Hogan.

RI MS JT/1/TYP/11/3646
LT Typescript Only

---

1. *Jane Hogan*: Jane Hogan (née Roberts, dates unknown), the sister of Margaret Ginty.
2. *money you so kindly lent me some months ago*: in October 1864, Tyndall received a letter from Jane Hogan thanking him for lending her £2 (see letter 2158, *Tyndall Correspondence*, vol. 8).
3. *Mrs. Ginty*: Margaret Ginty.
4. *lady*: not identified.
5. *Mr.*: William Ginty.

From Henry Wilkinson Cookson 14 February 1865 2233

S^t^. Peters College, | Cambridge. | Feb. 14, 1865.

Dear Sir

The Rede Lecture was given last year on the 10th of May at half past 2 P.M.[1] The Astronomer Royal[2] found that that was the most likely time to find the students at liberty. The boat races began on the 11th! And most of the students about the 10th were dinning at 2.—I will gladly assist you in fixing upon a convenient time.[3]

The subject which you mention would be as good as possible, but the windows in the Senate House cannot be closed and the Optical Lecture room would be too small for the audience.[4]

I can only say that it is not necessary to announce the subject at present unless you are quite prepared to do so.

Shall I announce the subject which you have mentioned or would you prefer to leave the matter open?

I have mentioned the matter to Prof[esso]r. Stokes who agrees with me as to the unsuitableness of the Optical Lecture room for so large a class. I am sure that he and every one else here will be glad to lend you every assistance and will rejoice that you undertake the office.

I am | Yours very faithfully | H. W. Cookson V.C.[5]

Prof[esso]r. Tyndall

RI MS JT/2/10
RI MS JT/2/13c/1281–2

1. *The Rede Lecture . . . half past 2 P.M.*: see letter 2231, n. 3.
2. *Astronomer Royal*: George Biddell Airy.
3. *fixing upon a convenient time*: see letter 2231, n. 1.
4. *The subject which . . . for the audience*: Tyndall's letter to Cookson, likely describing the logistics of giving a lecture on radiation, is missing.
5. *V.C.*: Vice Chancellor.

To Alfred Newton[1] [16 February 1865][2] 2234

Royal Institution of Great Britain | Thursday

Dear Sir,

I am much obliged to you for permitting me to see your sketch[3] of a Spitzbergenglacier.[4] It seems to me very accurate. At all events the air portion of the crevasses & Dirt Band is just such as might be inferred from the character of the ground.

Yours very truly | John Tyndall

The porter informs me that you will be in town again soon. I shall then hand you your sketch & thank you personally.

Alfred Newton Papers, Cambridge University Library Add.9839/1T/296

1. *Alfred Newton*: Alfred Newton (1829–1907), a British ornithologist and graduate of the University of Cambridge, who travelled extensively as a naturalist in Lapland, Iceland, the West Indies, and in 1864, Spitzbergen (*ODNB*).
2. *[16 February 1865]*: the date is written on the back of the letter, probably in Newton's hand: 'J. Tyndall, London. Feb. 16/17 /65. | Answered Feb 17/65'. Newton's reply is missing.
3. *your sketch*: this possibly refers to a written manuscript rather than a drawing that Newton would have sent to Tyndall, perhaps to have his opinion on what he wrote about the glacier before he presented some 'Notes on Spitzbergen' to the Cambridge Philosophical Society on 27 February 1865 (see *Phil. Mag.*, 30 (1865), pp. 376–7).
4. *Spitzbergenglacier*: a glacier on the island of Spitsbergen in the Norwegian archipelago Svalbard.

From Rudolf Clausius 19 February 1865 2235

Zürich, 19/2 65.

Lieber Tyndall,

Ich habe den Aufsatz von Akin im letzten Hefte des Phil. Mag. gelesen, und ich glaube, daß Du ganz Recht hast, wenn Du annimmst, daß die Sprache, welche Akin gegen Dich angewandt hat, Deine Freunde noch mehr indignirt, als Dich selbst. Du kannst auch ganz ruhig darüber sein, daß dieser Artikel Deinem Rufe in der wissenschaftlichen Welt keinen Schaden thut, denn es leuchtet aus dem ganzen Artikel überall ein gewisser krankhafter Zustand hervor, in welchem sich Akin befinden muß, und welcher ihn veranlaßt, alles schwärzer zu sehen, als es ist. Er überschätzt offenbar sein eigenes Verdienst im hohen Grade, und scheint jeden, der darin nicht mit ihm übereinstimmt, als seinen Feind zu betrachten. Besonders unüberlegt finde ich die Forderung, daß darum, weil er sich mit einem gewissen wissenschaftlichen Gegenstande beschäftigt, kein anderer sich damit beschäftigen dürfte. Auch der Umstand, daß die British Association ihm eine Unterstützung zu seiner Untersuchung bewilligt hat, kann ihn in keiner Weise zu einer solchen Forderung berechtigen, denn die British Association kann doch über die verschiedenen Theile der Wissenschaft nicht in der Weise verfügen, daß sie dieselben einzelnen Personen zur Bearbeitung übergiebt und dadurch andere Personen davon ausschließt.

Ich will nur hoffen, daß diese Sache Dir nicht zu viel Verdruß macht. Zum erfolgreichen Arbeiten in der Wissenschaft ist vor Allem Seelenruhe nothwendig, und die mußt Du Dir daher erhalten, und Dich durch diesen Streit nicht mehr als er es verdient, aufregen laßen.

Die Nachricht, daß Du zu meinem Buche eine Vorrede geschrieben hast, hat mich sehr erfreut. Ich wüßte keinen Namen in der Wissenschaft, mit dem ich meinen Namen so gern vereinigt sehe, wie mit dem Deinigen.

Der junge Wynne, welcher vor einiger Zeit von hier abgereist ist, ist hoffentlich glücklich in London angekommen. Er hat hier in Zürich bei allen, die ihn kennen gelernt haben, eine angenehme Erinnerung hinterlassen, und wir wünschen ihm von Herzen, daß es ihm in seiner zukünftigen Carriere immer gut gehe.

Mit herzlichen Grüßen von meiner Frau und mir

Dein | Clausius.

Zurich, 19/2 65.

Dear Tyndall,

I have read Akin's essay in the last issue of the Phil. Mag.[1] and I think that you are completely right when you assume that the language that Akin has used against you makes your friends even more indignant than yourself. You can be quite assured that this article will not harm your reputation in the scientific world, for throughout there shines forth a certain pathological state in which Akin must find himself, and which induces him to see things even more negatively than they really are. He seems to greatly overrate his own merits and to view anyone who does not agree with him as his enemy. I find his demand that since he is dealing with a certain scientific topic, no one else has the right to investigate it, particularly irrational. Also, the circumstance that the British Association has supported his investigations[2] cannot in any way justify such demand, for the British Association cannot decide over different parts of science in the sense of assigning them to certain individuals, while excluding others from them.

I only hope that this topic is not causing you too many annoyances. In order to work productively in science, one needs serenity above all: this you must maintain and not let yourself be irritated by this quarrel more than it deserves.

I was very pleased to hear that you wrote a foreword to my book.[3] There is no name in science that I would wish more to see associated with my own than yours.

I hope that young Wynne,[4] who left from here a while ago, has arrived safely in London. He left a pleasant impression in everyone who made his acquaintance here in Zurich, and we wish him from heart all the best in his future career.

With heartfelt greetings from my wife[5] and myself
Your | Clausius.

RI MS JT/2/10

1. *Akin's essay in the last issue of the Phil. Mag.*: see letter 2227, n. 8.
2. *British Association has supported his investigations*: see letter 2227, n. 10.
3. *you wrote a foreword to my book*: see letter 2222, n. 4.
4. *young Wynne*: Frank George Wynne (dates unknown), the third son of Tyndall's longtime friend George Wynne (1804–90), an Irish surveyor, engineer, and military officer, and his wife Anne Wynne (née Osborne, 1808–64). Clausius wrote to Tyndall of Wynne's coming to Germany in letters 2147 and 2191 of *Tyndall Correspondence*, vol. 8.
5. *my wife*: Adelheid Clausius.

From George Griffith 19 February 1865 2236

British Association | 19 Feb. 1865

My dear Sir,

I enclose the only document which Dr Akin sent to the Bath meeting of the Association.[1]

I will send you some information about his doings here by tomorrow's post.[2]

I am | very truly yours | George Griffith.

RI MS JT/1/TYP/1/63
LT Typescript Only

1. *I enclose . . . meeting of the Association*: see letter 2227, especially n. 5 and n. 7. Enclosure missing.
2. *by tomorrow's post*: see letter 2239.

To Alfred Newton [18 February 1865][1] 2237

Saturday

Dear Sir

I have read your note[2] with much interest: it excites in me the desire to see what you have seen & describe so well.

The ends of many glaciers are sensibly stationary in winter, though the motion continues in the upper parts. A thickening of the glacier is a consequence of this, as was pointed out by Prof. Forbes.[3]

But the cause you assign may be the true one.[4] The glacier may melt underneath, or the superficial melting may be carried to the bed, which may be of so porous a character as to convey the water by subterranean rivulets to the sea.

It is hardly possible to be more precise than this in the absence of observations.

Yours very truly | John Tyndall

This post will convey the sketch to you.[5]

Alfred Newton Papers, Cambridge University Library Add.9839/1T/297

1. *[18 February 1865]*: the date is written on the back of the letter, probably in Newton's hand: 'J. Tyndall, London. Feb. 18/19 /65. | Answered Feb 19/65'. Newton's reply is missing.
2. *your note*: letter missing. See letter 2234.
3. *Prof. Forbes*: James David Forbes.
4. *But the cause you assign may be the true one*: Tyndall's own theory of glacial motion—through the fracture and regelation of ice—is the topic of many letters in, and discussed in the introduction to, *The Correspondence of John Tyndall*, vol. 6.
5. *convey the sketch to you*: see letter 2234, n. 3.

To George Gabriel Stokes 19 February [1865][1] 2238

Royal Institution of Great Britain | 19th Feb[ruar]y

My dear Stokes,

You can save me some trouble if you would be good enough to answer the following question:—Has olive oil—fine Lucca oil—been examined spectroscopically and with reference to its fluorescent power? I ask the question because its absorption spectrum is very remarkable, and also its fluorescence.

Yours very sinc[ere]ly | John Tyndall

RI MS JT/1/T/1417
RI MS JT/1/TYP/4/1475

1. *[1865]*: the year is given by insertion of a note on both the handwritten and typescript letters indicating '1865' as a possibility.

From George Griffith 20 February 1865 2239

British Association for the | Advancement of Science. | Oxford.

20 Feb. 1865.

My Dear Sir,

You will have seen that the paper which I sent you yesterday[1] contains no intimation from Dr Akin[2] that he intended to give over scientific researches—After the meeting he wrote a letter to the President of the Association[3] which he sent to me—whether he intended that it should be forwarded to the President or whether it is a copy of a letter which he has sent to him, I cannot say. In this letter he states that he would be unable to continue his researches in the transmutation of rays in conjunction with me. He has published an extract from this letter in the Jan. number of the Phil. Mag.[4] I was the person to whom he ought to have communicated his inability to act, being the first named on the Committee. I have had no communications with him since last August. He is one of those unfortunate men who cannot help making themselves disagreeable. He came to Oxford in 1862, having previously stayed at Cambridge for some time. He *[has]* *[put]* no introductions, but by attending my lectures he made himself known to me. He spoke of Prof. Stokes as a friend & showed me some letters which he had received from Prof. Stokes. This gradually led me to listen to his proposal of making some experiments on ray transmutation. I saw by his conversation & writings that he was a man of considerable information on mathematical & physical subjects, but I never imagined that it was possible for anyone much less such a person as Dr Akin is in theoretical knowledge, to have been so utterly helpless in devising suitable experiments & in carrying out the simplest details in practice. I must not occupy your time with recounting his proposals—but should you visit Oxford I will show you one of his theories in the tangible form of a lamp—a modified Bunsen's burner. I introduced him to many of my friends here, but it was strangely impossible for him to "get on" with any one. I had very great difficulty in working with him, but in spite of this I continued the experiments; At the Newcastle meeting of the Association[5] he communicated the results of these to Section A,[6] which led to the appointment of a Committee consisting of Dr A. & myself—with a grant of £40.[7] The results of these experiments he has published in the Phil. Mag. for Feb.[8] I intended to bring them before the Association when I was at Bath,[9] but was prevented by my business.

I think that you were quite justified in getting rid of Dr Akin as you did. He is morally incapable of working in concert with any one. Before he came to England he had tried to fasten himself on Regnault[10] in Paris but failed to do so.

I heard your lecture at the Royal Institution[11] with great pleasure. If I can get a rock salt lens I hope to show these experiments to our Ashmolean Society[12]

I am | very faithfully yours | George Griffith

RI MS JT/2/10/331
RI MS JT/2/13c/1283–4

1. *paper which I sent you yesterday*: Griffith enclosed the paper with letter 2236, but the enclosure is missing. For the paper, see letter 2227, n. 7.
2. *D^r. Akin*: C. K. Akin.
3. *President of the Association*: William Armstrong (see letter 2227, n. 1).
4. *published an extract from this letter in the Jan. number of the Phil. Mag.*: see letter 2227, n. 8.
5. *Newcastle meeting of the Association*: see letter 2227, n. 5.
6. *communicated the results of these to Section A*: see letter 2227, n. 7.
7. *grant of £40*: for the announcement of the award, see *Brit. Assoc. Rep. 1863*, p. xxxix.
8. *Phil. Mag. for Feb.*: C. K. Akin, 'On the Conservation of Force', *Phil. Mag.*, 29 (1865), pp. 205–6.
9. *at Bath*: the thirty-fourth annual meeting of the BAAS was held in Bath from 14–21 September 1864 (*Brit. Assoc. Rep. 1864*).
10. *Regnault*: Henri Victor Regnault.
11. *your lecture at the Royal Institution*: on 20 January 1865, Tyndall delivered a Friday Evening Discourse at the RI, 'On Combustion by Invisible Rays' (*Roy. Inst. Proc.*, 5 (1862–6), pp. 329–35).
12. *Ashmolean Society*: a scientific society in Oxford founded in 1828.

From George Rush[1] 21 February 1865 2240

21st. Feb[ruar]^y. 1865.

Sir

The Committee of this Society[2] have requested me to beg you will do them the favor to allow your name to be added to the list of Vice-Presidents of this Institution; in anticipation of their obtaining your consent, you were unanimously elected at the last meeting, and I have the honor to request you will allow me to include your name in our forthcoming report.

I am | Sir | Your most obedient servant | Geo[rge] Rush | Hon[orary] Sec[retary].

J. Tyndall Esq[uire]. F.R.S.

RI MS JT/2/10/334
RI MS JT/2/13c/1285

---

1. *George Rush*: not identified, but at this time he was honorary secretary of the St. John's Wood Literary and Scientific Institution.
2. *this Society*: the St. John's Wood Literary and Scientific Institution, which operated out of 4 Blenheim Place between 1858 and 1915 (M. Brown, 'The Jews of Early St John's Wood', *Jewish Historical Studies*, 35 (1996), pp. 141–52, fn. 13). In the typescript of this letter, LT included '4 Blenheim Place' and 'Literary Society', which were probably on the envelope.

From George Griffith 24 February 1865 2241

British Association | Oxford. | 5 Park Villas | 24 Feb. 1865

Dear Dr. Tyndall,

I ought to have replied to your letter[1] before. How could you publish any letter that I might send you about Dr. Akin?[2] I suppose that you intend to reply[3] to his letter in the Phil. Mag.[4]

Last night I was asked by a friend what had become of him? I was told by my friend that one day last summer he came to my room at the museum.[5] I was away & Dr. Akin offered to show him the solar spectrum with my large spectroscope but utterly failed to do so!

I shall be glad to have his letter to Cayley[6] returned when you have finished with it.

Unless you are very desirous that I should do so, I would prefer not to write anything that is to be printed. It would probably lead to further trouble.

I am | Yours very truly | George Griffith

RI MS JT/2/10/332
RI MS JT/2/13c/1284

---

1. *your letter*: letter missing, but see letter from Griffith to Tyndall on 20 February 1865 (2239).
2. *Dr. Akin*: C. K. Akin.
3. *intend to reply*: Tyndall's reply to Akin in the *Phil. Mag.* is letter 2245 in this volume.
4. *his letter in the Phil. Mag.*: see letter 2225, n. 1.
5. *museum*: probably the original Ashmolean Museum at Oxford, which had connections with members of the Ashmolean Society founded in 1828, of which Griffith was elected a member on 3 November 1856 (*Abstracts of the Proceedings of the Ashmolean Society*, 3 (1856), p. 111; M. Price, 'The evolution of the Ashmolean Natural History Society 1880–1901', *Fritillary*, 7 (2018), pp. 84–100).
6. *Cayley*: Arthur Cayley (1821–95), a British mathematician who was Sadlerian Professor of mathematics at the University of Cambridge (*ODNB*).

To Rudolf Clausius 26 February 1865 2242

26th. February 1865

My dear Clausius

Accept my thanks for your letter:[1]—it is that of a true friend.

With regard to Dr Akin,[2] I shall endeavour to take your advice, and indeed at present I have got over the annoyance which his first attack caused me. You see stamped upon his countenance the marks of a quarrelsome and disagreeable disposition. He brought no letters of introduction to this country, tho he managed to introduce himself. He was for a time associated with Mr. Griffith of Oxford,[3] a most amiable man, but Akin could not agree with. Mr. Griffith introduced him to many of his friends, but Akin rendered himself disagreeable to all of [them]. Griffith pronounces him to be a man "morally incapable of working in concert with anybody." He tried to fasten himself on to Regnault[4] before he came to England, but failed to do so. All my difficulty with Akin has arisen out of my desire to be kind to him; but he is a man of that dark and brooding mind that can convert kindness into gall, as a wasp converts honey into poison.

His attack upon me[5] may be gratifying to a few; but with the general mass of the scientific public he has done himself far more harm than he has done me. I was intensely annoyed by it for a time. From childhood to manhood nobody had ever [spoiled] my name with anything untrue or unfair, and it harmed me much to find any human being capable of writing about me as Akin had done. If all my interviews were added together they would not amount to more than three hours' conversation, and yet it is this man whom nobody knows, and who knows nothing of me that has dared to slander me.

I have written a full & temperate reply to his attack.[6] I think it will settle the question. He has given me no peace for the last four months—but I will pay no further attention to him I will write to Francis[7] & ask him to forward you a copy of my reply immediately.

Kindest regards to Mrs Clausius[8] & love to little Johnny.[9]

Yours ever | J. Tyndall

RI MS JT/1/T/191

---

1. *your letter*: letter 2235.
2. *Dr Akin*: C. K. Akin. See letters 2222 and 2235.
3. *Mr. Griffith of Oxford*: George Griffith. See letters 2236, 2239, and 2241.
4. *Regnault*: Henri Victor Regnault.
5. *His attack upon me*: see letter 2225, n. 1.

6. *reply to his attack*: Tyndall's reply to Akin in the *Phil. Mag.* is letter 2245 in this volume.
7. *Francis*: William Francis (editor of the *Phil. Mag.*).
8. *Mrs Clausius*: Adelheid Clausius.
9. *Johnny*: Rudolf John Clausius.

From Charles Lyell 28 February 1865 2243

53, Harley S^t | London. W. | Febr[uar]^y 28 1865

My dear Sir,

As you have been working so successfully on heat[1] will you allow me to ask you a question.[2] In consequence of the eccentricity of the earth's orbit & the effects of precession the winter in one hemisphere is longer than in another; at some periods when the orbit is more eccentric than at present the difference is considerable. Humbold[3] remarked long ago[4] that the effect of this was a great loss of heat by radiation during such protracted winter. But I find from what Herschel[5] & others say that the quantity of light & heat annually received by a given hemisphere is always equal because during the shorter summer when the earth is moving more rapidly when nearer the sun it receives a larger quantity of radiant heat. It seems to me that in reasoning on this subject geologists & meteorologists have forgotten the greater store of heat which must be absorbed during the short but hotter summer.

I want to know whether they have any right to assume that during a lengthened winter the earth would lose more than the excess of heat which its crust receives from the sun. I think they have left this excess entirely out of account & by so doing have imagined a refrigeration of the hemisphere which happens to be exposed to the extremes of heat & cold in consequence of the coincidence of its winter with the aphelion at a time when the opposite hemisphere would have a very mitigated summer's heat & winter's cold because the summer coincided with the aphelion & the winter with the perihelion.

believe me | ever sincerely y[ou]rs | Cha[rles] Lyell

RI MS JT/1/L/50
RI MS JT/1/TYP/3/849

---

1. *working so successfully on heat*: at this time Tyndall was working on radiant heat, and would give the Rede Lecture at the University of Cambridge in May 1865 (see letter 2231, n. 1).
2. *ask you a question*: in 1864 the geologist James Croll (see letter 2264, n. 1) began a correspondence with Lyell about variations in the earth's orbit affecting climate, and this urged Lyell to add a new chapter to the tenth edition of his *Principles of Geology*, 'Vicissitudes in Climate, How Far Influence by Astronomical Changes' (C. Lyell, *Principles of Geology*,

10th edn, 2 vols (London: John Murray, 1867–8), vol. 1 (1867), pp. 268–304). On Lyell, Croll, and Tyndall, see S. Dry, *Waters of the World: The Story of the Scientists Who Unraveled the Mysteries of our Oceans, Atmosphere, and Ice Sheets and Made the Planet Whole* (Chicago: University of Chicago Press, 2019), pp. 31–9; J. R. Fleming, *Historical Perspectives on Climate Change* (Oxford: Oxford University Press, 2005), pp. 73–4; M. Hulme, *Exploring Climate Change through Science and in Society* (New York: Routledge, 2013), p. 63; and C. Summerhayes, *Earth's Climate Evolution* (Chichester: John Wiley & Sons, 2015). pp. 28–36.

3. *Humbold*: Alexander Von Humboldt (1769–1859), a world-famous Prussian explorer and naturalist, who went on expeditions to South America and Russia. His seven-volume *Personal Narrative of Travels to the Equinoctial Regions of the New Continent during the years 1799–1804*, written with Aimé Bonpland (1773–1858) and published in French between 1814 and 1829, was influential to many other explorers and naturalists including Charles Darwin, who read the English translation by Helen Maria Williams (London: Longman, Hurst, Rees, Orme and Brown, 1818–29) (*CDSB*).
4. *remarked long ago*: see A. Humboldt, 'Excentricity of the planetary orbits', in *Cosmos: Sketch of a Physical Description of the Universe*, trans. E. J. Sabine, 4 vols (London: Longman, Brown, Green, and Longmans; London: John Murray, 1849–58), vol. 3, part 2 (1852), pp. 333–7.
5. *Herschel*: see J. F. W. Herschel, 'On the Astronomical Causes which may influence Geological Phaenomena', *Transactions of the Geological Society of London*, 3 (1832), pp. 293–300.

From Edward Sabine February 1865[1] 2244

"DEAR TYNDALL,—Viewing your inquiry simply as an administrative question, Dr. Akin's[2] Report on the Transmutation of Spectral Rays in the British Association Reports for 1863[3] indicated a very wide field of experimental researches. In aid of these the General Committee made a grant of £45[4] to Mr. Griffith[5] and himself. A part of this grant was expended in apparatus to be employed by those gentlemen in the spring or early summer of 1864, the results of which were designed by Dr. Akin to be presented by himself to the Meeting of the British Association at Bath in the September of that year.[6] For this purpose Dr. Akin went to Oxford to join Mr. Griffith, but from some unexplained cause their purpose was frustrated, and 1864 passed without the experiments being made. The grant, however, did not lapse, as part of it had been expended in apparatus, which it appears to be still Dr. Akin's intention to employ (though not at Oxford where there is sunshine, or in London where there is no sunshine suitable for the purpose). Such, so far as I can collect them, is a statement of the facts.

"The object of the British Association in making such grants is to aid the progress of science, by enabling those, who may be so disposed, to make experiments, but by no means to retard its progress by discouraging others from approaching the same ground. One cannot read, I think, Dr. Akin's Report in 1863 without the impression that there is work enough in the subject for a dozen at least of active experimenters. He may have had an advantage (and properly used it is a great one) in meditating longer than others on the laurels that may be gathered, but he who desires to be foremost in the field must be 'prompt in action' where there are so many competitors,

"Always truly yours. | EDWARD SABINE."

J. Tyndall, 'On the History of Calorescence', *Phil. Mag.*, 29 (1865), pp. 218–31, on p. 222

---

1. *[February 1865]*: this letter is included within letter 2245, which was published in the March 1865 issue of the *Phil. Mag.*
2. *Dr. Akin's*: C. K. Akin.
3. *Report on the Transmutation of Spectral Rays in the British Association Reports for 1863*: see letter 2227, n. 7.
4. *grant of £45*: see letter 2227, n. 7.
5. *Mr. Griffith*: George Griffith.
6. *Meeting of the British Association at Bath in the September of that year*: see letter 2239, n. 9.

To the Editors of the *Phil. Mag.*[1] February 1865 2245

XXXII. *On the History of Calorescence.* |
By JOHN TYNDALL, *F.R.S.*, &c.[2]

On the 26th of May, 1859, I presented to the Royal Society a "Note on the Transmission of Radiant Heat through Gaseous Bodies."[3] The question had occupied me some time; but as the experimental difficulties were very great, I published the Note referred to in order to enable myself to vanquish those difficulties at my leisure. All the time at my disposal in 1859 and 1860 was devoted to the subject, and towards the end of 1860 I was so far advanced as to be able to prepare a memoir,[4] which was presented to the Royal Society on the 10th of January, and, being chosen as the Bakerian Lecture for that year, was read on the 7th of February, 1861.

In that memoir the comparative deportment of elementary and compound gases towards radiant heat is for the first time announced. I had attempted to make radiant heat an explorer of molecular condition, and had found that the simple gases possessed a power of transmission immensely greater than that of the compound ones. In the autumn of 1861 I pursued the subject, and by purifying more perfectly the elementary gases, rendered the differences between them and the compound ones still more vast. I then turned my attention to solids and liquids, and confirmed Melloni's[5] experiments on the diathermancy of lampblack;[6] I also tried to render the substance more transparent to invisible heat-rays by ridding it of the hydrocarbons which attach themselves to it during its formation. I next examined the element bromine and found it eminently diathermic; I tried sulphur dissolved in bisulphide of carbon and found it still more so. I finally operated on a solution of iodine in bisulphide of carbon, and found that a layer of it, sufficiently dense to intercept completely the light of the noonday sun, offered a scarcely sensible obstacle to the passage of the invisible calorific rays[7]

This fact is thus announced in a footnote at page 67 of the Philosophical Transactions for 1862:[8]—

> "A layer of bromine, sufficiently opake to intercept the entire luminous rays of a gas-flame, is highly diathermanous to is obscure rays. An opake solution of iodine in bisulphide of carbon behaves similarly. The details of these experiments shall be published in due time: they were publicly shown in my lectures many months ago.—June 13, 1862."

Turning to my published letters on Heat as a Mode of Motion," I find at page 357[9] one of these experiments described in the following words:—

> "I cannot use iodine in a solid state, but happily it dissolves in bisulphide of carbon. I have the densely coloured liquid in this glass cell. I throw the parallel electric beam upon the screen; this solution of iodine completely cuts the light off; but if I bring my pile into the path of the beam, the violence of the needle's motion shows how copious is the transmission of the obscure rays."

Turning back to page 307 of the same work, I find experiments on smoked rock-salt and black glass described as follows:—

> "Here is a plate of rock-salt coated so thickly with soot that the light, not only of every gas-lamp in this room, but the electric light itself is cut off by it. I interpose the plate of smoked salt in the path of the beam; the light is intercepted, but this rod enables me to find with my pile the place where the focus fell. I place the pile at this focus; you see no beam falling on the pile, but the violent action of the needle instantly reveals, to the mind's eye, a focus of heat at the point from which the light has been withdrawn."

I would ask the reader to picture any experimenter standing by a focus of invisible rays, the exposure to which, for an instant, of the face of my thermo-electric pile caused the heavy needles of a coarse galvanometer to dash against their stops. Could he escape the temptation to put his hand there? I did so fifty times, trying moreover to concentrate the radiation by pushing out my lens. The camera of my electric lamp was furnished with a concave reflector, at the centre of which stood the carbon-points whence issued the electric light. The rays were converged by a lens in front; and when the points were at their proper elevation, the focus of the lens coincided with that of the mirror. Causing the two foci to coincide, and converging the rays as much as possible, I cut off the light by the solution of iodine, and brought in succession my hand, my cheek, pieces of brown paper and of lead-foil into the dark focus. It was simply a substitution of these bodies for the face of my thermo-electric pile. But while the action on the skin was almost intolerable, I obtained neither the charring of the paper nor the fusion of the foil. I knew that a large portion of the radiant heat of the lamp was lodged in the lens in front of it, and in the sides of the cell containing my solution of iodine; and I determined, as soon as I could turn my thoughts to it, to supplant that lens and that cell by rock-salt ones.

At that time my experiments on the action of aqueous vapour as an absorbent of radiant heat were at variance with those of an eminent natural philosopher.[10] Respect for him, and the desire to place my results beyond the pale of doubt, determined my course of action. The autumn of 1862 was devoted to the preparation of a memoir "On the Relation of Radiant Heat to Aqueous Vapour,"[11] which was presented to the Royal Society on the 20th of November, and read on the 18th of December of that year. In the autumn of 1863 I completed my demonstration of the action of aqueous vapour on radiant heat, and at the same time executed further experiments on dissolved iodine, illustrative of its extraordinary transparency to heat-rays of ultra-red refrangibility.

My course was now cleared, and my thoughts reverted at once to the experiments which I had suspended more than two years before. I went to the Isle of Wight in March 1864 with my ideas perfectly defined and ripe for execution. While there, under circumstances which I have fully described in my article "On Negative Fluorescence,"[12] published in the January Number of the Philosophical Magazine, I received a note[13] from Dr. Akin,[14] a Hungarian gentleman who had arrived in this country in the early part of 1862, and who was personally unknown to me. His name had been mentioned to me by Professor Stokes; I had read an article from his pen published in the 'Saturday Review,'[15] and he now forwarded to me a proof of a Report presented by him to the British Association assembled at Newcastle in 1863.[16] In this Report three experiments were proposed: two of them were impracticable, but they nevertheless showed ability, while the third seemed to offer a promise of real success. These were not the experiments that I had proposed to make; I did not intend to experiment on the sun, for his beams are unattainable where I work; I intended to operate with the radiant source with which the constant practice of the previous eleven years had rendered me familiar, and to realize ideas of which the origin has been indicated above, but which, with the caution of a man who knows the difference between conception and performance, I did not publish in detail. The reader will now be in a position to appreciate the use which Dr. Akin has made of the following note[17] written to him at the time:—

"Clarendon Hotel, Chale, Isle of Wight, | Wednesday.

"MY DEAR SIR,—I have read the proof which you have been kind enough to send me with extreme interest. Being a truant from the Sections at Newcastle, I did not know that you had brought the subject forward there. Professor Stokes, however, informed me that you were engaged on the question, and I had similar information from the 'Saturday Review.'

"There is certainly a most singular coincidence in the thoughts which have passed through both our minds on some of the points in question. As to the possibility of converting the Herschellic into the Newtonic *rays* I do not entertain a doubt; and indeed the piece of work which I had set before me for attack on my return from this place was a series of experiments on this very subject.

"I have devised a very perfect means of sifting the Herschellic from the Newtonic rays, and with the latter alone I have strong hopes of being able to produce incandescence. But you have entered this field before me, and although the question lies directly in the way of my own inquiries, I should like to do in the matter what would be most pleasing to you. I shall be in town on Saturday next, at the Royal Institution at noon; could you not look in upon me there?

"Yours very sincerely, | "JOHN TYNDALL."

In the December Number of the Philosophical Magazine, p. 559,[18] Dr. Akin employs the following language with reference to this letter:—

"I must be allowed to express here a doubt whether Sir Humphry Davy,[19] for example, or any other predecessor of Professor Tyndall at the Royal Institution, having read in a public print that two persons were engaged in making researches on a certain subject with the aid and sanction of the British Association, would have chosen 'that very subject' for 'attack' some little time after." I, on the other hand, may be permitted to express the assured conviction that Sir Humphry Davy would never have given Dr. Akin the chance of garbling his language as he has here garbled mine. The reader will compare the words "attack" and "that very subject," as held in the claws of his inverted commas, with the same words as they occur in my note, and draw his own conclusions as to the temper of the quoter.

The fundamental error of Dr. Akin consists in his interpreting the attitude assumed by me in the foregoing note, as one which, under the circumstances, he had a right to expect. What most men would have regarded as an act of free courtesy on my part, Dr. Akin interprets as an acknowledgement of his inordinate claims. He does not seem to understand why, of my own free will, I conceded to him a position which I should unquestionably have refused had it been demanded as a right. As I have already stated to him, "there was nothing in the circumstances of the case to prevent me from working at the subject, restricting myself to a due reference to his labours." Had I been aware that my note had excited his "astonishment," I should have acted very differently towards him. Had he urged "the authority of the British Association," I should have informed him that I could not recognize its authority in such a matter. At the same time I should have known that the eminent leaders of that body would never countenance Dr. Akin in the attempt to shut me out of a field of inquiry which I had virtually entered before him, and which consideration for him as a stranger had alone caused me to relinquish. To render this assurance doubly sure, I wrote[20] to General Sabine,[21] who knows more about the constitution of the British Association than any other living man, and who, with the kindness which I have always experienced at his hands, promptly wrote to me[22] the following reply:—

"DEAR TYNDALL,—Viewing your inquiry simply as an administrative question, Dr. Akin's Report on the Transmutation of Spectral Rays in the British Association Reports for 1863 indicated a very wide field of experimental researches. In aid of these the General Committee made a grant of £45 to Mr. Griffith and himself.[23] A part of this grant was expended in apparatus to be employed by those gentlemen in the spring or early summer of 1864, the results of which were designed by Dr. Akin to be presented by himself to the Meeting of the British Association at Bath in the September of that year.[24] For this purpose Dr. Akin went to Oxford to join Mr. Griffith, but from some unexplained cause their purpose was frustrated, and 1864 passed without the experiments being made. The grant, however, did not lapse, as part of it had been expended in apparatus, which it appears to be still Dr. Akin's intention to employ (though not at Oxford where there is sunshine, or in London where there is no sunshine suitable for the purpose). Such, so far as I can collect them, is a statement of the facts.

"The object of the British Association in making such grants is to aid the progress of science, by enabling those, who may be so disposed, to make experiments, but by no means to retard its progress by discouraging others from approaching the same ground. One cannot read, I think, Dr. Akin's Report in 1863 without the impression that there is work enough in the subject for a dozen at least of active experimenters. He may have had an advantage (and properly used it is a great one) in mediating longer than others on the laurels that may be gathered, but he who desires to be foremost in the field must be 'prompt in action' where there are so many competitors,

"Always truly yours, | EDWARD SABINE."

I also wrote to Sir William Armstrong[25] on the same subject.[26] As President of the British Association at Newcastle, Sir William Armstrong does not think himself entitled now to speak, but as a private individual he states that the British Association "could not, even if they proposed anything so foolish, confer upon Dr. Akin any exclusive privileges in the matter."[27]

It was in answer to the express desire of Dr. Akin to see me that I informed him at the end of my note from Chale that I should be in town on the following Saturday. He came. I have since ransacked my memory for the details of our interview; and though I cannot recall the whole of them, the essential ones are perfectly clear to me, while the *result* is as vivid to my mind as the occurrence of yesterday. Without waiting to obtain the knowledge of Dr. Akin's character that ten minutes' conversation might have given, I remarked that although he and I had reached the same point by independent routes, he was the first to publish definitely his ideas, and that, therefore, I would not interfere with him. I stated that although my plans were prepared, I was perfectly willing to suspend them. Dr. Akin's reply was that he was delighted to find the subject taken up, and was only anxious to see it prosecuted to a successful issue. Of these words I have the most distinct recollection. I did not know at the time that the note written to him on the previous Wednesday had excited his "astonishment"; nor did I know that he had come, as he has since informed me, expecting that I would make him an offer that we should work together; but pleased by his apparent frankness, on the spur of the moment I used these words:—"We might perhaps attack it together." I use the phrase "spur of the moment" because it strictly represents the fact, but they who know me best will be the last to believe me capable of retreating from a position assumed even on the spur of the moment. And had Dr. Akin agreed to my proposal, no matter what consequent penalties it might have involved—and they, I doubt not, would have been great—I should have carried it out to the letter. But he did *not* agree to my proposition. He said that he had associated himself with Mr. Griffith at Oxford, and that he was compelled to fulfil his engagement with that gentleman.

Let me here remark that neither Dr. Akin nor myself professes to remember the whole of what occurred during this interview; and when he spoke of his being bound to Mr. Griffith, some conversation followed, the precise terms of which I do not recollect; but it drew from Dr. Akin a proposition, which I *do* remember perfectly well. I will allow him to supply what I have here forgotten. In a letter[28] written to me more than three months ago, he affirms that when he declared himself bound to Mr. Griffith at Oxford, I rejoined:—"Can we not make it a triple alliance?" His reply was, "No, leave us the sun at Oxford, but as regards experiments on artificial rays, I shall be happy to cooperate with you separately." Of the words "triple alliance" I have no recollection, but they are so like what I should use under the circumstances, that I do not doubt the correctness of Dr. Akin's memory regarding them. But I do remember his proposal to work with me in London and with Mr. Griffith at Oxford, and it caused me to pause. Neither in its tone nor in its terms did it please me, and to it, by neither word nor sign, did I ever show the slightest symptom of assent.

In his last published account of this conversation[29] Dr. Akin omits all reference to my proposal and his refusal of a "triple alliance," and represents me as accepting his counterproposition without reservation or qualification. "To this conclusion Professor Tyndall cheerfully agreed, observing that he had from the first intended to make experiments on the electric light; that he was consequently quite satisfied with the arrangement proposed." Though there seems *now* no shade of doubt upon Dr. Akin's mind regarding my acceptance of his offer, still I think there was a slight shade upon it four months ago. In a letter written to me on the 1st of last November, after reminding me of his offer to cooperate with me in London, leaving the sun to himself and Mr. Griffith at Oxford, he continues thus:—"If there was no 'closing' with any offer, it must have been on your part, and as regards this last-mentioned offer, which I made in answer to your own".[30] This passage, I submit, indicates a condition of mind and memory somewhat different from that which prompted the unqualified assertion "to this condition Professor Tyndall cheerfully agreed."

When I wrote to Dr. Akin from the Isle of Wight, my desire and intention was to allow him ample scope for the realization of his ideas, and during our conference this intention was revived. I resolved then and there not to interfere with Dr. Akin, to suspend the experiments I had planned, and leave the subject entirely in his hands throughout the summer. During the months of April, May, and June my assistant[31] was employed, under my direction, on an investigation which he afterwards published in the August Number of the Philosophical Magazine,[32] while I occupied my leisure with other matters.[33] The naked facts stand thus:—On Wednesday the 30th of March, I write to Dr. Akin stating that I had devised a series of experiments to be undertaken on my return to town. On the very day of my arrival the execution of these experiments is postponed, and not a word is heard of them until the end of October.

Meanwhile Dr. Akin had pursed his experiments at Oxford, and had sent in the following Report of his performances to the President of the Physical Section of the British Association. The document is dated from the Cavendish Club, London, 15th September, 1864.

"I am sorry to have to state that, owing to the scarcity of clear sunshine at Oxford last summer, and to Mr. Griffith being mostly unable to give his time to our work when the weather happened to be favourable to it, the experiments were left in August unfinished. I had the greater reason to feel disappointed at such a conclusion, as the methods and apparatus devised for the purposed appeared, from preliminary trials, competent to effect, in part at least, the proposed object, and as our preparations for the more decisive experiments were in a very forward state. Meanwhile I have received fresh testimony from several esteemed quarters to the importance of the research thus partially accomplished, which makes me feel yet deeper regret at seeing the matter after two years' labour, and nearly three years' thought and trouble—some of it unexpectedly galling—still in abeyance."

After the Bath Meeting of the British Association, Dr. Akin also wrote thus to the President of the Association:—

"I now beg leave to state that, after the experience of the last two years, it would be a hopeless undertaking for me to continue at Oxford the experiments begun there. Moreover I am not sure whether, after the end of the present year, I shall be able to give my attention any longer to scientific researches." Supposing Dr. Akin to be perfectly satisfied in his own mind that he had entered into a definite agreement with me, does it not seem odd that he should not have hinted at this engagement, either in his Report or in his letter? When he stated that he had abandoned hope in Oxford, it would surely have been natural to add that he had *not* abandoned it in London. The fact seems to be that Dr. Akin had no clear consciousness at the time that such an engagement existed, and that anger now supplies him with a power of memory which in his calmer moments he did not enjoy.

I went to Switzerland early in July, and from Pontresina[34] wrote to my assistant, describing the apparatus I had intended to employ, and desiring him to have it prepared for me. One portion of it consisted simply of the substitution of a rock-salt lens for the glass one of the electric lamp, and a rock-salt cell for the glass one that I had used in 1862. With this arrangement I executed the experiments "on luminous and obscure radiation" described in the November Number of the Philosophical Magazine.[35] Nay, abandoning the rock-salt altogether, and employing a glass lens of somewhat shorter focus than that made use of in 1862, with a battery of sixty cells, I obtained all the results with the electric lamp there described. In not the slightest particular does this arrangement differ from that which I had employed two years and three-quarters previously. We have the same source of rays, the same mode of convergence, and the same absorbent to intercept the luminous portion of the radiation. And yet it is from this, my own ground, which I had taken up practically before he had perused the "little German treatise which first taught him what had been done"[36] on the subject of fluorescence, that Dr. Akin would fence me out.

Experiments on combustion could not decide the question of ray-transmutation. "Intimately connected" they assuredly were,—so intimately, indeed, that the man who could neglect to pass immediately from the one to the other, would be unfit for the vocation of a natural philosopher. Was I, then, out of consideration for Dr. Akin, to keep my platinum-foil or my silver-leaf away from the focus of dark rays, lest it should become incandescent there? My duty to science ruled otherwise; and as no bond or promise existed to contravene that duty, I resolved not to let the autumn pass as Dr. Akin had the summer, without making some attempt to realize *my* ideas. He had already tried, failed, and announced his failure. Still, however, wishing to treat him with the utmost consideration, I resolved, while forwarding the work of science, to leave the chance of prior publication open to Dr. Akin. I determined to hold back whatever results I might obtain until he would be either able to precede me, or could so far merge his individual interests in the larger interests of science, as to be willing to see the results published. This he deems an unusual proceeding on my part, and so it assuredly was. But from first to last my proceeding towards him was unusual. In relation to any other man it would have been more than fair; in relation to him it has proved to be worse than useless.

Late in the day of the 29th of October, I obtained the incandescence of platinum foil by the dark rays of the electric light. The layer of iodine, however, was too thin to wholly intercept the light, while the battery had been enfeebled by long-continued previous action. On the following morning it was freshly charged, the number of cells being increased from fifty to sixty. Removing from the back of the camera the little reflector which I had made use of in 1862, I mounted it on a retort-stand, and with it converged the rays from the lamp. The mirror was silvered behind, and the reflected rays had therefore to pass to and fro across a plate of glass of considerable thickness. Notwithstanding this, however, and the employment of a much thicker layer of dissolved iodine, the platinum, when placed in the dark focus of the mirror, became vividly incandescent. Almost immediately after the performance of this experiment I found Dr. Akin in the library of the Royal Institution. Without a moment's hesitation I invited him to accompany me to the laboratory, and there almost simultaneously with myself he saw the result which I had obtained.[37]

Not one word of remonstrance or complaint escaped him on this occasion; not a syllable to indicate that he believed me in any way pledged to him. He made some critical remarks as to the experiment being made in daylight. He was moreover struck with the thickness of the layer of iodine employed, observing that the layer used by him at Oxford was a thin one, which did not cut off the light. He finally expressed himself delighted to see the experiment, and we quitted the laboratory together. I walked with him upstairs, and there at the eleventh hour expressed my desire not to interfere with him, and to leave, if he wished it, the subject still in his hands. His reply was, "It will all depend on the manner of publication." Knowing my own willingness to do him the amplest justice in any publication of mine, I responded, "You shall see my manuscript before I publish."

My intention when I parted from Dr. Akin was to give a brief account of the experiment at the first meeting of the Royal Society, but I had not definitely made up my mind. On the evening of the day to which I now refer I thought and spoke of associating myself with Dr. Akin for the further prosecution of the inquiry. On the following Monday, however, I received a letter from this gentleman which entirely changed the aspect of affairs, and in which he informed me that, on thinking over the experiment I had showed him, his mind had become greatly agitated. He referred to our interview in April, and to what he asserted had occurred between us since. Like his last article in this Journal,[38] that letter consisted of a single broken fibre of truth in a tissue of error. I then went through the assertions of Dr. Akin one by one, pointed out where they were exaggerated, where he had unduly intensified language, and where his statements were, as I believed, at direct variance with facts. His last paper furnishes some illustration of the intensification of language which I then sought to correct. In my reply to his intimation that I had appropriated his ideas, I stated that "my work would have been far more completely done by this time had he never existed."[39] Dr. Akin's version of this statement of a fact is this:—

> "Professor Tyndall, who is charitable enough to express an implied wish for my non-existence, will no doubt be gratified to learn that the experience I have met with at his hands has acted upon me in the manner he would seem to desire."[40]

It was sufficiently evident from Dr. Akin's correspondence that he was very much disappointed. He had been "greatly depressed and discouraged in mind"[41] at Newcastle. "An indelible mark of pain" had been left upon his memory by the treatment he received at Oxford; and now his misfortunes culminated on finding, in London, the goal at which he aimed attained by another. Still that other, though sorely irritated, was less anxious to avail himself of his success than to show Dr. Akin "how lightly he valued the whole matter when a question of good faith came into play." The letter which contained my corrections of Dr. Akin's misapprehensions concluded thus:—

> "And now for a few concluding words:—I cannot help expressing my deep regret that you did not write to me in a different manner. I should have gone almost any length to gratify you, had you adopted another tone; and I shall prove my sincerity in writing thus, by doing that which still remains in my power to do. I have to say then, that from this 3rd of November 1864 to the 3rd of November 1865 I shall not make known, publicly or privately, any experiments on 'ray-transmutation.' Nearly seven months ago I was, as you know, ready to take this subject up. Out of deference to you I did not do so; so that at the termination of the period referred to I shall have held back for nearly nineteen months. That you will be able to release me long before, by the actual performance of this experiment, is almost beyond a doubt; and in this event I hope to muster sufficient greatness of heart not to envy you."

Dr. Akin has the assurance to upbraid me with having, in the face of this promise, seized the first opportunity to publish my results. My promise was given in order to enable him to perform honest work, and not to indulge in aggression.

I halted that he might move on. The arrangement left open to him another summer, during which he might renew the investigation that had broken down at Oxford, with the assurance, moreover, that the object at which he aimed was no longer a doubtful one. Not satisfied with this arrangement, he continued to complain, and I was at length obliged to end a profitless and apparently interminable correspondence[42] with these words:—

> "I have now only to add, that when you have so far overcome the agitation of mind to which you so frequently refer, as to be able to look at this question with the eyes and mind of a philosopher, I shall be happy to communicate with you. But no possible good can arise from a continuance of this correspondence. Hence, as far as I am concerned, it is for the present at an end."

Dr. Akin then wrote to me his fifth letter,[43] and ended it by the statement that, although he could scarcely allow himself to hope much for the future, he wished, nevertheless, to say that whenever I should be pleased again to communicate with him he would be happy to hear from me. This letter was dated from Harley Street, November 9, 1864.

For nearly three weeks I heard nothing from or of Dr. Akin, but he meanwhile was not idle. Silently, and without a word of warning to me, he concocted his "Note on Ray-transmutation" for the December Number of the Philosophical Magazine.[44] After the completion of that article, and twenty-four hours previous to its publication, he wrote to me[45] commencing his letter with "Dear Sir," and ending it with "Believe me yours very faithfully." On the very day of its publication he sent me a second note[46] with somewhat similar termini; but neither of them breathed a whisper of his attack. During my correspondence with Dr. Akin I availed myself from time to time of the wisest advice within my reach; and the verdict of my advisers was now prompt and emphatic. From the legal, the scientific, and the general world, I heard the unanimous opinion that Dr. Akin's act of war cancelled every treaty subsisting between us, and rendered the act of publication a duty to science as well as to myself.

To sum up:—

1. Dr. Akin affirms that I have interfered with an investigation which he had originated and to which he had secured an exclusive right.

I assert, on the contrary, that my inquiries have been the direct result of labours commenced before Dr. Akin was heard of in this or any other field of scientific inquiry; and that even if they had not been, the assumption of any such right of exclusion is inadmissible.

2. Dr. Akin affirms that he closed with alacrity with a proposition of mine to work out, in conjunction with him, a certain subject.

But Dr. Akin admits that he did not "close" with my first proposition, on account of his tie to Mr. Griffith. He further admits that he declined the "triple alliance" between Mr. Griffith, Dr. Akin, and myself, which was next suggested by me. Finally, he produces no evidence to show that I proposed or accepted any other arrangement; and I categorically deny that any other arrangement was proposed or accepted by me.

3. Dr. Akin's words imply that I have availed myself of his ideas or suggestions, to anticipate his results. I answer thus:—

It has been shown in the forgoing article that early in the year 1862 I formed foci of invisible rays, and also illustrated in my public lectures the heat of such foci; that I employed for this purpose the condensed rays of the electric light, and intercepted the luminous part of the radiation by an absorbent of extraordinary properties discovered by myself.

Sir William Herschel[47] had tried to render the ultra-red rays of the solar spectrum visible by condensation, having first separated them, by prismatic analysis, from the visible rays.[48] But the idea of forming intense foci of invisible rays by concentrating the *total* radiation of a luminous source, and removing the visible rays by an absorbent, is, I believe, entirely mine.

4. In the autumn of 1863 Dr. Akin proposed to obtain an intense focus of invisible rays by converging the sun's beams and intercepting the luminous part of the radiation by absorbents which he did not name. In making this proposition, Dr. Akin had, without a word of acknowledgement, appropriated an idea which I had in substance enunciated and applied more than a year and half previously.

5. The only particular in which Dr. Akin was in advance of me—and this not in thought, but in publication—was that he was the first to express the belief that the dark rays of the sun, when detached from the luminous ones by my method, would be competent to raise platinum-foil to incandescence; and that if they did so we should have for the extra-red end of the spectrum phenomena analogous to those of fluorescence.

6. In consequence of the precision with which he had enunciated this idea, and also in part on account of his being a foreigner, working, as I conceived, against difficulties in this country, I somewhat Quixotically permitted him during the whole of last summer to pursue his idea undisturbed. He had tried, failed, and reported his failure to the British Association assembled at Bath, before my experiments on calorescence had commenced.

If I have not waited longer before publishing the results of my experiments, as I once proposed to do, it is because by his unwarranted and unprovoked attacks, Dr. Akin has forfeited all claim to the position in which I had voluntarily, but unwisely, placed him.

Royal Institution, February.

[The Editors have received a letter from Dr. Akin, in which he begs to enter a public and formal protest against the note appended to his communication in the last Number of the Magazine by W. F., and claims to have the last word in this discussion. As Dr. Akin commenced the controversy in the December Number, and in that of February has had every opportunity of placing all the evidence he deemed necessary before our readers, we cannot admit the justice of his claim, and abide by the decision already given.—THE EDITORS.]

J. Tyndall, 'On the History of Calorescence', *Phil. Mag.*, 29 (1865), pp. 218–31

---

1. *Editors of the Phil. Mag.*: at this time the editors of the *Phil. Mag.* were David Brewster, Robert John Kane (1809–90), an Irish chemist and educationist (*ODNB*), and William Francis.
2. The following footnote appeared here in the original: '* Communicated by the Author'.
3. *"Note on the Transmission of Radiant Heat through Gaseous Bodies"*: J. Tyndall, 'Note on the Transmission of Radiant Heat through Gaseous Bodies', *Roy. Soc. Proc.*, 10 (1859), pp. 37–9.
4. *memoir*: J. Tyndall, 'The Bakerian Lecture: On the Absorption and Radiation of Heat by Gases and Vapours, and on the Physical Connexion of Radiation, Absorption, and Conduction', *Phil. Trans.*, 151 (1861), pp. 1–36.
5. *Melloni's*: Macedonio Melloni (1798–1854), an Italian physicist who worked on heat and radiation. He was the 1834 recipient of the RS's Rumford Medal, 'For his discoveries relevant to radiant heat' (*CDSB*).
6. *experiments on the diathermancy of lampblack*: probably referring to: M. Melloni, *La thermochrôse ou la coloration calorifique démontrée par und grand nombre d'expériences, et considérée sous ses divers rapports avec la science de la chaleur rayonnante* (Naples: Impr. de J. Baron, 1850).
7. The following footnote appeared here in the original: '† The same à priori considerations which led to the discovery of the iodine, point also to red glass coloured by the element gold, instead of that coloured by the suboxide of copper, as most suitable for experiments on ray-transmission. The colouring matter of the former appears to be without sensible action upon the invisible heat-rays; and the rays transmitted through it are competent to raise platinized platinum to a white heat'.
8. *Philosophical Transactions for 1862*: J. Tyndall, 'On the Absorption and Radiation of Heat by Gaseous Matter. Second Memoir', *Phil. Trans.*, 152 (1862), pp. 59–98.
9. *page 357*: *Heat* (1863). While Tyndall quoted this passage as from p. 357, it is actually on p. 351.
10. *were at variance with those of an eminent natural philosopher*: Tyndall had a disagreement with one of his mentors from when he trained for his PhD in Germany, the physicist Gustav Magnus. Magnus did not accept Tyndall's experiments with rock-salt, which demonstrated the absorption of heat by water vapour. According to Tyndall, as he described in a letter to William Francis in 1867 (*Phil. Mag.*, 33 (1867), p. 425), part of his delay in finishing his experiments on radiant heat in the early 1860s was due to his desire to address the concerns of Magnus to his experimental results.

11. *"On the Relation of Radiant Heat to Aqueous Vapour"*: see letter 2229, n. 2.
12. *"On Negative Fluorescence"*: J. Tyndall, 'On the History of Negative Flourescence', *Phil. Mag.*, 29 (1865), pp. 44–55.
13. *note*: see letter 2075, *Tyndall Correspondence*, vol. 8.
14. *Dr. Akin*: C. K. Akin.
15. See 'Calcescence', *The Saturday Review of Politics, Literature, Science, and Art*, 17 (23 January 1864), pp. 105–6.
16. *Report presented by him to the British Association assembled at Newcastle in 1863*: see letter 2227, n. 5 and n. 7.
17. *following note*: published as letter 2076, *Tyndall Correspondence*, vol. 8.
18. *December Number of the Philosophical Magazine, p. 559*: C. K. Akin, 'Note on Ray-Transmutation', *Phil. Mag.*, 28 supplement (1864), pp. 554–60.
19. *Sir Humphry Davy*: Humphry Davy (1778–1829), a chemist and inventor born in Cornwall, England. He became directory of the RI laboratory in 1801 (until 1825) where he was also professor of chemistry (1802–12) and honorary professor (1813–23). He achieved fame as a lecturer and for his chemical research that isolated such elements as sodium, potassium, and chlorine. He is often attributed with the invention of the miner's safety lamp. He served as president of the RS from 1820–27 (*ODNB*).
20. *I wrote*: letter missing.
21. *General Sabine*: Edward Sabine.
22. *wrote to me*: published as letter 2197, *Tyndall Correspondence*, vol. 8.
23. *grant of £45 to Mr. Griffith and himself*: see letter 2239.
24. *Meeting of the British Association at Bath in the September of that year*: see letter 2239, n. 9.
25. *Sir William Armstrong*: see letter 2227, n. 1.
26. *I also wrote . . . the same subject*: letter 2227.
27. *"could not . . . privileges in the matter"*: see letter 2227, n. 10.
28. *letter*: see letter 2165, *Tyndall Correspondence*, vol. 8.
29. *last published account of this conversation*: C. K. Akin, 'Further Statements Concerning the History of Calcescence', *Phil. Trans.*, 29 (1865), pp. 136–51.
30. The following footnote appeared here in the original (the 'second letter' referred to is missing):

    '* Here is the extract *in extenso*: it is taken from Dr. Akin's second letter to me, the word "closing" being placed between inverted commas, because in my reply to Dr. Akin's first letter I had reminded him that he had not "closed" with my proposal:—

    Dear Sir,

    I have just received your letter of today, and intend to shortly reply to such portions as seem to me to require it.

    As regards the way in which your attention was turned to the subject which is in question, of course you are the best as well as the only authority, and I have no intention of challenging any statement on your part, deliberately made, and where there are no facts within my knowledge proving that your recollection misleads you. Such, however, I believe to be the case in reference to what occurred at the Royal Institution in April, in the course of our conversation. You then spoke of a "triple alliance" [a footnote appeared here in the original: 'That was after I had said that I was bound to Mr. Griffith'] (those were the words you used). I replied, 'Leave the sun to us at Oxford," but added that, as regarded experiments on artificial sources of rays (the expressions, I believe, were 'the other two experiments,' meaning those described in my paper), I should be happy to cooperate with you separately. If there was 'no closing' with my offer, it must have been on your part, and as regards this last-mentioned offer, which I made in answer to your own."

31. *my assistant*: William Fletcher Barrett (1844–1925), a physicist and psychical researcher who was an assistant to Tyndall at the RI from 1863–6. Following his time at the RI, he held various teaching positions at the International College, London, the Royal School of Naval Architecture, and the Royal College of Science for Ireland, and was elected FRS in 1899 (*ODNB*). Barrett and Tyndall later conflicted over spiritualism (see R. Noakes, 'The "Bridge Which Is between Physical and Psychical Research": William Fletcher Barrett, Sensitive Flames, and Spiritualism', *History of Science*, 42 (2004), pp. 419–64).
32. *he afterwards published in the August Number of the Philosophical Magazine*: W. F. Barrett, 'On a Physical Analysis of the Human Breath', *Phil. Mag.*, 28 (1864), pp. 108–21.
33. *occupied my leisure with other matters*: see letter 2227, n. 4.
34. *Pontresina*: a small town about 3 miles to the east of St. Moritz, in the Engadine, a long Alpine valley in southeastern Switzerland.
35. *in the November Number of the Philosophical Magazine*: see letter 2226, n. 3.
36. *"little German treatise which first taught him what had been done"*: in 'Further Statements' (see. n. 29), Akin states 'I may as well state now that the article on Fluorescence in Cornelius and Marbach's *Physikal. Lexicon* (the work being one intended for popular circulation and reference, not for perusal) became first known to me from E. Emsmann's paper on "Negative Fluorescence" in Poggendorf's *Annalen* in 1861' (p. 137).
37. The following footnote appeared here in the original: '* Dr. Akin had informed me that he was collecting materials for a "History of Force," and needed access to a good library. Deeming his work both useful and honourable, to enable him to prosecute it I introduced him to the libraries both of the Royal Institution and the Athenaeum Club'.
38. *his last article in this Journal*: C. K. Akin, 'On Calcescence', *Phil. Mag.*, 29 (1865), pp. 28–43.
39. *"my work would . . . he never existed"*: see Tyndall, 'On the History of Negative Fluorescence', p. 53.
40. *"Professor Tyndall, who . . . seem to desire"*: see Akin, 'Further Statements', p. 149.
41. *"greatly depressed and discouraged in mind"*: see Akin, 'Further Statements', p. 138.
42. *end a profitless and apparently interminable correspondence*: letter missing.
43. *fifth letter*: letter missing.
44. *December Number of the Philosophical Magazine*: see n. 18.
45. *wrote to me*: letter missing.

46. *second note*: letter missing.
47. *Sir William Herschel*: William Herschel (1738–1822), a German-born British astronomer and telescope maker, brother of astronomer Caroline Herschel (1750–1848), and father of John Herschel, a regular correspondent of Tyndall's. Among many contributions to astronomy, Herschel is most well-known for his discovery of the planet Uranus in 1781 (*ODNB*).
48. *tried to render . . . from the visible rays*: see W. Herschel, 'Experiments on the solar, and on the terrestrial Rays that occasion Heat; with a comparative View of the Laws to which Light and Heat, or rather the Rays which occasion them, are subject, in order to determine they are the same, or different', *Phil. Trans.*, 90 (1800), pp. 293–326, on pp. 317–9.

To Charles Lyell 2 March 1865 2246

2 March 1865. | Thursday

Dear Sir Charles

I hardly think that geologists are entitled to assume "a refrigeration of the hemisphere" which is exposed to the extremes of heat & cold, solely on account of the existence of such extremes.[1] You might I think fairly reason that the one would neutralize the other. But the existence of our atmosphere & the transport of water in the shape of snow from the equatorial regions to the polar ones, render the actual problem a complicated one. It is I think conceivable that through a series of Secondary actions, brought about by the protracted winter the mean temperature of the hemisphere might be lowered. This is a subject to which I intend to turn my entire attention by and by, meanwhile I fear any opinion of mine regarding it is but of little value.

Yours very truly | John Tyndall

RI MS JT/1/T/1054
RI MS JT/1/TYP/3/839

1. *I hardly think . . . of such extremes*: see letter 2243.

From George Griffith 5 March 1865 2247

British Association | 5 March 1865

Dear Dr. Tyndall,

I really cannot tell you who Dr. Akin[1] is—He seems to have spent his life in wandering about from one University to another—He has sat under Delauney,[2] Kummer,[3] and numerous Mathematicians, and when Helmholtz was here last spring I think that he recognized Dr. Akin as an attendant of his lectures at Heidelberg. I do not recollect that any indication of pleasure marked the recognition. I do not at all imagine he is a friend of Prof. Stokes, the letters which I referred to[4] were apparently answers to letters which Prof. Stokes had received from Dr. Akin.

Can you be tempted to pay Oxford a visit—if only for a Sunday—it would give Mrs Griffith[5] and myself great pleasure to have you in our house. You could see the apparatus which Dr Akin and I employed, I shall be glad to have your opinion on the possibility of success with it.

I am | very faithfully yours | George Griffith.

RI MS JT/1/TYP/1/64
LT Typescript Only

1. *Dr. Akin*: C. K. Akin.
2. *Delauney*: possibly Charles-Eugène Delaunay (1816–72), a French astronomer and mathematician (*CDSB*).
3. *Kummer*: Ernst Eduard Kummer (1810–93), a Prussian mathematician and Professor of Mathematics at the University of Berlin from 1855 (*ADB*).
4. *letters which I referred to*: see letter 2239.
5. *Mrs Griffith*: Griffith's wife is not identified.

To Juliet Pollock [7 March 1865][1] 2248

Tuesday

My dear Friend

I hope this will meet you before you take your flight from Euston farm.[2] I have just seen your husband[3] & he has just returned from Cambridge, full of pleasant thoughts, and proud of your son.[4] I have told him that I know no fellow (that was the word) who had better reason to be satisfied with the dispensations of Providence.

I am glad to hear that you thought of me—I think often of you, and it would destroy the principle of polarity if you did not sometimes think of me. And the more I think of you & of the long years that we have known each other, the more I like you—that is all!

Yours ever | John Tyndall

He carried his man today triumphantly at the Athenaeum Committee.[5]

RI MS JT/1/T/1187
RI MS JT/1/TYP/6/2024

1. *[7 March 1865]*: the date is given based on William Frederick Pollock's time as a member of the Athenaeum Committee, with the successful election of a member to the club in 1865 matching the content of this letter. See n. 5 below.

2. *Euston farm*: the Pollocks rented Easton Farm near the town of Freshwater on the western end of the Isle of Wight (they had done so in 1864 as well).
3. *your husband*: William Frederick Pollock.
4. *your son*: Frederick Pollock, the Pollocks' eldest son who apparently William Frederick had been visiting at Trinity College, University of Cambridge (he graduated with a BA in 1867 and MA in 1870).
5. *He carried his man today triumphantly at the Athenaeum Committee*: William Frederick Pollock was an Athenaeum Committee member for the Rule II elections in 1865 and 1867. In 1865, he put forward John William Kaye (1814–76) for the second Rule II election. Kaye, noted as being 'Secretary in the Office of Secretary of State for India . . . ' and the author of many histories, was elected by Committee on 7 March 1865. In 1867, Pollock put forward for the first Rule II election the Very Reverend Harvey Goodwin (1818–91), Dean of Ely, Fellow of Caius College, Cambridge, and an author. The election took place on 12 February 1867 and Godwin was unsuccessful (Jennie De Protani, Archivist, the Athenaeum Club, personal communication). Since Tyndall noted in the letter that Pollock 'carried his man triumphantly', this would match the 1865 date in which the election of member was successful.

From Edward Sabine 14 March 1865 2249

The Royal Society | Burlington House, London. W. | March 14. 1865

Dear Tyndall,

I feel personally extremely obliged to you for assisting so well in entertaining the Prince[1] on Saturday.[2] It made, I am persuaded, all the difference to him, between being bored on the one hand, and well amused on the other hand.

He is desirous of Hardy[3] for admission;[4] and I told him that I could submit some arrangement to that purpose.

One of our Ordinary Meetings would be in many respects preferable to a special morning meeting; but there we must have some suitable paper; and this is not an easy matter. Have you any thing in preparation? It might be a fitting occasion for [more] of experimental illustration than is our usual practice.

Always truly yours | Edward Sabine

RI MS JT/2/13c/1286

1. *Prince:* Albert Edward (1841–1910), 'Prince of Wales', eldest son of Queen Victoria and Prince Albert, who succeeded to the throne of the United Kingdom as King Edward VII when Queen Victoria died in 1901 (*ODNB*). He was FRS from 1863 and as a member of the Royal family, a patron of the RS.

2. *on Saturday*: Tyndall wrote in his journal entry for 15 March 1865 about the events of 11 March: 'On Saturday attended the president's Soiree at Burlington House: the Prince of Wales came early, and Sabine brought him to my room. Made a number of experiments, and spoke in a free manner. It was perfectly successful—nothing stiff, but all genial. The Prince lit a cigar at the dark focus, and laughed heartily at the performance' (RI MS JT/2/13c/1285).
3. *Hardy*: Thomas Hardy (1840–1928), an English novelist and poet who had a deep interest in science, especially astronomy (*ODNB*).
4. *for admission*: to become FRS (he was never elected).

From David Brewster 15 March 1865 2250

Dear Professor Tyndall

I take the liberty of sending you a copy of an unpublished pamphlet on the Invention & Introduction of the Dioptric Lights.[1]

Lord Stanley[2] has kindly undertaken to move for a Commission to inquire into my services as stated in the Pamphlet; and it is of great importance to me to show the value and national importance of the Dioptric Lights.

You would oblige me therefore greatly, if you could give me your opinion, such as Sir John Herschel has done, on that part of the subject.

I am, Dear Prof. Tyndall, | Ever Most Truly Yours | D Brewster

Edinburgh | 26 Rutland Street | March 15th 1865

RI MS JT/1/B/131
RI MS JT/1/TYP/1/162

1. *unpublished pamphlet on the Invention & Introduction of the Dioptric Lights:* this pamphlet was published in June 1865 as D. Brewster, *The History of the Invention of Dioptric Lights and their Introduction into Great Britain* (London: John Murray, 1865).
2. *Lord Stanley*: Edward Henry Stanley, 15th Earl of Derby (1826–93), a Conservative MP who in 1866 became the Secretary of State for Foreign Affairs in his father's (Prime Minister Edward George Geoffrey Smith-Stanley) government. He had the title Lord Stanley from 1851–69 (*ODNB*).

From David Brewster 18 March 1865 2251

My Dear Prof[esso]r Tyndall,

I trust my pamphlet has now reached you.[1] I am sure your knowledge of the Dioptric Lights is sufficient to entitle you to give an opinion not as my Friend, but in the interests of truth & justice.

The persecution I have suffered from the Scotch Board,[2] whom, as you will see, I compelled to introduce the new Lights, & from their Engineers, who had a family interest in keeping up the old Reflectors, is I believe unexampled in the History of Science.

I am, My Dear Prof. Tyndall, | Ever Most Truly Yours | D Brewster
Edinburgh | 26 Rutland Street | March 18th 1865

RI MS JT/1/B/132
RI MS JT/1/TYP/1/163

---

1. *my pamphlet has now reached you*: see letter 2250.
2. *Scotch Board*: the Commissioners of Northern Lighthouses, the lighthouse authority for Scotland. Brewster had been arguing for a new system of dioptric light, or a Fresnel lens (named after its inventor, the French physicist August-Jean Fresnel), which allowed for light to be projected over a longer distance.

From Mary Anne Coxe 18 March [1865][1] 2252

Kinellan | Edinburgh | March 18th

My dear Dr Tyndall

I am very certain that you, who always took so warm an interest in our dear Alick,[2] will grieve to hear that *[1 word illeg]* had *[1 word illeg]* accounts of his Death!—

We knew him to be very ill—& were in one sense prepared for the worst—still—unconsciously (one finds when all is over) one clings to hope & only resigns it with the knowledge that all is over!

He had volunteered for service at Bhutan[3] a most deadly climate. Very shortly his liver became so diseased that he had to leave in hope of improvement on a return to Darjeeling.[4] He just made out this journey—reaching the house of the Rev[eren]d. Mr Nesfield,[5] (to whose Sister[6] he was engaged to be married) when he died, a very few days after arrival. An abscess had formed in the liver—& he was too reduced in strength, to stand the treatment w[hic]h could alone have given a hope of his recovery. I need not tell you, this is a great blow to us all—for we loved him very dearly—& life was opening to him with every prospect of happiness. Still—there is much to comfort us in the circumstances of his death. He was surrounded to the last by most loving friends & all was done that could be done to soothe his last hours on earth. He had little suffering—& was conscious till within two hours of his Death. He was quite resigned & willing to die—his last words were those of prayer.———

A letter from Miss Nesfield—(written but a few hours before her brother's, announcing that all was over) impressed us all most favourably. It was so simple, natural & touching.———

He died at Darjeeling in Bengal—on the 3d February.———

This sad news came upon us just after ten days of great enjoyment in a visit from our dear old friend Charles Edward Stuart—Comte D'Albanie[7] the younger of the two brothers—who often reverted to the pleasant ev[enin]g they & we spent with you—I must now close for I do not feel up to writing on ordinary subjects—

Ever my dear Dr Tyndall | very much yours M A Coxe

RI MS JT/2/13c/1288

---

1. *[1865]*: the handwritten letter was inserted in Tyndall's journal for that year. Also, Tyndall wrote in his journal for 21 March 1865, 'My dear friend Aleck Cumming is dead' (RI MS JT/2/13c/1287).
2. *Alick*: Alick Cumming (d. 1865), nephew of Mary Anne Coxe, educated at East India Company Military Seminary in Addiscombe, Surrey. Tyndall referred to him several times in his journal during the 1850s (RI MS JT/2/13c/725, 746, 751, 808).
3. *volunteered for service at Bhutan*: between 1864 and 1865, British India and Bhutan fought a war over territory.
4. *Darjeeling*: a city in the Bengal region of British India.
5. *Rev[eren]d. Mr Nesfield*: possibly John Collinson Nesfield (1836–1919), a colonial educator in British India, serving at Krishnagarh and Presidency Colleges in Bengal (*The India List and India Office List for 1905* (London: Harrison and Sons, 1905), p. 576).
6. *Sister*: not identified.
7. *Charles Edward Stuart—Comte D'Albanie*: Charles Manning Allen (1799?–1880) and John Carter Allen (1795?–1872) were brothers who claimed to be the heirs of Charles Edward Stuart, the Young Chevalier. They went by the names Charles Edward Stuart and John Sobieski Stolberg Stuart, respectively, and used the title Count D'Albanie (*ODNB*).

From Richard Dawes 20 March 1865 2253

Deanery, Hereford, | March 20th 1865.

My dear Tyndall

A thousand thanks for your so kindly sending me a copy of the new edition of your book on Heat as a Mode of Motion.[1] I was looking for its publication and I assure you my dear Tyndall I shall always feel a most lively interest in what you are doing and in all that concerns you & Mrs Dawes[2] not less so. I only regret not being sufficiently in London to have an opportunity of attending your lectures and witnessing the experiments which so strikingly illustrate this department of science, and that we should so seldom have the pleasure of seeing you. My friend Herbert[3] tells me he saw you at the Athenaeum a short time ago and he wanted to know why the cook uses snow instead of eggs in making pancakes,[4] a sort of puzzle we don't understand.

We are just now going to re-lead one of the transepts[5] of our cathedral.[6] I think I must have it examined to see if the lead behaves here in the same way as at Bristol.

Mrs Dawes joins me in most kind and affectionate regards, and believe me, my dear Tyndall,

Always sincerely yours | R. Dawes.

P.S. A friend of ours, Mr Rich[ar]d. Wade,[7] a very fit person for the Athenaeum Club,[8] is to be ballotted for on Monday, April the 3rd. If you will give him a helping hand,[9] it will much oblige.

RI MS JT/1/TYP/1/326A
LT Typescript Only

1. *new edition of your book on Heat as a Mode of Motion*: *Heat* (1865).
2. *Mrs Dawes*: Mary Helen Dawes.
3. *my friend Herbert*: possibly George Herbert (1825–94), an English clergyman who was appointed to the Deanery of Hereford Cathedral after Dawes died in 1867 (J. A. Venn, *Alumni Cantabrigienses, 1752–1900*, 6 vols (Cambridge: Cambridge University Press, 1940–54), vol. 2, part 3, p. 337). In *The Rules & Regulations and List of Members. 1874* (London: Athenaeum Club), a 'George Herbert, Esq.' is shown as a member from 1849.
4. *snow instead of eggs in making pancakes*: indeed, snow could be used as an egg substitute in pancakes, as described in the popular English cookbook by Maria Rundell (1745–1828), *A New System of Domestic Cookery* (London: John Murray, 1806): 'when eggs are scarce, make the batter with flour, and small beer, ginger, &c. Or clean snow, with flour, and a very little milk, will serve as well as eggs' (p. 181). This cookbook went through numerous editions during the first half of the nineteenth century, and continued to include snow as an egg substitute for pancakes.
5. *transepts*: the transverse part of a cruciform church considered apart from the nave (*OED*).
6. *our cathedral*: Hereford Cathedral in Hereford, England, which for a long period had been under restoration (D. Whitehead, 'The Architectural History of the Cathedral since the Reformation', in G. Alymer and J. Tiller (eds), *Hereford Cathedral: A History* (London: Hambledon Press, 2000), pp. 241–85).
7. *Mr Rich[ar]d. Wade*: not identified. In *The Rules & Regulations and List of Members. 1874* (London: Athenaeum Club), a 'Richard B. Wade, Esq., F.R.G.S.' is shown as a member from 1865.
8. *Athenaeum Club*: a private members' club located at 107 Pall Mall in London, particularly for those with intellectual interests, and accomplishments in art, science, or literature. The club was founded in 1824, at the instigation of John Wilson Croker. The English chemist

Humphry Davy (see letter 2245, n. 19), who became the director of the RI laboratory in 1801 (until 1825) where he was also professor of chemistry (1802–12) and honorary professor (1813–23), was the first chairman and Michael Faraday the first secretary. Faraday became a member after a few months when he was able to resign the secretaryship. Tyndall became a member in 1860, Thomas Huxley in 1858, Joseph Hooker in 1851, and Thomas Hirst in 1866. On the club, see M. Wheeler, *The Athenaeum: More Than Just Another London Club* (New Haven, CT: Yale University Press, 2020).

9. *a helping hand*: in Tyndall's reply (letter 2258), he informed Dawes that 'before quitting London, I had the pleasure of voting for your friend at the Athenaeum'.

From Frédéric Zurcher[1] 26 March 1865 2254

Toulon 26 mars 1865. | à l'Enclos, près du fort Lamalgue.

Monsieur,

J'ai l'honneur de vous adresser avec cette lettre un volume dans lequel j'ai eu le plaisir de rappeler, malheureusement trop sommairement, quelques uns de vos excellents travaux. Je vous prie de bien vouloir en agréer l'hommage en témoignage de sentiments reconnaissants qu'inspirent à tout ami de la Science, votre dévouement, votre zèle pour son avancement et les belles découvertes que nous leur devons.

Notre ami, M. Félix Foucou, a eu la bonté de vous transmettre l'an dernier, en mon nom et en celui de mon beau frère, M. Margollé, *<4–5 words missing>* que nous devons à votre obligeance. Nous profitons avec plaisir de cette occasion pour les renouveler, et pour vous prier tous deux, d'agréer l'assurance de nos sentiments respectueux et dévoués.

F. Zurcher | Ancien Elève de l'Ecole Polytechnique

Je joins au volume

Toulon 26 March 1865. | at the Enclos, near Fort Lamalgue.[2]

Sir,

With this letter I have the honour of offering you a volume[3] in which I was pleased to cite, albeit too briefly, some of your excellent work. I hope that you will kindly accept this tribute to them as proof that all friends of Science are most grateful for your devotion to scientific advancement and for the great discoveries generated by your zeal.

Last year, our friend Mr. Félix Foucou was kind enough to send you, in my name and in the name of my brother in law, Mr. Margollé,[4] *<4–5 words missing>* that we owe to your courtesy. We are glad to renew them on this occasion, and the two of us send our most respectful and devoted good wishes.

F. Zurcher | Former Student of the Ecole Polytechnic

I am attaching [this letter] to the volume

RI MS JT/1/Z/1

1. *Frédéric Zurcher*: Frédéric Zurcher (1816–90), a French popularizer of science who wrote, with Elie Margollé, numerous books on a variety of scientific subjects including Les Glaciers (Paris: L. Hachette, 1868) (Archives Nationales (France), Base Léonore (Légion d'honneur), http://www2.culture.gouv.fr/).
2. *Enclos, near Fort Lamalgue*: l'Enclos; possibly the Domaine du Clos Lamalgue, a winery not far from Fort Lamalgue, a military fort built in the mid-eighteenth century in the Mediterranean port city of Toulon in the department of Var.
3. *volume*: probably E. P. Margollé and F. Zurcher, *Les Météores* (Paris: Librairie de L. Hachette et Cie, 1865), in which Tyndall's *Glaciers of the Alps* is cited (p. 86).
4. *Margolle*: Elie Phillipe Margollé (1816–84), a French popularizer of science who co-authored numerous books with Frédéric Zurcher, including F. Zurcher and E. Margollé, *Les Ascensions Célèbres aux plus hauteas Montagens du Globe* (Paris: L. Hachette, 1869); and F. Zurcher and E. Margollé, *Les Glaciers* (Paris: L. Hachette, 1870) (Archives Nationales (France), Base Léonore (Légion d'honneur), http://www2.culture.gouv.fr/).

From Félix Foucou 4 April 1865 2255

Paris. 4 avril 1865. | 5. rue de Martignac

Mon cher Monsieur Tyndall,

Je viens de réussir enfin à trouver un bon éditeur. Je vous envoie ci-contre la lettre en réponse à mes propositions à M. Hetzel. Ce monsieur est un éditeur très [répandu], qui fonde une bibliothèque d'éducation, et qui a compris la grande importance de votre livre. Tout le manuscrit et les Electros sont entre ses mains. Je vous enverrai dans peu de jours notre accord. J'ai stipulé que vous serez immédiatement remboursé des £4.10S déboursés par vous, puis que nous partagerions ensemble les bénéfices qui resteraient, après prélèvement des frais de l'éditeur.

L'impression commencera ce mois-ci et je vous enverrai les épreuves, au fur et à mesure, pour que vous les examiniez vous même avant le tirage.

J'espère que votre livre se vendra bien, et c'est aussi l'espoir de l'éditeur.

Si vous venez à Paris, ne manquez pas de nous consacrer au moins le diner.

Je vous annonce la naissance d'un petit garçon, qui s'appelle Mister Charley.

Votre bien sympathique | Felix Foucou

Professor Tyndall | Londres

Paris. April 4 1865. | 5 Martignac St.

My dear Mr. Tyndall,

I just succeeded, finally, in finding a good editor. I am sending you here attached the letter in response to my propositions to Mr Hetzel.[1] This gentleman is a much known editor, who founds a Library of Education, and who understood the great importance of your book.[2] The entire manuscript and the Electros[3] are in his hands. In a few days I will send you our agreement. I stipulated that you will be immediately reimbursed the £4.10S that you spent, and then we will share the profits that will remain after deduction of the editor's costs.

The printing will begin this month and I will send you the drafts as we go, so that you can examine them yourself before the printing.

I hope that your book will sell well, and it is also the wish of the editor.

If you come to Paris, be sure to at least have some dinner with us.

I am letting you know about the birth of a baby boy, whose name is Mister Charley.

Yours very truly | Felix Foucou

Professor Tyndall | London

RI MS JT/1/F/37

---

1. *Mr. Hetzel*: Pierre-Jules Hetzel.
2. *importance of your book*: the only book of Tyndall's to have been translated into French and published by Hetzel, as far as is known, was *Mountaineering in 1861* (1862), which appeared as: J. Tyndall, *Dans les Montagnes*, trans. L. Lortet (Paris: Hetzel, 1869). This letter's date in 1865 does not seem to work with the publication of the translation in 1869, as Foucou wrote that 'The printing will begin this month'. A possibility is that this letter was referring to the planned translation and publication of another of Tyndall's mountaineering books—J. Tyndall, *The Glaciers of the Alps* (London: John Murray, 1860)—but the plan fell through. Also, *Dans les Montagnes* not only includes the translation of *Mountaineering in 1861*, but a variety of other alpine writings by Tyndall, some of which were not first published in English until *Hours of Exercise in the Alps* in 1871.
3. *Electros*: *electro* is short for *electrotype*, 'a copy or duplicate of an object, *esp*. a printing plate, made by a process involving the electrolytic deposition of copper or other metal; a relief printing plate made in this way' (*OED*).

To Annette Roche 4 April 1865 2256

4th,[1] April, 1865.

My dear Miss Roche

Would you do me the great favour of applying the cheque which I enclose to my mother's[2] requirements? I do not know how she may be circumstanced[3] and I am anxious that money should be at hand if she should find it necessary to call upon it.

Yours most truly | John Tyndall

RI MS JT/1/TYP/11/3853
LT Typescript Only

1. *4th*: LT left a note here on the typescript: 'or possibly, 11th'.
2. *my mother's*: Sarah Tyndall. See also letters 2313, 2322, 2323, and 2337.
3. *circumstanced*: Tyndall arranged Annette Roche, a local friend, to help with managing money, accommodations, and suitable clothing for his mother and sister Emily Tyndall.

To Michael Faraday 6 April 1865 2257

Dublin Thursday, 6th, April. 1865

My dear Mr Faraday

Forgive me for not saying good bye to you before I came away.[1] I have been much worried of late, and have come over here to have a run to strengthen my nerves and enable them to bear worry. One gets sadly weakened in London, and both you and I when thus weakened are, I fear, far too sensitive for our own happiness. Hirst and I go to day to Belfast and thence tomorrow to the Giant's Causeway.[2] We afterwards mean to work along the coast to Galway;[3] and perhaps further south—returning to London I trust about the 25th.[4] Give my kindest regards to Mrs Faraday[5] and Miss Barnard.[6]

Yours very affectionately | John Tyndall

Post Office | Londonderry[7] | Ireland | up to Tuesday 11th April.[8]

Faraday Correspondence, 6:4543
RI MS JT/1/TYP/12/4122

1. *came away*: as Tyndall wrote in letter 2258 to Richard Dawes a day after writing to Faraday, he went to Ireland to 'give my brain a little rest', and as indicated in the present letter, returned to London toward the end of April (see n. 4).

2. *Giant's Causeway*: a group of columns of basalt on the coast of northern Ireland (now in Northern Ireland) that has been and continues to be a popular tourist attraction, and as of 1986 a World Heritage Site.
3. *Galway*: a harbor city in western Ireland, on the River Corrib.
4. *returning to London I trust about the 25th*: it appears that Tyndall returned to London on 27 April at the earliest (see letter 2263) and certainly before 29 April (see letter 2264).
5. *Mrs Faraday*: Sarah Faraday.
6. *Miss Barnard*: Jane Barnard.
7. *Londonderry*: Londonderry, a city in northern Ireland (now in Northern Ireland). Known colloquially as Derry, it was given the 'London' prefix in the seventeenth century.
8. *up to Tuesday 11th April*: Tyndall could be reached via the Londonderry post office until this date (he and Hirst headed to Londonderry after visiting the Giant's Causeway).

To Richard Dawes 7 April 1865 2258

7th April 1865.

My dear Mr Dean,

I have been over to Ireland to give my brain a little rest, it needs it much. But before quitting London, I had the pleasure of voting for your friends[1] at the Athenaeum.[2] They were carried I believe. I now write from a little place called Glenarm[3] within about 30 miles of Belfast. A friend[4] and myself are working round this portion of the coast to the Giant's Causeway.[5] The day after tomorrow we hope to be among its columns. Your last note to me[6] turning up this evening among a group of papers has prompted me to say to you that I did not neglect your request regarding the Athenaeum. Present to Mrs Dawes[7] my very warmest regards, and believe me ever

Yours affectionately | John Tyndall.

RI MS JT/1/TYP/1/327
LT Typescript Only

---

1. *your friend*: in letter 2253, Dawes wrote to Tyndall: 'A friend of ours, Mr Rich[ar]d. Wade, a very fit person for the Athenaeum Club, is to be ballotted for on Monday, April the 3rd. If you will give him a helping hand, it will much oblige'.
2. *Athenaeum*: see letter 2253, n. 8.
3. *Glenarm*: a coastal village in northeastern Ireland (now in Northern Ireland).
4. *friend*: Thomas Hirst.
5. *Giant's Causeway*: see letter 2257, n. 2.
6. *Your last note to me*: either letter 2253 or letter missing.
7. *Mrs Dawes*: Mary Helen Dawes.

To Thomas Henry Huxley 10 April [1865][1] 2259

Giants Causeway[2] | 10th. April

My dear Huxley,

On Tuesday evening last[3] after parting with you at the Athenaeum[4] Hirst and myself started off to Rugby[5] where we spent the night: thence next morning at a rattling pace to Holyhead.[6] We had a splendid passage to Dublin, and I strolled to and fro for four hours on the bridge of the steamer, watching the glimmer of the sea and letting the breeze do what it could to cool and cleanse me within and without. We stayed at night in Dublin and next day reached Belfast. Thence we proceeded by rail to a place called Larne[7] and here our footwork began. We walked from Larne by the coast to Glenarm,[8] a secluded little village reared up, as the guide book informed us, "by the fostering care of Lord Antrim"[9] whose castle is close by. We walked from Glenarm to Cushendall,[10] forsaking the coast and taking the high moors some way inland. From Cushendall we took a car to a place called Ballycastle,[11] thence yesterday (Sunday) we took a car to the Giants Causeway and here we are.

The latter part off yesterday we spent clambering over the Causeway and wandering from headland to headland along the coast: we had a guide but he was no climber and though but four and twenty and exceedingly fresh looking he complained of palpitation of the heart. The Causeway at a little distance is utterly mean, but actually standing amid its columns it is as utterly wonderful. The coast scenery is of the noblest character.

Though I have been away a good while still the amount of exercise which I have had has been so small that it would be too soon to speak of its effect. My brain was in a curious condition when I came away and its tone is by no means yet restored. It felt as if it had been physically bruised and wounded by worry and work, and I found my will, instead of being a cable of iron, a mere rope of sand. To think of myself standing in the open market place of science and having dirt thrown upon me is almost intolerable,[12] and still I see that almost every man who has raised himself above the general level has had to endure more or less of this. Science and its glories have ever been to me contemptible if measured against manhood and a chivalrous bearing; and my chief annoyance at the present moment is to find myself unable by an act of renunciation to publicly shew the contempt that I feel for this last achievement. I would willingly throw the bone to the <u>hungry wolf</u>[13] if it were in my power to do so. Five months ago I had written a letter with this intent, but withheld it because I knew it would be misinterpreted.

To day we go to Derry[14] and then round the north west and west coast to Donegal.[15] If you care to send me a line address it to the Post Office Donegal. <u>Please tell Reeks[16] that I shall commence my lectures at Jermyn Street[17] on</u>

Monday the 1st. of May. But I will write directly to him.[18] Give my kindest regards to your wife[19] and my love to your children[20]

Yours ever | J. Tyndall.

IC HP 8.45
Typed Transcript Only

---

1. *[1865]*: the year is given by reference to Tyndall's trip to Ireland in 1865 with Thomas Hirst.
2. *Giants Causeway*: see letter 2257, n. 2.
3. *Tuesday evening last*: 4 April 1865.
4. *Athenaeum*: see letter 2253, n. 8.
5. *Rugby*: a city in England northwest of London, on the way toward Birmingham.
6. *Holyhead*: a city on the west coast of England, from where Tyndall and Hirst travelled by sea to Dublin.
7. *Larne*: a coastal town in northeastern Ireland (now in Northern Ireland).
8. *Glenarm*: see letter 2258, n. 3.
9. *"by the fostering care of Lord Antrim"*: the guide Tyndall used was probably S. C. Hall and A. M. Hall, *Ireland: Its Scenery, Character, &c.*, 3 vols (London: Jeremiah How, 1843), vol. 3. In the description for Glenarm, we find: 'On entering, we pass to the left a very pretty school-house, which under the judicious and fostering care of Edmund McDonnel, Esq. [Lord Antrim] and his niece, Lady Louisa Kerr, will rescue the rising generation from ignorance' (p. 130).
10. *Cushendall*: a coastal village in northeastern Ireland (now in Northern Ireland), to the north of Glenarm.
11. *Ballycastle*: a coastal town in northeastern Ireland (now in Northern Ireland), to the north and west of Cushendall.
12. *To think of myself . . . is almost intolerable*: here Tyndall was referring to his dispute with C. K. Akin, the topic of many of the preceding letters in this volume (for example, 2222, 2227, 2239, 2242, and 2245). See also the introduction to this volume.
13. *I would willingly throw the bone to the hungry wolf*: in his published responses to Akin, despite the stress Akin's actions had given him, Tyndall appeared to resist attacking Akin even more negatively than he already was.
14. *Derry*: Londonderry, a city in northern Ireland (now in Northern Ireland). Known colloquially as Derry, it was given the 'London' prefix in the seventeenth century.
15. *Donegal*: a town in Ulster, Ireland.
16. *Reeks*: Trenham Reeks (1823/4–79), a British mineralogist and Registrar of the Royal School of Mines (1851–79) ('Trenham Reeks', *Nature* 20, (8 May 1879), pp. 38–9).
17. *my lectures at Jermyn Street*: Tyndall held the Chair of Natural Philosophy at the Royal School of Mines from 1859–68, and in this role lectured to students during the week and in some years gave evening lectures for working men. Here he was referring to a course of thirty lectures on 'Magnetism, Electricity, Sound, Light, and Heat', starting on 1 May

1865, on Mondays through Fridays at 3pm (*The Athenaeum*, 21 April 1865, p. 569). From its establishment in 1851 until 1863, it was known as the Government School of Mines and Science Applied to the Arts. See R. M. Clary, 'The Royal School of Mines: Henry de la Bache's Convergence of Professionalization and Public Advocacy', *Earth Sciences History*, 39 (2020), pp. 291–304.

18. *write directly to him*: letter missing.
19. *your wife*: Henrietta Huxley.
20. *children*: at this time, the Huxley children were Jessie Oriana Waller (née Huxley, 1858–1927), Marian Collier (née Huxley, 1859–87), Leonard Huxley (1860–1933), Rachel Eckersley (née Huxley, 1862–1934), Henrietta (Nettie) Roller (née Huxley, 1863–1940), and Henry Huxley (1865–1946). Another child was born in 1866, Ethel Collier (nee Huxley, 1866–1941). Their first-born son Noel Huxley had died in 1860 of tuberculosis at the age of 4 (Tyndall attended his funeral).

To Juliet Pollock 10 April [1865][1] 2260

Ireland | 10th April.

My dear Mrs Pollock,

I was obliged to run away to rest my brain and clear away its mists in order to prepare myself for the long course of lectures which I have to give at the School of Mines[2] in May and June. The west Coast of Ireland occurred to me, so accompanied by my friend Hirst I reached Dublin on Wednesday last.[3] On Thursday we reached Belfast. Thence to a little town called Larne,[4] and thence, mostly on foot, but partly by car to the Giants Causeway,[5] within ten minutes walk of which I now write. It is a noble coast: and the causeway, though quite mean when looked at from a little distance is a most wonderful thing when you stand amid its columns.

Today we hope to get on to Londonderry[6] and thence by some unknown means to the Donegal mountains[7] & the West Coast. I shall be in Donegal[8] on Thursday I hope, so if you care to send me a line—& a line from you would be very pleasant to me—you might address it to the

Post Office | Donegal | Ireland.

Give my kindest regards to your husband[9] and love to the boys.[10]

Yours ever | John Tyndall

RI MS JT/1/T/1163
RI MS JT/1/TYP/6/2038

1. *[1865]*: the year is given by reference to Tyndall's trip to Ireland with Thomas Hirst.
2. *long course of lectures which I have to give at the School of Mines*: see letter 2259, n. 17.

3. *reached Dublin on Wednesday last*: 5 April 1865.
4. *Larne*: a coastal town in northeastern Ireland (now in Northern Ireland).
5. *Giants Causeway*: see letter 2257, n. 2.
6. *Londonderry*: a city in Ireland (now in Northern Ireland). Known colloquially as Derry, it was given the 'London' prefix in the seventeenth century. It is perhaps notable that Tyndall had originally written 'Derry' and crossed this out to write 'Londonderry' instead.
7. *Donegal mountains*: various mountains within County Donegal, Ireland.
8. *Donegal*: a county in the north west of Ireland.
9. *your husband*: William Frederick Pollock.
10. *the boys*: Frederick Pollock, Walter Pollock, and Maurice Pollock.

From Johann Conrad Bohn 13 April 1865 2261

Giessen d. 13 April 1865

Hochgeehrter Herr Tyndall!

Gestatten Sie mir zunächst Ihnen meinen Dank auszusprechen für Ihre freundliche Vermittlung bei Aufnahme meines letzten kleinen Aufsatzes im Phil. Mag. Dann bitte ich Sie um die Erlaubniss Ihnen einige Bemerkungen vorzulegen, auf welche ich beim Studium Ihrer Abhandlungen Phil. Mag. [4] 28. 438, 508 gekommen bin, als ich von denselben einen Auszug für den in Giessen erscheinenden Jahresbericht machte.

1.) Sie fanden (p. 531)

Absorption der mit Kienruss überzogenen Steinsalzplatte 82,7%
" " reinen " 24,0

und daraus schliessen Sie

Absorption der Schichte Kienruss 58,7%

Den hier von Ihnen gemachten Schluss finde ich nicht *[gerechtfertigt]*; ich berechne aus Ihren Versuchen die Absorption der Kienrussschichte zu 77,3%

Wenn auf dem Wege dx die Intensität durch Absorption im Verhältniss von 1:1-adx gemindert wird so ist die aus einer Platte von der Dicke x tretende Quantität Strahlen durch e^{-ax} gegeben, wenn 1 die an der Vorderseite ankommende Menge ist.

Bezeichnet a' den Coefficienten für Kienruss, welchen a für Steinsalz angibt und y die Dicke der Kienrussschichte, so kommt nach Durchstrahlung dieser Schichte zu der Steinsalzplatte die Menge $e^{-a'y}$. Demnach tritt aus der berussten Steinsalzplatte die Menge

$$e^{-a'y}\ e^{-ax} = e^{-(ax+a'y)} = 0{,}173$$

Aus der reinen Steinsalzplatte tritt

$$e^{-ax} = 0{,}76$$

Also aus der Kienrussschicht tritt e-a'y = $\frac{173}{760}$ = 0,227

folglich werden 77,3% in der Kienrussschicht absorbiert.

2.) Ihre Resultate über die Absorption des W. in Gasen, Dämpfen und Flüssigkeiten enthalten auch die Absorption in den zum Verschluss dienenden Steinsalzplatten, welche Absorption, wie Sie zeigen verschieden ist, je nach der Wärmequelle. Ich hätte mir gewünscht diese durch Rechnung in Abzug bringen zu können.

Wenn x die Gesammtdicke der 2 Steinsalzplatten und a ihr Absorptionscoeff. ist, ferner d_1 d_2 d_3 . . . die Länge der durchstrahlten Gas-, Dampf- oder Flüssigkeitsschicht, so sind die durchgelassenen Wärmemengen

$T_1 = e^{-(ax+a'd_1)}$

$T_2 = e^{-(ax+a'd_2)}$ wo a' der Absorptionscoeff.

$T_3 = e^{-(ax+a'd_3)}$ der in der Röhre befindlichen Substanz ist.

. . .

Also $\frac{lgT_1 - lgT_2}{d_2 - d_1} = a'$ $\frac{lgT_2 - lgT_3}{d_3 - d_1} = a'$

. . .

Besitzt man also wenigstens 3 Versuche mit derselben Substanz bei verschiedener Dicke so kann man a', dann auch a und die Absorption der Steinsalzplatten selbst berechnen. Leider finde ich bei Ihren Angaben nur von der allerungünstigsten Substanz Wasser mehr als 2 Dicken; trotzdem jedoch dass diese Versuche mit Wasser sicher keine genauen Resultate liefern können, so stimmen doch die aus den geringsten Dicken abgeleiteten Werthe von a' ziemlich gut mit einander.

Pag. 527

Combinat. von Versuch	1 & 2 gibt a'	= *[*k*]*. T.84187
	1 & 3	= *[*k*]*. T.85559
	2 & 3	= *[*k*]*. T.86474
	1 & 4	= *[*k*]*. T.91129
	2 & 4	= *[*k*]*. T.92518
	3 & 4	= *[*k*]*. T.95108

Auf die Reflexionen an den Platten habe ich dabei keine Rücksicht genommen, was zunächst wohl erlaubt sein mag.

Vielleicht haben Sie noch einige Versuche vorräthig, welche die gewünschte Rechnung anzustellen gestatten oder finden Sie es interessant genug noch einige anzustellen.

Nehmen Sie meine Bemerkungen freundlich auf und wenn Sie mich sehr erfreuen wollen, so beehren Sie mich mit ein paar Zeilen Antwort.

In vorzüglichster Hochachtung | Ihr ganz ergebener | J C Bohn

Giessen, 13 April 1865

Esteemed Mr Tyndall!

Allow me first of all to express my thanks to you for your kind mediation in the acceptance of my last short essay in the Phil. Mag.[1] Then I ask for permission to present some remarks to you which I have arrived at while studying your papers in Phil. Mag. [4] 28. 438, 508.,[2] when preparing an abstract of them for the Annual Report appearing in Giessen.[3]

1.) You found (p. 531)

Absorption of the plate of rock salt coated with lamp-black	82.7%
" " " clean plate of rock salt	24.0

and from there you conclude

Absorption of the layer of lamp-black	58.7%

I do not find the conclusion made by you here *[justified]*; from your experiments, I calculate the absorption of the layer of lamp-black at 77.3%

If, along the path dx the intensity is reduced by absoprtion in the ratio 1:1-adx, then theen the quantity of rays emanating from a plate of thickness x will be given by e^{-ax}, if 1 is the amount reaching the front side.

If a' is the coefficient for lamp-black, which a indicates for rock salt and y the thickness of the layer of lamp-black, then, after irradiation of this layer, the quantitiy reaching the plate of rock salt will be $e^{-a'y}$. Accordingly, the quantity will emanating from the plate of rock-salt coated with lamp-black is:

$$e^{-a'y}\ e^{-ax} = e^{-(ax+a'y)} = 0.173$$

From the clean plate of rock salt emanates

$$e^{-ax} = 0.76$$

Thus, from the layer of lamp-black emanates $e^{-ay} = \frac{173}{760} = 0.227$

and consequently 77.3% are absorbed in the layer of lamp-black.

2.) Your results on the absorption of W[ater]. in gases, vapours, and liquids also contain the absorption in the plates of rock salt serving as seals, an absorption that, as you show, is different depending on the source of heat. I would have wished that I had been able to deduct this by calculation.

If x is the total thickness of the 2 plates of rock salt and their coeff[icient]. of absorption, and further $d_1\ d_2\ d_3 \ldots$ the length of the irradiated layer of gas, vapour, or liquid, then the quantities of heat let through are

$$T_1 = e^{-(ax+a'd_1)}$$
$$T_2 = e^{-(ax+a'd_2)}$$
$$T_3 = e^{-(ax+a'd_3)}$$

where a' is the coeff[icient]. of absorption of the substance found in the tube.

....

Then $\frac{lgT_1 - lgT_2}{d_2 - d_1} = a'$ $\qquad \frac{lgT_2 - lgT_3}{d_3 - d_1} = a'$

...

Thus, if one has at least 3 experiments with the same substance at different thicknesses, then one can calculate a', then also a, and the absorption of the plates of rock salt themselves. Unfortunately, with your data, I find only more than 2 thicknesses of the most unfavourable substance of all, water; however, despite the fact that these experiments with water certainly cannot provide precise results, the values of a' derived from the smallest thicknesses do agree quite well with one other.

Pag. 527

Combinat[ion]. of experiment	1 & 2 yields a'	= /k/. T.84187
	1 & 3	= /k/. T.85559
	2 & 3	= /k/. T.86474
	1 & 4	= /k/. T.91129
	2 & 4	= /k/. T.92518
	3 & 4	= /k/. T.95108

I did not give any consideration to the reflections on the plates while doing all this, which may be permitted initially.

Perhaps you have some more experiments available which permit doing the desired calculation, or do you find it interesting enough to do any more.

Accept my comments in a friendly way and if you wish to please me very much, then honour me with a few lines of reply.

Yours very fauithfully | Your most humble servant | J C Bohn

RI MS JT/1/B/104

1. *my last short essay in the Phil. Mag.*: J. C. Bohn, 'On the History of the Conservation of Energy, and of its application to Physics', *Phil. Mag.*, 29 (1865), pp. 215–8.
2. *your papers in Phil. Mag. [4] 28. 438, 508*: J. Tyndall, 'The Bakerian Lecture.—Contributions to Molecular Physics. Being the Fifth Memoir of Researches on Radiant Heat', *Phil. Mag.*, 28 (1864), pp. 438–58, 508–35.
3. *abstract of them for the Annual Report published in Giessen*: *Jahresbericht über die fortschritte der chemie und verwandter theile anderer wissenschaften. Für 1864* (Giessen: J. Ricker, 1865), pp. 24–31.

From Henry Russell[1] 21 April [1865][2] 2262

London April 21st

My Dear Mr. Tyndall

It was much too kind of you to send me your most useful and interesting brochure on radiation.[3] It gave me, I am sure, even greater pleasure than you think; and I shall read it not only with the greatest care, but with such jealous eyes that none but mine will be allowed to look at it.

With many thanks pray | believe me | Very sincerely yours | Cte Henry Russell.

RI MS JT/1/R/81

1. *Henry Russell*: Henry Patrick Marie, Count Russell-Killough (1834–1909), an Irish mountaineer known for numerous first ascents in the Pyrenees (R. Bailey, *The Man Who Married a Mountain: a Journey through the French Pyrenees* (London: Transworld Publishers Limited, 2005)).
2. *[1865]*: the year is given by reference to Tyndall's Rede Lecture at the University of Cambridge ('On Radiation') in May 1865 (see letter 2231, n. 1). Russell's words 'with such jealous eyes that none but mine will be allowed to look at it' indicate that Tyndall sent him a copy of his lecture beforehand.
3. *brochure on radiation*: the Rede lecture was published as *On Radiation*.

To Thomas Henry Huxley [26 April 1865][1] 2263

Sligo[2] Wednesday

Dear old Scipio,[3]

I see it all, and perhaps may again feel it all; but latterly I have felt "a law in my members warring against the law of my mind and bringing me into captivity to the law of sin which is in my members."[4] Poor old Paul,[5] a devilish clever fellow and no coward, was obliged to knock under in some particulars. And Achilles[6] himself had a thumbspace assailable. I have not found Irish whiskey a wholesome stimulant, neither has Hirst:[7] but I have strong hopes that when wholesome food is attainable all will be right. I pause this night in Dublin, and tomorrow evening I hope to be in London.[8]

Yours always old Lion | J. Tyndall.

IC HP 8.46
RI MS JT/1/TYP/9/2903
Typed Transcript Only

1. *[26 April 1865]*: LT wrote this date on the RI typescript. This letter was surely written during Tyndall's 1865 trip to Ireland with Thomas Hirst. In letter 2264 to James Croll, it is clear that Tyndall was back in London by at least 29 April (a Saturday), dating this Wednesday letter from just before returning to London to 26 April.
2. *Sligo*: a coastal town in northwestern Ireland.
3. *Scipio*: a nickname for Huxley; Publius Cornelius Scipio Africanus (236–183 BC) was a Roman General who defeated Hannibal at the battle of Zama in the 2nd Punic War.
4. *"a law in my members warring . . . in my members"*: here Tyndall is quoting Paul from Romans 7:23. Tyndall was likely referring to drinking alcohol.
5. *Paul*: a reference to Paul the Apostle from the New Testament.
6. *Achilles*: a reference to the hero of the Trojan War in Greek mythology.

7. *Hirst*: Thomas Hirst, who accompanied Tyndall on this trip to Ireland.
8. *tomorrow evening I hope to be in London*: this would be 27 April 1865 if he indeed returned to London the following day.

To James Croll 29 April 1865 2264

29th April 1865

Dear Mr Croll

I have thought ten times of writing to you about your manuscript:[1] and ten times over the subject has left my mind. I have been unwell & overtasked, and have been out of England[2] for several weeks. Up to the present moment I have not been able to look at the manuscript. If you can leave it with me I will try what I can do before I quit England for my summer tour; but at present I have barely sufficient strength to meet the imperative duties which lie before me.

Yours very truly | John Tyndall

BL Add MS 41077, ff. 19–19a

1. *your manuscript*: possibly J. Croll, 'On the Physical Cause of the Submergence and Emergence of the Land during the Glacial Epoch', *Phil. Mag.*, 31 (1866), pp. 301–5.
2. *out of England*: Tyndall spent three weeks in Ireland in April 1865 with Thomas Hirst (see letters 2257, 2258, 2259, and 2260).

From Johann Conrad Bohn 2 May 1865 2265

Giessen d 2 Mai 1865

Hochgeehrter Herr Tyndall!

Gestatten Sie mir meinen verbindlichsten Dank für Ihre freundliche Beantwortung meiner Briefe auszusprechen. Mit Bedauern vernehme ich dass Ihre Gesundheitszustände augenblicklich nicht so befriedigend sind, um Ihnen eine alsbaldige Wiederaufnahme und Fortführung Ihrer so interessanten und wichtigen Untersuchungen über die Wärme zu gestatten.

Obgleich der Jahresbericht, an welchem ich Mitarbeiter bin, keine Kritik zu geben beabsichtigt, war es doch bei *[seiner]* Redigirung zu allen Zeiten üblich, wichtigere Arbeiten, wenn die Gelegenheit sich darbot, mit Anmerkungen von Seiten der Berichterstatter zu begleiten. Ich erlaube mir Ihnen mitzutheilen, dass ich folgende Anmerkungen, zu dem Bericht über Ihre Abhandlung zu machen gedenke. Ihrer Tabelle der Absorption des W. in

Flüssigkeitsschichten verschiedener Dicke, beabsichtige ich folgende Note anzuhängen:

„Wird durch Absorption die Intensität der Strahlen auf dem Wege dA im Verhältniss 1:1-adx geschwächt, so verhält sich die Intensität der durch die Dicke x hindurchgelassen Strahlen zur Intensität der *[eintretenden]* Strahlen wie e^{-ax}:1. Es ergeben sich hieraus für die Intensitäten I_1, I_2, I_3 . . . der durch die Dicken x_1, x_2, x_3, . . . derselben Substanz gelassenen Strahlen die Gleichungen:

$$\frac{\log l_1 - \log l_2}{x_2 - x_1} = \frac{\log l_1 - \log l_3}{x_3 - x_1} = \frac{\log l_m - \log l_n}{x_n - x_m}.$$

Tyndall's Angaben entsprechen dieser theoretischen Anforderung nicht."

Dieser Note lege ich keine andere Bedeutung bei, als dass Sie Zeugnisgaben sind, von der Aufmerksamkeit, mit welcher ich ihre Abhandlung gelesen habe und von dem Interesse das mir dieselbe erregt hat. Diese Note könnte auch ganz wohl wegbleiben und es würde mich keine Ueberwindung kosten, sie zu unter drücken.

Eine zweite Note gedenke ich nach folgenden Worten des Textes meines Berichtes zu geben: Fazit: „Während eine reine Steinsalzplatte 24,0 Procent der von einer Wasserstoffflamme ausgegangenen Strahlen verschluckte, absorbirte dieselbe mit Kienruss überzogene Platte 82,7 Procent, woraus folgen soll, dass der Kienruss 58,7% absorbirt."

Anmerkung: „Von 100 auffallenden Strahlen lässt die Kienrussschicht $100e^{-ax}$ durch, wenn x ihre Dicke und a die Absorptionsconstante des Russes ist. Von den $100e^{-ax}$ auf das Steinsalz fallenden Strahlen, lässt die Steinsalzplatte $100e^{-ax} \cdot e^{-by}$ durch, wenn y die Dicke, b die Absorptionsconstante des Steinsalzes ist. Die Versuche ergaben $100e^{-ax}.e^{-by} = 17{,}3$ und $100e^{-by} = 76{,}0$. Also ist $e^{-2x} = 0{,}227$ und der Kienruss absorbirt hiernach 77,3 Procent, nicht 58,7 wie Tyndall rechnet."

Ich habe mich bemüht diese Noten so zu halten, dass ich kein Urtheil darin ausspreche ob die mathematische Theorie der Erscheinungen auf Ihre Versuche anwendbar ist und glaube nichts über weitere Untersuchungen präjudiert zu haben. Es würde mir zu einer besonderen Befriedigung gereichen, wenn Sie gegen die Form dieser Anmerkungen nichts einzuwenden hätten, und ich bin, falls Sie eine Aenderung wünschen sollten gerne dazu bereit.

Was die Reflexion der Wärmestrahlen angeht und ihren Einfluss auf die Resultate, so scheint mir als würden nur die absoluten Werthe der Absorptionscoefficienten durch die Berücksichtigung der Zurückwerfung eine kleine Modification erleiden, nicht wesentlich aber der Sinn meiner Anmerkungen getroffen werde.

Sei R der von einer Ihrer Steinsalzplatten reflectirten Antheile. Wenn der Brechungsexponent des Steinsalzes = 1,5 genommen wird und die Absorption in derselben zu 24% angenommen, so ist bei normaler Incidenz der Werth von R = 0,0612. Ist y die Dicke der Flüssigkeitsschicht und b die Absorptionsconstante, so wird durch Ihre Röhre folgende Zahl von Strahlen gelassen (wenn 1—R = T)

$$T^2e^{-by} + T^2e^{-3by}R^2 + T^2e^{-5by}R^4 + T^2e^{-7by}R^6 + \ldots = T^2e^{-by} \cdot \frac{1}{1-R^2e^{-2by}}$$

Für eine andere Dicke y, berechnet sich dann $T^2e^{-b_1y_1}\frac{1}{1-R^2e^{-2b_1y}}$ wegen der Kleinheit von $R^2[(0{,}0612)^2]$ und da auch $e^{-2by} < 1$, bleibt die von mir, ohne Berücksichtigung der Reflexion abgebildete Bedingung $\frac{I}{I_1} = e^{y-y_1}$ nahezu richtig. Und ähnlich ist für die zweite Anmerkung die Reflexion nicht von bedeutendem Einfluss.

Wenn Sie Veranlassung zu einem experimentellen Studium der Frage nach dem Einflusse der Reflexion finden, so wird das bei mir das grösste Interesse erregen und ich wünsche schon deswegen lebhaft dass Sie bald Musse und Kraft dazu finden mögen.

Für Ihre gütige Absicht mich durch Uebersendung Ihres Buches on Heat zu beehren, bin ich Ihnen auf grösste zu Dank verpflichtet. Die Gelegenheit zur Uebermittlung wäre sehr bequem, wenn Sie dasselbe entweder an Mr. D. Nutt, Strand 270 oder an Mr. Williams & Norgate, 14 Henrietta Street, Covent Garden zur Uebermittlung an die J. Ricker'sche Buchhandlung in Giessen, übergeben wollten.

Ich verbleibe mit der ausgezeichnetsten Verehrung | Ihr ergebenster | Dr Bohn

Giessen, 2 May 1865

Esteemed Mr Tyndall!

Allow me to express my most sincere thanks for your friendly replies to my letters.[1] With regret I learn that your state of health is momentarily not so satisfactory as to permit a speedy resumption and continuation of your interesting and important investigations on heat.

Although the annual report in which I am a collaborator does not intend to give criticism, during its redaction it was customary at all times to accompany more important works with annotations by the reporters. I allow myself to inform you that I plan to make the following comments on the report about your article.[2] I intend to attach the following note to your table on the absorption of W[ater]. in fluid layers of varying thickness:

"If the intensity of the rays decreases in a ratio 1:1-αdx due to absorption on the path dA, then the intensity of the rays transmitted through a thickness x relates to the intensity of incoming rays as e^{-ax}:1. Hereby one obtains the following equations for the intensities I_1, I_2, I_3 . . . of the rays that pass through thicknesses x_1, x_2, x_3 . . . of the same substance:

$$\frac{\log I_1 - \log I_2}{x_2 - x_1} = \frac{\log I_1 - \log I_3}{x_3 - x_1} = \frac{\log I_m - \log I_n}{x_n - x_m}$$

Tyndall's results do not comply with this theoretical requirement."

I do not attach any importance to this note other than it being a testimony for the attention with which I have read your article and for the interest that it has awakened in me. This note could also be entirely omitted and it would cost me no effort to suppress it.

I intend to write a second note after the following words of my report: Conclusion:

"While a plate of pure rock salt absorbed 24.0 percent of the rays emitted by a hydrogen flame, the same plate absorbed 82.7 percent when coated with lamp-black, which would imply that the lamp-black absorbs 58.7%".

Note: "Of 100 incoming rays, the layer of lamp-black permits the passage of $100e^{-ax}$, where x is the thickness and a the constant of absorption of the soot. Of the $100e^{-ax}$ rays that reach the rock salt plate, it permits the passage of $100e^{-ax} \cdot e^{-by} = 17.3$ and $100e^{-by} = 76{,}0$. Accordingly, $e^{-2x} = 0.227$ and the lamp-black absorbs 77.3 percent, not 58.7 as Tyndall calculates."

I have endeavoured to keep these notes such that I do not pass judgement whether the mathematical theory of the phenomena can be applied to your experiments, and I believe that I have not caused any prejudice against further investigations. It would give me great satisfaction if you had no objections to make against the form of these comments and should you wish it, I would be glad to modify them.

Regarding the reflection of heat rays and their influence on the results, it seems to me that if the reflection were considered, only the absolute values of the coefficients of absorption would undergo slight modification, but not the essential point of my annotations.

Let R be a proportion of those reflected by one of your rock salt plates. If the exponent of refraction of rock salt is taken to be = 1.5 and its absorption is assumed as 24%, then the value of R = 0.0612 at normal incidence. If y is the thickness of the fluid layer and b the constant of absorption, then the following number of rays will be allowed through your tube (if 1—R = T)

$$T^2e^{-by} + T^2e^{-3by}R^2 + T^2e^{-5by}R^4 + T^2e^{-7by}R^6 + \ldots = T^2e^{-by} \cdot \frac{1}{1-R^2e^{-2by}}$$

For a different thickness y, one calculates $T^2e^{-b_1y_1}\frac{1}{1-R^2e^{-2b_1y}}$ due to the smallness of $R^2[(0.0612)^2]$ and because also $e^{-2by} < 1$, then the condition I propose without considering reflection, $\frac{I}{I_1} = e^{y-y_1}$, remains approximately true. And similarly, reflection is not a significant influence for the second comment.

If you find the opportunity to investigate the question of the influence of reflection experimentally, it would arouse my greatest interest and for that reason alone I fervently wish that you might soon find inspiration and energy to do so.

I am greatly indebted to you for your kind intention to honour me with the sending of your book on Heat.[3] The most convenient opportunity for sending it would be if you gave it either to Mr D. Nutt,[4] Strand 270 or to Misters Williams

& Norgate,[5] 14 Henrietta Street, Covent Garden for delivery to J. Ricker's[6] bookshop in Giessen.

I remain, with the most excellent admiration | Your humble servant | Dr Bohn

RI MS JT/1/B/105

1. *your friendly replies to my letters*: letters from Tyndall to Bohn are missing.
2. *report about your article*: see letter 2261, n. 3.
3. *your book on Heat*: *Heat* (1865).
4. *Mr D. Nutt*: Alfred Trübner Nutt (1856–1910), a British publisher based in London (*ODNB*).
5. *Misters William & Norgate*: the London-based publishing firm Williams and Norgate, founded by Edmund Sydney Williams (1817–91) and Frederick Norgate (1817–1908) ('In Memorium: Mr. Edmund Sydney Williams', *The Publisher's Circular*, 55 (12 September 1891), p. 281; *ODNB*).
6. *J. Ricker's*: Joseph Ricker, a book publisher and seller in Giessen, Germany (R. Schmidt,-*Deutsche Buchhändler. Deutsche Buchdrucker. Beiträge zu einer Firmengeschichte des deutschen Buchgewerbes*, 6 vols (Berlin: Verlag der Buchdruckerei Franz Weber, 1902–8), vol. 5 (1908), pp. 820–23).

To George Gabriel Stokes 6 May [1865][1] 2266

Royal Institution of Great Britain | 6th May

My dear Stokes

I should like to show my audience at Cambridge[2] the proof of a spectrum by daylight. Could you lend me a screen on which to project it?

Yours very truly | J. Tyndall | Prof. Stokes | &c &c &c

CUL SC—7656

1. *[1865]*: the year is given by reference to Tyndall's Rede Lecture in 1865.
2. *audience at Cambridge*: Tyndall delivered the annual Rede Lecture at the University of Cambridge in May 1865 (see letter 2231, n. 1).

To James Croll 9 May 1865 2267

9th May 1865

Dear Sir

I detained your manuscript[1] for two days in order to make myself acquainted in a general way with its contents. It was however posted to you yesterday.

The philosophic tone of the manuscript pleases me, and there are numerous remarks in it to which I would heartily subscribe. But I must frankly state my opinion that what you say of gravity needs reconsideration.

Gravity is in precisely the same category as all the other forces: we cannot compare gravity with heat; but we can compare it with the molecular attractions & repulsions which produce the motion of heat—and every difficulty urged against gravity may be urged against them.

If I had time I would write to you at greater length: but I would now simply say that in my opinion there is nothing inconsistent whatever with the principle of the conservation of energy in our present ideas of gravity

Yours very truly | John Tyndall[2]

BL Add MS 41077, f. 20–21v

J. C. Irons (ed), *Autobiographical Sketch of James Croll with Memoir of his Life and Work* (London: Edward Stanford, 1896), pp. 147–8

1. *manuscript*: see letter 2264, n. 1.
2. On the reverse side of the last page of the letter is a list of references to scientific articles, presumably in Croll's hand.

To William Frederick Pollock [9 May 1865][1] 2268

Tuesday

My dear Pollock,

The arrangement is delightful.[2]

I must keep quiet—I must avoid dinners—let us vote for the Bull.[3] My head is just within working limits and the least thing would push it beyond them.

Yours ever | John Tyndall.

RI MS JT/1/TYP/6/2201
LT Typescript Only

1. *[9 May 1865]*: the date is given by relation to letter 2269, dated to Thursday, 11 May [1865], thus dating this Tuesday letter to 9 May 1865.
2. *arrangement is delightful*: Pollock had probably asked Tyndall about arrangements for his Rede Lecture at the University of Cambridge ('On Radiation') on 16 May 1865 (see letter 2231, n. 1).
3. *vote for the Bull*: the Bull Hotel in Cambridge, built in 1828. In letter 2269, Tyndall thanked Pollock 'for having asked Fred to get a room for me at the Bull'.

To William Frederick Pollock 11 May [1865][1] 2269

11th May.

My dear Pollock,

Many thanks for having asked Fred[2] to get a room for me at the Bull.[3] I hope it will be a quiet one, as I must sleep if I would avoid breaking down.

Make your plans entirely irrespective of me. I cannot promise either to breakfast or dine anywhere. My health is so shaky that I shall in all probability rigidly confine myself to the Bull.

We must have a stroll through the gardens[4] together. I have long wished to see them in your company

Yours ever | John Tyndall.

RI MS JT/1/TYP/6/2200
LT Typescript Only

1. *[1865]*: the year is given since Tyndall was describing plans to stay in Cambridge in order to deliver the annual Rede Lecture at the University of Cambridge ('On Radiation') on 16 May 1865 (see letter 2231, n. 1).
2. *Fred*: Frederick Pollock.
3. *the Bull*: the Bull Hotel in Cambridge, built in 1828.
4. *gardens*: probably the botanic gardens at the University of Cambridge.

From Rudolf Clausius 15 May 1865 2270

ZÜRICH, 15/5/65.

Lieber Tyndall,

Ich muss Dich um Entschuldigung bitten, dass ich Deinen freundlichen Brief vom 26sten Febr. so lange unbeantwortet gelassen habe. Ich habe leider die Angewohnheit, wenn nicht eine specielle Veranlassung zum Schreiben vorliegt, das Schreiben fort und fort aufzuschieben, bis ich endlich selbst über die Länge der Zeit erstaunt bin. Deine Antwort an Akin für deren Zusendung ich Dir bestens danke, habe ich mit grossem Interesse gelesen, und habe mich daraus noch mehr, wie früher, überzeugt, dass Akin nicht nur von finsterem unerträglichem Character und in hohem Grade in Bezug auf seine eigenen Verdienste verblendet ist, sondern dass er auch unaufrichtig ist, und die Sache total entstellt hat. Wenn Du etwas in der Sache versehen hast, so besteht es meiner Ansicht nach gewiss nicht darin, dass Du zu wenig Rücksicht auf Akin genommen hast, sondern höchstens darin dass Du zu viel Rücksicht auf ihn genommen hast. Er konnte von Anfang an nichts weiter beanspruchen als dass dasjenige, was er publicirt hatte, erwähnt würde. Ich glaube übrigens dass alle unparteiischen Leser den Eindruck behalten haben, dass Akin mit seinen Ansprüchen vollständig im Unrechte ist.

Akin hat zu Anfang des März auch an mich geschrieben, und hat mir unter anderem den Wunsch ausgesprochen, dass ich ihm in Bezug auf das, was er über Kirchhoff geschrieben hat, meine Ansichten mittheilen möge. Er ist mir aber durch sein Benehmen gegen Dich so zuwider geworden, dass ich ihm gar nicht geantwortet habe.

Dieses Jahr wird die Versammlung der schweizerischen Naturforschergesellschaft zu Genf an den Tagen 21, 22, und 23 August abgehalten werden, so dass die Ankunft der Gäste am Sonntag den 20sten August erwartet wird. Ich gedenke mit Frau und Kindern nach dem Genfer See zu reisen und wir wollen in irgend einer Pension in oder bei Montreux Wohnung nehmen, und von da will ich dann nach Genf gehen. Wenn Du auch wieder nach der Schweiz kommen und entweder uns in Zürich besuchen oder mit uns am Genfer See zusammentreffen könntest, würden wir uns herzlich freuen. Ich denke, die Versammlung in Genf wird eine recht zahlreiche und belebte werden, weil dieses Mal das fünfzigjährige Jubilaeum der schweizerischen naturforschenden Gesellschaft, deren Stiftung auch in Genf stattgefunden hat, gefeiert wird. Dir wird nach den Arbeiten und Aufregungen des letzten Jahres gewiss eine Erholung in den Bergen wohlthun. Soviel ich gehört habe, kommt Frankland auch nach der Schweiz, und dieser Umstand kann vielleicht die für uns sehr erfreuliche Wirkung haben, Deine Reiselust nach der Schweiz zu vermehren.

In der Hoffnung, Dich in diesem Jahre wieder zu sehen und mit freundlichen Grüssen von meiner Frau und mir,

Dein | Clausius.

ZURICH, 15/5/65.

Dear Tyndall,

I have to ask your forgiveness for not having answered your kind letter from February 26th for such a long time.[1] Unfortunately, I have the habit of endlessly postponing answering letters if there is no specific reason for writing, up to the point that even I am surprised by how long it has taken me. I want to thank you for sending me your response to Akin.[2] I read it with great interest and am now more than ever convinced that Akin not only has a glum, unbearable personality and is highly deluded with regard to his own merits, but also that he is insincere and has misrepresented things completely. If you really had done him wrong, in my view it is certainly not because you were not sufficiently considerate to him, but that at most you were too considerate. From the beginning, there was nothing he could demand except that what he had published be mentioned. But I believe that all impartial readers will have had the impression that Akin was absolutely wrong with his demands.

Akin wrote to me as well at the beginning of March and among other things expressed the wish that I communicate to him my opinion about what he had written regarding Kirchhoff.[3] But I find his behaviour toward you so repugnant that I did not even reply.

This year, the meeting of the Swiss Society of Natural Sciences will be held in Geneva[4] on August 21st, 22nd and 23rd, so that participants are expected to arrive on Sunday, August 20th. I am planning to travel with my wife and children[5] to Lake Geneva; we want to take accommodations in a small guesthouse in or near Montreux, and I want to go to Geneva from there. If you could come once again to Switzerland and visit us in Zurich or meet at Lake Geneva, we would be very glad. I think the meeting in Geneva will be well attended and animated, since we will be celebrating the fiftieth anniversary of the Swiss Society of Natural Sciences, which was founded in Geneva. A chance to relax in the mountains will certainly do you good after all the work and agitation of the last year. As far as I have heard, Frankland is also coming to Switzerland, which might help to increase your desire to travel to Switzerland, which would be a great delight to us.

In the hope of seeing you again this year and with best wishes from my wife and myself,

Your | Clausius.

RI MS JT/1/TYP/7/2275–6
LT Typescript Only

---

1. *kind letter*: letter 2242.
2. *Akin*: C. K. Akin.
3. *Kirchhoff*: Gustav Kirchhoff.
4. *meeting of the Swiss Society of Natural Sciences will be held in Geneva*: the 1865 meeting of the Société Helvetique des Sciences Naturelles (Swiss Society of Natural Sciences, now Swiss Academy of Natural Sciences, founded in 1815; 'Versammlung der schweizerischen Naturforschergesellschaft' in German) was held in Geneva from 21–3 August. Tyndall did attend (see letter 2298).
5. *wife and children*: at this time, Clausius and his wife Adelheid had three children (of an eventual six): daughters Mathilde (1861–1907) and Helene (1863–1919), and son Rudolf John Clausius.

To [Susan] Durant[1] [before 16 May 1865][2] 2271

Thursday m[orni]ng

Dear Miss Durant

I received your note[3] last night; but too late to reply to it.

This is to be a full day with me in my laboratory; and I cannot say when I shall be able to get loose from my experiments.

Hence I cannot say "yes" to your kind invitation, though it would give me real pleasure to do so.

Remember me kindly to the Baron[4]—We appeared to differ when we last met, but we were one at heart.

I have been trying to find out the real agent by which Archimedes burned the Roman ships at Syracuse[5]—If burned at all, they were burned with darkness. And if they never were burned they might be burned with darkness. The brothers Davenport,[6] for example, plus their ropes & cabinets could be burnt up bodily by an agent totally invisible to them. Mr. Grote[7] may feel an interest in this—not in the "brothers", but in the case of Archimedes. To *[1 word illeg]* mystery I find it possible to detach the sun's heat from the light, and to burn up anything by the former.

Yours very truly | John Tyndall

Royal Archives/VIC/ADDW/3/136 by the permission of Her Majesty Queen Elizabeth II

---

1. *[Susan] Durant*: possibly Susan Durant (1827–73), a British sculptor. Durant trained under Baron Henri de Triqueti (1803–74) in Paris, and was acquainted with the Royal Family, receiving several commissions from Queen Victoria (*ODNB*).
2. *[before 16 May 1865]*: the date is given since Tyndall had referred to Archimedes and his burning ships at Syracuse in his Rede lecture at the University of Cambridge on 16 May 1865 (see n. 5).
3. *your note*: letter missing.
4. *Baron*: possibly Henri de Triqueti (1803–74), a French sculptor
5. *Archimedes burned the Roman ships at Syracuse*: in his Rede lecture at the University of Cambridge on 16 May 1865 (see letter 2231, n. 1), Tyndall shared: 'The sun's invisible rays transcend the visible ones in heating power, so that if the alleged performances of Archimedes during the siege of Syracuse had any foundation in fact, the dark solar rays would have been the philosopher's chief agents of combustion. On a small scale we can readily produce with the purely invisible rays of the electric light all that Archimedes is said to have performed with the sun's total radiation' (*On Radiation*, p. 21).
6. *brothers Davenport*: Ira Erastus Davenport (1839–1911) and William Henry Davenport (1841–77), American magicians.
7. *Mr. Grote*: George Grote (1794–1871), an English classical historian and politician, known for his 12-volume *A History of Greece* (1846–56) (*ODNB*).

From Gerard F. Cobb[1] 16 May 1865 2272

May 16, 65.

My dear Professor Tyndall,

I must add my personal testimony to the voice of general acclamation, and assure you of the very great and universal satisfaction your lecture[2] gave.

"Radiation" will form the stock subject of conversation in academic circles for some time to come.

I never like filling a fellow creature's mind with unnecessary apprehensions, but really you will have to keep strict watch for the future, for I am sure some secret agent of our many Insurance Societies will be quietly 'putting you down' if you go about the country teaching people such a new & easy method of '<u>invisible</u> incendiarism'.[3]

Our first impression on entering the Senate House door was that we were going to have a lecture on the Matterhorn,[4] and I shall never see that noble form whether in art or nature without calling to mind the relative scale of visible & invisible heat,—that "Matterhorn of heat"[5] as it will henceforth be called. I only hope the courage which could risk an eye in the cause of science, will ultimately be rewarded by the successful ascent of that peak, as it has no doubt been strengthened and developed by successive attempts at it.

Knowing as I do your occupation, I shall never forgive you if you think of adding to them by an acknowledgement of this scrawl—July 1 to August 25 will see me at Rydal;[6] should anything happen to bring you there be assured of a hearty welcome from

Y[ou]rs. most sincerely | Gerard F. Cobb.

My kind remembrances to Prof[esso]r Huxley—

P.S. I think you may satisfy yourself of having done one thing yesterday, that never mortal man did before, and that is smoked a cigar in the Senate House. Note it as a fact unique in the history of the world. Your future biographer will epitomize you as the one man who ascended the Matterhorn, & smoked a cigar in the Senate House.

RI MS JT/2/10/342
RI MS JT/2/13c/1291–2

---

1. *Gerard F. Cobb*: Gerard Francis Cobb (1838–1904), composer and Junior Bursar of Trinity College at the University of Cambridge (*ODNB*).
2. *your lecture*: Tyndall delivered the annual Rede Lecture at the University of Cambridge ('On Radiation') on 16 May 1865 (see letter 2231, n. 1).
3. *'invisible incendiarism'*: Cobb was referring to Tyndall's discussion in his lecture of imagining that which had never been observed nor detected: atoms, molecules, vibrations, and waves (*On Radiation*, pp. 46–7).
4. *Matterhorn*: one of the high peaks of the Alps, located on the border between Switzerland and Italy. Tyndall attempted to climb the Matterhorn in August 1860 and July 1862 (*Hours of Exercise in the Alps*, pp. 27–52, 153–65), and on 28 July 1868, his party was the first to traverse the mountain (*Hours of Exercise in the Alps*, pp. 271–94).
5. *"Matterhorn of heat"*: in describing the distribution of heat in the spectrum of the electric light, Tyndall lectured: 'when this [thermo-electric pile] is brought up to the violet end of the spectrum of the electric light, the heat is found to be insensible. As the pile gradually

moves from the violet end towards the red, heat soon manifests itself, augmenting as we approach the red. Of all the colours of the visible spectrum the red possesses the highest heating power. On pushing the pile into the dark region beyond the red it attains a maximum. Moving the pile still forward, the thermal power falls, somewhat more suddenly than it rose. It then gradually shades away, but for a distance beyond the red greater than the length of the whole visible spectrum, signs of heat may be detected. Drawing, as Sir William Herschel did, a datum line, and erecting along it perpendiculars, proportional in length to the thermal intensity at the respective points, we obtain the extraordinary curve which exhibits the distribution of heat in the spectrum of the electric light. In the region of dark rays beyond the red curve shoots up in a steep and massive peak—a kind of Matterhorn of heat, which dwarfs by its magnitude the portion of the diagram representing the luminous radiation' (*On Radiation*, pp. 20–1). The diagram described is given as the frontispiece to the published lecture. Michael Reidy has noted of this diagram that 'Tyndall's mountain metaphor serves to create a mental image of what is physically unseeable through one of the most well-known and colossal mountains in Europe. Moreover, as one of the first climbers to attempt to reach the Matterhorn's summit . . . Tyndall could use his own experiences to convey his authority on the subject' (M. S. Reidy, 'John Tyndall's Vertical Physics: From Rock Quarries to Icy Peaks', *Physics in Perspective*, 12 (2010), pp. 122–45, on p. 139).

6. *Rydal*: the village and lake of Rydal near Ambleside in the Lake District of northwest England. The Pollocks spent time here in 1865 (see letter 2307, n. 1).

From William Frederick Pollock 17 May [1865][1] 2273

59 Montague Square W.[2] | Wednesday Night | 17th May

My dear Tyndall

I enclose your Ticket for the Adelphi[3] on Saturday—Fazio[4] begins at 8—We will meet you at the Theatre—

Fred[5] has sent a glowing account of your Senate House Lecture[6]—Of course it would be a great success—Hurrah for the Pollycules[7]—

Yours ever | W. F. Pollock

Juliet[8] thanks you for her note.[9]

RI MS JT/1/P/255
RI MS JT/1/TYP/6/2140

---

1. *[1865]*: the year is given by reference to Tyndall's Rede Lecture (see n. 6).
2. *59 Montague Square W.*: the Pollocks' residence in Marylebone, London.
3. *Adelphi*: a theatre in London, founded in 1806 under its first name, Sans Pareil.
4. *Fazio*: *Fazio*, an 1815 poetical tragedy by Henry H. Milman (1791–1868), an English historian and poet (*ODNB*), was first performed on stage in Surrey in 1816 as *The Italian Wife*. It first opened in London at Covent Garden in 1818, and at the Adelphi on 8 May 1865.

5. *Fred*: Frederick Pollock.
6. *Senate House Lecture*: Tyndall delivered the annual Rede Lecture at the University of Cambridge ('On Radiation') on 16 May 1865 (see letter 2231, n. 1).
7. *Pollycules*: presumably a comical play on the word 'molecules' and the name 'Pollock'.
8. *Juliet*: Juliet Pollock.
9. *her note*: the previous letter from Tyndall to Juliet Pollock is 2260, dated 10 April [1865], so it is possible this refers to a more recent letter that is missing.

From Henry Wilkinson Cookson 19 May 1865 2274

S^t^. Peters College, | Cambridge | May 19, 1865.

My dear Sir

The Council of the Senate have to day approved a Grace[1] to be offered to the Senate to confer on you the honorary degree of L.L.D. if you give your consent.

The grace would be proposed on the 24th instant and the degree might be most conveniently conferred on the 30th.[2]

It will give Mrs Cookson[3] and myself very great pleasure to see you at our house at that time.

Degrees will be conferred on the same day on the head masters of Eton and Harrow[4] and the Prize Exercises will be recited in the Senate house.

I am, my dear Sir | Yours very faithfully | H. W. Cookson V.C.[5] | Professor Tyndall

RI MS JT/2/13c/1291

---

1. *approved a Grace*: Tyndall was not present for the grace ('Graces', *Cambridge Chronicle and Journal*, 27 May 1865, p. 4).
2. *conferred on the 30th*: following his Rede Lecture on 16 May 1865 (see letter 2231, n. 1), Tyndall received an honorary LL.D (Doctor of Law) from the University of Cambridge on 30 May 1865 ('University Intelligence', *London Evening Standard*, 31 May 1865, p. 6).
3. *Mrs Cookson*: Emily Valence Cookson (née Ainslie), daughter of Gilbert Ainslie (1793–1870), an English clergyman and Master of Pembroke College (*Alum. Cantab.*). Emily and Cookson were married on 23 January 1855 (J. Foster (ed), *Collectanea genealogica*, Part II (London, July 1881), p. 45).
4. *head masters of Eton and Harrow*: the degree of Doctor of Divinity was conferred on Edward Balston (1817–91), Headmaster of Eton from 1862–8, and Henry Montagu Butler (1833–1918), Headmaster of Harrow from 1860–85 ('Graces', *Cambridge Chronicle and Journal*, 27 May 1865, p. 4).
5. *V.C.*: Cookson was Vice Chancellor of the University of Cambridge in 1863 and 1864. When Tyndall received his degree, the Vice Chancellor was James Cartmell (see letter 2275, n. 10).

To Rudolf Clausius 20 May 1865 2275

20th. May 1865

My dear Clausius.

Since I last wrote to you[1] I have been a little unwell. My head some time ago became so weak and painful that I went away to Ireland[2] for three weeks accompanied by my friend Hirst. I am much better now, but I do not expect to be quite strong again until I get renewed in the mountain air either of Switzerland or Norway. In six weeks my duties in London will be ended and soon afterward I shall be among the mountains.[3]

De la Rive[4] has been here for a few days and he pressed me strongly to go to Geneva;[5] your letter[6] certainly greatly augments the inducement to go there, and if I can manage it I will do so. Last year however I gave up the British Association[7] because I had stayed so long in Switzerland I could not afford the time to go to Bath. This year I am pledged to go to the British Association in Birmingham,[8] and I almost fear that attending both it and the meeting at Geneva will involve the loss of too much time. However the pleasure of going to Geneva would be so great that I should not wonder if I was tempted to do so.

Last Tuesday I lectured before the University of Cambridge in the Senate House.[9] The Vice Chancellor[10] had appointed me Rede Lecturer, an endowment dating back many centuries. The duties consist in giving one lecture, and this year I gave one on radiation. I feared the room, which is very large and ill-constructed for hearing, and not at all constructed for experiments. I thought the Undergraduates, who are rather a wild lot, might be inclined to be noisy. Nothing could be better than their reception of me. The galleries were crowded with them, and they listened for an hour and a half with unbroken attention, and ended in a storm of applause. The little incident was very pleasant.

I had strong hopes that you might be induced to attend the Birmingham meeting of the British Association—Is not this possible? The next in September. I will meet you in London, and the hospitable Birmingham people have arranged to entertain a number of the scientific élite free of all expense; and you of course are among the number: Could you not come?

This Akin affair[11] has troubled me a good deal: had it not been for him I should have accomplished a capital piece of work before this time. But I will pay no more attention to him. His suspicious & venomous temper has caused him to quarrel with many besides me. I do not think he will succeed in England—it is not the place for such a man. William Thomson made a long communication to the Royal Society a few nights ago[12] in which he harped a little upon the glaciers: These discussions have become wearisome & contemptible.

How is little Johnny[13] getting on?

He will be able to fight the battles of his godfather[14] at a future day, and to show the world that whatever that godfather's failings may have been they were not such as the man Akin has ascribed to him. With kindest regards to Mrs Clausius[15]

Dein | J. Tyndall | Hirst is engaged in the translation of your notes & memoirs.[16]

RI MS JT/1/T/192

---

1. *last wrote to you*: Tyndall's previous extant letter to Clausius is 2242.
2. *away to Ireland*: Tyndall spent three weeks in Ireland in April 1865 with Thomas Hirst (see letters 2257, 2258, 2259, and 2260).
3. *I shall be among the mountains*: Tyndall and Thomas Hirst left London for the Alps on 10 July 1865, and returned to London by 26 August.
4. *De la Rive*: Auguste de la Rive.
5. *to go to Geneva*: see letter 2270, n. 4.
6. *your letter*: letter 2270.
7. *however I gave up the British Association:* the thirty-fourth annual meeting of the BAAS was held in Bath in September 1864 (see letter 2239, n. 9). In a 16 September 1864 letter to Juliet Pollock, Tyndall wrote: 'I have made myself an exile from Bath, notwithstanding all its seductions' (see letter 2139, *Tyndall Correspondence*, vol. 8).
8. *British Association in Birmingham*: the thirty-fifth annual meeting of the BAAS was held in Birmingham from 6–13 September 1865 (*Brit. Assoc. Rep. 1865*). Tyndall served as one of the vice-presidents for Section A—Mathematics and Physics.
9. *I lectured before the University of Cambridge in the Senate House*: Tyndall delivered the annual Rede Lecture at the University of Cambridge ('On Radiation') on 16 May 1865 (see letter 2231, n. 1).
10. *Vice Chancellor*: James Cartmell (1810–81), Master of Christ's College, Cambridge from 1849–81, and Chaplain to Queen Victoria (1851–81). He served as Vice Chancellor of the University in 1849, 1865, and 1866 (*Alum. Cantab.*).
11. *Akin affair*: here Tyndall was referring to his dispute with C. K. Akin, the topic of many of the preceding letters in this volume (for example, 2222, 2227, 2239, 2242, and 2245). See also the introduction to this volume.
12. *long communication to the Royal Society a few nights ago*: W. Thomson, 'On the Elasticity and Viscosity of Metals', *Roy. Soc. Proc.*, 14 (1865), pp. 289–97.
13. *Johnny*: Rudolf John Clausius.
14. *his godfather*: in 1864, Tyndall agreed to be godfather to Clausius's first son (see letters 2147, 2153, 2191, 2192, and 2205, *Tyndall Correspondence*, vol. 8).
15. *Mrs Clausius*: Adelheid Clausius.
16. *translation of your notes & memoirs*: Hirst and Tyndall both translated papers by Clausius for publication in the *Phil. Mag.*

From Henry Wilkinson Cookson 22 May 1865 2276

S[t]. Peters College, | Cambridge. | May 22, 1865.

My dear Sir

The ceremonial on the 30th. instant[1] will begin punctually at 12 o'clock. The Great Eastern train which leaves London at 8 A.M. will bring you here before 10, so that you will have ample time to prepare for the Congregation.

As I shall consider you my guest I will order a tailor to provide a scarlet gown for the occasion.

I find that your lecture[2] has given great satisfaction.

I am | yours very faithfully | H. W. Cookson V.C.[3]

P.S. I may mention now that you have only to inscribe your name in the Registrar's book, and to appear in the Senate house in a scarlet gown at 12 o'clock. The writing of your name is for a mere record.

Professor Tyndall

RI MS JT/2/13c/1293

1. *ceremonial on the 30th. instant*: see letter 2274, n. 2.
2. *your lecture*: Tyndall delivered the annual Rede Lecture at the University of Cambridge ('On Radiation') on 16 May 1865 (see letter 2231, n. 1)
3. *V.C.*: see letter 2274, n. 5.

To William Spottiswoode [between 19 and 30 May 1865][1] 2277

Thinker and Climber!

Take your choice:—I am informed that I am to be placed behind a mob of mathematicians.[2] It shows my humility to endure this.

I shall come to see the boys soon: but I fear not on Sunday. I am engaged upon most slavish work

J. T.

RI MS JT/1/T/1344

1. *[between 19 and 30 May 1865]*: the month and year are given on the presumption that Tyndall was referring to his being awarded an honourary LL.D (Doctor of Law) from the University of Cambridge on 30 May 1865, of which he was informed in a letter of 19 May 1865 from Henry Wilkinson Cookson (see letter 2274).
2. *behind a mob of mathematicians*: Tyndall was commenting on the number of mathematicians who were to be honoured at the ceremony.

To Henry Cole[1] 31 May 1865 2278

31st May <u>1865</u>

Dear Sir

I shall be happy to take part in the duty of the Committee[2] which you have named to me, so far as its labours are compatible with my duty at the School of Mines[3]—I lecture there from 2 to 3 on all days except Saturday.

It is also probable that I shall be leaving England towards the end of June or the beginning of July.[4]

Yours very truly | John Tyndall | Henry Cole Esq[ui][r][e]. C.B.[5] &c.

RI MS RD/10/B/04/C Accessions

---

1. *Henry Cole*: Henry Cole (1808–82), an English civil servant who served as Executive Commissioner at the Paris exhibitions in 1855 and 1867. He was also one of the principal managers of the International Exhibition of 1862 in London (*ODNB*).
2. *duty of the Committee*: Cole wrote in *Reports on the Paris Universal Exhibition, 1867*, 7 vols (London: Eyre and Spottiswoode, 1868), vol. 1, p. viii: 'Her Majesty the Queen was pleased to issue a Commission, dated 27th April 1865, addressed to His Royal Highness the Prince of Wales, to the Lord President of the Council, and a numerous body of noblemen and gentlemen representing various interests in the United Kingdom. After reciting the intention of the Emperor of the French to hold the Exhibition, the Commission expressed Her Majesty's wish that the Exhibition should afford full and suitable representations of the Industry and of the Fine Arts in the United Kingdom, its Colonies and Dependencies throughout the world. The Commission recited that estimates were to be submitted to the Commons' House of Parliament for the purpose of affording the necessary assistance through the Department of Science and Art, to be administered as the Lord President of the Council and the Vice President of the Committee of Council on Education for the time being might direct; and the Commission appointed the Commissioners to advise upon the best mode by which the products of Industry and Fine Arts might be procured and sent to the Exhibition, and gave power to appoint additional commissioners and to name and appoint Jurors. The functions of the Commissioners were deliberative, and the responsibility for all executive measures rested solely upon the Lord President of the Council. The Commissioners held seven meetings, and their proceedings will be set forth in a separate report to Her Majesty. The names of the associate Commissioners, the Jurors, and associate Jurors will be found in the Appendix (pp. 10 and 51)'. The exhibition ran from 1 April to 31 October 1867. Tyndall was in Paris with Hirst from 19–26 April (Brock and MacLeod, *Hirst Journals*, pp. 1801–2), and he served as a juror for Apparatus and Processes for Heating and Lighting (*Reports on the Paris Universal Exhibition,* 1867, vol. 1, p. 30).
3. *my duty at the School of Mines*: see letter 2259, n. 17.

4. *leaving England towards the end of June or the beginning of July*: see letter 2275, n. 3.
5. *C.B.*: Companion class of The Most Honorable Order of the Bath.

To Thomas Archer Hirst 31 May 1865 2279

14 Waverley Place[1] | 31st. May 1865.

My dear Tom

All is happily over. I started yesterday morning—reached Cambridge before 10; walked through some of the Museums and Chapels with the wife and little daughter[2] of the Vice Chancellor.[3] "Robed!" at a quarter to 12—walked down the street in a blaze of scarlet to the Senate House.[4] There was a colonial bishop[5] to be doctored and his turn came first. Then it was mine.[6] The Vice Chancellor sat on a Dais at the head of the room. The public orator[7]—a fine fellow—came, and standing at my left side put out his right hand and took mine. Then we walked towards the Vice Chancellor—halted about 6 paces from him, and then a Latin speech, referring to the doings of your friend was pronounced. It was admirable I am told: but I could only catch its meaning here and there. The undergraduates received both it and me with extreme cordiality. The speech ended I walked up to the Vice Chancellor, who taking me by the hand pronounced the words conferring the degree. The galleries cheered vociferously and all was over.

<*3–4 words excised*> with the Vice Chancellor and met Carlyle there. Afterwards I went home with him and his wife[8] and had two capital games of Croquet with them. Thus ended my visit. I quitted Cambridge by the 5.15 train.

Give my kind regards to Mr and Mrs Wright[9]

Yours affect[tionately]. | John

RI MS JT/1/HTYP/534
LT Typescript Only

---

1. *14 Waverley Place*: Tyndall lived with Hirst at 14 Waverly Place, St. John's Wood, London until January 1868.
2. *wife and little daughter*: not identified.
3. *Vice Chancellor*: see letter 2275, n. 10.
4. *Senate House*: Tyndall delivered the annual Rede Lecture at the University of Cambridge ('On Radiation') on 16 May 1865 (see letter 2231, n. 1).
5. *colonial bishop*: Robert Machray (1831–1904), an Anglican bishop and missionary who was the first Primate of the Church of England in Canada, in Rupert's Land (*ODNB*; 'University Intelligence', *London Evening Standard*, 31 May 1865, p. 6).

6. *Then it was mine*: Tyndall received an honorary LL.D (Doctor of Law) from the University of Cambridge (see letter 2274, n. 2).
7. *public orator*: William George Clark (1821–78), an English literary and classical scholar who was Public Orator at the University of Cambridge from 1857–69 (*Alum. Cantab.*).
8. *his wife*: Jane Carlyle.
9. *to Mr and Mrs Wright*: probably their mutual friend Richard Pears Wright (1829–92), an English mathematician and educator who taught engineering and surveying at Queenwood College from 1851–63 and later was teacher of mathematics at University College School, London (1870–92); and his wife Fanny Wright (née Bellamy), married in 1852.

From Robert Bunsen 1 June 1865 2280

Mein theuerster Freund.

Nach einer längeren Abwesenheit auf den Balearen und in Spanien hierher zurückgekehrt, habe ich Ihr schönes Werk über die Wärme zu meiner größten Freude hier vorgefunden. Mit tausend Decanats- und Senats-Geschäften überhäuft und in meinen Vorlesungen und allen Arbeiten durch meine lange Abwesenheit recht arg in Rückstand gerathen, wollte ich mir doch, bevor ich Ihnen schrieb, die Freude nicht versagen, das Werk erst durchzu*[studieren]*. Nur so ist es gekommen, daß ich schon über 14 Tage hier bin und Ihnen erst jetzt meinen freundlichsten und herzlichsten Dank ausspreche, für die unverdiente Anerkennung u große Freundschaft, die Sie mir durch Ihre freundliche Dedication haben zutheil werden lassen. Ich habe das treffliche Werk <mit> dem höchsten Genuß gelesen und daran, was Inhalt und Form anbelangt, mehr noch als in der ersten Auflage Belehrung geschöpft. Kirchhoff und Helmholtz, denen ich die beiden Exemplare geschickt, laßen nicht minder herzlich danken.

Ich bin vor der Reise mit einer recht dornenvollen Arbeit über die Verbrennungstemperaturen der Gase beschäftigt gewesen, bei der recht hartnäckige Schwierigkeiten zu überwinden gewesen sind. Wenn man den Druck bestimmt, den ein durch seine Maße gleichzeitig in einem verschloßenen Gefäß verbrennendes Gas ausübt, so gewinnt man eine 2te Gleichung für die Verbrennungstemperatur, mit Hülfe deren sich berechnen läßt, der wie vieltste Theil des Gasgemischs seine Verbrennlichkeit bei den hohen Temperaturen der Flamme verloren hat. Es ergiebt sich dabei, daß oft mehr als 3/4 eines Knallgasvolumens seine Verbindbarkeit bei der Flammentemperatur gänzlich verloren hat.

In herzlicher Freundschaft | Ihr | R W Bunsen
Heidelberg den 1 Juni 1865.

My dearest friend,

Having returned after a long absence on the Balearic Islands[1] and in Spain, I have found here to my greatest joy your beautiful work on heat.[2] Despite a thousand affairs for the deanship and the senate and having fallen behind terribly in my lectures and all works due to my long absence, I did not want to deny myself the joy of studying the work before writing to you. It is only because of this that even though I been here for more than 14 days already, I am only now expressing my most friendly and heartfelt thanks for the undeserved recognition and great friendship that you offered me in your friendly dedication.[3] I have read the splendid work with the greatest pleasure and in regard to content and form have drawn even more instruction from it than from the first edition. Kirchhoff[4] and Helmholtz, to whom I sent both copies, are no less grateful.

Before the journey, I was occupied with quite a thorny project on the combustion temperatures of gases, in which quite persistent difficulties had to be overcome.[5] If one determines the pressure exerted by a gas that burns <u>simultaneously</u> through its [mass] in a closed container, one obtains a 2nd equation for the combustion temperature, with which one can calculate what fraction of the gas mixture has lost its combustibility due to the high temperatures of the flame. Hereby, one finds that often more than 3/4 of the volume of oxyhydrogen has entirely lost its reactivity at that flame temperature.

In warm friendship | Your | R W Bunsen

Heidelberg 1st June 1865.

RI MS JT/1/B/150

1. *Balearic Islands*: an archipelago off eastern Spain, in the Mediterranean.
2. *work on heat*: *Heat* (1865).
3. *dedication*: the dedication in *Heat* (1865) is: 'To his friend and teacher Robert Bunsen this book is dedicated by John Tyndall'.
4. *Kirchhoff*: Gustav Kirchhoff.
5. *I was occupied . . . to be overcome*: the history of combustion is long and complicated. French chemist Antoine-Laurent Lavoisier (1743–94) discovered in 1772 that the products of burned sulfur or phosphorus outweighed the initial substances, which led him to postulate that the increased weight was due to their having combined with air. Later, Lavoisier concluded that the 'fixed' air that had combined with the sulfur was identical to a gas obtained by English chemist Joseph Priestley (1733–1804). Lavoisier called the gas 'oxygen'. Soon after, studies of gases by English chemist John Dalton (1766–1844), and the first table of atomic weights that Dalton compiled, as well as many new gases discovered by other scientists, were important in supporting not only Lavoisier's theory of combustion but his whole new system of chemistry based on accurate measurement. The discoveries of nitrogen and hydrogen in the latter half of the eighteenth century, added to the earlier

discoveries of carbon dioxide and carbon monoxide, and the discovery that the composition of air is remarkably constant though it is a mixture, all supported Lavoisier's theory. From 1815–9 the English chemist Humphry Davy (see letter 2245, n. 19) experimented on combustion, including measurements of flame temperatures, investigations of the effect on flames of rarefied gases, and dilution with various gases; he also discovered catalytic combustion—the oxidation of combustibles on a catalytic surface accompanied by the release of heat but without flame.

To Hermann Helmholtz 1 June 1865 2281

1st. June 1865

My dear friend

I forward you an invitation from the executive Committee of the British Association which meets at Birmingham next September.[1] I trust you will be able to accept the invitation and to give your numerous English friends the opportunity of seeing you once more. The meeting promises to be a very good one; and Birmingham possesses many objects of interest to a man like yourself.

Believe me | Yours ever sincerely | John Tyndall

BBAW NL Helmholtz, Nr. 477 BI_5r
RI MS JT/1/T/482

1. *Committee of the British Association which meets at Birmingham next September*: the thirty-fifth annual meeting of the BAAS was held in Birmingham from 6–13 September 1865 (*Brit. Assoc. Rep. 1864*). Helmholtz accepted the invitation (see letter 2290).

From Rudolf Clausius 6 June 1865 2282

ZÜRICH, 6/6/65.

Lieber Tyndall,

Dein lieber Brief vom 20sten Mai hat einerseits Bedauern und andererseits Freude bei uns erregt. Bedauern, weil er die Nachricht enthält dass Du unwohl gewesen bist, und Dich auch jetzt noch nicht ganz wohl befindest, und Freude, weil Du uns darin Hoffnung machst, dass Du diesen Sommer wieder nach der Schweiz kommen wirst. Ich glaube auch dass eine Erholung von den angestrengten und aufregenden Arbeiten Dir nothwendig ist, und ein Aufenthalt in den Bergen Dir wohl thun wird. Wenn Du Dich entschliessen wolltest, bei uns auf einige Wochen Deinen Wohnsitz aufzuschlagen, wo

Du ganz ungenirt leben, und nach Belieben kleinere oder grössere Excursionen machen könntest, so würde uns Dein Besuch ungemein freuen. Solltest Du aber vorziehen, gleich in die hohen Berge zu gehen, so sei so gut mir gelegentlich mitzutheilen, wohin ich einen Brief an Dich adressiren muss, um Dir zu schreiben, wann wir nach dem Genfer See reisen, und welchen Ort wir zu unserem Wohnsitze gewählt haben. Bis jetzt kennen wir die verschiedenen Pensionen noch zu wenig um uns zu entscheiden. Vor Mitte August kann ich leider meiner Geschäfte wegen nicht von Zürich losmachen.

Zu der Ernennung zum Rede Lecturer in Cambridge, die *[1 word illeg]* gewiss sehr ehrenvoll ist, gratulire ich Dir herzlich. Dass die Zuhörerschaft Dich nicht gestört, sondern mit grösster Aufmerksamkeit bis zu Ende angehört hat, und dann in Beifall ausgebrochen ist, wundert mich nicht. Ich würde es im Gegentheile sehr sonderbar gefunden haben, wenn es anders gewesen wäre.

Zur Versammlung der British Association zu kommen ist mir für dieses Jahr nicht gut möglich, obwohl ich an den grossen Annehmlichkeiten die ein solcher Besuch für mich haben würde keinen Augenblick zweifele. Für dieses Jahr will ich meine Reisen auf die Schweiz und Deutschland beschränken, hoffe aber in einem der folgenden Jahre einmal wieder nach England reisen, und dann einer Versammlung der British Association beiwohnen zu können.

Dass Hirst so freundlich ist, die noch unübersetzten Partien meines Buches zu übersetzen, freut mich sehr, denn darin liegt eine sichere Garantie für die Vortrefflichkeit der Übersetzung. Grüsse ihn doch freundlich von mir, und sage ihm in Voraus meinen herzlichen Dank. Vielleicht dürfen wir hoffen, ihn diesmal auch in der Schweiz zu sehen, was leider im vorigen Jahre nicht der Fall war.

Du hast vielleicht in den Comptes rendus gelesen, dass man mich zum Correspondenten der Pariser Academie gewählt hat. Ich war ganz überrascht davon indem ich, als ich die Nachricht von der geschehenen Wahl empfing, noch nicht einmal wusste, dass ich vorgeschlagen war. Natürlich war die Überraschung eine sehr angenehme.

Der kleine John ist einige Zeit krank gewesen, indem er ebenso wie die beiden anderen Kinder, eine Kinderkrankheit, die Masern, durchgemacht hat. Jetzt sind sie aber alle wieder hergestellt, und John ist heute zum ersten Male wieder in's Freie gekommen.

Mit herzlichen Grüssen von meiner Frau und mir und der nochmaligen Versicherung dass ein längerer Besuch von Dir uns ausserordentlich freuen würde,

Dein | Clausius.

ZURICH, 6/6/65.

Dear Tyndall,

Your kind letter from May 20th[1] made us sad and joyful at the same time. Sad, because you write that you were ill and that you are still not quite well, and joyful because you made us hope that you will come to Switzerland again this summer. I believe you need to recover from the hard and stressful work and a stay in the mountains will do you good. It would be a tremendous joy to us if you decided to set up residence with us for a few weeks, where you can be free of all care and make long or short excursions to the mountains at your pleasure. Should you prefer to go to the high mountains, please be so good as to inform me on occasion where I can send you a letter to let you know when we go to Lake Geneva and which place we have chosen as our abode. So far, we do not know the guesthouses sufficiently to be able to decide. Unfortunately, I cannot leave Zurich before mid-August because of my work.

My sincerest congratulations on being selected as Rede Lecturer in Cambridge,[2] which *[1 word illeg]* certainly is a great honour. I am not surprised that the audience did not disturb you, but listened to you with great attention and then broke into applause. On the contrary, I would have found it very strange had it been any different.

I will not be able to attend the meeting of the British Association this year,[3] although I have no doubt that it would be very pleasant to be there. This year, I want to limit my travels to Switzerland and Germany, but I hope to eventually come to England again in one of the following years and to be able to attend a meeting of the British Association then.

I am very glad to learn that Hirst is so kind as to translate those parts of my book[4] that have not been translated yet, as this is a sure guarantee for the excellence of the translation. Please send him my best wishes and thank him for me in advance. Perhaps we may hope to see him this year in Switzerland, which unfortunately was not the case in the previous year.

You have perhaps read in the Comptes rendus[5] that I was elected a Correspondent of the Parisian Academy.[6] I was quite surprised when I received the news of the election, as I did not even know that I had been nominated. Naturally, it was a very pleasant surprise.

My little John[7] was ill for a quite some time—like two other children,[8] he had measles. They have all recovered now and today, John went outside again for the first time.

With best wishes from my wife[9] and myself and the repeated assurance that a longer visit from you would make us extraordinarily glad.

Your | Clausius.

RI MS JT/1/TYP/7/2277–8
LT Typescript Only

1. *letter from May 20th*: letter 2275.
2. *Rede Lecturer in Cambridge*: see letter 2231, n. 1.
3. *meeting of the British Association this year*: see letter 2275, n. 8.
4. *translate those parts of my book*: the English translation of Clausius's *Abhandlungen über die mechanische Wärmetheorie* (Braunschweig: Vieweg und Sohn, 1864) is: R. Clausius, *The Mechanical Theory of Heat, with its Applications to the Steam-Engine and to the Physical Properties of Bodies*, trans. T. Hirst (London: John van Voorst, 1867). This included an introduction by Tyndall.
5. *read in the Comptes rendus*: for Clausius's election as 'd'un Correspondant dans la Section de Mécanique', see 'Nominations', *Comptes rendus hebdomadaires des séances de l'Académie des sciences*, 60 (1865), p. 1126.
6. *Parisian Academy*: l'Académie des sciences, or French Academy of Sciences, founded in 1666.
7. *little John*: Rudolf John Clausius.
8. *two other children*: see letter 2270, n. 5.
9. *wife*: Adelheid Clausius.

From William Jardine [1] 7 June 1865 2283

8 Kings Bench Walk | Temple[2] | 7th. June 1865.

Dear Sir

I am much obliged by your communication about the Reader.[3] I sent a notice of the last meeting[4] and hope to find time in a day or two to draw up some short account of the objects & foundation[5] of the Mathem[atica][1] Society[6]

I sent the Insurance record which I hope reached you safely.

A member of the Society Mr. M. Jenkins[7] informs me that he has investigated geometrically the properties of the Envelope of the Tripodal line (if I may use this term) as enunciated some time ago by Steiner,[8] and some additional properties. He has not seen the solutions by Cremona[9] nor have I and I thought it worth while to ask you whether you think a paper embodying Mr. Jenkins solutions would be suitable for our next meeting on the 19th inst.[ant] [10] or whether the subject is already exhausted

There are several names to be proposed at the next meeting for membership. It seems to me that it is a pity to keep them waiting till after the Vacation

Are there any other papers that would insert an account of the Society?

I remain | Yours truly | William Jardine

RI MS JT/1/J/24

---

1. *William Jardine*: William Jardine (1841–73), a mathematician who was educated at the University of Cambridge (22nd Wrangler in 1864), and served as the London Mathematical Society's secretary for most of its first year (1865). In November 1865, Jardine accepted an Indian Civil Service appointment in Lahore (*Alum. Cantab.*; A. C. Rice, R. J. Wilson, and J. H. Gardner, 'From Student Club to National Society: The Founding of the London Mathematical Society in 1865', *Historia Mathematica*, 22 (1995), pp. 402–21).
2. *8 Kings Bench Walk | Temple*: in London.
3. *your communication about the Reader*: letter missing. In November 1864, Tyndall and other members of the X Club (see letter 2527, n. 4) had purchased *The Reader*, a journal of science and literature originally established by the English social reformer and author Thomas Hughes (1822–96; *ODNB*) and English astronomer and journal editor Norman Lockyer (1836–1920; *ODNB*). They hoped to turn the journal to good use in promoting and publishing the works, society meetings, and correspondence of scientific men. This project was short-lived, however (see *Tyndall Correspondence*, vol. 8; and M. Baldwin, *Making Nature: The History of a Scientific Journal* (Chicago: University of Chicago Press, 2015), pp. 21–47).
4. *notice of the last meeting*: the previous meeting of the London Mathematical Society was on 15 May 1865. Jardine presented his paper 'An Account of a Proof of Pohlke's Fundamental Proposition of Axonometry' (which does not appear to have been published in *Proceedings of the London Mathematical Society*).
5. *short account of the objects & foundation*: see 'The London Mathematical Society', *The Reader*, 5 (17 June 1865), p. 687.
6. *Mathem[atica]l Society*: the London Mathematical Society, founded in January 1865. See A. C. Rice, R. J. Wilson, and J. H. Gardner, 'From Student Club to National Society: The Founding of the London Mathematical Society in 1865', pp. 402–21.
7. *M^r^. M. Jenkins*: Morgan Jenkins (1841–1913), a mathematician who was educated at the University of Cambridge (34th Wrangler in 1864) and served as the London Mathematical Society's secretary from November 1865 (taking over for Jardine) to 1894 (*Alum. Cantab.*; A. C. Rice, R. J. Wilson, and J. H. Gardner, 'From Student Club to National Society: The Founding of the London Mathematical Society in 1865', pp. 402–21).
8. *Steiner*: Jakob Steiner (1796–1863), a Swiss mathematician (*HLS*).
9. *Cremona*: Antonio Luigi Gaudenzio Giuseppe Cremona (1830–1903), an Italian mathematician and professor of geometry first at the University of Bologna and after 1867 at the Istituto Tecnico Superiore of Milan (*CDSB*).
10. *paper embodying Mr. Jenkins solutions . . . next meeting on the 19th inst.[ant]*: the London Mathematical Society next met on 19 June 1865, with Jenkins presenting his paper 'The Regular Hypocycloidal Tricusp'.

From Cecilia Anne Barlow[1] 9 June 1865 2284

5 Berkeley St.[2] | June 9th. /65.

Dear Dr. Tyndall,

Let me first congratulate you on your honors;[3] and next I must tell you how very sorry we were when we heard that you had called last Sunday and had not been admitted—but you see we keep very early dinner hours, as best suited to an invalid, and our servants say not at home to visitors after a certain hour, without consulting us.

I was so much obliged to you for your letter from Ireland,[4] but could not answer it then, not knowing where a letter would find you.

I hope that you will soon be starting on your annual health-restoring trip.[5]

We are off to Brighton to-day[6]—my husband[7] feels the heat and noise of London very much, and is very languid. I trust that the sea-breezes may invigorate him.

Good bye—take care of yourself in your hazardous expeditions,[8] for the sake of your friends if not for your own.

My husband's very kind remembrances

Ever sincerely yours | C. A. Barlow.

RI MS JT/1/TYP/1/145
LT Typescript Only

---

1. *Cecilia Anne Barlow*: Cecilia Anne Barlow (née Law, d. 1868), the daughter of an MP who made his fortune in India. She married John Barlow in 1824.
2. *5 Berkeley St.*: the Barlow's home in Piccadilly, London.
3. *your honors*: Tyndall received an honorary LL.D (Doctor of Law) from the University of Cambridge (see letter 2274, n. 2).
4. *your letter from Ireland*: letter missing. Tyndall spent three weeks in Ireland in April 1865 with Thomas Hirst (see letters 2257, 2258, 2259, and 2260).
5. *annual health-restoring trip*: see letter 2275, n. 3.
6. *We are off to Brighton to-day*: John Barlow wrote to Tyndall from their seaside rental in Brighton, Belgrave House (see letter 2286).
7. *my husband*: John Barlow (1798–1869), an Anglican clergyman and secretary of the RI from 1842–60. Together with Michael Faraday, he undertook the day-to-day running of the institution. Barlow lived near the RI and entertained many of its wealthy and aristocratic supporters as well as fellow scientists, including Tyndall, who was a frequent guest.
8. *hazardous expeditions*: mountaineering in the Swiss Alps.

To George Gabriel Stokes 13 June 1865 2285

13th June 1865

My dear Sir

I have read M. de. Schlagintweit's[1] paper[2] more than once. It seems to have been written somewhat hastily. Its style needs improvements and its really valuable result is, I think, worthy of a more thorough discussion of the observations on which it rests than that given in the paper. I would recommend its reconsideration by the author. If he succeeds in establishing the result communicated at page 4a of his manuscript he will in my opinion furnish a paper well worthy of a place in the Philosophical Transactions.

I am dear Sir | Yours very truly | John Tyndall

Prof. Stokes | Sec[retary]. R.S. | &c &c &c

RS RR/5/240 Tyndall 13 June 1865

1. *M. de. Schlagintweit's*: Hermann Schlagintweit (1826–82), a member of a German family of five brothers, all explorers (*ADB*).
2. *paper*: there is no paper by Schlagintweit published in the *Phil. Trans.* in 1865, 1866, or 1867. An abstract of a paper received by the RS on 27 December 1864 and communicated on 9 March 1865 was published as: H. Schlagintweit, 'Numerical elements of Indian meteorology.—Series II. Insolation, and its connexion with atmospheric moisture', *Roy. Soc. Proc.*, 14 (1865), pp. 111–12.

From John Barlow[1] and Cecilia Anne Barlow[2] 17 June 1865 2286

Belgrave House, | Marine Parade | Brighton[3] | June 17th. 1865

My dear Tyndall,

Your kindness has always supplied me with a new Book or a new Memoir for interesting study during the periods of my visits to this place. And now I have to thank you for the Rede lecture.[4]

It is indeed most acceptable: for, as I infer from the Contents, it will present a lucid résumé of your researches during the last five or six years.

I trust however that you will soon cease for a time from giving any Lectures, however interesting, and that you will enjoy your holidays.

Ever your sincere friend | John Barlow.

The same hand under another name, will now tell you that we got down here very well and are comfortably located in very convenient lodgings. The quiet and freshness of the sea breezes are very enjoyable, and will I hope, in time, do good to my poor invalid.[5] We have never felt any heat since we came here. Good bye. I trust that your summer excursion[6] will be very pleasant, and that you will be very prudent.

Ever sincerely y[ou]rs. | C. A. Barlow.

RI MS JT/1/B/19
RI MS JT/1/TYP/1/146

1. *John Barlow*: John Barlow (see letter 2284, n. 7).
2. *Cecilia Anne Barlow*: Cecilia Anne Barlow (see letter 2284, n. 1).
3. *Belgrave House, | Marine Parade | Brighton*: a seaside home the Barlows were renting on the southern England coast.
4. *Rede lecture*: see letter 2231, n. 1.
5. *my poor invalid*: as a visit to the coast might suggest, John Barlow was probably ill.
6. *summer excursion*: see letter 2275, n. 3.

To Thomas Archer Hirst [19][1] June 1865 2287

Monday June 1865

My dear Tom,

The very address on the top of your note[2] is refreshing. You speak of glorious weather—to me here in London it is intolerable—& deprives me of every vestige of energy. I long to escape to a cooler region—long to get far away from London & forget it wholly for a time. I will see what answer can be sent to Lubbock.

You are right he is now Sir John:[3] they will probably quit Lamas[4] & and go to High Elms.[5] Change! Change! & Change!—The world is so ephemeral.

Yours ever | John

Mr Key[6] was here to obtain your address: I dined with Williamson[7] yesterday as he wished to talk with me about a project of getting both you & me to join their staff[8]—For me I see no way to it—but it would be fine if you could find a corner there

RI MS JT/1/T/904
RI MS JT/1/HTYP/534

1. *[19]*: the day is given based on Hirst's journal entry about trying to visit the Lubbocks in Lamas (see n. 2). He went there on 18 June 1865, a Sunday. Hirst's letter that Tyndall referred to was written from Lamas, and this reply written on a Monday implies 19 June.
2. *your note*: letter missing, but presumably Hirst had gone to Lamas to visit the Lubbocks. In his journal entry for 25 June 1865, a Sunday, he wrote: 'On Sunday last I went alone to Lamas but found the children only Mr and Mrs Lubbock had gone to High Elms on their last visit to Sir John who died the day after' (Brock and MacLeod, *Hirst Journals*, p. 1736).

3. *he is now Sir John*: Lubbock succeeded his father as 4th Baronet in 1865.
4. *Lamas*: Lubbock's home in Chislehurst, Kent, named after the Lubbock ancestral home of Lamas, Norfolk, England.
5. *High Elms*: Lubbock's new home near Down House, the home of Charles Darwin in the English village of Downe, Kent.
6. *Mr Key*: Thomas Hewitt Key (1799–1875), an English classical scholar and joint headmaster of University College School in London from 1832–75 (*ODNB*). Hirst taught mathematics at University College School from 1860 until July 1864, when he resigned to devote more time to research, although he took up another position the following year (see n. 8).
7. *Williamson*: probably Alexander William Williamson (1824–1904), an English chemist and Professor of Practical Chemistry at University College, London from 1849–87 (*ODNB*).
8. *project of getting both you & me to join their staff*: in August 1865, Hirst was appointed Professor of Mathematical Physics at University College, London, and remained with the college until 1873, when he became Director of Studies at the Royal Naval College, Greenwich.

From Rudolf Clausius 6 June 1865 2288

ZÜRICH, 19/6/65

Lieber Tyndall,

Unsere letzten beiden Briefe haben sich gekreuzt. Ich hoffe, dass Du den meinigen richtig empfangen hast. Für Deine freundliche Gratulation danke ich Dir herzlich, und freue mich, dass ich sie gleich erwiedern kann in Bezug auf die Ehre welche Dir vom Universitäts-Senat zu Cambridge erwiesen ist. Ich weiss zwar nicht ganz was unter L.L.D zu verstehen ist, aber ich denke mir, D. bedeutet Doctor, und ich weiss dass der Doctortitel in England eine viel höhere Bedeutung hat als in Deutschland. Ich wünsche Dir vor allem Gesundheit um an den Dir erwiesenen Ehren, an denen es nie fehlen wird, auch rechte Freude zu haben.

Ich schicke Dir gleichzeitig mit diesem Briefe eine Abhandlung von mir, welche eben erschienen ist. Ausserdem adressire ich noch 3 andere Exemplare an Dich, eins für Hirst, den ich freundlich zu grüssen bitte, eins für die Royal Society, was Du wohl so gut bist zu überreichen, und ein drittes für W. Siemens, dessen Adresse ich nicht weiss. Du kannst es ihm vielleicht mit der Stadtpost zuschicken, oder sonst gelegentlich zukommen lassen.

Vor Kurzem habe ich aus Lüttich in Belgien, von einem mir bisher unbekannten Herrn Folie, Professor der Mathematik und Mechanik an der Gewerbeschule und Bergschule zu Lüttich, einen Brief erhalten, worin er bei mir anfragt, ob ich meine Zustimmung dazu gebe, dass er mein Buch über die mech. Wärmetheorie in's Französische übersetzt. Ich habe natürlich meine Zustimmung sehr gern gegeben. Eine gleiche Anfrage habe ich schon früher aus Italien erhalten. Ein Ingenieur Luchesini zu Turin, den ich ebenfalls nicht persönlich kenne, will mein Buch in's Italienische übersetzen, und zwar soll es, wie wenigstens der ursprüngliche Plan war, in Absätzen im Nuovo Cimento gedruckt werden, und nach der Vollendung kann es dann auch im Ganzen als besonderes Werk verkauft werden. Hiernach scheint es, dass ich in nicht zu langer Zeit die Freude haben werde, mein Buch auch in England, Frankreich und Italien verbreitet zu sehen.

Mit herzlichen Grüssen von meiner Frau und mir, und in der Hoffnung Dich in diesem Sommer wieder zu sehen

Dein | Clausius.

P.S. Wenn Du einmal den jungen Wynne siehst, so grüsse ihn doch von uns. Er und sein Vater haben aus Dover sehr freundlich an mich geschrieben, was noch mehr dazu beigetragen hat, ihn uns lieb zu machen.

ZURICH, 19/6/65

Dear Tyndall,

Our last two letters[1] must have crossed. I hope you have received mine without a problem. Thank you very much for your kind congratulations; I am happy that I can reciprocate them in regards to the honor that was bestowed upon you by the Cambridge University Senate.[2] I do not really know how to interpret L.L.D. but I think that D. stands for Doctor and I know that this title has a much higher significance in England than in Germany. More than anything, I wish you the health to properly enjoy the honours—which will never be lacking—that are bestowed upon you.

Together with this letter, I am also sending you one of my papers,[3] which has just been published. Moreover, I am sending you three additional copies: one for Hirst—whom I ask you to greet kindly from me—; one for the Royal Society which you will hopefully be so good as to hand over to them; and the third one for W. Siemens,[4] whose address I do not know. Perhaps you can send it to him with the local post or forward it to him in another way on occasion.

Recently I have received a letter from a Mr. Folie,[5] who is unknown to me and who is a Professor of mathematics and mechanics at a vocational and mining school in Liège, Belgium. He asked me whether I would authorise that he translates my book on mechanical theory of heat[6] into French.[7] Obviously, I was very glad to give him my consent. I had already received a similar request from Italy. An engineer named Luchesini[8] from Turin, whom I do not know either, wants to translate my book into Italian. It is supposed, at least that was the original plan, to be published in parts in the Nuovo Cimento,[9] and after this is done they will also sell the whole as a special edition. So it appears that before long I will be able to enjoy seeing my book distributed in England, France and Italy.

With best wishes from my wife[10] and myself, in the hope to see you again in the summer.

Your | Clausius

P.S. If you happen to see young Wynne,[11] send him our best wishes. He and his father sent me a very friendly letter from Dover, which endeared him to us even more.

RI MS JT/1/TYP/7/2279–80
LT Typescript Only

1. *two letters*: letters 2285 and 2294.
2. *honor that was bestowed upon you by the Cambridge University Senate*: Tyndall received an honorary LL.D (Doctor of Law) from the University of Cambridge (see letter 0063, n. 2).
3. *one of my papers*: probably R. Clausius, 'Ueber verschiedene für die Anwendung bequeme Formen der Hauptgleichungen der mechanischen Wärmetheorie', *Poggendorff's Annal.*, 125 (1865), pp. 353–400.
4. *W. Siemens*: Charles William Siemens (1823–83), formerly Karl Wilhelm, a German-British electrical engineer and businessman (*ODNB*).
5. *Mr. Folie*: François-Jacques-Philippe Folie.
6. *mechanical theory of heat*: R. Clausius, *Abhandlungen über die mechanische Wärmetheorie* (Braunschweig: Vieweg und Sohn, 1864).
7. *into French*: R. Clausius, *Théorie Mécanique de la Chaleur*, 2 vols, trans. F.-J.-P. Folie (Paris: Eugène Lacroix, 1868–9).
8. *Luchesini*: Alessandro Lucchesini (dates unknown), an Italian inventor.
9. *Nuovo Cimento*: Italian translations of Clausius's book have not been identified. *Il Nuovo Cimento* was a physics journal established in 1855 by Italian chemists Carlo Matteucci (1811–68) and Raffaele Piria (1814–65), as a continuation of the journal *Il Cimento* (est. 1844).
10. *wife*: Adelheid Clausius.
11. *young Wynne*: see letter 2235, n. 4.

From Auguste de la Rive 19 June 1865 2289

Genève, le 19 juin, 1865

Mon cher Tyndall,

Je vous remercie des lignes que vous avez ajoutées à la lettre de M. Boyle qui m'invite à la réunion de Birmingham. J'ai répondu à M. Boyle qu'il m'est impossible, à mon grand regret, de m'y rendre, ayant été déjà absent tout le printemps & ne pouvant recommencer à la fin de l'été un voyage en Angleterre.

Mais en venant vous remercier, je tiens à vous rappeler que nous vous attendons à Genève pour le 20 Août, votre chambre sera prête chez moi à Genève & vous vous trouverez logé chez moi avec Clausius, Schoenbein, Deville, Dufour & quelques autres de nos amis communs. Nous tenons beaucoup à vous avoir & j'espère bien que vous ne nous manquerez pas. Notre réunion sera terminée le 24, de sorte que vous aurez *[amplement]* le temps de vous trouvez à celle de Birmingham.

Je crois que notre réunion de Genève sera intéressante; nous y attendons de France, d'Allemagne & d'Italie plusieurs des principales notabilités scientifiques; vous comprenez combien nous tenons à ce que nos amis d'Angleterre ne nous fassent pas défaut.

Vous devez bien engager M. Huxley à venir avec vous; Lyell m'a à peu près promis, ainsi que Bentham & Sharpie. Tâchez d'engager encore quelques uns de vos amis scientifiques à nous honorer de leur présence; vous nous rendrez un vrai service; mais avant tout, venez vous-même, c'est l'essentiel.

Tous vos amis de Genève se joignent à moi pour vous presser de venir à notre réunion d'Août en vous envoyant leurs compliments affectueux.

Votre tout dévoué et affectionné | Aug. de la Rive

Geneva, 19 June, 1865

My dear Tyndall,

Thank you for the lines that you added to Mr. Boyle's[1] letter, who invites me to the Birmingham meeting.[2] I answered to Mr. Boyle that, to my great regret, it is impossible for me to come, since I have been away for the entire spring & I cannot start a trip to England at the end of the summer.

But while I am thanking you, I would like to remind you that we are waiting for you in Geneva for the 20th of August.[3] Your room will be ready at my house in Geneva & you will be staying at my place with Clausius, Schoenbein,[4] Deville,[5] Dufour[6] & some of our other friends in common. We are very keen on seeing you, & I hope that you will not miss us. Our meeting will end on the 24th, such that you would have plenty of time to attend the one in Birmingham.

I think that our meeting in Geneva will be interesting; we have several scientific notables visiting from France, Germany & Italy. You understand how much we wish that our friends from England will not fail us.

You should persuade Mr. Huxley to come with you; Lyell nearly promised me, as well as Bentham[7] & Sharpie.[8] Try to encourage some more of your scientific friends to honour us with their presence—that would be a great service that you could provide us with; but above all, the most important thing is for you to come.

All of your friends in Geneva are joining with me to ask you to come to our reunion in August, while sending you the warmest of wishes.

Your devoted and affectionate | Aug. de la Rive

RI MS JT/1/TYP/1/363
LT Typescript Only

---

1. *Mr. Boyle's*: George David Boyle (1828–1901), later Dean of Salisbury (*ODNB*). He served as Local Secretary for the BAAS meeting in Birmingham (*Brit. Assoc. Rep. 1865*, p. xxxiiii).
2. *Birmingham meeting*: the thirty-fifth annual meeting of the BAAS was held in Birmingham from 6–13 September 1865 (*Brit. Assoc. Rep. 1865*). Tyndall served as one of the vice-presidents for Section A—Mathematics and Physics.
3. *Geneva for the 20th of August*: an invitation to Tyndall to attend the 1865 meeting of the Société Helvetique des Sciences Naturelles (see letter 2270, n. 4).
4. *Schoenbein*: Christian Friedrich Schönbein (1799–1868), a German-Swiss chemist, who discovered the principle of the fuel cell in 1838 (*ADB*).
5. *Deville*: Henri Étienne Sainte-Claire Deville (1818–81), a French chemist who worked at the Chemistry Laboratory, Ecole Normale Supérior, Paris. He developed a process for producing aluminum (*CDSB*).
6. *Dufour*: either Guilliame Henri Dufour (1787–1875), a Swiss army officer and engineer (*HLS*); or Charles Dufour (1827–1902), a Swiss mathematician, astronomer, and meteorologist (*HLS*).
7. *Bentham*: George Bentham (1800–84), an English botanist most well-known for, both with Joseph Dalton Hooker, *The Handbook of the British Flora* (London: Lovell Reeve, 1858) and *Genera Plantarum*, 3 vols (London: Reeve & Co., Williams & Norgate, 1862–83), a monumental taxonomic classification of plants (*ODNB*).
8. *Sharpie*: likely a mis-transcription of 'Sharpe'; probably Edmund Sharpe (1809–77), an English architect, architectural historian, and railway engineer, who starting in 1863 designed and built a horse-drawn tramway in Geneva (*ODNB*; J. M. Hughes, *Edmund Sharpe: Man of Lancaster* (Stockport: J. M. Hughes, 2010), pp. 578–82).

From Hermann Helmholtz 19 June 1865 2290

Heidelberg, 19.6.65

Bester Freund

ich sage Ihnen meinen besten Dank für die Übersendung Ihres Buches on Heat. Es hat etwas lange bei Bunsen gelegen, welcher in irdischen Geschäftsangelegenheiten ebenso unzuverläßig ist, als er in wißenschaftlichen Dingen genau und zuverläßig zu sein pflegt. Sonst hätten Sie schon früher meinen Dank erhalten. Bunsen ist jetzt Decan der philosophischen Facultät und hat als solcher wieder eine Menge der heitersten Verwechselungen ausgeführt. Neulich waren zwei Examinanden, der eine gut bestanden, der andere sollte zurückgewiesen werden. Dem ersteren verkündete Bunsen in sehr ernstem Tone zu dessen größter Überraschung, daß die Facultät mit seinen Leistungen ganz unzufrieden sei. Der junge Mann war ganz vernichtet, glücklicher Weise ergab sich bei der Rückgabe der Dissertation, daß es gar nicht gemeint sei. Man darf ihm also dergleichen nicht an rechnen.

Auch für die Übersendung der Einladung zur British Association danke ich vielmals. Ich hoffe kommen zu können, und habe deshalb schon an M^r^. Boyle in Birmingham geschrieben, und die Einladung des Committee angenommen. Ich hoffe, daß ich Sie dort finden werde glücklich und heil zurückgekehrt von ihren alljährlichen Bergbesteigungen. Im letzten Winter habe ich auch einmal eine populäre Vorlesung über Gletschereis gehalten, und Ihre Versuche etwas modificirt. Ich habe sie schon mit einigen anderen zusammen zum Druck gegeben, und hoffe Ihnen ein Exemplar nach England mitbringen zu können. Es wäre mir lieb, einige Puncte dieser Theorie noch mit Ihnen besprechen zu können.

Wißenschaftliche Neuigkeiten von hier sind noch keine wieder zu melden. Bunsen hat sehr interessante Versuche über chemische Verbindungen in hoher Temperatur angestellt, die er aber noch weiter fortsetzen will, ehe er sie veröffentlicht. Ich selbst schreibe zu meiner Physiologischen Optik, um die endlich fertig zu machen.

Viele Grüße | Ihr treuer | H. Helmholtz

Heidelberg, 19.6.65

Dearest friend

I express my best thanks for sending me your book on Heat.[1] It has long remained with Bunsen, who in worldly affairs is as unreliable as he is reliable and precise in scientific matters. You would otherwise have received my thanks much earlier. Bunsen is now dean of the Faculty of Philosophy and as such has once again been involved in a series of the most amusing confusions.[2] The other day, there were two examinees; the one passed well, while the other one was to be rejected. To the former, to his greatest surprise, Bunsen announced in a very earnest tone that that the faculty was very dissatisfied with his performance. The young man was devastated; fortunately, as the dissertation was returned, it became clear that he was not meant. So the like should not be counted against Bunsen.

I also thank you very much for sending me the invitation to the British Association.[3] I hope to be able to come and have therefore written already to M^r^. Boyle[4] in Birmingham and accepted the committee's invitation. I hope to see you there, happy and safely returned from your annual mountain climbs. I gave a popular lecture about glacier ice[5] last winter and slightly modified your experiments. I have already sent this lecture to print,[6] together with some others and hope to bring you a copy for you to England.[7] I would be glad to also be able to discuss points of this theory with you.

There are no further scientific news to tell you from here. Bunsen has conducted some very interesting experiments on chemical compounds at high temperatures, but he wants to pursue them further before publishing them.[8] I myself am working on my Physiological Optics[9] in order to finally finish it.

Regards | your faithful | H. Helmholtz.

RI MS JT/H/42

---

1. *your book on Heat*: *Heat* (1865).
2. *amusing confusions*: Bunsen had developed quite the reputation and become the subject of many anecdotes collectively known as 'Bunseniana'. Some of these idiosyncrasies are recorded in: R. E. Oesper, 'Robert Wilhelm Bunsen', *Journal of Chemical Education*, 4 (1927), pp. 430–9.
3. *sending me the invitation to the British Association*: letter missing. The thirty-fifth annual meeting of the BAAS was held in Birmingham from 6–13 September 1865 (*Brit. Assoc. Rep. 1865*). Although Helmholtz accepted the invitation, he was unable to attend (see letter 2331).
4. *Mr. Boyle*: see letter 2289, n. 1.
5. *popular lecture about glacier ice*: Helmholtz delivered this popular lecture in Heidelberg in February 1865. On 24 February 1865, he had also delivered a lecture on ice for the Heidelberg Association for Natural History and Medicine, an abstract of which was published as: H. Helmholtz, 'Ueber Eigenschaften des Eises', *Verhandlungen des Naturhistorisch-medizinischen Vereins zu Heidelberg*, 3 (1865), pp. 194–6. A note from Helmholtz about this lecture was published as: H. Helmholtz, 'On the Regelation of Ice', *Phil. Mag.*, 32 (1866), pp. 22–3.
6. *this lecture to print*: the popular lecture was published as H. Helmholtz, 'Eis und Gletscher', in *Populäre Wissenschaftliche Vorträge* (Braunschweig: Friedrich Vieweg und Sohn, 1865), pp. 93–133. This was published in English much later as: H. Helmholtz, 'Ice and Glaciers: A Lecture Delivered at Frankfort-on-the-Main, and at Heidelberg, in February, 1865', in C. W. Eliot (ed), *Scientific Papers: Physics, Chemistry, Astronomy, Geology, with Introductions, Notes and Illustrations* (New York: P. F. Collier & Son, 1910), pp. 221–59.
7. *a copy for you to England*: Tyndall did receive a copy and discussed Helmholtz's lecture in J. Tyndall, 'Professor Helmholtz on Ice and Glaciers', *Phil. Mag.*, 30 (1865), pp. 393–407. This is letter 2330 in this volume.
8. *chemical compounds at high temperatures*: see R. Bunsen, 'Flammenreactionen', *Annalen der Chemie und Pharmacie*, 133 (1866), pp. 257–96; published in English as: R. Bunsen, 'Flame Reactions', *Phil. Mag.*, 32 (1866), pp. 89–107.
9. *Physiological Optics*: H. Helmholtz, *Handbuch der Physiologischen Optik*, 3rd edn (Leipzig: Leopold Voss, 1867).

From Heinrich Gustav Magnus 26 June 1865 2291

Berlin 26.6.65

Verehrter Freund!

Von einem kleinen Ausfluge auf's Land, wo wir unsere silberne Hochzeit feierten, zurückgekehrter fand ich den Brief von Mr George D Boyle, den Sie so gut waren mit freundlichen Worten zu begleiten. Es ist ein großer Wunsch von mir die Versammlung in Birmingham zu besuchen. Die einzige Versammlung der Art die ich in England mitgemacht habe war die 1849 ebenfalls in Birmingham, und die Erinnerung daran, namentlich an die frohen Stunden die ich mit Faraday dort verlebt, gehört zu den aller angenehmsten. Ich bin indeß bis jetzt noch nicht im Stande zu übersehen ob ich werde von hier fort können und vermag daher auch noch keine Pläne zu machen. Sie würden mich verbinden wenn Sie so gut wären dies dem Herrn Boyle mitzutheilen. Daß die Hofnung Sie wieder zu sehn nicht wenig dazu beiträgt daß die Waage meines Entschlußes für Birmingh. neigt, brauche ich Ihnen wohl nicht zu sagen. Sollte es wirklich dazu kommen daß ich zu jener Zeit nach England gelange, so schreibe ich Ihnen oder Hr. Boyle so bald als irgend möglich.

Mit den herzlichsten Grüßen von Frau und Töchtern

Ihr | G Magnus.

Berlin 26.6.65

Dear friend!

Returned from a short trip to the countryside where we celebrated our 25th wedding anniversary,[1] I found the letter from Mr George D Boyle,[2] which you were so kind to accompany with a few friendly words. It is a great wish of mine to attend the meeting in Birmingham.[3] The only similar meeting in which I participated in England was the one in 1849, also in Birmingham,[4] and my memories of it, especially of the pleasant hours I spend with Faraday, are among the most pleasant. I am, however, still not able to tell if I can depart from here and therefore do not want to make plans yet. I would feel obliged if you were so kind to inform Mr Boyle about this. I do not have to tell you that the hope to see you again significantly shifts the balance of my decision toward Birmingham. Should it really happen that I get to England at that time, I will write to you or Mr Boyle as soon as at all possible.

With the kindest greetings from my wife[5] and daughters[6]

Yours | G Magnus.

RI MS JT/1/M/30

1. *25th wedding anniversary*: Magnus married Bertha Humblot, a French Huguenot, in 1840.
2. *Mr George D Boyle*: see letter 2289, n. 1.
3. *meeting in Birmingham*: see letter 2289, n. 2.
4. *1849, also in Birmingham*: the BAAS held its nineteenth annual meeting in Birmingham in 1849 (*Brit. Assoc. Rep. 1849*).
5. *my wife*: Bertha Magnus (née Humblot, 1820–1910).
6. *daughters*: Anna (1841–75) and Christine (1842–1936), who married her cousin Viktor von Magnus in 1866.

From Auguste De la Rive 1 July 1865 2292

Genève, le 1er Juillet, 1865.

Mon cher Tyndall,

J'ai reçu de vous un petit opuscule dont je vous remercie infiniment & une invitation de vous pour m'engager à aller à la réunion de l'association Britannique qui aura lieu à Birmingham au commencement du mois de Septembre. Je regrette bien d'avoir été obligé de répondre par un refus à cette obligeante invitation; mais ayant beaucoup voyagé ce printemps, il m'est impossible de m'absenter de nouveau cet automne.

Par contre je compte toujours sur vous pour le 20 Août; votre chambre est prête chez nous à Genève; Vous vous trouverez avec Clausius, Deville & beaucoup d'autres amis soit Suisses, soit Allemands. Je tiens bien à ce que vous arriviez pour le Dimanche 20, parce que je réuni chez moi le soir de ce jour là tous les membres de la Société. Vous ne nous manquerez pas au moins; nous avons absolument besoin de vous & nous avons si peu de nos amis d'Angleterre qui nous ont donné l'espoir de venir à Genève, qu'il faut nécessairement que ceux qui, comme vous, nous ont à peu près promis, ne nous fassent pas défaut. J'ai fait préparer tout ce qu'il faut pour répéter vos belles expériences sur la chaleur obscure de l'arc voltaïque; mais il faut que vous soyez là pour les faire vous-même. Vous nous apporterez bien aussi quelques autres communications scientifiques; vous n'avez qu'à ouvrir votre portefeuille pour cela. Vous devriez bien engager M. Frankland à venir avec vous; dites lui bien de ma part & de celle de tous mes collègues, le plaisir et l'honneur qu'il nous ferait en se rendant à Genève pour notre réunion du 20, 21, 22 & 23 Août. Je lui aurais écrit directement si j'avais l'honneur de le connaître; mais passant par vous, mon invitation n'aura que plus de chance d'être agrée. Si vous aviez quelques autres amis d'Angleterre disposés à venir à notre réunion, vous seriez bien aimable de les engager à le faire.

Au revoir donc à bientôt, très cher et excellent ami, & recevez l'assurance de mes sentiments affectueux & bien dévoués. | Aug. de la Rive.

Geneva, 1 July, 1865.

My dear Tyndall,

I received a small booklet[1] from you, for which I am infinitely grateful, & an invitation from you to come to the reunion of the British Association that will take place in Birmingham at the beginning of September.[2] I deeply regret having to turn down this kind invitation; but having traveled a lot during the spring, it is impossible for me to be absent again this autumn.

On the other hand, I am still counting on you for the 20th of August.[3] Your room is ready at our house in Geneva. You will find yourself with Clausius, Deville[4] & many other friends,[5] whether they are Swiss or German. I am very keen on your arrival on Sunday the 20th, since at the evening of that day at my house I am reuniting all the members of the Society. At least you will not fail us, we definitely need you, & we have so few of our friends from England who gave us hope of coming to Geneva, that it would be necessary that those who, like you, more or less promised, do not fail us. I have prepared everything that is needed to repeat your beautiful experiments on the obscure heat of the voltaic arc;[6] but you need to be there to do them yourself. You would also bring us other scientific works; all you have to do for that is open your portfolio. You should encourage Mr Frankland to come with you; tell him from me & from all of my colleagues, about the pleasure and honour that he would bring us by coming to Geneva for our reunion on the 20th, 21st, 22nd & 23rd of August. I would have written to him directly if I would have had the honour of knowing him; but going through you, my invitation has a higher chance of being accepted. If you have some other friends in England who wish to come to our reunion, you be kind to ask them to do so.

Goodbye, or rather, see you soon, very dear and excellent friend, & accept the assurance of my affectionate & devoted sentiments. | Aug. de la Rive.

RI MS JT/1/TYP/1/364
LT Typescript Only

---

1. *booklet*: not identified, but perhaps Tyndall's Rede lecture (see letter 2231, n. 1).
2. *invitation from you . . . beginning of September*: the thirty-fifth annual meeting of the BAAS was held in Birmingham from 6–13 September 1865 (*Brit. Assoc. Rep. 1865*).
3. *20th of August*: an invitation to Tyndall to attend the 1865 meeting of the Société Helvetique des Sciences Naturelles (see letter 2270, n. 4).
4. *Deville*: see letter 2289, n. 5.
5. *other friends*: see letter 2289, n. 4 and n. 6.
6. *your beautiful experiments on the obscure heat of the voltaic arc*: probably Tyndall's eight lectures on 'Voltaic Electricity' delivered at the RI in the spring of 1855 (*Roy. Inst. Proc.*, 2 (1854–8), p. 56). He also delivered a Friday Evening Discourse at the RI on 1 June 1865,

'On the Currents of the Leyden Battery' (*Roy. Inst. Proc.*, 2 (1854–8), pp. 132–5). See also J. Tyndall, 'The Polymagnet', *Phil. Mag.*, 9 (1855), pp. 425–30.

From John Herschel 1 July 1865 2293

Collingwood[1] July 1./65

Dear Sir

Pray accept my best thanks for the copy of your interesting lecture on the heat rays of the Electric Spectrum[2] The experiments are exceedingly curious and not the least curious point about them is the perfect transparency to heat of your Iodine Solution while so very opake[3] to light! Wishing you continued success in this fertile line of enquiry.

Believe me yours very truly | J F W Herschel | Prof. Tyndall

RI MS JT/1/H/105
RI MS JT/1/TYP/2/538

1. *Collingwood*: Herschel's home near Hawkhurst in Kent.
2. *copy of your interesting lecture on the heat rays of the Electric Spectrum*: Tyndall delivered the annual Rede Lecture at the University of Cambridge ('On Radiation') on 16 May 1865 (see letter 2231, n. 1).
3. *opake*: alternative spelling of *opaque*.

From John William Draper 24 July 1865 2294

My Dear Sir

I received this morning your very acceptable gift of a copy of y[ou]r work on Heat[1] It must have been delayed a long time somehow as I see that you sent it last month

I was of course already familiar with it *[1 word illeg]* It cannot but extend a most acceptable influence on the progress of Chem[istry] and apart from the inestimable value of your own original researches it must reflect great credit on y[ou]r authority as the as the[2] most*[1 word illeg]* and satisfactory treatise we yet have on the subject—Accept my warmest thanks both for your gift and the instruction & pleasure I have derived from it

After devoting a great many years to scientific researches I at length gave them up. finding how difficult it is to learn the recognition of first priority in them. I have turned more to historical pursuits and shall have the pleasure of sending you by an early opportunity a work I have recently published "Thoughts on the Future Civil Policy of America"[3]—a little later I will send you a larger book on which I am now engaged "History of the Am[erican] Civil War"[4]

Yours truly | J W D | Univ[ersity] N.Y.[5] | July 24, 1865 | Prof Tyndall FRS &c

Library of Congress, John William Draper Family Papers

---

1. *y[ou]r work on Heat*: *Heat* (1865).
2. *as the*: Draper mistakenly wrote 'as the' twice.
3. *"Thoughts on the Future Civil Policy of America"*: J. W. Draper, *Thoughts on the Future Civil Policy of America* (New York: Harper & Brothers, 1865).
4. *"History of the Am[erican] Civil War"*: J. W. Draper, *History of the American Civil War*, 3 vols (New York: Harper & Brothers, 1867–70).
5. *Univ[ersity] N.Y.*: New York University, where Draper was Professor of Chemistry (1840–81) and President of the its Medical College (1850–73).

From Adam Sedgwick[1] 26 July 1865 2295

Cambridge July 26. | 1865

My dear Professor Tyndall

I do not know when your Rede Lecture on Radiation[2] was sent to me. Since the day when to my great delight (spite of my deaf ears & half blind eye I was delighted) I sat under you in our Senate House,[3] I have been a *[2–3 words illeg]* & a great vagabond. But on my last return to Cambridge I found a presentation copy of your Lecture[4] on my table and it is my pleasant duty to thank you for it. I have read it with much pleasure & its vibrations have set the chords of my fancy in merry movement. It is a noble essay—the facts & inferences—the exquisite mechanism—the beauty of its handling—the tests of exquisite sensibility—the logical inventions took when which little thread of phenomena are granted into a glorious pattern of *[*learning*]* workmanship. I declare the apprehension of such things is sought to make the poor brain retreat out of the limits of the *[*precranium*]*—It is delightful to see the results of long continued & continued labour exhibited by one who is in the grand work of discourse.

I congratulate you on the noble field in which you are working—& I hope you will go on gathering year by year fresh materials to show forth the glory of the Creation & to contribute to the good of man's estate.

I remain | Truly & grateful yours | A Sedgwick

RI MS JT/2/13c/1293–4

1. *Adam Sedgwick*: Adam Sedgwick (1785–1873), a highly influential British geologist and priest. He was educated at Trinity College, University of Cambridge, earning his BA in 1808 and his MA in 1811. He was ordained as a priest in 1818, the same year he became Woodwardian Professor of Geology at Cambridge, a position he held for the rest of his life. Elected FRS in 1821, one of Sedgwick's most important contributions to science was the determination of different time periods in geologic history, the Cambrian and the Devonian. A mentor to a young Charles Darwin in geology, Sedgwick became an opponent to Darwin's theory of evolution by natural selection (*ODNB*).
2. *your Rede Lecture on Radiation*: Tyndall delivered the annual Rede Lecture at the University of Cambridge ('On Radiation') on 16 May 1865 (see letter 2231, n. 1).
3. *Senate House*: location of Tyndall's Rede Lecture at the University of Cambridge.
4. *a presentation copy of your Lecture*: *On Radiation*.

From Johann Conrad Bohn 11 August 1865 2296

Sehr verehrter Herr Tyndall!

Ihr Buch „on Heat" welches Sie so freundlich waren schon im Anfang des Monats Mai einem Londoner Buchhändler zur Besorgung an mich zu übergeben, ist mir erst vor wenigen Tagen zugestellt worden. Entschuldigen Sie daher, dass ich bis heute zögerte Ihnen für das gütige Geschenk meinen verbindlichsten Dank auszusprechen. Ich habe eine grosse Freude an dem was ich bisher darin gelesen, (und es ist, da ich gerade einige Musse hatte, ein gutes Stück) empfunden. Sie wissen die bekanntesten Dinge auf eine neue Art schlagend vorzuführen und zu erläutern, Sie gewinnen ihnen eine neue Seite für die Betrachtung ab und verstehen zu dem Bekannten noch etwas Belehrendes Neues hinzuzufügen, so dass auch Jener, der sich fachgemäss mit den behandelten Gegenständen beschäftigt und für welchen das Buch nicht zunächst bestimmt ist, nicht nur Vergnügen empfindet, sondern auch etwas zu lernen vermag. Sie haben vortrefflich Ihre zahlreichen und so interessanten eigenen Untersuchungen zu verwerten verstanden und es ist Ihnen gelungen, scheinbar weit aus einander liegende Gegenstände durch das Band der Theorie der Wärme in befriedigender Weise zu verbinden. Und nicht gering schätze ich die Klarheit und <Bestimmtheit> mit welcher Sie Ihre Lehren vortragen und die Eindringlichkeit, mit welcher Sie dieselben darlegen, so dass auch der gebildete Mann, dem physikalische Untersuchungen und Betrachtungen sonst ferne zu liegen pflegen, Theilnahme an denselben gewinnt und reichen Nutzen aus Ihren Vorträgen ziehen lernt.

Ich glaube mich nicht zu täuschen, wenn ich annehme, dass auch in Deutschland Viele mit Vergnügen und Nützen Ihr Buch lesen und kaufen werden. Leider ist die englische Literatur (schon wegen der hohen

Preise), wenn es nicht gerade Romane in „Tauchnitz Edition" sind, schwer in Deutschland zu verbreiten, trotz dem, dass meine Landsleute vielleicht mehr ausländische Bücher zu lesen verstehen und auch wirklich lesen, als andere <Nationen>. Ich glaube desshalb dass eine Uebersetzung Ihres Buches ins Deutsche Vielen eine angenehme Sache sein werde. So viel mir bekannt ist nach der ersten Ausgabe eine Uebersetzung von Moigno in die französische Sprache gemacht worden. Wenn Sie zu mir das Vertrauen haben können, dass ich ein treuer Dollmetsch Ihrer Gedanken sein könnte, so würde ich mich mit Vergnügen der Arbeit der Uebersetzung unterziehen und Sie um Ihre freundliche Erlaubnis und Zustimmung zu meinem Plane bitten. Die Ansichten der Buchhändler sind zwar zu <weilen> unberechenbar verschieden von denen der Männer der Wissenschaften; allein ich hoffe ganz sicher einen Verleger für die Uebersetzung finden zu können. Das muss man freilich gleich im Voraus sich sagen ein hoher Lohn wird nicht geboten werden, namentlich wird das etwaige Honorar nicht mit englischem Maassstab zu messen sein und würde in L. Sterl. ausgedrückt eine sehr ärmliche Zahl geben.—Ich habe einstweilen mit keinem Buchhändler gesprochen oder verkehrt über diese Angelegenheit.—Wenn Sie jedoch die Güte haben wollen mich zur Uebersetzung und Herausgabe Ihres Buches zu ermächtigen, so werde ich die nötigen Schritte sofort thun.

Erweisen Sie mir, hochverehrter Herr Tyndall, die Freundlichkeit, mich bald mit einer Antwort auf diese meine Anfrage zu beehren, und zwar bitte ich wiederholt: lassen Sie dieselbe eine Zustimmung sein.

Genehmigen Sie die Versicherung meiner vorzüglichsten Verehrung und meines ergebensten Dankes

Ihr Dr J Bohn | Giessen d. 11 August 1865.

Dear Mr Tyndall!

Your book "On Heat"[1] that you were so kind to give to a bookseller in London[2] at the beginning of the month of May to be forwarded to me, has only been delivered to me a few days ago. Therefore, please excuse that I hesitated until today to give you my most sincere thanks for the kind present. I was very pleased with what I have read so far (and it is a considerable portion, as I currently have some leisure). You know how to present and explain the most well-known things in a new, convincing way; you find a new perspective from which to look at them and you understand how to add new and edifying aspects to the familiar, so that those who occupy themselves professionally with the subjects under discussion, and for whom the book is not primarily intended, will not only find pleasure but will also be able learn something. You make superb use of your own numerous and interesting investigations and in a satisfying manner you were able to tie together seemingly disparate subjects with the ribbon of the theory of heat. And not little do I appreciate the clarity and resolution with which you present your lessons and the depth with which you elaborate them, so that learned men, who otherwise tend to be far from physical investigations and considerations, participate in them and learn to reap rich benefit from your lectures.

I believe that I am not deceived if I assume that in Germany as well, many will read and purchase your book with pleasure and profit. Unfortunately, English literature is difficult to diffuse in Germany (if alone due to the high prices), unless it is novels in "Tauchnitz editions",[3] even though my compatriots are probably more capable of understanding foreign books, and actually do read them, than other nations. I therefore think that a translation of your book into German would be a pleasant thing to many. As far as I know, Moigno[4] did a translation into French after the first edition.[5] If you had the confidence that I could be a truthful interpreter of your thoughts, I would gladly undertake the task of translation and ask you for your kind permission and approbation of my plan. The opinions of booksellers occasionally differ unpredictably from those of men of science; I hope certainly to be able to find a publisher for the translation. Of course, one has to tell oneself quite clearly beforehand: a high wage will not be offered; in fact, the royalties will not correspond to English standards and expressed in £. Sterl[ing]. will give but a miserable sum.—For the time being, I have not spoken or exchanged thoughts with any bookseller about this matter.—However, if you were so kind as to authorise me to translate and publish your book, I will immediately take the necessary steps.[6]

Be so kind, highly admired Mr Tyndall, to honour me soon with an answer to my request, and allow me to plead again: let it be an assent.

Please accept the assurance of my highest admiration and my most sincere thanks

Your Dr J Bohn | Giessen 11 August 1865.

RI MS JT/1/B/106

1. *Your book "On Heat"*: *Heat* (1865).
2. *so kind to give to a bookseller in London*: see letter 2265.
3. *"Tauchnitz editions"*: English-language editions of British works published for European readers by the German publisher Bernhard Tauchnitz (1816–95). See A. Bowden and W. B. Todd, *Tauchnitz International Editions in English, 1841–1955: A Bibliographical History* (New York: Bibliographical Society of America, 1988).
4. *Moigno*: François-Napoléon-Marie Moigno.
5. *translation into French after the first edition*: J. Tyndall, *La Chaleur considérée come un mode de mouvement*, trans. F. Moigno (Paris: Étienne Giraud, 1864).
6. *to authorise me to translate and publish your book, I will immediately take the necessary steps*: Tyndall agreed to a German translation of *Heat* (1865), first to be published by Eduard Vieweg and then Theodore Fischer. However, as many letters later in this volume show, Tyndall decided against this agreement due to Bohn's inferior translation skills. A German

translation eventually appeared as: J. Tyndall, *Die Wärme, betrachtet als eine Art der Bewegung*, trans. H. Helmholtz and G. Wiedemann (Braunschweig: Friedrich Vieweg und Sohn, 1867). While Helmholtz and Wiedmann are credited with the translation, it was in fact their wives Anna and Clara, respectively, who carried out the work of translating the book (see letter 2410).

From Juliet Pollock 15 August [1865][1] 2297

59 Montagu Square[2] | August 15.th

My dear Mr. Tyndall

We should like to hear from you;—& so, not in a wholly disinterested spirit, I write to you. I shall make no comments on the Matterhorn accident[3] about which enough has been said. . but I will tell you something of our own way of life in the yellow leaf[4] of the season, when the city is stripped of its honours, and the thorough fares are given up to curious kind of creatures; specimens of life that crawl from unknown depths to the surface, giving rise to speculations in one's mind which are not fruitful. Wonderful dresses, odd wigs; evidently false whiskers, & patched finery walk about. Where were they when they didn't walk about?—And empty London is more noisy than full London—seems so from the broken continuity of the sounds,—is in fact as noisy as a country village.—We escape to our favourite refuge in the Regents Park[5] where only the Parson remains to greet us, but which is still green and pleasant and full always of agreeable associations.

Walter[6] has come home, looking well and in capital spirits, and in spite of ill health, & lost time, has taken the first place in the examination[7] for the half year & brought home the prize! His English verse translations from Latin & Greek poets are very good indeed, & I suppose mainly contributed to his distinction. It has pleased us very much and I know it will please you.

Fred[8] came to us last Saturday, & next Saturday we all go to St. Julians[9] together. Afterwards my Husband[10] and self and Moss[11] visit the Romillys[12] & the Speddings,[13] and Walter goes to the Speddings independently, while we are with the Romillys. There is a first rate piece of criticism by James Spedding[14] on Kate Terry's performance[15] in The 12.th Night (Shakespeare's 12.th Night) in the current no. of Fraser.

Having finished my vols. of Plato's dialogues,[16] I am now occupied with Lecky.[17] I admire & like the book.[18] I do not agree with Geo[rge]. Elliot's strictures upon it.[19] Its moderate tone will make it a useful book; it will be acceptable to some, to many, whose stomachs would revolt at a more strongly seasoned dish, & it will work into the system, & work good.—I do not think it is true that no result is to be obtained from it. I think the author arrives at an evident conclusion: this is it. Theology has been the curse of mankind: that is, the fixed creeds which mankind has endeavoured to establish concerning things of which we really know nothing. The creeds of Xtianity are the more perverse when you reflect that the great founders especially condemned the dogma and the form. There is another conclusion arrived at: the great contending power that has driven Theology out of its idolatries & dark caverns, out of its Temples & its terrorism, is Physical science.—

It has done a good deal, it is doing more, & will do enough; but the onward movement is better gradual than precipitate, & we must not cast away too much; we must keep reverence and worship; for man must have something more than man to look to. We must not let go the hope & trust that belong to the idea of divinity; we must not neglect Christ's teaching as we find it; unperplexed by the additions of superstition & ignorance or fraud & arrogance. I cannot say that Lecky's statements are new to me, I knew his facts long ago; but they are very well assembled and his thought is neither too bold nor too timid. I had thought with him; but it is satisfactory to see one's thought turned out into a shapely mould.—But note this, M^r^. Tyndall:—very very few of my most intimate friends have known, or know, or ever will know these thoughts of mine; only you, & perhaps three others.—& so you see progress is arrested by timidity; a timidity, which as society is at present constituted, is a necessity. I would never willingly say what I did not think; but I am not bound to declare all that I do think: in fact I am bound in discretion to withhold it.

Polly[20] is gone off on a holiday for a fortnight to her friends in Wales & Moss is left to Jane[21] & me.. What a long letter I have written.

I am ever truly yours | Juliet Pollock

RI MS JT/1/P/192

---

1. *[1865]*: the year is given by reference to the 'Matterhorn accident" (see n. 3). Also, a book and a review of it that are mentioned were both published in 1865 (see n. 18 and n. 19).
2. *59 Montagu Square*: the Pollocks' residence in Marylebone, London.
3. *the Matterhorn accident*: on 14 July 1865, four members of the group accompanying Edward Whymper (1840–1911) on their successful first ascent of the Matterhorn fell to their deaths during the descent. They were Francis William Bouverie Douglas, Lord Douglas (b. 1847); Charles Hudson (b. 1828); Douglas Hadow (b. 1846); and Michel Croz (b. 1830); Whymper and two Zermatt guides, Peter Taugwalder father and son, survived the accident. For Tyndall's account of hearing of the tragedy while he was himself in the Alps, see *Hours of Exercise in the Alps*, pp. 252–4. For Whymper's account, see E. Whymper, 'The Matterhorn Accident', *Times*, 8 August 1865, p. 9; and E. Whymper, *The Ascent of the Matterhorn* (London: John Murray, 1880), pp. 284–98.
4. *yellow leaf*: possibly a reference to a line from Shakespeare's *Macbeth* (V.iii.24–5): 'I have lived long enough. My way of life | Is fall'n into the sere, the yellow leaf'.

5. *Regents Park*: a large public park in central London.
6. *Walter*: Walter Pollock.
7. *the first place in the examination*: at Eton College, a boarding school for boys in Windsor, England.
8. *Fred*: Frederick Pollock.
9. *S^t. Julians*: the Herries family home in the town of Sevenoaks in Kent, England, southeast of London.
10. *my Husband*: William Frederick Pollock.
11. *Moss*: Maurice Pollock.
12. *Romillys*: the family of Edward Romilly (1804–70), an English amateur cricketer and a Member of Parliament from 1832–5, of Porthkerry in southern Wales.
13. *Speddings*: the Pollocks were good friends with James Spedding (1808–81), an English literary editor and biographer known especially for editing the works of Francis Bacon (*ODNB*). The family home was at Greta Bank, an estate located in the town of Keswick in the county of Cumberland in northwestern England.
14. *piece of criticism by James Spedding*: J. Spedding, 'Miss Kate Terry in Viola', *Fraser's Magazine*, 72 (August 1865), pp. 261–8.
15. *Kate Terry's performance*: Kate Terry (1844–1924), an English actress who performed from an early age until 23 (obituary notice, *Times*, 7 January 1924, p. 14). She played both roles of the twin brothers Viola and Sebastian in Shakespeare's *Twelfth Night* at London's Royal Olympic Theatre in June 1865.
16. *my vols. of Plato's dialogues*: dialogues about moral and philosophical problems as written by the Greek philosopher Plato (428/427 or 424/423–348/347 BC). Pollock's volumes were possibly: W. Whewell, *The Platonic Dialogues for English Readers*, 3 vols (London: Macmillan and Co, 1859).
17. *Lecky*: William Edward Hartpool Lecky (1838–1903), an Irish historian later known for his eight-volume *History of England during the Eighteenth Century* (1878–90) (*ODNB*).
18. *book*: W. E. D. Lecky, *History of the Rise and Influence of the Spirit of Rationalism in Europe*, 2 vols (London: Longman, Green, Longman, Roberts, & Green, 1865).
19. *Geo[rge]. Elliot's strictures upon it*: G. Eliot, 'The Influence of Rationalism', *Fortnightly Review*, 15 May 1865, pp. 43–55.
20. *Polly*: probably a governess for the Pollocks' son Maurice, then aged 8 years old (born 28 March 1857).
21. *Jane*: probably another governess for Maurice Pollock.

To Michael Faraday 23 August 1865 2298

Geneva, | 23rd. Aug. 1865.

My dear Mr. Faraday,

During the last five or six weeks I have thought many times thought of writing to you, but not until the conclusion of my excursion[1] do I put my thought into execution. Many months ago I promised our friend DelaRive[2] to attend the meeting of the Swiss Savants at Geneva.[3] It is the Jubilee of the Society—a kind of Swiss British Association. Dumas,[4] Schoenbein,[5] Wöhler,[6] Dove,[7] Clausius and many others celebrities are here, and the meeting thus far has been a very pleasant one. It will end to-day, tomorrow I dine with Marignac,[8] and the next day I turn my face homewards.

Through some mistake I suppose on Smith's[9] part I have had no letters sent to me either to Zermatt or to Geneva, so that I am entirely ignorant of the condition of my friends at home I thought to remedy matters by writing to Mr. Barrett[10] from Zermatt, asking him to forward my letters to Geneva and to write to me himself: but I have had no reply. The post in this latter case, is I suppose the delinquent.

I went to Zermatt soon after the occurrence of that terrible accident.[11] I had been informed that Lord Francis Douglas's[12] Mother[13] suffered much from the idea of her son not having been found. I therefore resolved to make an effort to regain the body. I proposed to climb to the point from which they fell, fix irons there in the rocks, attach ropes to the irons, and descend by the ropes along the precipices, examining the mountain right and left of the line of ascent. Nearer than Geneva ropes were not to be found; so I sent a man[14] to that city, to select and purchase two thousand feet of rope of the best quality. It was bought and carried to Zermatt. I also engaged a man[15] who had been accustomed to blast rocks, and who was (also) an able mountaineer. His business was to bore holes in the cliffs, fix iron and steel jumpers there, to which the ropes were to be attached. All my arrangements were made. As far as *ideas* went, my plans were complete; so I waited at Zermatt for fine weather. I waited eighteen days, but the weather was abominable throughout. Heavy snow had fallen and covered the rocks of the Matterhorn; the said snow had been partially melted and refrozen, thus covering the rocks in thick envelopes of ice, and rendering them a hundred times more dangerous than ordinary. Day after day moreover heavy clouds swathed the mountain, and discharged themselves at intervals in hail and snow against its sides. Not until hope deferred had made the heart sick did I quit the place, and I do not know that I should have quitted it so soon, had it not been for my engagements with our friend DelaRive to attend the Jubilee of the Helvetic Natural Society.

Will you present my kindest regards to Mrs. Faraday[16] and Miss Barnard.[17] I hope to find you all in good health on my return.

Yours ever | John Tyndall

Faraday Correspondence, 6:4561
RI MS JT/1/TYP/12/4123–4

1. *my excursion*: Tyndall and Thomas Hirst left London for the Alps on 10 July 1865, and returned to London by 26 August.
2. *DelaRive*: Auguste de la Rive.
3. *meeting of the Swiss Savants at Geneva*: Tyndall attended the 1865 meeting of the Société Helvetique des Sciences Naturelles (see letter 2270, n. 4).
4. *Dumas*: Jean-Baptiste Dumas (1800–84), a French chemist who received the RS Copley Medal in 1843 'for his late valuable researches in organic chemistry, particularly those contained in a series of memoirs on chemical types and the doctrine of substitution, and also for his elaborate investigations of the atomic weights of carbon, oxygen, hydrogen, nitrogen and other elements' (*CDSB*).
5. *Schoenbein*: see letter 2289, n. 4.
6. *Wöhler*: Friedrich Wöhler (1800–82), a German chemist who taught at the Polytechnic Schools in Berlin and Kassel before becoming Ordinary Professor of Chemistry at the University of Göttingen (1836–82) (*ADB*).
7. *Dove*: Heinrich Wilhelm Dove (1803–79), a Prussian meteorologist and experimental physicist. He was educated at the University of Berlin and became an ordinary professor at Berlin's Friedrich-Wilhelms-Universität in 1845. In his research, he focused on meteorology and was awarded the RS's Copley Medal in 1853 for his contributions to understanding the distribution of heat in the atmosphere (*ADB*).
8. *Marignac*: Jean Charles Galissard de Marignac (1817–94), Professor of Chemistry at Geneva from 1841–78 (*CDSB*; *HLS*).
9. *Smith's*: Thomas Smith (d. 1898), a hall porter at the Royal Institution from 1861–89 (*Faraday Correspondence*, vol. 6, p. lix).
10. *Mr. Barrett*: see letter 2245, n. 31.
11. *that terrible accident*: a reference to an accident on the Matterhorn (see letter 2297, n. 3).
12. *Lord Francis Douglas's*: see letter 2297, n. 3.
13. *Mother*: Caroline Margaret Douglas (née Clayton, 1821–1904), Marchioness of Queensbury, widow of Archibald William Douglas (1818–58), 8th Marquis of Queensbury.
14. *man*: in *Hours of Exercise in the Alps*, Tyndall wrote: 'None of the Zermatt guides would second me, but I found one of the Lochmatters of St. Nicholas willing to do so' (p. 253).
15. *man*: in *Hours of Exercise in the Alps*, Tyndall wrote: 'I had seen the road-makers at work between St. Nicholas and Zermatt, and was struck by the rapidity with which they pierced the rocks for blasting. One of these fellows could drive a hole a foot deep into hard granite in less than an hour. I was therefore determined to secure in aid of my project the services of a road-maker' (p. 253).
16. *Mrs. Faraday*: Sarah Faraday.
17. *Miss Barnard*: Jane Barnard.

From Michael Faraday 28 August 1865 2299

The Green, | Hampton Court,[1] | Monday, 28 August / 65.

Welcome to London, my dear Tyndall. I was in the house on Saturday[2] when you were there, but did not know until afterwards. Glad to hear you are well.

I write very inconsistently, and know not from line to line what I have said. I received your letter[3] the same day (Saturday). You must have had a most grievous time of it.[4] Many thanks from all of us to you for it. If you meet with Wilson,[5] who has left me his name but not his address, say to him as you think I would say, or send me his address.

Yours, | M. Faraday | Dr. Tyndall

Faraday Correspondence, 6:4562
RI MS JT/1/TYP/12/4158

1. *The Green, | Hampton Court*: Faraday's house in London since 1858.
2. *on Saturday*: 26 August 1865.
3. *your letter*: letter 2298.
4. *You must have had a most grievous time of it*: a reference to an accident on the Matterhorn (see letter 2297, n. 3).
5. *Wilson*: probably James Maurice Wilson.

From François-Napoléon-Maire Moigno 30 August [1865][1] 2300

Mon cher Mr Tyndall

Aussitôt après votre passage, j'ai commencé la traduction de votre charmante lecture <u>Rede</u>, je l'ai poursuivie sans un instant de repos le samedi et le dimanche entre les offices; j'ai fini le lundi et aujourd'hui cette incomparable dissertation est presque entièrement imprimée. Elle fera une petite brochure semblable à celles que vous avez emportées. Seulement il manquera quelques pages, *[*et*]* pour les remplir, je vous prierais *[*instamment*]* de me faire envoyer immédiatement par le retour du *[*courrier*]* s'il est possible, par Mr Longmann les clichés ou blocs de la deuxième leçon de la chaleur pages 411, 412, 414.

Je *[1 word illeg]* comptant les *[*frais*]* que Mr Longmann *[*fera*]* pour les clichés. Il me tarde de vous envoyer l'épreuve de ma traduction, car je la crois très *[*digne*]* de vous. Les clichés, je vous en *[1 word illeg]*. La chaleur *[*fera*]* les *[1 word illeg]* d'un nombre considérable de *[*savants*]* français.

Je *[1 word illeg]* de pouvoir aller à Birmingham pour l'association. D'ici là j'aurai à vous demander un tout petit service.

Votre humble et dévoué serviteur et ami | L'abbé F. Moigno

30 août, 2 rue d'Erfurt

My dear Mr Tyndall,

Immediately after your visit,[2] I began the translation of your marvellous <u>Rede</u> lecture.[3] I worked on it without a moment of rest on Saturday and Sunday in between the offices; I finished on Monday, and today this incomparable essay is almost completely printed. It will be a small brochure similar to those that you took away with you. It only misses a few pages, and to complete them, I beg you earnestly to send me immediately by a return mail, if possible, by Mr Longmann,[4] the pictures or *[*blocks*]* of the second lesson on Heat, pages 411, 412, 414.[5]

I *[1 word illeg] [*counting*]* the expenses that Mr Longmann *[*will make*]* for the pictures. I cannot wait to send you the proofs of my translation, since I believe it is worthy of you. The pictures, I will *[1 word illeg]* to you. The Heat does *[1 word illeg]* a considerable number of French savants.

I *[1 word illeg]* to be able to go to Birmingham for the association.[6] Until then, I will have to ask you a little favour.

Your humble and devoted servant and friend | Father F. Moigno

30 August, 2 rue d'Erfurt[7]

RI MS JT/1/M/132

---

1. *[1865]*: the year is given by reference to the thirty-fifth annual meeting of the BAAS held in Birmingham from 6–13 September 1865 (*Brit. Assoc. Rep. 1865*).
2. *your visit*: Tyndall presumably visited Moigno on his pass through Paris during his return to London from his annual alpine excursion.
3. *translation of your marvellous <u>Rede</u> lecture*: J. Tyndall, *Sur la radiation*, trans. F. Moigno (Paris: E. Giruad, 1865). Tyndall delivered the annual Rede lecture at Cambridge in May 1865 (see letter 2231, n. 1).
4. *Mr Longmann*: either Thomas Longman (1804–79) or William Longman (1813–77), brothers who ran the firm that published most of Tyndall's books, including *Heat* (1863).
5. *second lesson on Heat, pages 411, 412, 414*: Moigno had translated the French edition of Tyndall's *Heat* (1863), which appeared as: J. Tyndall, *La Chaleur considérée come un mode de mouvement*, trans. F. Moigno (Paris: Étienne Giraud, 1864).
6. *Birmingham for the association*: see n. 1.
7. *2 rue d'Erfurt*: in Paris.

To William Frederick Pollock 1 September 1865 2301

1st September 1865.

My dear Pollock,

I have already stated to you my theory of "that boy's"[1] success.[2]—Blood depend upon it carries it in intellectual as in physical matters. There may be exceptions but this is the rule; and had his mother's[3] brain—not to mention his father's—been other than it is his success would have been other than it is. I hope to see you this evening.

Yours ever | John Tyndall.

RI MS JT/1/TYP/6/2044
LT Typescript Only

1. *"that boy's"*: Walter Pollock.
2. *success*: Walter had taken first place in an examination at Eton College. In letter 2297, Juliet Pollock wrote, 'Walter has come home, looking well and in capital spirits, and in spite of ill health, & lost time, has taken the first place in the examination for the half year & brought home the prize!'.
3. *mother's*: Juliet Pollock.

From Johann Conrad Bohn 7 September 1865 2302

Giessen d. 7. Sept. 1865

Hochverehrter Herr Tyndall!

Nehmen Sie meinen verbindlichsten Dank entgegen für die Bereitwilligkeit, mit welcher Sie mir die Erlaubniss zur Uebersetzung Ihres Werkes über die Wärme gestatteten. Ich bedaure dass Ihnen Herr Vieweg nicht ganz als Verleger zu passen scheint; hätte ich früher darum gewusst, so würde ich nach einem anderen mich umgesehen haben; ich that erst geraume Zeit nachdem ich meinen ersten Brief an Sie geschrieben hatte Schritte um einen Verleger, denn ich wollte Antwort von Ihnen und allenfallsige Wünsche hinsichtlich dieses Punktes abwarten. Allein da Sie ausser England waren (ich lass in der Zeitung dass Sie in Genf an der Vers. d. N. theilnahmen) so verzögerte sich Ihre Rückäusserung so lange, dass ich voreilig mit dem Buchhändler mich einliess. Ich bin nun nicht mehr frei und es wäre mir unangenehm die Abrede mit Vieweg mit mehr oder weniger Gewalt rückgängig machen zu müssen. Für das <u>Buch</u> selbst ist kein besserer Verleger zu finden als Vieweg, er gibt die beste Ausstattung an Papier, Druck u.s.w. Der Nachtheil eines unvortheilhaften Vertrages trifft also mich allein und ich will ihn eben tragen. Nur werde ich noch Verabredung über die Stärke der Auflage treffen und für eine nothwendig werdende zweite Auflage mir vorbehalten die Bedingungen zu machen; möglich dass dann bei dieser Gelegenheit ich der Verkürzung die ich gegenwärtig zu erleiden habe, wieder beikomme.

Es ist Ihnen bekannt, dass der Uebersetzer immer einige Fragen an den Autor zu stellen hat, und indem Sie so gütig waren die Uebersetzung zu gestatten haben Sie zugleich implicite Ihre Bereitwilligkeit erklärt, bis zu bescheidenem Umfange—die Mühseligkeit der Erledigung solcher Fragen zu übernehmen. Ich komme schon heute mit einer solchen Frage.

Da das englische Maass bei uns wenig gekannt und auch keinerlei Vortheile in sich gewährt, so rechne ich alles auf das französische, welches ja demnächst vollständig für den bürgerlichen Verkehr bei uns eingeführt werden soll, und für wissenschaftliche Dinge bei uns schon lange gilt, um. In (76)–(82) machen Sie die Berechnung des Aeq. d. W. in Mayers Art.

Sie setzen den Druck einer Atmosphäre auf d Quadratzoll = 15 lb. Das gibt 1,05464 Kilogr. auf dem Quadrat Centimetr. Regnault (nach de Relat . . .) gibt als spec. Gew. des Quecksilbers 0°, 13,396, also der Druck des Quad. Centim. bei $0{,}760^{mm}$ (Barometer wird = 13,596 x 0,96 = 1,033296 Kil. merklich verschieden von dem welches Sie annahmen.

Ferner geben Sie als Gewicht eines Cubikfusses Luft 1,29 Unz oder das Liter gleich 1,417053 Grammes, während Regnault 1,29318 Gr. (0°, 760^{mm}.) angibt. Nehme ich an Sie rechneten die Luft unter dem stärkeren Druck 1,054. . während Regn. 1,035. . hat, so würde doch Ihre Luft bei 0° 760^{mm} 1,3202 wiegen. Die Grundlagen Ihrer Rechnung sind also schon von den Regnault'schen Angaben ziemlich verschieden. In der That, mit der genauen Regnault'schen Angabe berechnet sich (wenn ich nicht irrte) das Aeq. zu 416,77 Kilogr M. wahrend Joule's Mittelwerth 423,55 beträgt.—Die in (92) von Ihnen ausgesprochene nahe Uebereinstimmung von dem revidirten Mayer W Aeq. mit dem von Joule finde ich also nach meiner Rechnung *[*nicht bestehen*]*.

Wollen Sie mir gefälligst sagen ob Sie Ihre Angaben beibehalten zu sehen wünschen (und auf welche Autorität hin oder ob diess mit Stillschweigen zu übergehen ist)—oder ob ich meine nach Regnault's Daten geführte Rechnung nehmen darf. §(92) wäre dann etwas abzuschwächen,—ohne dass dadurch ja etwas Wesentliches geändert würde.

Entschuldigen Sie die Belästigung welche ich Ihnen gebe, sie ist aber nothwendige Folge Ihrer vorausgehenden Freundlichkeit oder wie wir sagen: wer A spricht muss auch B sprechen oder gebe dem Teufel den kleinen Finger und er will die Hand und den Arm.

Ich finde viel Vergnügen daran Ihr Buch zu übersetzen, das ist auch ein Theil des Lohnes für meine Mühe, mehr als Herr Vieweg gibt.

Werden wir nicht das Vergnügen haben Sie in Hannover bei der Naturforscher Versammlung zu sehen? Sie beginnt am 18^{n} Spt.

Mit dem herzlichsten Gruss | Ihr dankbar ergebenster | Dr Bohn

Giessen, 7. Sept. 1865

Esteemed Mr Tyndall!

Please accept my most sincere thanks for the willingness with which you granted me the permission to translate your work on heat.[1] I regret that Mr Vieweg[2] does not seem to fully please you as publisher; had I known about this earlier, I would have looked for a different one; I took steps to find a publisher only after considerable time had passed since sending you my first letter, as I wanted to wait for your answer and possible wishes in this matter. However, as you were outside of England (I read in the newspaper that you participated at the meeting of natural scientists in Geneva[3]) your reply[4] was delayed for so long that I rashly made an agreement with the bookseller. I am now no longer at liberty and would feel uncomfortable to have to more or less violently break off the deal with Vieweg. For the <u>book</u> itself, there cannot be a better publisher than Vieweg; he offers the best quality in paper, print, etc. The disadvantage of an unflattering contract, then, concerns only me and I am willing to bear it. I will now only discuss the size of the print run and make sure that I reserve the right to state my conditions in case a second edition becomes necessary; it is possible that on that occasion I may recover the cutback that I presently have to endure.

You know that the translator always has questions to pose to the author, and in being so kind as to authorise the translation, you have at the same time implicitly expressed your willingness, to a modest degree, to assume the labour of settling these questions. Already today I come to you with one such question.

As the English measures are only little known to us and do not provide any advantages, I am converting everything to the French ones, which after all shall soon be fully introduced here for civilian uses, and have long been in use for scientific matters. In (76)–(82)[5] you calculate the equivalent of W[ater]. according to Mayer.[6]

You set the pressure of an atmosphere on a square inch = 15 lb. This gives <u>1.05464</u> kilograms on a square centimetre. Regnault[7] (following de Relat . . . [8]) calculates the specific Weight. of mercury 0°, 13.396, that is, the pressure on a square. centimeter. at 0.760^{mm} (the barometer will = 13.596 x 0,96 = <u>1.033296</u> Kil[ograms]. significantly different from that which you assumed.

Further, you indicate the weight of a cubic foot of air to be 1.29 ounces or the litre equal to <u>1.417053</u> grams, while Regnault states <u>1.29318</u> Gr. (0°, 760^{mm}.). If I assume correctly that you calculated the air under the larger pressure 1.054. . while Regn[ault]. has 1.035. . then your air would weigh <u>1.3202</u> at 0° 760^{mm}. The foundations of your calculations thus are quite different from Regnault's data. In fact, using the exact data of Regnault's one obtains (if I am not mistaken) for the equivalent. 416.77 kilograms M. while Joule's[9] mean value is <u>423.55</u>. According to my calculations, then, I do not find the close correspondence of Mayer's revised equivalent of W[ater] with that of Joule that you mention in (92).[10]

Would you please kindly tell me if you wish to see <u>your</u> values retained (and on which authority or if this is to be passed over tacitly)—or if I may take my calculations that are done in accordance with Regnault's data. §(92) should be then tempered to a slight degree,—without anything fundamental being changed thereby.

Please excuse the annoyance that I am causing you, but it is a necessary consequence of your previous kindness or as we say: he who says A should also say B or give the devil an inch and he will take a mile.

I very much enjoy translating your book; this also is part of the wage for my labour, more than what Mr Vieweg pays.

Will we not have the pleasure to see you in Hannover for the meeting of natural scientists?[11] It begins on Sept 18th.

With the warmest regards | Your gratefully devoted | Dr Bohn

RI MS JT/1/B/107

1. *permission to translate your work on heat*: see letter 2296, n. 6.
2. *Mr Vieweg*: Eduard Vieweg.
3. *conference of natural scientists in Geneva*: see letter 2270, n. 4.
4. *your reply*: letter missing.
5. *(76)–(82)*: referring to sections 76 through 82 in Tyndall's *Heat* (1865), pp. 72–3.
6. *Mayer*: Julius Mayer.
7. *Regnault*: Henri Victor Regnault.
8. *Relat . . .* : H. V. Regnault, *Relation des expériences pour déterminer les principales lois et les données numériques qui entrent dans le calcul des machines à vapeur*, 3 vols (Paris: Didot, 1847–70). Bohn was referencing the first (1847) and second (1862) volumes, as the third was not published until 1870.
9. *Joule's*: James Prescott Joule (1818–89), an English physicist and brewer who studied the relationship between work and heat (*ODNB*).
10. *(92)*: referring to section 92 in Tyndall's *Heat* (1865), pp. 75–6.
11. *in Hannover for the meeting of natural scientists*: the 1865 annual meeting of the Society of German Natural Scientists and Physicians (Versammlung Deutscher Naturforscher und Ärzte) was held in Hanover from 18–24 September.

From Michael Faraday 7 September 1865 2303

Green, | Hampton Court,[1] | 7 Sept. 1865.

My dear Tyndall,

Many thanks for your experiment:[2] it is beautiful. I made it last night, but thinking there might be a little pyrotechny about it, waited until dark before I set it off, and so lost the opportunity of remarking as closely as we might have done the particular facts of the evolution of <u>force</u>.

Having kept the results of the combustion until this morning, I can only refer the phenomena &c. to the power and *<1 word excised>* of Brodie's[3] strange form of *<1 word excised>*[4], or the oxide of carbon, which I have shown once or twice to foreign philosophers. But whatever the arrangement, it is very neatly made.

My wife's[5] letter[6] to you is gone, and I write irrespective of her matter to express to you the pleasure your kindness has given you. If I am right about Brodie's share in the *<1 word excised>,* and you meet him at Birmingham,[7] remember me kindly to him.

I write as you see, confirming[8] myself in every line, but believe me
ever yours, | M. Faraday
Dr. Tyndall, | &c. &c. &c.

Faraday Correspondence, 6:4564
RI MS JT/1/TYP/12/4159

1. *Green, | Hampton Court*: Faraday's house in London since 1858.
2. *your experiment*: referring to his experimental researches on the radiance and absorption of heat.
3. *Brodie's*: Benjamin Collins Brodie (1817–80), second baronet, an English chemist and Aldrichan Chair at Oxford University from 1865–72 (*ODNB*).
4. <1 word excised>: LT wrote in the margin of her typescript that this word might have been 'carbonine'.
5. *My wife's*: Sarah Faraday.
6. *letter*: letter missing.
7. *at Birmingham*: see letter 2289, n. 2.
8. *confirming*: LT wrote in the margin of her typescript that this word might also have been 'confessing'.

From Johann Conrad Bohn 9 September 1865 2304

Giessen 9 Sept. 1865

Hochverehrter Herr Tyndall!

Ihr soeben mir zugekommener Brief vom 7^{n} Sept. zeigt eine freundliche Fürsorge für mein Privatinteresse, insoweit es den Lohn für die begonnene Uebersetzung Ihres Buches betrifft, für welche ich Ihnen zu ergebenstem Dank verpflichtet bin.—Gestatten Sie mir Ihnen den Verlauf der Unterhandlung mit dem Verleger darzustellen und daran meine Ansicht über die Lage der Dinge zu knüpfen.

Beim Lesen Ihres Buches wurde in mir der Wunsch erregt dasselbe zu übersetzen, theils weil es eine angenehme Arbeit zu sein versprach, die mir obendrein noch Geld einbringt, theils weil ich es nur für Vortheil halte, wenn ein solch' ausgezeichnetes Werk auch in Deutschland eine grössere Verbreitung findet. Ich schrieb nun (im ersten Drittel August's) an Sie jenen Brief, der noch nicht in Ihre Hand gekommen zu sein scheint. Bald nachher erfuhr H. v. Liebig, der mir sehr freundlich gesinnt ist von meinem Vorhaben und erbot sich gegen einen Verwandten von mir er wolle mir für das Buch den Verleger alsbald schaffen. Liebig schrieb nun an seinen Freund Vieweg und bezeichnete als Minimum des Honorars 2 Louis d'ors p. Bogen (1⅚ Liv. Sterl.), welche Summe ich gegen meinen Verwandten als jene bezeichnet hatte, welche ich erwarten zu können glaube, wenn meine Absicht zur Ausführung käme. Herr von Liebig schickte mir 2 Briefe die ihm Vieweg geschrieben. Der <erste> erklärt mit Vergnügen nehme V. das Buch zum Verlag, der zweite sagt: „<Es> handle sich wohl um eine Bearbeitung des engl. Originals, nicht um eine blosse Uebersetzung, denn sonst wäre das verlangte Honorar doppelt so gross als es gewöhnlich gegeben werde.<"> Liebig rieth mir bei Zusendung dieser Briefe eine Bearbeitung zu machen. Hierauf trat ich selbst mit Vieweg in Verkehr. Es erfolgte einiges Hin- und Herschreiben über die Behandlungsweise des Buches; ich erklärte mich bereit zu einer blossen Uebersetzung, aber auch, und zwar eigentlich noch lieber, zu einer Bearbeitung. Nur meinte ich die Bearbeitung könne nicht der Art gemacht werden wie H. Vieweg zu denken scheine. Meiner Ansicht nach müssten Ihre Vorlesungen möglichst genau ohne Zusatz und ohne Weglassung übersetzt werden, denn das seien Kunstwerke und alles was ein Anderer dazu oder davon thäte würde diese Kunstwerke verletzen. Allein so wie Sie selbst in Anhängen (Appendix) noch weitere Erörterungen einzelner Fragmente gegeben hätten, so scheine mir dass in Zusätzen welche nach Ihren Anträgen kämen für den Uebersetzer als Bearbeiter noch Raum und Gelegenheit bliebe. Da liesse sich zunächst das Einschlagende jener Arbeiten mittheilen, welche seit Abfassung Ihres Buches veröffentlicht wurden, da liesse sich etwas über die Literatur beibringen (was in Deutschland meist gewünscht wird), da liesse sich angeben was wissenschaftlichen Werth oder einen solchen für die Geschichte der Wissenschaften, von den Meinungen Ihrer Gegner habe; Sie seien über manche der im Buche behandelten Fragen in wissenschaftlichem Streit und es erschiene mir ganz angemessen einen Bericht über die Lage der Streiter zu geben (ich dachte dabei z.B. an Ihre Discussionen mit Magnus, an die Ansichten über den Leidenfrostischen Versuch die anders sind als die Ihren etc.), endlich könnte in jenen Zusätzen auch einiges <weniges>, was ich selbst über die betreffenden Fragen denke Platz finden.

Vieweg antwortete hierauf er zöge eine blosse Uebersetzung vor, namentlich auch glaube er nicht dass Ihre Zustimmung zu einer Beobachtung erlangt werden könne, welche gelegentlich in Widerspruch mit Ihren wissenschaftlichen Ansichten tritt. Für die Uebersetzung bot er dann 8 Thlr = 1⅙ Liv. Sterl für den Bogen in der Grösse u. Ausstattung von Müllers Physik<.> Er führte an, dass er das gleiche Honorar zahle für die Bearbeitung von Percy's Metallurgie und von Murchisons Fieber und typhoi Krankheiten.—Ich besitze keine Erfahrung im Verkehre mit Verlegern und *[1 word illeg]* H. Vieweg, der mir von so guter Seite genannt war für einen *[3 words illeg]*. Ich schrieb ihm also, im guten Glauben dass er mich nicht unbillig behandeln wolle, stimme ich seinem Vorschlag zu und ersuche ihn nur zur Formulirung eines Vertrages. Hierauf antwortete er 8 Th sei in der That alles was er geben könne und motivirt diess durch einen voraussichtlich nicht grossen Absatz u.s.w. Ein Vertrag sei nicht nöthig; die gewechselten Briefe böten Anhaltspunkt genug. Ich solle nur trachten Sie irgendwo aufzufinden und Ihre Zustimmung zu erlangen.

Aus dem Vorstehenden können Sie nun ersehen, wie viel ich H. Vieweg gegenüber gebunden bin.

Was die Ausstattung und den Vertrieb des Buches angeht, so kann es meiner Meinung nach in keinen besseren Händen sein, als Viewegs. Nur zwei oder drei Verleger in Deutschland kommen ihm darin nahe (Trübner—Fischer, Cassel). Dieser Umstand gäbe also einen Grund ab, dem Buche zu wünschen dass es bei Vieweg erscheine. Zum anderen würde es mir wegen Herrn von Liebig, der in dieser Angelegenheit wie auch sonst sehr gütig für mich war, nicht angenehm sein, ohne ganz genügenden Grund einen anderen Verleger, als dem von ihm proponirten den Vorzug zu geben.—Allein die ganzen Verhandlungen sind keineswegs in unabänderlicher Weise geschlossen.

Sie haben sich das Recht der Uebersetzung in fremde Sprachen gewahrt und damit auch das Recht an die Ertheilung Ihrer Erlaubniss Bedingungen zu knüpfen, wie es Ihnen gefällt. Wenn Sie auf einem bestimmten Verleger bestehen, oder auf einem bestimmten Honorar für den Uebersetzer, so ist meine Verabredung mit H. Vieweg natürlich hinfällig, da die unter der Voraussetzung Ihrer Zustimmung geschah, ja Herr Vieweg noch ganz besonders hervorhob, dass wenn diese Ihre Zustimmung nicht zu erlangen wäre, oder eine andere deutsche Uebersetzung von Ihnen bereits autorisirt sei, er (Vieweg) mir gegenüber keine Verbindlichkeit habe.

Meine Wünsche hinsichtlich des Buches betreffen zunächst ein besseres Honorar für meine Arbeit, dann auch die Art der Ausstattung. Mein Geschmack (eine logische Begründung desselben ist nicht möglich) würde die Ausstattung ähnlich dem englischen Originale oder dem Chemische Briefe von Liebig (Winter's Verlag) jenem grossen Formate von Müllers Physik vorziehen.

Wollen Sie nun, verehrter Herr Tyndall, in dieser Angelegenheit weiter zu thun, so scheint mir folgender Vorschlag annehmbar:

Sie schreiben mir einen Brief, den ich dem Herrn Vieweg vorlegen kann. In demselben ertheilen Sie die Erlaubniss zur Herausgabe einer von mir gemachten Uebersetzung und knüpfen aber an diese Erlaubniss jene

Bedingungen, welche Ihnen passend erscheinen. Also hinsichtlich der Ausstattung des Buches, der Stärke der Auflage, des Maximalpreises für ein Exemplar, hinsichtlich des Honorares für den Uebersetzer u.s.w. Kann und will nun Herr Vieweg auf Ihre Bedingungen eingehen, so wird mich diess freuen und ein vortheilhafter Vertrag zu Stande kommen, vortheilhaft für das Erscheinen und die Ausstattung und Vertreibung des Buchs, (und wie ich wiederhole liegt mir diess so sehr am Herzen wie mein persönlicher Vortheil, und von diesem Gedanken geleitet gab ich so schnell meine Zustimmung zu H. Viewegs Anerbieten) und für mich als Uebersetzer. Will oder kann H. Vieweg diess nicht thun, dann bin ich aller Verbindlichkeit gegen ihn ledig und eben so bin ich dann auch Herrn von Liebig gegenüber vollkommen gerechtfertigt. Alsdann kann die Unterhandlung mit Theod. Fischer in Cassel oder sonst einem Druckhändler frei beginnen. Nicht unerwähnt will ich lassen, dass ich den Buchhändler Ricker in Giessen frug ob er eventuell gesonnen sei eine deutsche Uebersetzung Ihres Buches zu verlegen, dass dieser aber nach einigen Tagen Bedenkzeit es ablehnte, da er in Deutschland kein genügendes Publikum glaubte.

Ich finde es ganz besonders liebenswürdig von Ihnen, dass Sie auch auf den Vortheil Ihres Uebersetzers bedacht sind und nur Ihr erster Schritt in dieser Richtung hat mich ermuthigt Sie in vorstehendem mit diesen materiellen Kleinlichkeiten und Erbärmlichkeiten zu belästigen, welche allerdings leider nur von zu grossem Einfluss auf unser Leben sind. Vieweg ist gegenwärtig irgendwo in einem Bade. Es *[gilt]* also nicht sehr mit ihm zu verhandeln.

Darf ich Sie bitten mich Herrn Dr. Debus freundlichst in Erinnerung zu bringen. Er war mein erster Lehrer in der Chemie in Marburg 1849 und unser Tischgenosse in *[Rappin]*.

Nachträglich zu meinem gestrigen Briefe bemerke ich dass nach den Regnault'schen Zahlen das Aeq. sich zu 416,. . berechnet wenn man, wie Sie es in ihrem Buche Herrn Dulong's Werth <für> K = 1,421 nimmt. Eine abermalige Rechnung ergab mir für

Druck auf [1er Cant.] = 1,033296 Kilogr.
Gew.von 1 liter Luft (0°,760mm) = 1gr,29318 } Regnault's exacte Angaben

$\alpha = 0{,}003665$

Spec. Warme d. Luft (zw 0 u 200°) 0,2375

Und

K = 1,41 (nach der Schallgeschwindigkeit etc. . . .)

das Aequivalent = 424,045

So würde es dann mit H. Joule's Mittelwerth 423,55 sehr nahe stimmen und wenn Sie zustimmen, dass diese Rechnungdaten in der Uebersetzung benützt werden, so könnte §92 ganz unverändert bleiben.

Mit den herzlichsten Grüssen verbleibe ich Ihr dankbar | ergebenster | Bohn

Giessen 9 Sept. 1865

Esteemed Mr Tyndall!

Your letter from Sept. 7th,[1] which has just reached me, shows a kind consideration for my private interests in as far as the wage for the already begun translation of your book[2] is concerned. I am sincerely grateful to you for this.—Allow me to describe to you the course of the negotiations with the publisher[3] and following this, to add my views on the state of affairs.

While reading your book, I felt the wish to translate it, partly because this promised to be a pleasant occupation that moreover would bring in some money for me, partly because I can only see it as advantageous if such an outstanding work found wider diffusion also in Germany. Thus, I wrote to you (in the first third of August) the letter that apparently has not yet reached you.[4] Soon afterwards, Mr v. Liebig,[5] who is kindly disposed towards me, heard about my plan and made the offer to a relative of mine that he would directly procure a publisher for the book. Liebig wrote to his friend Vieweg[6] and specified 2 Louis d'ors per sheet (1⅚ Lb. Sterling) as a minimum wage, an amount that I had mentioned to my relative as what I thought I might expect if my intentions were executed. Mr von Liebig sent me 2 letters that Vieweg had written to him. The first one states that V[ieweg]. would be pleased to accept the book for publication, the second one says: "presumably it is an <u>edition</u> of the Engl[ish]. original, not a mere <u>translation</u>; otherwise, the requested royalties would be twice as high as what is customarily given.["] When sending the letters, Liebig advised me to do an <u>edition</u>. Following this, I contacted Vieweg myself. There followed some writing back and forth about the manner of treatment for the book; I expressed my willingness to do a mere translation but also—and in fact preferably—to do an edition. I only opined that the edition could not be done in the way that Mr Vieweg seemed to think. That in my opinion, your lectures should be translated as accurately as possible, without addition or omission, as they are works of art and anything that someone else added or took away would violate these works of art. That only in the manner that you yourself would give additional explanations of individual passages, in <u>appendices</u>—it appeared to me—would there be room and opportunity for the translator, as editor, to provide <u>additions</u> as proposed by you. That here, one would have the opportunity to present the key points of those works that were published since the composition of your book; one could add something about the literature (something that is often desired in Germany); one could indicate what—of the opinions of your opponents—is of value for science or the history of the sciences, that you are in scientific dispute about some of the questions treated in the book,[7] and that I would deem it quite appropriate to report on the positions of opponents (I thought, for instance, about your discussions with Magnus,[8] about his views on the Leidenfrost[9] experiment[10] which differ from your own, etc.); finally, that there might be space in these additions for a few of my own thoughts on the respective questions.

To this, Vieweg answered that he would prefer a mere translation and that he did not think that your consent could be obtained to include observations that occasionally contradicted your scientific views. He then offered 8 Thaler[11] = 1⅙ Lb. Sterling per sheet for the translation, with the dimensions and features of Müller's[12] Physik.[13] He added that he paid the same wage for the edition of Percy's[14] Metallurgie[15] and of Murchison's[16] Fever and Typhoid Diseases.[17]—I do not have experience in dealing with publishers and *[1 word illeg]* Mr Vieweg, who has been recommended to me by a reliable party as a *[3 words illeg]*. Thus, I wrote to him that in good faith that he did not wish to treat me unfairly I accepted his proposal and requested him to draw up a contract. To this, he answered that 8 Thalers was indeed all he could give and justified this with the likelihood of low sales, etc. A contract would not be necessary, our correspondence offering sufficient points of reference. I should only strive to locate you and obtain your approval.

From the above, you can see how much I am bound to Mr Vieweg.

Concerning the quality and distribution of the book, in my opinion it could not be in better hands than Vieweg's. In these aspects, only two or three publishers in Germany come close to him (Trübner[18]—Fischer, Cassel[19]). This circumstance would be one reason to wish the book that it be published with Vieweg. Additionally, it would be uncomfortable for me because of Mr von Liebig, who has been very kind to me in this matter as well as in general, to give preference to a publisher other than the one that he proposed, without sufficient reason.—However, the negotiations are by no means unalterably concluded.

You have retained the right for translation into foreign languages and thereby also the right to attach whatever conditions you please to the granting of your approval. If you insist on a particular publisher or on a specific wage for the translator, then my arrangement with Mr Vieweg is naturally invalid, since it happened on the condition of your approval and Mr Vieweg stressed very particularly that if your approval could not be obtained, or if you had already authorised another German translation, he (Vieweg) would have no obligations towards me.

My wishes in relation to the book concern firstly a better wage for my work, and then also the type of execution. My taste (logical reasoning for which is not possible) would prefer features similar to the English original or Liebig's Chemische Briefe (Winter's Publishers)[20] rather than the large format of Müller's Physics.

If you, dear Mr Tyndall, would now like to proceed in this matter, I consider the following proposal to be acceptable:

You send me a letter that I can present to Mr Vieweg. In it, you give the authorisation to publish a translation made by me and you attach to this authorisation those conditions that seem appropriate to you. That is, with regards to the quality of the book, the number of copies, the maximum price per copy, the wage for the translator, etc. If Mr Vieweg can and wants to accept your conditions, I will be glad and a favourable contract will be made; favourable for the publication and the features and distribution of the book (and I repeat that this is as close to my heart as my personal benefit, and guided by this thought I quickly gave my approval to Mr Vieweg's proposal) and for me as translator. If Mr Vieweg cannot or does not want to do this, I will be free from all obligations towards him and at the same time, I will also be completely justified against Mr von Liebig. I may then begin negotiations with Theod. Fischer in Cassel or with any other publisher. I do not want to neglect mentioning that I asked the bookseller Ricker in Giessen[21] if he might potentially be disposed to publish a German translation of your book, but that after a few days' reflection he declined because he did not believe that it would find a sufficient readership in Germany.

I find it very kind of you that you also care for the welfare of your translator and it is because of your first step in this direction that I felt motivated to importune you in the above with these trivialities and this paltriness, which, however, unfortunately have too great an influence on our lives. At present Vieweg is at a spa somewhere. Thus, it is not really an option to negotiate with him.

May I please ask you to greet Dr Debus[22] from me. He was my first chemistry teacher in Marburg[23] in 1849, and a table companion in *[*Rappin*]*.[24]

As an addendum to my letter from yesterday,[25] I observe that according to Regnault's[26] numbers[27] the equivalent is calculated as 416. . . . if one takes Mr Dulong's[28] value for K = 1.421 as you do in your book. A repeated calculation I did produced for

Pressure on [1er Cant.]=1.033296 Kilogram
Weight of 1 liter air (0°,760mm)=1gr.29318 } Regnault's exacte Angaben

$\alpha = 0.003665$

Specific heat of air (between 0 and 200°) 0.2374

And

K= 1.41 (according to the speed of sound etc. . . .)

the equivalent = 424.045

This would coincide largely with Mr. Joule's[29] mean value of 423.55 and if you approve that these numerical data be used in the translation, then §92[30] could remain unchanged.

With the kindest greetings I remain your | most devoted | Bohn

RI MS JT/1/B/108

1. *Your letter from Sept. 7th*: letter missing.
2. *translation of your book*: see letter 2296, n. 6.
3. *publisher*: Eduard Vieweg.
4. *I wrote to you (in the first third of August) the letter that apparently has not yet reached you*: letter 2296.
5. *Mr v. Liebig*: Justus Liebig (1803–73), the noted German chemist and educator. He helped found the field of organic chemistry and made substantial contributions to agricultural

chemistry, plant and animal physiology and the chemistry of foodstuffs, as well as helping to introduce the laboratory-based teaching method (*ADB*).

6. *Vieweg*: Eduard Vieweg.
7. *scientific dispute about some of the questions treated in the book*: probably a reference to Tyndall's dispute with James Joule, William Thomson, and Peter Tait regarding what he believed to be the German physicist Julius Mayer's rightful claim to priority for the discovery of the mechanical equivalent of heat, as well as his dispute with C. K. Akin over priority of discoveries in relation to heat. On these, see the introductions to *Tyndall Correspondence*, vol. 8, and this volume.
8. *Magnus*: Heinrich Magnus.
9. *Leidenfrost*: Johann Gottlob Leidenfrost (1715–94), a German doctor (*ADB*).
10. *experiment*: the Leidenforst effect, as it is now known, is the 'phenomenon observed by Leidenfrost, in which liquid, when placed in contact with a surface whose temperature exceeds some critical value (the Leidenfrost point), forms a layer of vapour which insulates it from the surface' (*OED*).
11. *Thaler*: German silver coins, probably referring to the Vereinsthaler (see this volume's Note on Money).
12. *Muller's*: Johann Heinrich Jakob Müller (1809–75), a German physicist, mathematician, and professor at the University of Freiburg whose research interests included optics and radiant heat (*CDSB*).
13. *Physik*: J. Müller, *Pouillet's Lehrbuch der Physik und Meteorologie*, 2 vols (Braunschweig: F. Vieweg und Sohn, 1842–3). Published in English as: J. Müller, *Principles of Physics and Meteorology* (London: Hippolyte Bailliere, 1847).
14. *Percy's*: John Percy (1817–89), an English metallurgist (*ODNB*).
15. *Metallurgie*: Percy's *Metallurgy: the art of extracting metals from their ores, and adapting them to various purposes of manufacture* (London: John Murray, 1861) was translated into German and published as: J. Percy, *Die Metallurgie: Gewinnung und Verarbeitung der Metalle und ihrer Legirungen, in praktischer und theoretischer, besonders chemischer Beziehung* (Braunschweig, F. Vieweg und Sohn, 1863).
16. *Murchison's*: Charles Murchison (1830–79), a British physician (*ODNB*).
17. *Fever and Typhoid Diseases*: Murchison's *A Treatise on the Continued Fevers of Great Britain* (London: Parker, Son, and Bourne, 1862) was translated into German and published as: C. Murchison, *Die Typhoiden Krankheiten: Flecktyphus, Recurrirender Typhus, Ileotyphus, und Febricula* (Braunschweig: F. Vieweg und Sohn, 1867).
18. *Trübner*: Nikolaus Trübner (1817–84), a German publisher and bookseller (*ADB*).
19. *Fischer, Cassel*: Theodore Fischer.
20. *Chemische Briefe (Winter's Publishers)*: J. Liebig, *Chemische Briefe* (Heidelberg: C. F. Winter, 1851).
21. *Ricker in Giessen*: see letter 2265, n. 6.
22. *Dr Debus*: Heinrich Debus.
23. *my first chemistry teacher in Marburg*: Bohn was at the University of Marburg from 1850–51.

24. *[Rappin]*: possibly Rappin, a town in northeastern Germany.
25. *my letter from yesterday*: probably letter 2302, dated 7 September 1865 (although 'yesterday' would have been 8 September).
26. *Regnault's*: Henri Victor Regnault.
27. *numbers*: see letter 2302, n. 8.
28. *Mr Dulong's*: Pierre Louis Dulong (1785–1838), a French physicist known for his study of the elasticity of steam and conduction of heat. He is especially noted for his research with Alexis Thérèse Petit, which resulted in the Law of Dulong and Petit (*CDSB*).
29. *Mr. Joule's*: see letter 2302, n. 9.
30. *§92*: referring to section 92 in Tyndall's *Heat* (1865), pp. 75–6.

From Pieter van der Burg[1] 20 September 1865 2305

Nimègue 20 Sept. 1865

Monsieur!

Dans cette ville, qui compte 24 000 habitants, il y a une société de physique, laquelle a déjà 25 ans d'existence. Depuis 16 ans j'ai eu la conduite des séances, et chaque hiver j'ai parlé dans huit séances, en faisant un cours suivi et réglé de quelque sujet.

En lisant vos 12 leçons „<u>La chaleur considérée comme un mode de mouvement</u>" etc. je fus saisi d'une admiration qui allait toujours croissant. Je me suis maintenant proposé de les traiter et des les expliquer cet hiver, et de faire en même tems, autant que possible, toutes les épreuves, avec lesquelles vous avez expliqué, d'une manière si excellente, la nature de la chaleur à vos auditeurs.

Pour atteindre à ce but il me faut cependant une pile thermo-électrique et un galvanomètre, qui aient l'un et l'autre la même sensibilité délicate que ceux que vous avez employés vous-même.

Que l'appareil soit beau, c'est ce que je ne prétends pas: mais j'ignore absolument quels sont les frais de la structure, et s'ils ne surpasseront pas la somme, dont notre société peut disposer, vu qu'elle n'est nullement bien dotée.

J'ose donc vous demander humblement, si vous voulez avoir la bonté de me faire faire ces deux instruments, et de m'écrire quel en serait le prix.

Il est vrai, que je ne suis point connu de vous; mais je m'empresse de vous informer, que je vous ai toujours suivi avec grand intérêt dans toutes vos recherches scientifiques: les Annales de Poggendorff me fournissaient toujours les matériaux nécessaires, et c'est avec des sentiments d'une reconnaissance profonde, que je vous avoue d'avoir appris de vous bien des choses intéressantes.

Veuillez aussi recevoir mes félicitations de ce que vous venez d'avoir eu le dessus dans le combat dernier.

Moi je n'ai pas fait grand chose pour les sciences: je n'ai écrit que quelques ouvrages populaires. Un de ceux-ci Eléments de la Physique a déjà eu 4 éditions, et environ 11 000 exemplaires en ont été répandus dans notre pays.

Agréez Monsieur l'assurance de mon sincère estime.

votre très humble serviteur | P. van der Burg

Professeur de mathématiques et de Physiques à l'Ecole supérieure de Nimègue (Pays-Bas)

Membre de la société des arts et des sciences d'Utrecht.

Nijmegen[2] 20 Sept. 1865

Sir!

In this city, which has 24,000 inhabitants, there is a physics society[3] that has been running for 25 years. I have been leading sessions there for sixteen years, and each winter I have spoken at eight sessions, giving an organised lecture on some subject.

In reading your 12 lessons "Heat considered as a mode of motion" etc.[4] I was overcome by an ever-growing admiration. I have now offered to lecture on your lessons this winter, and at the same time, as much as possible, to do all the experiments that you performed to explain the nature of heat to your audience in such an excellent way.

However, in order to do so I need a thermo-electrical battery and a galvanometer that have the same sensitivity as the ones you used yourself.[5]

That the device should be beautiful does not concern me: but I would have it *[fully functional]*, no matter the cost, so long as it is within the means of our society, as it is not al all well-funded.

And so I dare to ask you, humbly, if you would be kind enough to have these two instruments made for me and to write to tell me what the price would be.

It is true that you do not know me; however, I have always followed your scientific research with great interest: Poggendorf's Annals[6] always published the necessary materials, and it is with feelings of profound gratitude, that I confess that I have learned from you many interesting things.

Also let me congratulate you for having come out on top in your last confrontation.[7]

I have not done much for the sciences; I have only written a few popular books.[8] One of those, Eléments de la Physique,[9] has already had 4 editions, and roughly 11,000 copies were sold in our country.

Please accept, dear Sir, the assurance of my sincere esteem.

Your very humble servant | P. van der Burg

Professor of mathematics and physics at the School of Nimègue (Netherlands)

Member of the Utrecht society for arts and science

RI MS JT/1/V/2

1. *Pieter van der Burg*: Pieter van der Burg (1808–89), a Dutch teacher at various schools until he joined the gymnasium (l'Ecole supérieure) in Nijmegen in 1850. Van der Burg also wrote popular science textbooks (E. Zuidema, 'Burg (Pieter van der)', in P. C. Molhuysen and P. J. Blok (eds), *Nieuw Nederlandsch Biografisch Woordenboek*, 10 vols (Leiden: A. W. Sijthoff's Uitgevers-Maatschappij, 1911–37), vol. 3 (1914), pp. 181–2).
2. *Nijmegen*: a city in the Netherlands (Nimègue in Dutch).
3. *physics society*: in the biographical entry cited in n. 1, it is stated that van der Burg was very involved with a scientific society in Nijmegen called 'Physica'.
4. *Heat considered as a mode of motion*: *Heat* (1863).
5. *thermo-electrical battery . . . you used yourself*: a thermoelectric battery converts heat into chemical energy to charge, and when discharged it produces electricity. A galvanometer is an electromechanical instrument used for detecting and indicating an electric current. Tyndall discussed both of these devices in the first lecture (*Heat* (1863), pp. 13–36).
6. *Poggendorff's Annals*: *Poggendorff's Annal.*, or *Annalen der Physik und Chemie*, was a German scientific journal edited by the physicist Johann Christian Poggendorf (1796–1877) between 1824–76. Many of Tyndall's scientific papers were translated and published in the journal.
7. *last confrontation*: probably a reference to Tyndall's dispute over priority of discoveries in relation to heat with C. K. Akin (see the introductions to *Tyndall Correspondence*, vol. 8, and this volume).
8. *a few popular books*: including *Belangrijke onderwerpen uit de natuur-en werktuigkunde: een leesboek voor de scholen*, a book on physics and mechanics for schools which went through many editions, and a translation of a physics textbook by the French publisher Adolphe Ganot (1804–87), *Leerboek der proefondervindelijke en toegepaste natuurkunde* (Gouda: G. B. van Goor, 1859).
9. *Eléments de la Physique*: probably P. van der Burg, *Eerste grondbeginselen der natuurkunde* (Gouda: G. B. van Goor, 1854).

To Juliet Pollock 21 September 1865 2306

Royal Institution of Great Britain | 21st. Sept. 1865

My dear Mrs. Pollock

After wandering from London to Geneva and from Geneva to Bale[1] your excellent letter[2] reached my hands this morning. I have been assured by the "Direction des Postes"[3] that I myself gave directions to have my letters forwarded from Geneva to Bale: which statement I can only explain by the fact that another John Tyndall[4] was in the neighbourhood at the time & that he unconsciously in having his own letters re-directed had mine also.

Let me say that I use the word "excellent" advisedly: and let me also add that I shall not deliver you up to the Inquisition.[5]

I write this in the hope of its finding you at Greta Bank[6]—Kindest greetings to all my friends there.

Yours ever | John Tyndall

RI MS JT/1/T/1143
RI MS JT/1/TYP/6/2046

---

1. *from London to Geneva and from Geneva to Bale*: Tyndall and Thomas Hirst left London for the Alps on 10 July 1865. After some climbing and waiting around for several weeks to attempt to recover a body from the Matterhorn (see letter 2316), he visited Auguste de la Rive in Geneva and attended there the meeting of the Société Helvetique des Sciences Naturelles (Swiss Society of Natural Sciences, now Swiss Academy of Natural Sciences, founded in 1815) from 21–3 August 1865. His purpose in Basel (*Bale* in French) has not been identified. Tyndall and Hirst were back in London by 26 August.
2. *your excellent letter*: letter 2297.
3. *"Direction des Postes"*: postal management (French).
4. *another John Tyndall*: not identified.
5. *not deliver you up to the Inquisition*: probably a reference to Pollock's discussion in letter 2297 regarding reading William Edward Hartpool Lecky's *History of the Rise and Influence of the Spirit of Rationalism in Europe*, 2 vols (London: Longman, Green, Longman, Roberts, & Green, 1865), and her concerns about theology, creeds, and dogma.
6. *Greta Bank*: an estate located in the town of Keswick in the county of Cumberland in northwestern England, a Spedding family home. The Pollocks were good friends with James Spedding (1808–81), an English literary editor and biographer known especially for editing the works of Francis Bacon (*ODNB*). Juliet Pollock remarked in letter 2297 that 'next Saturday we all go to St. Julians together. Afterwards my Husband and self and Moss visit the Romillys & the Speddings'.

From Juliet Pollock 24 September 1865 2307

Rydal Lodge | Ambleside[1] | Sept[embe]r. 24.th 1865

My dear Mr. Tyndall

I am glad the letter[2] came into your hands at last for it is an uneasy feeling to have a scrap of one's own soul wandering about, drifting one knows not whither: unfriended. Yours[3] was forwarded to me by Sarah Spedding[4] for we are not yet at G. B.[5] and not due there till the 9th October. We are capitally lodged here with a roomy comfortable house which we have taken the whole of, though larger than we want, to avoid any unpleasant intrusion. It was the parsonage house, and retains more of that than of the ordinary furnished apartments character. How much I wish you could have come to stay with us.

How pleasant it would have been to ramble and scramble with you up these interesting mountain *[sides]*, or to row with you on Rydal Water?[6]

—Who is the man with the poker in Punch?[7]—

We expect Fred[8] home—to this mountain home of ours, this evening—fresh from Switzerland, will he think little of our beauties that we think so much of? this is the size of my sketch book[9] on which I note down remembrances of favourite scenes, and over leaf, I give you the size of my colour box[10] is not this a pretty case of packing, with eight colours & a palette in it?—

Walter[11] is with us till next Thursday. His holiday has been pleasant to him & he has been tolerably strong. Moss[12] was the delight of the Romillys at Porthkerry[13] and a source of general entertainment at the Windermere Hotel.[14] One day we found him on our return from walking, established outside the porch with a fat old traveller, discussing all his experiences of farming &c with him, and presently "Boots"[15] appeared with a newspaper (the Manchester Guardian) which Mr. Moss had ordered for himself!—meaning to sit and read it outside his Inn, like the other loungers. It was rather a blow to him that we should request him to come in to tea with Polly,[16] and a serious grievance that we prohibited such proceedings for the future. The air with which he pointed out to me an article that he thought very good upon Austria & Prussia was first rate—a bit of high comedy.

And now Mr. Tyndall—before I take leave of you, I hope you are well. I thought you a little depressed when we last met, and I do not like to think you so..

I hope you are happy, & that your gases are behaving as they ought,[17] and that you have no more trouble on your mind than the strain put upon it by the investigation of Nature & her laws and her seeming caprices. I thought of you at Porthkerry when the moon came out and swelled with liberal largess on the sea a broad full smile—she is fond of that white bearded old Neptune I am sure, & has little right to reproach you for your inconstancies.

Good bye—I shall like to hear from you when you have leisure to write and I am ever

Most truly yours | Juliet Pollock

RI MS JT/1/P/173

1. *Rydal Lodge | Ambleside*: Rydal Lodge was a country home situated in the village of Rydal near Ambleside in the Lake District of northwest England, now operated as a bed and breakfast. Juliet Pollock's husband William Frederick recorded for 1865: 'In the autumn we spent some pleasant days with the Edward Romillys, at their beautiful place of Porthkerry on the Bristol Channel, and were afterwards at Rydal Lodge, taken by us for some time' (*Remembrances*, vol. 2, pp. 138–9).

2. *letter*: letter 2297.
3. *Yours*: letter 2306.
4. *Sarah Spedding*: possibly the mother of James Spedding (see letter 2297, n. 13); or an unmarried sister of Thomas Story Spedding (1800–70), a Cumberland landowner and elder brother of James.
5. *G. B.*: Greta Bank (see letter 2306, n. 6).
6. *Rydal Water*: a small body of water near Ambleside in the Lake District of northwest England.
7. *Who is the man with the poker in Punch?*: Pollock was referring to a caricature published in the 23 September 1865 issue of the satirical magazine *Punch* (on p. 113) that was probably one Tyndall's earliest appearances in its pages. The caricature, given as the frontispiece to this volume, depicts a variety of members of the BAAS, which just held its thirty-fifth annual meeting in Birmingham from 6–13 September, demonstrating their scientific practice in humorous fashion, including Thomas Huxley, the comparative anatomist Richard Owen (1804–92), the mathematician Charles Babbage (1791–1871), Charles Lyell, the geologist George Scrope (1797–1876), David Brewster, and Roderick Impey Murchison. The 'man with a poker' is Tyndall himself, pricking Joseph Beete Jukes (1811–69) with the hot poker (a reference to their dispute over coal measures), with the text below stating that Jukes 'said that he had got something to say on coals. He proceeded to make a statement', while Tyndall 'offered to haul MR. JUKES over the coals for that remark'. Jukes was a British geologist known primarily for his work aboard HMS *Fly* from 1842–6. He published, among other works, *Narrative of the Surveying Voyage of H.M.S. Fly*, 2 vols (London: T. & W. Boone, 1847) (*ODNB*). At the BAAS meeting in Birmingham, on 8 September, Jukes had delivered an evening lecture, 'Probabilities as to the Position and Extent of the Coal-Measures beneath the Red Rocks of the Midland Counties', to which Tyndall must have had some disagreements.
8. *Fred*: Frederick Pollock.
9. *this is the size of my sketch book*: Juliet may have been indicating that her sketch book was the same size as her stationary.
10. *colour box*: a portable box for holding an artist's paints, brushes, etc. (*OED*). Juliet indeed drew a box in the top left of the backside of the letter, and inside it wrote the words 'is not this a pretty case of packing, with eight colours & a palette in it?—'.
11. *Walter*: Walter Pollock.
12. *Moss*: Maurice Pollock.
13. *Romillys at Porthkerry*: see letter 2297, n. 12.
14. *Windermere Hotel*: a hotel in Windermere, Lake District.
15. *"Boots"*: possibly the hotel boot boy.
16. *Polly*: see letter 2297, n. 20.
17. *your gases are behaving as they ought*: referring to his experimental researches on the radiance and absorption of heat.

To Juliet Pollock [28 September 1865][1] 2308

My dear M^rs^. Pollock

I wish I could join you—so strong indeed was this wish on receiving your letter[2] that I pondered for ten minutes about it. But alas it cannot be. I have now broken into my work, and it holds me captive.

Whether it is that I had not my due amount of exercise in Switzerland,[3] or that this hot weather in London is relaxing I do not find my vigour up to high water mark. This also caused me to think of Fairfield and Helvellyn.[4] Beyond this paltry lassitude I am all right, and I hope to shake it off in due time.

I spent three hours with Carlyle on Sunday evening. He was asleep when I called, and I was kindly and cordially received by his wife.[5] She had him stirred up, and before sitting down to tea we had ten minutes walk in the fresh air. We afterwards talked for two hours, and then he led me through a thousand intricacies—apparently through all the slums and lanes of his part of Babylon,[6] till finally we emerged on the Brompton Road.[7] He placed me safely at Hyde Park Corner,[8] and then turned homewards.

I have never seen him better, full of fire, fit to ignite the tinder of a soul so dry as mine. The intellectual power of the man even when speaking on a purely scientific subject is enormous. He grasped what I had to say before it was half uttered.

Only imagine my book[9] being called a "truly harrowing book"—I send you the scrap where the expression occurs: it is from a very nice review in the Spectator.[10]

I spent a day and a half pleasantly at Birmingham,[11] and went thence to Wrottesley.[12] There I played croquet, and was very kindly treated. I met a most charming lady there, the wife of M^r^. Arthur Wrottesley,[13] bright, cheerful, & winning—almost fascinating. Grove[14] was there at the time. He is to be the president of the British Association next year.

I spent an evening at Hampton Court[15] some time ago. Our friends were all well, still that expression of weariness hangs about M^r^. Faraday—I fear it will not be easily removed.

Yours ever my friend | John Tyndall

RI MS JT/1/T/1172
RI MS JT/1/TYP/6/2053

---

1. *[28 September 1865]*: the handwritten and typescript letters include LT's addition of 28 September as the date. The year is given by reference to the book review in *The Spectator* (see n. 11), and reference to visiting Birmingham (presumably for the meeting of the BAAS; see n. 12).
2. *your letter*: letter 2307.

3. *not my due amount of exercise in Switzerland*: Tyndall's travels in the Alps during the summer of 1865 did not yield any significant mountain ascents.
4. *Fairfield and Helvellyn:* mountains in the Lake District of northwest England.
5. *his wife*: Jane Carlyle.
6. *his part of Babylon*: the Carlyle's lived at 5 Cheyne Row in Chelsea, central London since 1834.
7. *Brompton Road*: a street in Chelsea, London.
8. *Hyde Park Corner*: an area of London where several streets converge, at the southeast corner of Hyde Park.
9. *my book*: *Heat* (1865).
10. *nice review in the Spectator*: 'Heat' (review), *The Spectator*, 38 (23 September 1865), pp. 1064–6. The words 'truly harrowing book' appear on p. 1065, in the second column. The author of this anonymous review was by the British barrister Bernard Cracroft (1828–88) of Trinity College, Cambridge. He published this review in his *Essays: Political and Miscellaneous*, 2 vols (Philadelphia: J. B. Lippincott and Co., 1868), vol. 2, pp. 174–85.
11. *at Birmingham*: the thirty-fifth annual meeting of the BAAS was held in Birmingham from 6–13 September 1865 (*Brit. Assoc. Rep.* 1865).
12. *Wrottesley*: Wrottesley Hall in Wolverhampton, near Birmingham.
13. *wife of Mr. Arthur Wrottesley*: Arthur Wrottesley (1824–1910), third Baron Wrottesley, the eldest son of John Wrottesley, married Augusta Elizabeth Denison (1841–87) in 1861.
14. *Grove*: William Grove.
15. *Hampton Court*: Michael Faraday's house in London since 1858.

From Jacques Louis Soret 29 September 1865 2309

Genève 29 Sept 1865

Mon cher Tyndall

Votre lettre du 15 Sept m'est parvenue hier seulement parce que Dimanche jour de fête fédérale la poste ne fonctionnait point. Je me suis rendu immédiatement auprès du Directeur de la Poste qui était déjà occupé de votre affaire. Je n'ai pu en savoir le résultat qu'après le départ du courrier d'aujourd'hui *[ensorte]* que lorsque vous recevrez cette lettre, vous serez sans doute déjà possesseur de votre courrier qui vous attendait à Bâle. Vous devez recevoir 5 lettres, un journal, et de plus une lettre du Directeur de la Poste de Genève qui vous explique pourquoi vous n'avez rien trouvé ici. Voilà donc une affaire en règle.

Je comptais vous écrire en tout cas pour vous demander un conseil. Je dois faire cet hiver un cours public d'une dizaine de leçons sur les Glaciers. La salle où j'aurai ces séances est trop mal disposée et le public trop nombreux pour qu'il soit possible de faire beaucoup d'expériences. C'est en partie ce qui m'a décidé dans le choix de ce sujet qui n'en *[exige]* pas beaucoup. Pourtant je voudrais en faire quelques'unes sur le regel de la glace. D'abord rien ne sera plus facile que de montrer que des morceaux de glace superposés se soudent les uns aux autres; mais j'aimerais beaucoup répéter quelques'unes des expériences que vous avez faites, en moulant de la glace brisée par exemple faire une sphère [ou] une coupe par compression. Si cela vous est possible je vous serais très reconnaissant de m'indiquer comment il faut opérer; de me dire en particulier si une très forte pression est nécessaire, car je crains de ne pas trouver facilement de bonne presse hydraulique ici. Dites moi ainsi quelle disposition il faut adopter pour les mouler et quelle dimension il faut leur donner.

Croyez vous qu'il y ait quelqu'autres expériences possible, par exemple sur la structure de la glace? Peut-on arriver facilement à donner la structure veinée par compression?

Vous m'obligerez beaucoup en me donnant quelques renseignements à cet égard.

J'ai étè très reconnaissant de votre promesse de m'envoyer votre photographie et pour que vous ne l'oubliez pas je vous envoie la mienne que j'ai fait faire dernièrement.

Madame Soret a été très sensible à votre souvenir et me chargé de ses compliments pour vous. Recevez mon cher Tyndall l'expression de mon sincère attachment

L. Soret

Geneva 29 Sept 1865

My dear Tyndall,

Your letter from 15 Sept.[1] reached me only yesterday because Sunday was a federal holiday and the post office was not running. I immediately went to the Director of the Post Office, who was already dealing with your case.[2] It was only after the departure of today's post that I was able to learn the result, so by the time you receive this letter, you will no doubt be in possession of the mail that awaited you in Basel.[3] You should receive 5 letters, a journal, and also a letter from the Director of the Geneva Post Office, which will tell you why there was nothing for you here.[4] And so, the matter is settled.

I intended to write to you in any case to ask for your advice. This winter I must do a public course of about ten lessons on Glaciers. The room where I will hold these sessions is too poorly laid out and the crowd too large to all for many experiments. This is partly why I chose this subject, which does not require too many. Yet I would like to do a few on the refreezing of ice. To begin, there will be nothing easier than showing that stacked pieces of ice coalesce with each other; but I would very much like to repeat some of the experiments that you have done,[5] showing, for example, how to use compression to make a sphere or a section out of broken ice. If possible, I would be very grateful if you could tell me how it should be done. I particular, if you could tell me if a very high pressure is required, because I fear that it will not be easy to find a good hydraulic press here. Also tell me how to arrange the ice for moulding and what dimensions to make the shards.

Do you think that there might be other possible experiments, for example on the structure of ice? Would it be easy to create a veined structure through compression?

You would greatly oblige me if you could give me some information on the matter.

I was very grateful for your promise to send me your photograph; and so that you will not forget, I am sending you one that I had made recently.

Mrs Soret[6] was very moved that you remembered her and instructed me to send you her compliments. Yours most sincerely, my dear Tyndall,

L. Soret

RI MS JT/1/S/88

1. *Your letter from 15 Sept.*: letter missing.
2. *your case*: not identified.
3. *Basel*: a city in northwestern Switzerland, located on the Rhine river (*Basel* in French, *Bale* in English).
4. *receive 5 letters, a journal, and another letter*: letters missing; journal not identified.
5. *experiments that you have done*: referring to experiments Tyndall conducted in the late 1850s for his own theory of glacial motion, through the fracture and regelation of ice. See the introduction to *Tyndall Correspondence*, vol. 6, as well as: J. Tyndall, 'Observations on Glaciers,' *Roy. Soc. Proc.*, 8 (1856), pp. 331–8; J. Tyndall and T. H. Huxley, 'On the Structure and Motion of Glaciers', *Phil. Trans.*, 147 (1857), pp. 327–46; J. Tyndall, 'On Some Physical Properties of Ice,' *Phil. Trans.*, 148 (1858), pp. 211–29; J. Tyndall, 'On the Veined Structure of Glaciers; with observations upon White Ice-seams, Air-bubbles and Dirt-bands, and remarks upon Glacier Theories', *Phil. Trans.*, 149 (1859), pp. 279–307; J. Tyndall, 'Remarks on Ice and Glaciers', *Phil. Mag.*, 17 (1859), pp. 91–6; and J. Tyndall, *Glaciers of the Alps* (London: John Murray, 1860).
6. *Mrs Soret*: Clémentine Soret (née Odier).

From Juliet Pollock 1 October [1865][1] 2310

Rydal Lodge | Ambleside[2] | Oct. 1.

My dear Mr. Tyndall

I enclose two letters from Walter[3] & I ask a favour of you. I ask you to go to see him & to send me your true impression of his condition. If you were to find him really ill, you would not fail to see the Headmaster[4] & get his leave to take him up to London for De Mussy's[5] advice. But I trust he will not be ill, & that you may find that his indisposition, the result of fatigue, & possibly of this sudden change of weather, is passing off.—

He must not play at football—it has knocked him up before & it was very unwise of him to attempt it. If you could be at Eton[6] either after six in the evening or before twelve in the morning you would be able to see Walter at once without any tiresome waiting about. I feel that I am taxing your time but I rely on your friendship & I know that there is nothing you could ask of me that I would not try to do.—

Your own nature has with all its masculine force such an amount of feminine tenderness in it, that I know you will understand the anxiety that Walter's notes have excited in me which is great in proportion to the distance that divides me from him. I cannot run down to Eton from Rydal.—as from Montagu Square.[7]—

I do not apologize; I know that you will understand me.—

I remain ever my dear Mr. Tyndall | Truly your friend | Juliet Pollock | Thanks for your letter.[8]

RI MS JT/1/P/199

1. *[1865]*: the year is given by relation to letters 2311, 2312, and 2315.
2. *Rydal Lodge | Ambleside*: see letter 2307, n. 1.
3. *Walter*: Walter Pollock. Letters missing.
4. *Headmaster*: the headmaster of Eton College at this time was Edward Balston (1817–91).
5. *De Mussy's*: Henri Guéneau de Mussy (1814–92), a French physician who lived and practiced in London from 1848–71 (*Faraday Correspondence*, vol. 6, p. 790).
6. *Eton*: Eton College, a boarding school for boys in Windsor, England.
7. *Montagu Square*: a square in Marylebone, London, where the Pollocks had their primary residence at no. 59.
8. *your letter*: letter 2308.

To Juliet Pollock 4 October 1865 2311

4th Oct. 1865

My dear Mrs. Pollock

I went down to Eton[1] yesterday and after some investigation discovered Walter.[2] The boy was in his gown, he received me cheerily and led me to his little room. At first I rallied him on writing in a desponding tone, and afterwards I observed him. He is all right just now.[3] De Mussy[4] could do nothing for him. He assured me that his lassitude had departed and that he had written to tell you so. But even while thus assuring me there was something in the tone of his voice too gentle for a boy of his years. He lacks rude physical vigour—a vigour which medicine will by no means impart. The spirit of the boy is entirely predominant over his clay,[5] and though this makes him wonderfully loveable, the clay must be taken care of. Walter I am persuaded is at this moment no weaker than when you last saw him, but I am equally persuaded that to prevent him from becoming weaker you must somewhat alter his life at Eton. The room which he occupies was built by men who knew absolutely nothing of the circulation of the blood, absolutely nothing of the influence of oxygen or carbonic acid on that blood; and such rooms have been left unaltered by men who ought to know something of both. Let me say that I wonder it did not occur to his father[6] that that room is unfit for Walter. It is exceedingly small—a mere den, and it is heated by pipes of warm water. Now he can sleep with his window open, but in winter he must close his window and endure the suffocating air of that place. I say emphatically that that room would imperil the constitution of a more robust boy than Walter, and that he, on no account, ought to be permitted to remain there through the winter. These are my "true impressions"

Yours ever | John Tyndall

RI MS JT/1/T/1144
RI MS JT/1/TYP/6/2057

---

1. *Eton*: Eton College, a boarding school for boys in Windsor, England.
2. *Walter*: Walter Pollock.
3. *He is alright just now*: in letter 2310, Pollock had asked Tyndall to check in on an ill Walter, as she 'cannot run down to Eton from Rydal.—as from Montagu Square'.
4. *De Mussy*: see letter 2310, n. 5.
5. *clay*: the human body (living or dead) as distinguished from the soul; the earthly or material part of man (*OED*).
6. *father*: William Frederick Pollock.

From Juliet Pollock 5 October 1865 2312

Rydal Lodge | Ambleside[1] | Oct 5.th 1865

My dear Mr. Tyndall

I thank you inwardly more than I can mark outwardly on paper; for your goodness; your account of Walter[2,3] is the more welcome to me because it is evidently true; exactly true.

I feel too, that there is an absence of physical vigour in my child with nerves too susceptible; and a nature so tender, though certainly not deficient in courage, that it seems rather a woman's than a boy's.

I am very much afraid that we have not the power of procuring him a better room, though we both feel with you that this confined space is bad for him.—

We are just returned from a two days visit to Coniston to the James Marshall's.[4]

The place is beautiful and James Marshall is an interesting man, when you know him; and his daughter Julia is a noble frank hearted girl with a fine intellect.—

We had most beautiful weather and we enjoyed it though I was suffering from a cold which was a heavy drag upon me.

I have read with satisfaction the preface of your lecture in French[5] & I thank you for sending it to me.

What a pity that you could not come to us here!

Good bye. I am ever | Most truly yours | Juliet Pollock

RI MS JT/1/P/174

---

1. *Rydal Lodge | Ambleside*: see letter 2307, n. 1.
2. *your account of Walter*: see letter 2311.
3. *Walter*: Walter Pollock.
4. *Coniston to the James Marshall's*: Monk Coniston Hall was a family estate of James Garth Marshall (1802–73), an English Liberal Party politician who served as Member of Parliament for Leeds from 1847–52, in Coniston, Cumbria, England (N. C. Smith, *The Letters of Sydney Smith*, 2 vols (Oxford: Clarendon Press, 1953), vol. 2, p. 721). Marshall married Mary Alicia Pery Spring (1812–75) in 1841 (H. R. Forster, *The Pocket Peerage of Great Britain and Ireland* (London: David Bogue, 1852), p. 235), and they had two children: Victor Alexander Ernest Garth Marshall (1841–1928) and Julia Mary Garth Marshall (1846–1907). The Pollocks possibly stayed in Tent Lodge, a separate house on the estate, where Alfred Tennyson and his wife honeymooned in 1850.
5. *preface of your lecture in French*: probably the French translation of Tyndall's Rede Lecture at the University of Cambridge earlier that year (see letter 2231, n. 1): J. Tyndall, *Sur la radiation*, trans. F. Moigno (Paris: E. Giruad, 1865). Gauthier-Villars published a new edition in 1884.

To Sarah Tyndall — 9 October 1865 — 2313

9th Oct. 1865

Dear Mother.

About a week ago I received a note[1] from Emma[2] asking me to send her some money to enable her to go to Tramore.[3]

I enclosed this note to you[4] and asked you to give me some explanation of it: no such explanation has as yet reached me.

This morning I have received a note[5] from Mrs. Doyle,[6] informing me that she has in compliance with Emma's wishes, taken her to Tramore: she says she has no money and writes to me for some.

I think I have some reason to complain of this treatment—not that Emma has gone to Tramore if it were thought necessary to her health, but that I should be applied to for money in this way.

This year if my recollection be right I have sent you more than a hundred pounds. I also pay to Dr. Phillips[7] twenty pounds a year for looking after my sister. I also at your request, pay to your brother John McAssey[8] six or seven and twenty pounds a year. It is only a few days ago since I requested you to let me know what yearly sum would be sufficient for your wants.[9] You informed me[10] that eighty pounds would be sufficient and I at once sent you a twenty pound cheque as the allowance for the first quarter.

Nothing that you have asked from me have I ever refused and nothing that I can afford shall ever be wanting to make you comfortable; but whatever money is needed to your comfort must be applied for by yourself. Why did you not commission Mrs. Doyle to write to me? Did she take this step without your consent, and merely because it was Emma's wish that she should do so? I beg for an explanation of this affair and I think I have a right to demand it.

If you thought it necessary either for her health of your peace that she should go to[11]

RI MS JT/1/T/1484

1. *note*: letter missing.
2. *Emma*: Emily Tyndall.
3. *Tramore*: a seaside and resort town in County Waterford, Ireland.
4. *I enclosed this note to you*: letter missing.
5. *note*: letter missing.
6. *Mrs. Doyle*: Abby Doyle.
7. *Dr. Phillips*: presumably Tyndall's mother's physician.
8. *John McAssey*: John McAssey, Sarah's older brother.
9. *I requested you to let me know what yearly sum would be sufficient for your wants*: letter missing.
10. *You informed me*: letter missing.
11. *she should go to*: letter ends; fragment only.

To the Editor of the *PMG*[1] 11 October 1865 2314

To the EDITOR *of the* PALL MALL GAZETTE.

SIR,—An eminent and respected scientific friend[2] has drawn my attention to your exceedingly clever article entitled "Prayers against Cholera."[3] The position there taken is a strong one; for, granting the entire freedom of the human will, that it, unlike natural phenomena, is uncontrolled by its antecedents, it follows that, within certain limits, arbitrary changes may be wrought in the order and sequence of those phenomena. And if the possibility of such changes, even in the smallest particular, be conceded, the abstract possibility, or in other words the *conceivability*, of a change upon a grand scale follows as a matter of course. Hence (you would argue) the petitioning of the Almighty against cholera or cattle plague is rescued from the charge of necessary absurdity.

But you will, I think, admit that the value of this argument is not bounded by the limits of nineteenth century Christendom; that it would apply equally well to the beliefs of ancient heathens and modern savages, who saw and see in almost every change of the aspects of nature the hand of an arbitrary deity. It justifies equally the mildest and the most extravagant belief in spontaneous interference. Who, indeed, in such a case, is to draw the line between mildness and extravagance? Once upon a time we prayed against the ravages of small-pox—with what effect? You may answer (and rightly answer) that you do not know. But you will, at all events, admit that the prayer, as a preventive or remedial agent, proved no match for vaccination. Would the suppliant voice of a whole nation have atoned for the bad engineering, or caused a suspension of the laws of hydraulic pressure, in the case of the Bradfield reservoir?[4] I think not. The great majority of sane persons at the present day believe in the necessary character of natural laws, and it is only where the antecedents of a calamity are vague or disguised that they think of resorting to prayer to avert it. Such unhappily is the case with the cholera and the cattle plague.[5] With regard to them we are in a state of darkness similar to that of the ancient pagan world with regard to natural phenomena generally, and hence the disposition to resort to pagan methods of meeting these scourges.

JOHN TYNDALL. | *October* 11, 1865.

PMG, 11 October 1865, p. 4
RI MS JT/8/3

---

1. *Editor of the PMG*: the *PMG* began its run in 1865, and its first editor was Frederick Greenwood (1830–1909), an English journalist, editor, and novelist. When the *PMG* changed ownership in 1880 as well as its political sway, Greenwood resigned as editor and would soon take on the same role with the new *St. James Gazette* (*ODNB*).
2. *eminent and respected scientific friend*: Edward Sabine.
3. *"Prayers against Cholera."*: 'Prayers Against the Cholera', *PMG*, 9 October 1865, p. 1. A second article on the subject was 'On Prayer and the Cholera', *PMG*, 17 October 1865, p. 10, to which Tyndall also replied (see letter 2318).

4. *case of the Bradfield reservoir*: on the night of 11 March 1864, the new Dale Dyke Reservoir, built to provide clean water for the growing city of Sheffield, burst upon being filled for the first time and brought devastation as the flood weaved through the Loxley valley near the village of Bradfield and then into Sheffield, in the end killing 238 people (R. A. Buchanan, 'The Causes of the Great Sheffield Flood of 1864', *History of Technology*, 26 (2005), pp. 113–30).
5. *cattle plague*: rinderpest, an infectious viral disease of cattle and other species of ungulate. See T. M. Romano, 'The Cattle Plague of 1865 and the Reception of "The Germ Theory" in Mid-Victorian Britain", *Journal of the History of Medicine and Allied Sciences*, 52 (1997), pp. 51–80.

To Juliet Pollock 11 October 1865 2315

11th Oct. 1865

Read this to cheer your mother's heart:[1] the boy[2] is all right, but that is a d–d[3] bad room.[4]

J. T.

RI MS JT/1/T/1145
RI MS JT/1/TYP/6/2064

1. *to cheer your mother's heart*: see letters 2310, 2311, and 2312.
2. *boy*: Walter Pollock.
3. *d—d*: damned.
4. *room*: at Eton College, a boarding school for boys in Windsor, England.

To Annette Roche 13 October 1865 2316

Royal Institution, 13th, Oct, 1865.

My dear Miss Roche

I am exceedingly thankful to you for your letter.[1] It would relieve my mind very much if you would allow me to make you my banker:[2] but I shrink from asking you to take so much trouble. Perhaps the quarterly bills might be sent to myself and I could send a cheque for the amount of them.[3]

I know it will be necessary to see how the money sent to Leighlin[4] is expended, and my mother[5] I fear has not the energy to look after the matter.

You write as if she were straitened for money: would you please go to her and see how she stands at present?—Learn what are her debts and what money she has in her possession to meet them. Would you also let me know the sum that you think amply sufficient to secure her every comfort. Would 100 pounds a year do it: if so the sum shall be forthcoming: If you would allow me I would send the quarterly cheques to you.

But I am strongly of opinion that it would be best to remove Emma[6] from my mother's neighbourhood. It will augment her peace and probably lengthen her days:—I think if I wrote earnestly to my mother on that point that she would listen to my advice. Could you give me any suggestion as to a suitable place for her?[7]

I was under the impression that Mrs Doyle[8] was attentive to both my mother and sister and I was resolved not to forget her attention. But she has been so unfaithful[9] to the trust reposed in her that were I in Leighlin I should at once remove her. I have written to her[10] saying that I hand the matter over to Dr Phillips[11] who will remove her if he thinks fit.

My Swiss trip this year was not so successful as usual, I first went, in poor health, to the Canton of Glarus[12]—staying two or three days at Stachelberg,[13] Thence to Altdorf[14] over the Klausen Pass.[15] Thence by Wasen[16] over the Susten Pass.[17] I soon picked up a fair share of strength. At a little place called Gadmen,[18] not far from Meyringen,[19] my friend[20] and myself passed a night and next morning as he was smoking his pipe outside a guide came up to him, asking whether he knew me. My friend answered in the affirmative: and the man forthwith commenced to give an account of my death on the Matterhorn.[21] I had been smashed to pieces on those terrible precipices. In fact the rumour had spread through Switzerland, and even as far as Paris that I was of the party. I had then heard nothing of the accident,[22] and the error regarding myself caused me to think it a mere canard. On coming to Hof[23] however though the accounts were vague I could see that a disaster of some kind had occurred. Of course I soon after learned the details.

From Hof I went through Grindelwald[24] to Mürren,[25] a village perched on the heights above Lauterbrunnen.[26] The spot commands a noble view of the range of mountains which separates the Canton of Valais[27] from that of Berne.[28] Near to this place moreover poor Mrs Arbuthnot had been killed.[29] I climbed the mountain (the Schilthorn[30]) on whose shoulder the bolt struck her. I should scarcely have anticipated danger in such a position. The cloud must have moved up the valley and while grazing the shoulder of the mountain fired the shot which killed the poor young Lady.

I intended to cross the mountain to the AEggischhorn[31] but was beaten back by bad weather. Then I drove to Kandersteg,[32] crossed the Gemmi[33] and made my way to the AEggischhorn. There I stayed a couple of days, the weather being still bad. Afterwards I went to Zermatt. Considerable quantities of snow had fallen upon the Matterhorn, the sun had partially melted this snow and the frosts at night had refrozen the melted mass, so that the rocks were all encased in an enamel of ice. The danger of climbing amongst them was then augmented 100 fold. But I thought it a very unsatisfactory thing to leave the body of Lord Frances Douglas[34] bleaching there without making some effort to recover it. The Zermatt guides would not second me: they manifestly feared the undertaking. Well I engaged a good man[35] from a neighbouring village, who undertook to second me in every way. Having no trustworthy rope near at hand I sent this man all the way to Geneva, whence he brought 2000 feet of rope. I also engaged a man[36] who had been accustomed to the blasting of rocks and who had constructed for me various iron and steel implements for piercing the rocks. All was prepared. I had sent my rope, implements and tent to the base of the mountain and waited only for a change of weather. My idea was to climb to the point at which they slipped, fix the irons stanchions there, attach the rope to them and let myself down by the rope—looking right and left for the body. But the weather was inexorable. I waited 20 days, and then in disgust went off to Geneva. I had promised to join my friend De la Rive[37] there on the 20th of August: The Swiss savants held their Jubilee there this year[38] and thus with this most agreeable scientific gathering my trip ended.

Yours very truly | John Tyndall | I write by this post to Dr Phillips[39]

RI MS JT/1/TYP/11/3855–6
LT Typescript Only

---

1. *your letter*: letter missing.
2. *if you would allow me to make you my banker*: in order to use the money Tyndall sent to support his mother back in Ireland (see letters 2322 and 2337).
3. *them*: LT wrote on the typescript that this word could also have been 'these'.
4. *Leighlin*: Leighlinbridge, County Carlow, Ireland, where Tyndall was born and his mother lived.
5. *my mother*: Sarah Tyndall.
6. *Emma*: Tyndall's older sister, Emily Tyndall, was suffering from a mental illness which manifested as an extreme form of religiosity.
7. *a suitable place for her*: letters in the tenth volume of *The Correspondence of John Tyndall* refer to Emily being under the care of Dr. John Tyndall, a family relative, in Gorey, east of Leighlinbridge.
8. *Mrs Doyle*: Abbey Doyle.
9. *she has been so unfaithful*: Elizabeth Dawson Steuart defended Abbey Doyle in letters 2349, 2355, 2356, and 2383.
10. *I have written to her*: letter missing.
11. *Dr Phillips*: presumably Tyndall's mother's physician, who, according to Elizabeth Dawson Steuart, was incompetent (see letters 2355 and 2356).
12. *Canton of Glarus*: a canton in central eastern Switzerland.
13. *Stachelberg*: a small alpine village in the Swiss canton of Bern, known for its baths.

14. *Altdorf*: a town in the Swiss canton of Uri.
15. *Klausen Pass*: a Swiss alpine pass connecting Altdorf in the canton of Uri with Linthal in the canton of Glarus.
16. *Wasen*: an alpine village in the Swiss canton of Bern.
17. *Susten Pass*: a Swiss alpine pass connecting Innertkirchen in the canton of Bern with Wassen in the canton of Uri.
18. *Gadmen*: a town in the Swiss canton of Bern, now merged with Innertkirchen.
19. *Meyringen*: Meiringen, a town in the Swiss canton of Bern.
20. *my friend*: Thomas Hirst.
21. *Matterhorn*: see letter 2272, n. 4.
22. *accident*: see letter 2297, n. 3.
23. *Hof*: in *Hours of Exercise in the Alps*, Tyndall referred to this location as *Imhof*, noting that 'At Imhof the rumour became more consistent' (p. 252).
24. *Grindelwald*: a town in the Swiss canton of Bern.
25. *Mürren*: a village in the Lauterbrunnen valley.
26. *Lauterbrunnen*: a village in the Swiss canton of Bern.
27. *Canton of Valais*: a canton in southwestern Switzerland.
28. *that of Berne*: the canton in central western Switzerland, the country's second largest.
29. *Mrs Arbuthnot had been killed*: on 21 June 1865, Alice Charlotte Arbuthnot (née Pitt-Rivers, b. 1841), first wife of William Arbuthnot (1838–93), was killed by lightning on the Schilthorn during a sudden thunderstorm ('Obituary', *Gentleman's Magazine, Or Monthly Intelligencer*, 80 (August 1865), p. 257).
30. *Schilthorn*: an alpine peak in the Swiss canton of Bern, above Lauterbrunnen.
31. *AEggischhorn*: Eggishorn, a mountain of the Bernese Alps in the Swiss canton of Valais.
32. *Kandersteg*: a town in the Swiss canton of Bern.
33. *Gemmi*: a Swiss alpine pass connecting Kandersteg with Leukerbad in the canton of Valais.
34. *Lord Frances Douglas*: see letter 2297, n. 3.
35. *good man*: see letter 2298, n. 14.
36. *man*: see letter 2298, n. 15.
37. *De la Rive*: Auguste de la Rive.
38. *Swiss savants held their Jubilee there this year*: see letter 2270, n. 4.
39. *I write by this post to Dr Phillips*: letter missing.

From Johann Conrad Bohn 17 October 1865 2317

<Giessen> d. 17 Oktober 1865

Hochverehrter Herr Tyndall:

Ihre freundlichen Briefe vom 21.[n] & 23.[n] September habe ich bis heute noch nicht beantwortet, da ich eine Rückäusserung von Herrn Vieweg erst heute erhalten habe (Er war von Hause so lang abwesend)<.> Ich habe ihm auf Ihren Rath hin den Vorschlag gemacht 850 Exemplar zu drücken; sei seine Meinung richtig und die Verbreitung des Buches eine geringe so sei diese Zahl genügend; sei Ihre Ansicht der grösseren Verbreitung begründet, so falle der Grund weg ein geringes Honorar zu zahlen und ich müsste dann bei einer nothwendig werdenden 2^{n} Auflage die gebührende Entschädigung finden. Vieweg schreibt nun er müsse mehr als 850 Ex. drucken; für den guten Vertrieb müsse er an mehr als 1000 Buchhandlungen Exemplare als Neuigkeit versenden; es sei ihm leichter 1200 Ex zu verkaufen als die 750 wenn er *[nur]* so viel drucke. Er schlägt vor die Ausgabe 1500 Ex. stark zu machen; würden diese 1500 verkauft, so sei er bereit das Honorar um die Hälfte zu erhöhen; würden sie nicht verkauft, so würde es bei der ersten Abrede bleiben.—Er beruft sich auf seine buchhändlerische Erfahrung u.s.w. Wie die Sachen einmal stehen, scheint es mir am Gerathensten diesem Vorschlag zuzustimmen; doch werde ich noch ein paar Tage warten, da die Sache keine Eile hat. Ich vergass zu bemerken, dass Vieweg noch vorschlägt, bei etwa nothwendig <werdenden> späteren Auflagen sollen dieselben Bedingungen gelten.— Den mir von Ihnen zugeschickten Brief des Leipziger Buchhändlers, behalte ich einstweilen, da ich denke Sie haben denselben nicht nöthig: er steht auf Verlangen natürlich jederzeit zu Ihrer Verfügung. Erlauben Sie mir nun, Sie mit einigen Anfragen zu belästigen und Sie um freundliche Beantwortung derselben zu bitten.

Ad pag. 139. Würde es nicht richtiger sein, die Anordnung in der Ueberschrift des Kapitels so zu ändern, dass das erst genannte „Application of etc.... nicht länger in erster Linie, sondern als Zweites stehe, vor „Energy of molecular force". Darf ich so abändern?

Ad pag 147. Die Zahlen der Tabellen geben die Wärmemenge an, welche beim Abkühlen der Substanzen um 1° bei Temperaturen zwischen 98° & 15° von der Gewichtseinheit ausgegeben werden; entsprechen nicht der Abkühlung von 98° auf 15° also um 83°. Darf ich so abändern?

Ad pag 148. Darf ich die Beziehung so abändern, dass ich schreibe

Phosphor (Krystallinisch-gelb) ... 0,1887
" (amorph— roth) 0,1700?

Ad pag 187. Zeile 2 von oben. Sind die hiergemeinten Temperaturgrade in Celsius oder Fahrenheit?

Ad p 187 Sind es englische oder Pariser Zolle nach denen die Regenmengen gemessen sind?

Ad pag 201. Note. Wie ist Principal Forbes zu übersetzen? Director?

Ad pag 202. Englische Zolle oder Pariser?

Ad pag 213. Zeile 10 von oben. Darf ich zur grösseren Deutlichkeit nur die Worte „in the same time" eingeschaltet denken und demnach übersetzen die Wärme verbreitet sich in der gleichen Zeit in Kupfer weiter ...

Ad pag 234 oben. Um gleiche Gewichte Eisen u. Wismuth um gleiche

Anzahl Temperaturgrad zu erwärmen bedarf es Wärmemengen die sich wie 0,1138:0,0308 verhalten. Hier aber handelt es sich eigentlich um gleiche Volume Eisen und Wismuth. Soll ich demgemäss die Zahlen abändern?

Ad pag 236. Ist fine lint feiner Hanf oder vielleicht Charpie?

Ad pag 239–240. Wenn ich eine Unze = 1⁄12 Pfund annehme, so wäre Ihre Rechnung, wie ich die Sache verstehe, nicht genau. Wenn die Uebertragung der moving force bedeuten soll dass die ganze lebendige Potenz, (vis viva) auf die 100 k[i]l [o]gr[am] Kugel übergegangen gedacht werden soll, so würde deren Geschwindigkeit sich berechnen nach $100 \cdot x^2 = \frac{2}{12}\,(1600)^2$, woraus $x = 65, \ldots$

Endlich finde ich Schwierigkeiten die Note über die Gletscher zu übersetzen. Ich glaube sie beruht hauptsächlich darin, dass ich nicht verstehe, was Sie a tributary of the glacier nennen.Was ist tributary? Auch trunkstream ist nur nicht ganz deutlich, ist es was im Deutschen der Gletscherstock genannt wird? Ich bitte Sie über diesen letzten Punkt mir gefälligst etwas deutlichere Auskunft zu geben; da mir auch die genauere Anschauung eines Gletschers fehlt, so könnte ich sonst leicht schlechte Uebersetzg liefern. Wenn Sie mir angeben wollen wo Ihre Abhandlungen über die Gletscher (Transaction. .) erschienen sind, so wird deren Studium mich vielleicht am besten fördern.

Die anderen Fragen habe ich so eingerichtet, dass Ihre Antwort mit Yes oder no schon verständlich sein wird; also keine grosse Mühe Ihnen durch sie erwächst.

Ich bin überzeugt, Sie entschuldigen die Belästigung die ich Ihnen mache durch mein Bestreben die Uebersetzung des Originales würdig zu machen. Ich bin nun bald mit der Hälfte des Buches fertig und hoffe meine Arbeit bis zu Neujahr 1866 beendet zu haben.

Sie um Fortdauer Ihrer freundlichen Gesinnung bittend verbleibe ich mit herzlichsten Grüssen

Ihr ergebenster | Dr Bohn

<Giessen>, 17 October 1865

Esteemed Mr Tyndall:

I have not yet replied to your kind letters from September 21st and 23rd[1] as I received an answer from Mr Vieweg[2] only today (he was absent from home so long). Following your advice, I made him the proposal to print 850 copies;[3] should his opinion be correct and the distribution of the book be limited, this number should be sufficient; if your view of a wider distribution is justified, then there is no reason to pay a reduced salary and I should, in case a 2nd edition becomes necessary, receive the appropriate compensation. Vieweg now writes that he has to print more than 850 copies; to promote good sales, he needs to send specimen copies to more than 1000 booksellers; it would be easier for him to sell 1200 cps than the 750 if he only printed that many. He proposes to make an edition of

1500 cps; should those 1500 be sold, he would be willing to increase the royalty by half; should they not be sold, the previous agreement would remain.—He appeals to his experience as bookseller, etc. As things stand, it seems to me the most advisable to agree with this proposal; but I will still wait for some days as there is no hurry. I forgot to mention that in addition, Vieweg proposes to maintain the same conditions if later editions become necessary.—I will keep the letter from the bookseller from Leipzig[4] that you sent me for now, as I think that you do not need it: of course, it is available to you any time upon demand. Please allow me now to importune you with some questions and to ask you for your kind answer to the same.

Ad pag. 139. Would it not be more correct to change the order in the title in such a way that "Application of etc. . . . , which is mentioned first, is no longer on the first line but the second, before "Energy of molecular force". Can I change thus?

Ad pag 147. The values in the tables indicate the quantity of heat that is released when substances of the weight unit cool down <u>by 1°</u> at temperatures between 98° & 15°; they do <u>not</u> correspond to a cooling from 98° to 15°, that is, by 83°. Can I change thus?

Ad page 148. Can I change the relation so as to write

Phosphorus (crystalline-yellow) . . . 0.1887
" (amorphous-red) 0.1700?

Ad page 187. Line 2 from the top. Are the degrees of temperature meant here in Celsius or Fahrenheit?

Ad p 187 Are English or Parisian inches used to measure the amount of rain?

Ad page 201. Note. How can Principal Forbes[5] be translated? Director?

Ad page 202. English inches or Parisian?

Ad page 213. Line 10 from the top. For better clarity, may I only think of the words "in the same time" as switched on and translate accordingly, the heat diffuses in copper <u>in the same time</u> . . .

Ad page 234 top. In order to warm up the same <u>weights</u> of iron and bismuth by the same number of degrees of temperature one requires quantities of heat that relate as 0.1138:0.0308. However, here it is a quesiton of equal <u>volumes</u> of iron and bismuth. Should I change the figures accordingly?

Ad page 236. Is fine lint fine hemp or perhaps caddis?

Ad page 239–240. If I assumed that 1 ounce =$\frac{2}{12}$ pounds, then your calculations would not be precise, as I understand the matter. If the transmission of the moving force is to mean that one has to imagine that the whole living force (vis viva) has moved to the 100 lb. cannon-ball, then its speed would be calculated according to $100 \cdot x^2 = \frac{2}{12}\,(1600)^2$, yielding $x = 65$. . . .

To finish, I have difficulties translating the note on glaciers.[6] I think the difficulties arise mainly because I do not understand what it is that you call a <u>tributary</u> of the glacier. What is tributary? Also <u>trunkstream</u> is not entirely clear, is it what is called a Gletscherstock in German? I ask you to kindly give me some clearer information on this last point; as I lack a more precise idea of a glacier, I might otherwise easily produce a bad translation. If you would like to indicate to me where your articles on glaciers have appeared (Transaction. .[7]), perhaps their study will be of greatest assistance to me.

I have organised the other questions in such a way that any answer with Yes or no will be comprehensible; therefore, they should not cost you too much effort.

I am convinced that you will forgive the burden that I cause you in my endeavour to make the translation worthy of the original. I will soon have completed half of the book and hope to finish my work by the new year 1866.

Pleading for the continuation of your friendly consideration, I remain with cordial greetings

Your most devoted | Dr Bohn

RI MS JT/1/B/109

---

1. *your kind letters from September 21st and 23rd*: letters missing.
2. *Mr Vieweg*: Eduard Vieweg.
3. *the proposal to print 850 copies*: Bohn was translating *Heat* (1865) into German (see letter 2296, n. 6).
4. *letter from the bookseller from Leipzig*: letter missing.
5. *Principal Forbes*: James David Forbes.
6. *note on glaciers*: in *Heat* (1865), 'Appendix to Chapter VI: Abstract of a Discourse on the Mer-de-Glace' (pp. 200–09). 'Trunk stream' appeared on p. 200.
7. *Transaction. .*: a reference to *Phil. Trans.*

To the Editor of the *PMG*[1] 19 October 1865 2318

CORRESPONDENCE. | PRAYER AND THE CHOLERA. | *To the* EDITOR *of the* PALL MALL GAZETTE.

SIR,—I have read with interest the letters of your correspondents "E. W." and "H. W. W.,"[2] and your own thoughtful second article on the influence of national prayer.[3] It gives me true pleasure to exchange ideas with such earnest men on so important a subject.

In answer to your correspondent "E. W." I would first say that when I affirm necessity I merely affirm the result of knowledge and experience. Science shows that certain consequents follow certain antecedents with such undeviating uniformity that the association between antecedent and consequent has become inseparable in thought. We explain the known and predict the unknown on the assumption of this inseparability. On the same ground of experience, the ideas of prayer and of a change in the course of natural phenomena refuse to be connected in thought. I believe that water will wet, that iron will sink in it, that fire will burn, that the sun will rise to-morrow, and hold that no prayer at the present day will alter such facts. Both you and your correspondents probably entertain the same opinion. You do not seem disposed to pray for undoubted miracles. Where the antecedent is perfectly clear, you prepare yourselves for the consequent. Now, as a matter of fact, in cases of national supplication the antecedents are often very clear to one class of the community, though very dark to another and a larger class. This explains the fact, that while the latter are ready to resort to prayer, the former decline doing so. The difference between both classes is one of knowledge, not of religious feeling. I turn to the account of the Epping cholera case[4] and learn that the people drank poisoned water. To alter, by prayer, the consequences of this or any similar fact—to deprive, by petition, even a single molecule of miasmatic matter of its properties—would in the eye of science be as much a miracle as to make the sun and moon stand still. For one of these results neither of us would pray; on the same grounds I refuse to pray for either.

With regard to the efficacy of prayer, I grant all manner of *possibilities*, or, more correctly, *conceivabilities*. Whenever we have undoubted evidence of the smallest phenomenon, we can, in imagination, expand it to the largest. I jump over a hillock, and can therefore imagine a man jumping over Mont Blanc.[5] Certain bodies are repelled by the poles of a magnet, and I can imagine this force of repulsion so augmented as to urge projectiles in war. But though I can *conceive* both, I *believe* in neither the jumping of the mountain nor the projectile force of diamagnetism. Similarly, a good-natured man grants my request when I ask him for a share of his umbrella. I can, in imagination, expand this fact to the infinite, and ask Omnipotence to ward off the rain from my paddock. That He *may* do so is conceivable, but experience renders it unbelievable. The people of England are already more or less conscious of this; and the practice of national propitiation is, I believe, doomed, which requires the great ability of this journal—not to direct the spontaneous fervour of a smitten and a threatened people to the throne of grace—not even to prove that it is the bounden duty of the nation to engage in supplication—but to show that such an act is not intrinsically absurd.

Both your correspondents seem to think that scientific discovery may be a result of a prayer. If this be believed, I will only say that the bearing of theology towards science at the present day is as unpardonable as it is unaccountable.

You speak very frequently of the combinations of material law possible to man. Have you ever analyzed these combinations? You stretch forth your arm and move your inkstand. Is this an act of volition pure and simple? It is not. The external motion of your arm is derived immediately from a motion *within* your arm—it *is*, in fact, this motion in another shape. While you were pushing your inkstand a certain amount of oxidation occurred in the muscles of your arm, which oxidation, under normal circumstances, produces a certain definite amount of heat. To move the inkstand, a quantity of that heat has been consumed, which is the exact thermal equivalent of the work done. You could do the same work with the same amount of heat from an ordinary fire. The force employed is the force of your food which is stored up in your muscles. The motor nerves pull the trigger and discharge this force. You have here a series of transformations of purely physical energy with one critical point involved in the question, "*What causes the motor nerves to pull the trigger?*" Is the cause physical or super-physical? Is it a sound or a gleam, or an external prick or pressure, or some internal uneasiness that stimulates the nerves to unlock the muscular force, or is it free will? Whatever the true answer to the question may be, *your* safety consists in affirming boldly that free will must be the cause of the nervous action. But this, your only line of retreat, you have deliberately closed by saying that, whether the will is free or not, "the physical results are the same." By thus dispensing with free will you cause human actions to take their place in the chain of physical sequence, and our combinations of material laws no more justify your conclusions regarding prayer to a free Deity than does the combination of the molecules of water to form frost-flowers upon your window-pain.

Finally, I object to any philosophy, or theology, which selects a special series of natural phenomena as the subject of national supplication and shrinks from the same act with reference to other phenomena. In reply to my question whether the suppliant voice of a whole nation would have altered the laws of hydraulic pressure in the case of the Bradfield reservoir,[6] you reply, "Undoubtedly not." Why not? I would earnestly ask. You advocate prayers for fair weather and for rain. Now the absence or presence of rain depends upon laws of gaseous pressure which are just as immutable as those of water-pressure, and the only reason that I can see for the assumption that the one is the object of Divine interference, and the other not, is that one of them is 770 times heavier than the other. Your position puts one in mind of the remark of Galileo,[7] that Nature abhorred a vacuum only to a height of thirty-two feet.[8] "Divine intervention is thinkable," you virtually say, "but only in the case of bodies of small specific gravity." "Stupendous interferences," or "results of gigantic magnitude," are not to be expected, but small shiftings reasonably may. Again I ask you, in all earnestness, how came you to know this? To me it appears that you are unwittingly taking dangerous liberties with the established laws of the universe. These laws abolish your distinctions of large and small, and make it as great a miracle to suspend the gravity of a straw as to extinguish the force which holds the solar system together.

With these remarks, and with thanks to you for the opportunity of making them, I would willingly refer the final adjudication of the case between us to the coming time.—Your obedient servant, | JOHN TYNDALL.

PMG, 19 October 1865, pp. 3–4

---

1. *Editor of the PMG*: see letter 2314, n. 1.
2. *letters of your correspondents "E. W." and "H. W. W.,"*: both letters were published together in the correspondence section as 'Prayer and the Cholera', *PMG*, 14 October 1865, p. 4.
3. *your own thoughtful second article on the influence of national prayer*: 'On Prayer and the Cholera', *PMG*, 17 October 1865, p. 10. The first article was: 'Prayers Against the Cholera', *PMG*, 9 October 1865, p. 1.
4. *Epping cholera case*: cholera broke out on a farm near the town of Epping, northeast of London, in late September and October 1865, resulting in nine deaths (see J. N. Radcliffe, 'On the Outbreak of Cholera at Theydon Bois, Essex, in 1865, with special reference to the propagation of cholera by water as a medium', *Transactions of the Epidemiological Society of London*, 3 (1869), pp. 85–98).
5. *Mont Blanc*: the highest mountain in the Alps, on the border between France and Italy, which Tyndall first summited on 13 August 1857 (see *Tyndall Correspondence*, vol. 6).
6. *case of the Bradfield reservoir*: in letter 2314, Tyndall had asked: 'Would the suppliant voice of a whole nation have atoned for the bad engineering, or caused a suspension of the laws of hydraulic pressure, in the case of the Bradfield reservoir?'. For information about the reservoir incident, see letter 2314, n. 4.
7. *Galileo*: Galileo Galilei (1564–1642), an Italian astronomer and natural philosopher known for his support of heliocentrism (*CDSB*).
8. *Nature abhorred a vacuum only to a height of thirty-two feet*: while Galileo was consulted regarding why pumped water would not ascend inside a tube higher than 32 feet—he thought it was due to the force of vacuum inside the tube ('forza del vacuo')—it was his student, the Italian physicist and mathematician Evangelista Torricelli (1608–47) who determined the reason why (the force of atmospheric pressure), leading to his invention of the mercury barometer. The phrase 'Nature abhors a vacuum' originated with the Greek philosopher Aristotle (in his *Physics*), which Galileo wrote as 'perchè dunque a sè medesimo repugna esser uno spazio vacuo, vieta la natura il far quello' (G. Galilei, *Le Opere di Galileo Galilei: Prima Edizione Completa*, 20 vols (Firenze: Societa Editrice Fiorentina, 1855), vol. 13 (*Discorsi e Dimostrazioni Matematiche intorno a due nuove scienze*, 1855), p. 17. In G. Galilei, *Dialogues Concerning Two New Sciences*, trans. H. Crew and A. de Salvio (London: Macmillan, 1914), p. 13, the phrase was translated into English as 'Since nature abhors a vacuum, she prevents that from which a vacuum would follow as a necessary consequence'). See also C. S. Maffioli, "La ragione del vacuo'. Why and How Galileo Measured the Resistance of Vacuum', *Galilaeana: The Journal of Galilean Studies*, 8 (2011), pp. 73–104; and J. B. West, 'Torricelli and the Ocean of Air: The First Measurement of Barometric Pressure', *Physiology*, 28 (2013), pp. 66–73.

From Charles Lyell 27 October 1865 2319

53, Harley St | London. W. | Oct. 27— | 1865

My dear Dr Tyndall

I have kept the copy of your Rede lecture[1] which you dedicated to me & have transferred to it the marks I had made from the copy I bought[2] which is a good geological investment. As you will have use for a spare copy I return it with many thanks

sinc[ere]ly y[ou]rs | Cha[rles] Lyell

RI MS JT/1/L/51
RI MS JT/1/TYP/3/850

---

1. *your Rede lecture*: Tyndall delivered the annual Rede Lecture at the University of Cambridge ('On Radiation') on 16 May 1865 (see letter 2231, n. 1).
2. *copy I bought*: *On Radiation*.

From James Joseph Sylvester[1] 27 October 1865 2320

27th Oct. 1865

My dear Tyndall,

I have just received the enclosed[2] from Smith[3]—Will it not be simpler at this stage for you to enter into direct confidential communication with him—You would certainly hold a most commanding position at Oxford[4] if you made up y[ou]^r^ mind to go on with the matter and which would give you unrivalled opportunities for promoting the study of Physical Science in G[rea]^t^ Britain.

Why should not Griffith[5] act under you, so as to leave you more time at y[ou]^r^ disposal, and at the same time to derive for himself a step and some pecuniary advantage from the vacancy?

Yours very truly, | and in haste | J. J Sylvester

RI MS JT/1/S/293
RI MS JT/1/TYP/4/1510

---

1. *James Joseph Sylvester*: James Joseph Sylvester (1814–97), an English mathematician of Jewish descent. He was appointed to the professorship of natural philosophy at the nondenominational University College London in 1838, after which he accepted a professorship at the University of Virginia in 1841, although he stayed there only six months. After returning to England the following year, he met Arthur Cayley, a fellow mathematician,

and together they made significant advances in the theory of invariants. He found work as an actuary before obtaining the professorship of mathematics at the Royal Military Academy, Woolwich (1855–70). Through the 1850s Sylvester was a prolific mathematical researcher, often publishing in the *Phil. Mag.* (*ODNB*).

2. *enclosed*: not identified.
3. *Smith*: Henry John Stephen Smith (1826–83), an Irish mathematician who held the Savilian Chair of Geometry at Oxford since 1861 (*ODNB*).
4. *a most commanding position at Oxford*: upon the death of the British clergyman and physicist Robert Walker (1801–65), reader of experimental philosophy at Oxford (*ODNB*), Tyndall was invited to replace him. Tyndall declined this offer because he did not wish to leave London nor the RI (*Ascent of John Tyndall*, pp. 193–4; *X Club*, p. 131). See also letters 2324 and 2325. The position was competitive, and in the end would go to the physicist Robert Bellamy Clifton (1836–1921), who studied under George Gabriel Stokes and previously was Professor of Natural Philosophy at Owens College, Manchester (*ODNB*). See R. Fox, 'The Context and Practices of Oxford Physics, 1839–77', in R. Fox and G. Gooday (eds), *Physics in Oxford, 1839–1939: Laboratories, Learning, and College Life* (Oxford: Oxford University Press, 2005), pp. 41–4.
5. *Griffith*: George Griffith.

From Gustav Heinrich Wiedemann 30 October 1865 2321

Braunschweig, d 30ten Oktober 1865

Mein lieber Tyndall!

Meinen herzlichsten Dank für die Uebersendung Ihres schönen Werkes! Es freut mich recht sehr, daß unsere Genfer Wünsche schon erfüllt sind und Vieweg eine Übersetzung ihres Buches herausgiebt. Er war erst von anderer Seite aufgefordert worden; eine Bearbeitung desselben zu verlegen. Da aber neben dem Inhalt gerade die treffliche *[1 word illeg]*, welche Sie dem Gegenstand in Ihren Vorlesungen gegeben haben, auch im Deutschen beibehalten werden muß, so scheint mir eine recht geschickte Uebersetzung viel angemessener, und diese wird hoffentlich Professor Bohn liefern. Da nun Vieweg selbst Prof. Bohn weniger kennt, und für die Güte der Übersetzung eine größere Garantie haben möchte, so wünscht er, daß ich die letzte Correctur noch einmal durchlese, und möchte dazu Ihre Autorisation haben. Sie wissen, wie <gern> ich mich der kleinen Mühe unterziehe, wenn ich dadurch beitragen könnte, daß Ihr Werk einem größeren deutschen Publicum in ganz angemessener Weise zugänglich gemacht würde.

Es hat mich ungemein erfreut, Sie in Genf nach so langer Zeit einmal wiederzusehen. Könnten wir uns doch in Zukunft öfter begegnen und zuweilen in Ruhe über allerlei Dinge plaudern. Ich glaube, unsere wissenschaftliche

Anschauung ist in sehr vielen Dingen die Gleiche. Wie sehr würde es mich freuen, dies auch durch einen häufigeren persönlichen freundschaftlichen Verkehr bestätigt zu sehen. Vielleicht liegt Ihnen meine neue Heimat Carlsruhe, wohin ich zu Ostern übersiedeln werde, mehr auf dem Wege und Sie machen Gelegentlich Ihrer Schweizer Reisen einen Abstecher dorthin.

Beiliegend meine letzte magnetische Arbeit, vielleicht interessiert Sie [darin] das verhältnißmäßig einfache Resultat.

Wissen Sie vielleicht etwas über die Legung des transatlantischen Kabels im nächsten Jahre? Sollte dieselbe in meine Ferienzeit ([nach] Mitte Juli) fallen und wäre es sonst ausführbar, so hätte ich eigentlich sehr große Lust, dabei zu sein. Glauben Sie, daß die Einwilligung dazu nur mit größeren Schwierigkeiten zu erlangen wäre?

Meine Frau grüßt Sie bestens; auch Sie würde sich sehr freuen, Sie einmal wieder sehen zu können.

In steter aufrichtiger Ergebenheit | Ihr | G. Wiedemann.

Braunschweig, 30th October 1865

My dear Tyndall!

My most heartfelt thanks for sending me your fine work![1] I am really quite pleased that our Genevan wishes[2] have already been fulfilled and that Vieweg[3] is publishing a translation of your book.[4] He was first asked by a different party to publish an adaptation of the same.[5] But since aside from the content it is especially the splendid *[1 word illeg]* that you gave to the subject in your lectures that must be maintained in the German version, it appears to me that a quite skilful translation would be much more appropriate, and I hope that Professor Bohn[6] will deliver the same. Since Vieweg himself does not know Prof. Bohn very well and would like a greater guarantee for the quality of the translation, he wishes that I should read through the final proof one more time and would like your authorisation for this. You know how gladly I would undertake this small burden if by doing so I could contribute to making your work accessible to a greater German audience in a suitable manner.

I was uncommonly pleased to see you again in Geneva after such a long time.[7] If only we could meet more often in the future and chat about all sorts of things in peace. I believe that our scientific views are the same in many aspects. How pleased would I be to see this confirmed through a more frequent, personal and amicable communication. Perhaps my new home Carlsruhe, wither I will relocate at Easter, will be more on the way and you will make a detour there on the occasion of your Swiss travels.

Enclosed is my latest work on magnetism,[8] perhaps you are interested in the relatively simple result therein.

Perhaps you know something about the laying of the transatlantic cable next year?[9] Should this fall within my vacation time ([after] the middle of July) and should it be otherwise feasible, I actually would have a great desire to be there. Do you think that the permission for this would only be obtainable with great difficulties?

My wife[10] sends her best regards; she would likewise be very pleased to see be able to see you again.

In perpetual, sincere faithfulness | Your | G. Wiedemann.

RI MS JT/1/W/40

1. *Your fine work*: *Heat* (1865).
2. *Genevan wishes*: Tyndall attended the 1865 meeting of the Société Helvetique des Sciences Naturelles (Swiss Society of Natural Sciences, now Swiss Academy of Natural Sciences, founded in 1815; 'Versammlung der schweizerischen Naturforschergesellschaft' in German) held in Geneva from 21–3 August.
3. *Vieweg*: Eduard Vieweg.
4. *translation of your book*: see letter 2296, n. 6.
5. *asked by a different party to publish an adaptation of the same*: see letter 2296.
6. *Professor Bohn*: Johann Conrad Bohn.
7. *see you again in Geneva after such a long time*: see n. 2.
8. *my latest work on magnetism*: G. Wiedemann, 'Magnetische Untersuchungen. Ueber den Magnetismus der Salze der magnetischen Metalle', *Poggendorff's Annal.*, 126 (1865), pp. 1–38.
9. *the laying of the transatlantic cable next year*: work on re-laying the Atlantic telegraph cable after the failed attempt of 1858 was completed on 27 July 1866.
10. *My wife*: Clara Wiedemann.

To Annette Roche [October 1865][1] 2322

Royal Institution, Friday.

My dear Miss Roche

I have been desperately busy of late—so busy that my head has become so weary that I am forced to rush away for two days walk in the country. I have been trying to find out something regarding the P & O. company,[2] but I live so much apart from the great world that I am not up in those matters. When I return, which will be on Monday, I will do what I can to get information for you.

Thank you a thousand times for all your kindness to my mother and to me.[3]

Yours ever truly | John Tyndall

RI MS JT/1/TYP/11/3854
LT Typescript Only

1. *[October 1865]*: the month and year are given based on relation to letter 2323, also dated to [October 1865] based on a note written on the typescript by LT: 'After or before this comes letter to mother 9th Oct. 1865' (letter 2313).
2. *P & O. company*: P. & O. (originally the Peninsular and Oriental Steam Navigation Company) was a prominent British shipping company.
3. *kindness to my mother and to me*: Tyndall had arranged for Roche, a local friend, to help with managing money, accommodations, and suitable clothing for his mother Sarah Tyndall and sister Emily Tyndall. See letters 2256 and 2316.

To Annette Roche [October 1865][1] 2323

Royal Institution. Saturday.

Dear Miss Roche

Thank you a thousand times for your letter[2]—the thought of what you hint to me is simply shocking—If her expenses were trebled I should meet them with gladness of heart.

With the half notes which I send by this post I forward a cheque for an additional fifteen pounds—it is made payable to you—and you I know will kindly cash it and hand it over to her.

I cannot tell you how much obliged to you I am. Draw upon me for any amount for my mother.[3]

I have been making enquiries about the P & O company.[4] A friend[5] of mine has forwarded your queries to Admiral Hall,[6] and I expect an answer from him daily.

Yours most truly | John Tyndall

RI MS JT/1/TYP/11/3854
LT Typescript Only

1. *[October 1865]*: the month and year are given based on a note written on the typescript by LT: 'After or before this comes letter to mother 9th Oct. 1865' (letter 2313).
2. *your letter*: letter missing.
3. *I cannot tell . . . for my mother*: Tyndall had arranged for Roche, a local friend, to help with managing money, accommodations, and suitable clothing for his mother Sarah Tyndall and sister Emily Tyndall. See letters 2256 and 2316.
4. *P & O company*: see letter 2322, n. 2.
5. *friend*: not identified.
6. *Admiral Hall*: William Hutcheon Hall (1797?–1878), a British naval officer and director of the Peninsular and Oriental Steam Navigation Company (*ODNB*).

To Henry Bence Jones — 3 November 1865 — 2324

3rd. Nov. 1865.

My dear Friend

I have acted, not in the colourless manner which your words to me as a friend were calculated to prompt, but in full conformity with your official wishes as expressed in your official letter.[1] That is to say, I have communicated to my friends at Oxford[2] my final resolution not to forsake the Royal Institution.[3]

Your friend | John Tyndall

Dr Bence Jones.

RI MS JT/1/TYP/3/795
LT Typescript Only

---

1. *your official letter*: letter missing, and thus Bence Jones's words that Tyndall described as being in a 'colourless manner' are not identified.
2. *my friends at Oxford*: probably George Griffith and Henry John Stephen Smith (see letter 2320).
3. *not to forsake the Royal Institution*: see letter 2320, n. 4.

To Juliet Pollock — 3 November 1865 — 2325

My dear friend

I came to the Rubicon[1] and found its waters cold; I looked back and saw the warm homesteads among which I have spent the last dozen years of my life. And at the door of one of them I saw a face which said silently "don't cross". And I, not having the heart of a Cesar[2] in my body, didn't cross. So here I am and here I intend to remain,[3] to eat your mutton and drink your brandy, world without end. Amen!

J. T.

RI MS JT/1/TYP/6/2066
LT Typescript Only

---

1. *Rubicon*: in allusive use, with reference to crossing or passing the Rubicon: a boundary, a limit; *esp*. one which once crossed entails irrevocable commitment; a point of no return (*OED*). When in 49 BC Julius Caesar and his army crossed the Rubicon, a small stream in northeast Italy that marked the boundary of Italy and Gaul, it led to war.
2. *Cesar*: Julius Caesar (100–44 BC), a Roman politician and military general.
3. *intend to remain*: to remain at the RI rather than accept a position at Oxford (see letter 2320, n. 4).

From John William Draper 6 November 1865 2326

University New York[1] | Nov 6th 1865

My Dear Sir,

I don't know whether I owe this favor to you—this Pall Mall Gazette which I have received today. If so, please accept my thanks.[2] I am already greatly indebted to you for your "Rede Lecture"[3] brought by Prof Youmans.[4] Your handsome allusions to my experiments[5] are very gratifying and the Lecture truly a clear statement of your own important contributions to the subject of Radiation.

The newspaper not only interests but also amuses me, for I am at the present moment passing through a literary trouble *[1 word illeg]* out of the same subject. If you have my book on the Intell[ectual] Dev[elopment] of Europe, the American edition[6] you will find on p. 412 twelve lines from the bottom the point of offence They add to this also the allusions conveyed in a few lines on the top of p 4.

So you see that even in America where there is so much freedom of thought and liberty of speech how strong an opposition on this point may arise It is wonderful how malignant is the bitterness

And yet I take encouragement from the fact that a book I published recently on the Civil Policy of America[7] and which contained these same paragraphs has had the most extraordinary success an edition of three thousand copies & it having been disposed of in little more than a month and a second larger one rapidly following. There is therefore a great *[&]* very influential and thoughtful class who accept *[1 word illeg]* news

Why don't you find a little time to investigate the distribution of heat and Chemical force in the Interference spectrum found by a grating or ruled surface In that the maximum of heat is found in the <u>yellow</u> space. I published something in relation to it in 1845 in the Philosophical Magazine[8] The photographs I obtained shewing the maximum of chemical action for Iodide Bromide & Chloride of Silver was very curious I had compounds made of them

I have given up Physics & Chemistry finding as I grow older History more to my taste Such papers as that on Phosphorescence published in the Phil Mag 1851[9] cost a world of labor yet it was read by nobody When Becquerel[10] wrote his excellent memoir[11] on the same subject more recently he did not know any thing about it. Physics no one reads, History every one reads

However don't forget to send me a copy of whatever <u>you</u> write and I will promise to read it thoroughly

Yours truly | J W D | Professor Tyndall

Library of Congress, John William Draper Family Papers

1. *University New York*: see letter 2294, n. 5.
2. *accept my thanks*: in letter 2352, dated 11 January 1866, Tyndall confirmed that he indeed sent Draper the *PMG* article.
3. *"Rede Lecture"*: *On Radiation*, the published version of Tyndall's 1865 Rede Lecture at the University of Cambridge (see letter 2231, n. 1).
4. *Prof Youmans*: Edward Youmans.
5. *allusions to my experiments*: in *On Radiation*, on pp. 6 and 27, Tyndall was likely referring to a number of articles Draper published in *Phil. Mag.*, of which Tyndall served on the editorial board from 1850s onwards (L. Brake and M. Demoor (eds), *Dictionary of Nineteenth-Century Journalism in Great Britain and Ireland* (London: British Library; Ghent: Academia Press, 2009), pp. 493–4, 642).
6. *Intell[ectual] Dev[elopment] of Europe, the American edition*: J. W. Draper, *History of the Intellectual Development of Europe* (New York: Harper and Brothers, 1863), in which Draper applied the methods of science to writing history.
7. *Civil Policy of America*: see letter 2294, n. 3.
8. *1845 in the Philosophical Magazine*: J. W. Draper, 'On the Interference Spectrum, and the Absorption of the Tithonic Rays', *Phil. Mag.*, 26 (1845), pp. 465–78.
9. *Phil Mag 1851*: J. W. Draper, 'On the Chemical Action of Light', *Phil. Mag.*, 4 (1851), pp. 314–20.
10. *Becquerel*: probably Alexandre-Edmond Becquerel (1820–91), a French physicist who studied electricity and magnetism, and discovered the photovoltaic effect (*CDSB*).
11. *memoir*: probably E. Becquerel, *Recherches sur divers effets lumineux qui résultent de l'action de la lumière sur les corps* (Paris: Mallet-Bachelier, 1859).

To Frances Power Cobbe[1] 7 November 1865 2327

My dear Miss Cobbe

Our minds—that is yours and mine—sound the same note as regards the influence of prayer on the ceremony of nature. With clearness and precision you have stated the question. In fact had I known that you had written upon the question[2] I might have copied your words & put my name to them.[3]

I intend to keep your book, but I have desired my publisher to send you a book of mine[4] in exchange—this is fair is it not?

The book so far as I have read it is full of strength. Of course I could not have written it all. Your images are too concrete & your personification of the mystery of mysteries too intense for me. But as long as you are tolerant of others—which you are—the shape into which you mould the power of your soul must be determined by yourself alone.

Believe me | Yours most truly | John Tyndall

John Tyndall to Frances Power Cobbe, 7 November 1865, mssCB 798, Box 2, Frances Power Cobbe Correspondence, The Huntington Library, San Marino, California

1. *Frances Power Cobbe*: Frances Power Cobbe (1822–1904), a leading women's rights campaigner and later founder of the National Anti-Vivisection Society and the British Union for the Abolition of Vivisection (*ODNB*).
2. *written upon the question*: probably F. P. Cobbe, *Broken Lights: An Inquiry into the Present Condition & Future Prospects of Religious Faith* (London: Trübner & Co., 1864), in which she argued that the 'faith of the future' will be a 'practical' Theism, where 'men will think more of justice and love and less of creeds and sacraments'.
3. *copied your words & put my name to them*: Tyndall was referring to his recent letters to the *PMG* regarding prayer and cholera (see letters 2314 and 2318, and *Ascent of John Tyndall*, pp. 212–4). Tyndall had previously published comments on the power of prayer in a chapter entitled 'Reflections' in *Mountaineering in 1861*, pp. 33–40. As Tyndall biographer Roland Jackson has noted, this was an odd topic to discuss in an otherwise mountaineering-focused book (*Ascent of John Tyndall*, p. 213).
4. *book of mine*: not identified, but given the year possibly *Heat* (1865).

To Rudolf Clausius 13 November 1865 2328

<u>13th. Nov. 1865</u>

My dear Clausius

It has occurred to me that you have expected some thought on the subject of Regelation; and hence before publishing the enclosed article[1] I would ask you kindly to give me your opinion of it. I hope you are not too busy to give the subject half an hour of your precious time.

On my return from Geneva[2] my health improved rapidly and I am now quite strong. Had I been well at the time I should have gone with you to pay a visit to your wife[3] and little Johnny.[4]

I had some conversation regarding you at Oxford on Wednesday last[5]—, and was pleased to observe the high opinion of you which is entertained by the scientific men of that university.

I am still hard at work at Radiant Heat[6] and have observed some good results, an account of which I will send you in due time.

Professor Bohn[7] of Giessen has undertaken the translation[8] of my little book on Heat,[9] and Wiedemann has kindly offered to look over the last proofs. I suppose it will be published in the Spring.

Hirst is working hard in his new position:[10] he evidently likes his work and this will contribute to the goodness of his results.

Your <u>Bernerkungen</u> on the work of St. Robert[11] has just reached me. He will hardly be content with what you say.

The Archbishop of Canterbury[12] has been writing a prayer against cholera and Rinderpest;[13] I wrote a letter[14] to one of the leading newspapers against such prayers: it is wonderful what a commotion has been caused by the letter. Those things do not interest the people of Germany, but they immensely interest the people of England.

Give my kind regard to your wife & love to the children.[15]

Believe me dear Clausius | Yours ever | John Tyndall

RI MS JT/1/T/193

---

1. *enclosed article*: probably J. Tyndall, 'Professor Helmholtz on Ice and Glaciers', *Phil. Mag.*, 30 (1865), pp. 393–407. This is letter 2330 in this volume.
2. *my return from Geneva*: Tyndall attended the 1865 meeting of the Société Helvetique des Sciences Naturelles (see letter 2270, n. 4).
3. *wife*: Adelheid Clausius.
4. *little Johnny*: Rudolf John Clausius.
5. *at Oxford on Wednesday last*: Tyndall had probably gone to Oxford to discuss the possibility of his taking a position there, which he declined (see letter 2320, n. 4).
6. *work at Radiant Heat:* J. Tyndall, 'Sixth Memoir on Radiation and Absorption—Influence of Colour and Mechanical Condition on Radiant Heat', *Phil. Trans.*, 156 (1866), pp. 83–96.
7. *Professor Bohn*: Johann Conrad Bohn.
8. *translation*: see letter 2296, n. 6.
9. *my little book on Heat*: *Heat* (1865).
10. *his new position*: in August 1865, Thomas Hirst was appointed Professor of Mathematical Physics at University College, London, and remained with the college until 1873, when he became Director of Studies at the Royal Naval College, Greenwich..
11. *Bernerkungen on the work of St. Robert*: R. Clausius, 'Ueber die Bestimmung der Disgregationeines Körpers und die wahre Wärmecapacität', *Archives des Sciences Physiques et Mathématiques*, 24 (1865), pp. 117–24. Clausius had read this paper on 22 August 1865 at the annual meeting of the Société Helvetique des Sciences Naturelles in Geneva (see letter 2270, n. 4). The paper was translated into English and published the following year as: R. Clausius, 'On the determination of the Disgregation of a Body, and on the True Capacity for Heat', *Phil. Mag.*, 31 (1866), pp. 28–33. Tyndall was referring to Paul Ballada de Saint Robert (1815–88), an Italian physicist, naturalist, and mountaineer (C. C. Gillispie and R. Pisano, *Lazare and Sadi Carnot: A Scientific and Filial Relationshop*, 2nd edn (New York: Springer, 2014), pp. 319–24).
12. *Archbishop of Canterbury*: Charles Thomas Longley (1794–1868), Archbishop of Canterbury from 1862–8 (*ODNB*).
13. *prayer against cholera and Rinderpest*: see letter 2314, n. 3. For rinderpest, see same letter, n. 5.
14. *I wrote a letter*: Tyndall's letter to the *PMG* on prayer and cholera is letter 2314.
15. *the children*: see letter 2270, n. 5.

From Rudolf Clausius 16 November 1865 2329

ZÜRICH, 16/11/65.

Lieber Tyndall,

Deinen lieben Brief mit dem gedruckten Aufsatze habe ich gestern Abend erhalten. Der Vortrag von Helmholtz auf den der Aufsatz sich bezieht, war mir noch ganz unbekannt. Ich habe heute bei meinem Buchhändler nach der Brochüre gefragt aber der weiss auch nichts davon. Die Brochüre muss also noch gar nicht im Buchhandel verschickt sein. Ich kann also über die von Helmholtz auseinandergesetzten Ansichten nur in soweit urtheilen als Du selbst sie in Deinem Aufsatze citirt hast.

Deinen Aufsatz habe ich mit grossem Vergnügen gelesen und ich stimme in allen Puncten mit Deinen Ansichten überein. Auch ich glaube dass in James Thomson's Erklärung zwar eine mögliche Ursache des Zusammenfrierens gegeben ist, aber nicht die einzig mögliche, und dass viele Erscheinungen vorkommen, sowohl bei den Gletschern, als auch in anderen Fällen, wo jene Erklärung nicht ausreicht. Besonders schlagend finde ich Deine Versuche, wo Eisstücke in warmem Wasser bei blosser Berührung zusammenfrieren.

Ich begreife nicht, weshalb Helmholtz die Wirkung des Eises durch blossen Contact nicht anerkennen will. Es ist bekannt dass wenn in kaltes Wasser ein Stück Eis gebracht wird dadurch das Frieren erleichtert wird, so dass das neu entstehende Eis sich an dem schon vorhandenen Eise ansetzt, was ganz damit übereinstimmt, dass das Eintreten der Crystallisation in einer Salzlösung dadurch befördert wird, wenn man einen Crystall von derselben Art in die Flüssigkeit bringt. Wenn man hiernach zugeben muss, dass Wasser in Berührung mit schon vorhandenem Eise leichter friert, als ohne die Berührung, weshalb will man dann nicht zugeben, dass eine dünne Wasserschicht, welche sich zwischen zwei Eisstücken befindet, also von zwei Stellen mit Eis in Berührung ist, noch leichter frieren kann, als wenn es nur an einer Stelle mit Eis und an der anderen Stelle mit flüssigem Wasser in Berührung ist?

Ich glaube daher, dass Du in Deinen Auseinandersetzungen ganz recht hast, und dass besonders die in diesem Aufsatze angeführten neuen Versuche sehr dazu beitragen können, auch andere davon zu überzeugen.

Eine kleine Stelle ist mir nicht ganz klar. Auf S.2 in dem ersten Absatze steht: I did not seek to show by a direct cabinet experiment that such changes of form do not occur when ice is subjected to pressure. This has been recently done by Professor Helmholtz. Hier verstehe ich nicht was unter "such changes of form" gemeint ist. Vielleicht wäre es gut, wenn Du noch ein Paar Worte zur Erklärung hinzufügtest.

Ausserdem kommen zwei Stellen vor, die ich für Druckfehler halte. S.1, unter Absatz von unten: "In fact the corroborated", statt "he corroborated"; S.4 vorletzter Absatz: "coulple".

Dass Dein Buch in's Deutsche übersetzt wird, freut mich sehr. Es war davon die Rede gewesen, statt einer wörtlichen Uebersetzung eine Bearbeitung zu machen; aber Wiedemann und ich haben uns übereinstimmend dahin ausgesprochen, dass eine Uebersetzung zweckmässiger sei, weil die Form in welche Du die Sache gebracht hast, meisterhaft ist, und es sehr schade wäre, wenn davon etwas verändert würde.

Von Deinem Streite mit der frommen Welt hatte ich noch nichts gehört, weil ich für gewöhnlich keine englischen Zeitungen lese. Jedenfalls thust Du ganz recht daran, solche Veröffentlichungen, die den Aberglauben fördern, entgegenzutreten. Du wirst dabei gewiss alle Vernünftigen auf Deiner Seite haben.

Von der Uebersetzung meines Buches habe ich seit langer Zeit keinen Correcturbogen erhalten. Ich denke mir, dass man in der Druckerei vielleicht gerade stark mit anderen Sachen beschäftigt ist.

Dass Hirst in seiner neuen Thätigkeit Gefallen findet, höre ich mit Vergnügen, und zweifele auch nicht daran, dass er in seinen Arbeiten den besten Erfolg haben wird. Grüsse ihn bestens und nimm selbst ebenfalls die herzlichsten Grüsse von meiner Frau und mir.

Dein | Clausius.

ZURICH, 16/11/65.

Dear Tyndall,

Last night I received your kind letter[1] with the printed article.[2] Helmholtz's lecture[3] to which the article refers, was entirely unknown to me. Today I asked my bookseller for the brochure, but he has not heard about it either. Thus, the brochure must not have been sent to the book trade yet. Regarding Helmholtz's opinions, I can therefore only pass my judgment in as far as they are quoted in your article.

I enjoyed reading your article with the greatest pleasure and I agree in all points with your views. I also believe that James Thomson's statement presents <u>one</u> cause for the [adhesion on freezing], but certainly not <u>the only possible one</u>, and that there are many phenomena, in glaciers as well as in other cases, where that explanation is insufficient. I find your experiments, in which pieces of ice freeze together in warm water simply upon touching each other, particularly impressive.

I do not really understand why Helmholtz does not want to acknowledge the effect of ice through mere contact. It is well known that if a piece of ice is put into cold water, freezing will be facilitated so that the newly created ice grows on already existing ice. This entirely corresponds with the fact that the formation of crystals in a salt solution can be promoted by placing a crystal of the same kind into the liquid. If one thus has to admit that water freezes more easily when interacting with existing ice than without this interaction, why does one not want to admit that a thin layer of water, which is situated between two pieces of ice and is, therefore, in contact with ice in two places, can freeze even faster than water that has ice on only one side and water on the other?

Thus, I believe that your arguments are absolutely correct and in particular that the new experiments that are discussed in this article will greatly contribute to convincing others.

One small point is not entirely clear to me. The first paragraph on p. 2 says: I did not seek to show by a direct cabinet experiment that such changes of form do not occur when ice is subjected to pressure. This has been recently done by Professor Helmholtz. Here, I do not understand what is meant by "such changes of form". I think it would be useful if you added a few words of explanation.

Beside that, there are two places that I think are misprints. On p. 1 in the bottom paragraph: "In fact the corroborated", instead of "he corroborated";[4] on p. 4, the penultimate paragraph: "coulple".[5]

I am glad that your book[6] will be translated into German.[7] There had been a discussion about doing an adaptation instead of a literal translation. But Wiedemann and I agreed that it would be more appropriate to translate it, because the way in which you present the topics is so masterful that trying to change it would be a great loss.

I had not heard about your dispute with the pious world[8] because I do not usually read English newspapers. By all means, you are completely right to confront such publications that promote superstitions. The rational of mind will certainly be on your side.

As far as translation of my book[9] is concerned, I have not received any proofs in a long time. I suppose the printers are perhaps very busy with other things at the moment.

I am glad to hear that Hirst enjoys his new occupation[10] and do not doubt that his work will be very successful. Please send him my best wishes and accept yourself my wife's[11] and my warmest greetings.

Your | Clausius.

RI MS JT/1/TYP/7/2281–3
LT Typescript Only

1. *your kind letter*: letter 2328.
2. *printed article*: J. Tyndall, 'Professor Helmholtz on Ice and Glaciers', *Phil. Mag.*, 30 (1865), pp. 393–407. This is letter 2330 in this volume.
3. *lecture*: see letter 2290, n. 5 and n. 6.

4. *"In fact the corroborated", instead of "he corroborated"*: 'In fact he corroborated' appeared in the published article. Clausius had been sent a preprint that was then corrected before final publication. See letter 2328.
5. *"coulple"*: the word 'couple', nor this misspelling, do not appear in the published article.
6. *your book*: *Heat* (1865).
7. *translated into German*: see letter 2296, n. 6.
8. *the pious world*: see letter 2327, n. 3.
9. *translation of my book*: see letter 2282, n. 4.
10. *Hirst enjoys his new occupation*: see letter 2328, n. 10.
11. *my wife's*: Adelheid Clausius.

To the Editors of the *Phil. Mag.*[1] 17 November 1865[2] 2330

LVI. Professor HELMHOLTZ *on Ice and* Glaciers. |
To the Editors of the Philosophical Magazine and Journal.

GENTLEMEN,

I HAVE been recently favoured by Professor Helmholtz with a *brochure* entitled "Popular Scientific Lectures,"[3] from the preface of which I learn with great satisfaction that other discourses, of a cast similar to those here placed before the public, may be expected to follow. The present pamphlet contains four lectures,—the first, "On the Relation of the Natural Sciences to Science in general;" the second, "On Goethe's[4] labours in Natural Science;" the third, "On the Physiological Origin of Musical Harmony;" and the fourth, "On Ice and Glaciers." The pamphlet is in German, and it is much to be desired that some competent person should undertake its translation into English.[5]

I turned with natural interest to the last-mentioned discourse to see how my notions and experiments on the formation and motion of glaciers[6] were regarded by so eminent a man. And as it can hardly be doubted that the subject will also interest many of the readers of the Philosophical Magazine, I will here endeavour to give a summary of the scientific portion of the lecture.

Professor Helmholtz refers the cold of the upper regions of the atmosphere to the causes generally assigned; but he adds a remark important at the present moment, when the origin of the hot wind called Föhn[7] in Switzerland is the subject of so much discussion. This wind, as Helmholtz justly observes, may not only be a cold wind upon the mountain summits, but a *wet* one. It is deprived of its moisture upon the heights, and it is warmed by its subsequent fall into the valleys. The heat and dryness of the Föhn are therefore no proof that its origin is Sahara, and not the Atlantic Ocean.

It will probably be remembered that I deduced the formation of glaciers, and their subsequent motion through valleys of varying width and flexure, from the fact that when two pieces of ice are pressed together they freeze together at their places of contact. This fact was first mentioned to me verbally by its discoverer, Mr. Faraday. Soon afterwards, and long before I had occasion to reflect upon its cause, the application of the fact to the formation and motion of glaciers flashed upon me. Snow was in the yard of the Royal Institution at the time; stuffing a quantity of it into one of the steel moulds which I had previously employed to demonstrate the influence of pressure on magnetic phenomena, I squeezed the snow, and had the pleasure of seeing it turn out from the mould as a cylinder of translucent ice. I immediately joined Mr. Faraday, and expressed the conviction that his little outlying experiment would be found to constitute the basis of a true theory of glaciers. It became subsequently known to me that the Messrs. Schlagintweit[8] had made a similar experiment with snow; but they did not connect with it the applications which suggested themselves to me, and which have since been developed into a theory of glacier-motion.

A section of the mould used in the experiment above referred to is given in fig. 1. AB is the solid base of the mould; CDEF a hollow cylinder let into the base; P is the solid plug used to compress the snow. When sufficiently squeezed, the bottom, AB, is removed, and the cylinder of ice is pushed out by the plug. The mould closely resembles one of those employed by Professor Helmholtz.

Fig. 1.

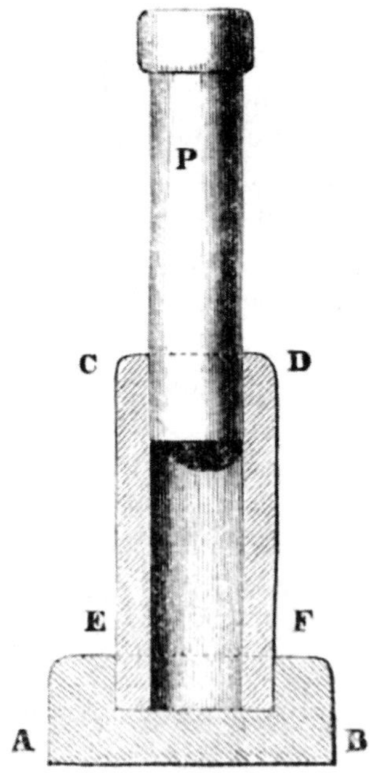

The subsequent development of the subject by the moulding of ice into various forms by pressure is too well known to need dwelling upon here. In applying these results to glaciers, I dwelt with especial emphasis upon the fact that while the power of being moulded by *pressure* belonged in an eminent degree to ice, the power of yielding by stretching to a force of *tension* was wanting. On this point Prof. Helmholtz speaks as follows:—"Tyndall in particular maintained, and proved by calculation and measurement, that the ice of a glacier does not stretch in the smallest degree when subjected to tension—that when sufficiently strained it always breaks;" and he adds in another place, that the property thus revealed establishes "an essential difference between a stream of ice, and one of lava, tar, honey, or mud."

In the beautiful experiments of M. Tresca[9] recently executed, the power of ice to mould itself under pressure has been very strikingly illustrated. Professor Helmholtz also, in the presence of his audiences at Heidelberg and Frankfort,[10] illustrated this property in various ways. From snow and broken fragments of ice he formed cakes and cylinders; and uniting the latter, end to end, he permitted them to freeze together to long sticks of ice. Placing, moreover, in a suitable mould a cylinder of ice of the shape represented in fig. 2, he squeezed it into the cakerepresented in fig. 3. In fact he corroborated, by a series of striking experimental devices of his own, the results previously obtained by myself.

Fig. 2.

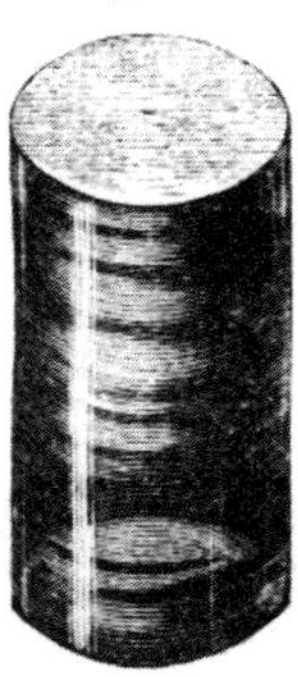

Fig. 3.

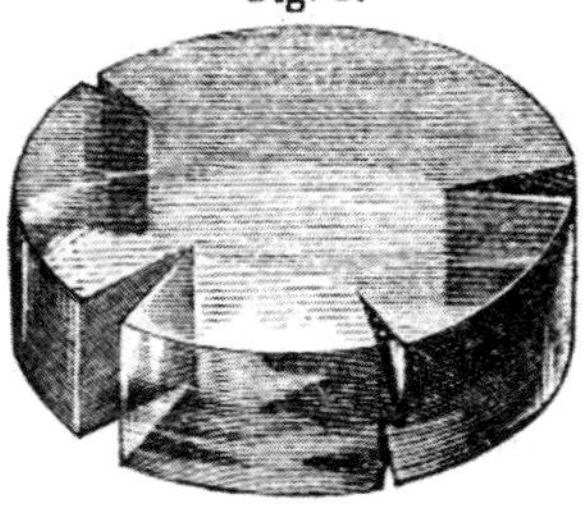

With regard to the application of these results to the phenomena of glaciers, Professor Helmholtz, after satisfying himself of the insufficiency of other hypotheses, thus finally expresses his conviction:—"I do not doubt that Tyndall has assigned the essential and principal cause of glacier-motion, in referring it to fracture and regelation."

It is perhaps worth stating that the term "regelation" was first introduced in a paper published by Mr. Huxley and myself[11] more than seven years after discovery of the fact by Faraday, and that it was suggested to us by Dr. Hooker,[12] now director of the Royal Gardens at Kew. As already remarked, the formation and motion of glaciers, and other points of a kindred nature, had been referred to regelation long before I occupied myself with the cause of regelation itself. This latter question is not once referred to in the memoir in which the regelation theory was first developed*.[13] The inquiries, though related, were different. In referring the motion of glaciers to a fact experimentally demonstrated, I referred it to its *proximate* cause. To refer that cause to *its* physical antecedents formed the subject of a distinct inquiry, in which, because of my belief in the substantial correctness of Mr. Faraday's explanation, I took comparatively little part.

Five persons however, mingled more or less in the inquiry—viz. Professor Faraday, Principal Forbes,[14] Professor James Thomson, Professor William Thomson, and myself†.[15] Professor James Thomson explained regelation by reference to an important deduction, first drawn by him‡,[16] and almost simultaneously by Professor Clausius§,[17] from the mechanical theory of heat. He had shown it to be a consequence of this theory that the freezing-point of water must be lowered by pressure; that is to say, water when subjected to pressure will remain liquid at a temperature below that at which it would freeze if the pressure were removed. This theoretic deduction was confirmed in a remarkable manner by the experiments of his brother, Professor William Thomson||.[18] Regelation, according to Professor Thomson's theory, was thus accounted for:—"When two pieces of ice are pressed together, or laid the one upon the other, their compressed parts liquefy. The water thus produced has rendered a latent a portion of the heat of the surrounding ice, and must therefore be lower then 0° C in temperature. On escaping from the pressure this water refreezes and cements the pieces of ice together." I always admitted that this explanation dealt with "true cause." But considering the infinitesimal magnitude of the pressure sufficient to produce regelation, in common with Professor Faraday and Principal Forbes I deemed the cause an insufficient one. Professor James Thomson, moreover, grounded upon the foregoing theory of regelation a theory of glacier-motion, in which he ascribed the changes of form which a glacier undergoes to the incessant liquefaction of the ice at places where the pressure is intense, and the refreezing, in other positions, of the water thus produced¶.[19] I am endeavoured to show at the time that this theory was inapplicable to the facts. Professor Helmholtz has recently subjected it to the test of experiment, and the conclusions which he draws from his researches are substantially the same as mine.

Thus, then, as regards the incapacity of ice to stretch in obedience to tension, and its capacity to be moulded to any extent by pressure—as regards the essential difference between a glacier, and a stream of lava, honey, or tar—as regards the sufficiency of pressure and regelation to account for the formation of glaciers, and of fracture and regelation to account for their motion—as regards, finally, the insufficiency of the theory which refers the motion to liquefaction by pressure, and refreezing, the views of Professor Helmholtz and myself appear to be identical.

But the case is different with regard to the cause of regelation itself. Here Professor Helmholtz, like M. Jamin*,[20,21] accepts the clear and definite explanation of Professor James Thomson as the most satisfactory that has been advanced; and he supports this view by an experiment so beautiful then it cannot fail to give pleasure even to those against whose opinions it is adduced. But before passing to the experiment, which is described in the Appendix to the lecture, it will be well to give in the words of Professor Helmholtz the views which he expresses in the body of his discourse.

"You will now ask with surprise," he says, "how it is that ice, the most fragile and brittle of all known solid substances, can flow in a glacier like a viscous mass; and you may perhaps be inclined to regard this as one of the most unnatural and paradoxical assertions that ever was made by a natural philosopher. I will at once admit that the inquirers themselves were in no small degree perplexed by the results of their investigation. But the facts were there, and could not be dissipated by denial. How this kind of motion on the part of ice was possible remained long an enigma—the more so as the known brittleness of ice also manifested itself in glaciers by the formation of numerous fissures. This, as Tyndall rightly maintained, constituted an essential difference between the ice-stream, and a stream of lava, tar, honey, or mud.

"The solution of this wonderful enigma was found—as is often the case in natural science—in an apparently remote investigation on the nature of heat, which forms one of the most important conquests of modern physics, and which is known under the name of the mechanical theory of heat. Among a great number of deductions as to the relations of the most diverse natural forces to each other, the principles of the mechanical theory of heat enable us to draw certain conclusions regarding the dependence of the freezing-point of water on the pressure to which the ice and water are subjected."

Professor Helmholtz then explains to his audience what is meant by latent heat, and points out that, through the circulation of water in the fissures and capillaries of a glacier, its interior temperature must remain constantly at the freezing-point of water.

"But," he continues, "the temperature of the freezing-point of water can be altered by pressure. This was first deduced by James Thomson, and almost simultaneously by Clausius, from the mechanical theory of heat; and by the same deductions even the magnitude of the change may be predicted. For the pressure of every additional atmosphere, the freezing-point sinks 0°·0075 C. The brother of the gentleman first named, William Thomson, the celebrated Glasgow physicist, verified experimentally the theoretic deduction by compressing a mixture of ice and water in a suitable vessel. The mixture became colder and colder as the pressure was augmented, and by the exact amount which the mechanical theory of heat required."

"If, then, by pressure a mixture of ice and water can be rendered colder without the actual abstraction of heat, this can only occur by the liquefaction of the ice and the rendering of heat latent. And this is the reason why pressure can alter the point of congelation * * * *.

"In the experiment of William Thomson just referred to, ice and water were enclosed in a solid vessel from which nothing could escape. The case is somewhat different when, as in the case of a glacier, the water of the compressed ice can escape through fissures. In this case the ice is compressed, but not the water which escapes. The pressed ice will become colder by the quantity corresponding to the lowering of its freezing-point by the pressure. But the freezing-point of the uncompressed water is not lowered. Here, then, we have ice colder than 0° C. in Contact with water at 0° C. The consequence is, that round the place of pressure the water will freeze and form new ice, while, on the other hand, a portion of the compressed ice continues to be melted (während dafür ein Theil des gepressten Eises fortschmiltz[22]).

"This occurs, for instance, when two pieces of ice are simply pressed together. By the water which freezes at the points of contact they are firmly united to a continuous mass. When the pressure is considerable and the chilling consequently great, the union occurs quickly, but it may also be effected by a very slight pressure if sufficient time be afforded. Faraday, who discovered this phenomenon, named it the *regelation of ice**.[23] Its explanation has given rise to considerable controversy: I have laid that explanation before you which I consider to be the most satisfactory."

In the Appendix, Professor Helmholtz returns to the subject thus handled in the body of his discourse. "The theory of the regelation of ice," he observes, "has given rise to a scientific discussion between Faraday and Tyndall on the one hand, and James and William Thomson on the other. In the text of this lecture I have adopted the theory of the latter, and have therefore to justify myself for so doing." He then analyzes the reasonings on both sides, points out the theoretic difficulties of Faraday's explanation, shows what a small pressure can accomplish if only sufficient time be granted to it, draws attention to the fact that when one piece of ice is placed upon another the pressure is not distributed over the whole of the two appressed surfaces, but is concentrated on a few points of contact. He also holds with Professor James Thomson, that in an experiment devised by Principal Forbes even the capillary attraction exerted between two plates of ice is sufficient, in due time, to produce regelation. To illustrate the slow action of the small differences of temperature which here come into play Professor Helmholtz made the following experiment, to which reference has been already made.

"A glass flask with a drawn-out neck was half filled with water, which was boiled until all the air above it was driven out. The flask was then hermetically sealed. When cooled, the flask was void of air, and the water within it freed from the pressure of the atmosphere. As the water thus prepared can be cooled considerably below 0° C. before the first ice is formed, while when iced is in the flask it freezes at 0° C. [why?], the flask was in the first instance placed in a freezing-mixture until the water was changed into ice. It was afterwards permitted to melt slowly in a place the temperature of which was +2° C., until the half of it was liquefied.

"The flask thus half filled with water with the disk of ice swimming on it was placed in a mixture of ice and water, being quite surrounded by the mixture. After an hour the disk within the flask was frozen to the glass. By shaking the flask the disk was liberated, but it froze again as often as the shaking was repeated. The flask was permitted to remain for eight days in the mixture, which was preserved throughout at a temperature of 0° C. During this time a number of very regular and sharply-defined ice-crystals were formed, and augmented very slowly in size. This is perhaps the best method of obtaining beautifully formed crystals of ice.

"While, therefore, the outer ice which had to support the pressure of the atmosphere slowly melted, the water within the flask whose freezing point, on account of a defect of pressure, was 0°·0075 C. higher, deposited crystals of ice. The heat abstracted from the water in this operation had, moreover, to pass through the glass of the flask, which, together with the small difference of temperature, explains the slowness of the freezing process."

A single additional condition in connexion with this beautiful experiment I should like to have seen fulfilled—namely, that the water in which the flask was immersed, as well as that within it, should be purged of its air by boiling. It is just possible that the point of congelation may not be entirely independent of the presence of air in the water.

The revival of the subject by Professor Helmholtz has caused me to make a few experiments on the moulding and regelation of ice. The following experiment illustrates both:—A quantity of snowy powder was scraped from a block of clear ice and placed in a boxwood mould having a shaped like the foot of a claret-glass. The ice-powder was compressed by an hydraulic press, and a clear mass of ice of the shape shown in section at the bottom of fig. 4 was the result. In another mould the same powder was squeezed so as to form small cylinders, three of which are shown separate in fig. 4. A third mould was then employed form a cup of ice, which is shown at the top of fig. 4. Bringing all the parts into contact, they were cemented through regelation to form the claret-glass sketched in fig. 5 from which several draughts of wine might be taken if the liquid were cooled sufficiently before pouring it into the cup of ice.

Fig. 4.

Fig. 5.

There are brass shapes used for the casting of flowers and other objects which answer admirably for experiments on the regelation of ice. One of them was purchased for me by Mr. Becker.[24] Ice-powder squeezed into it, regelated to a solid mass and came from the mould in the sharply defined form sketched in fig. 6.

Fig. 6.

I placed a small piece of ice in warm water and forced it underneath the water by a second piece. The submerged morsel was so small that the vertical pressure was almost infinitesimal. It froze, notwithstanding, to the under surface of the superior piece of ice. Two pieces of ice were placed in a basin of warm water, and allowed to come together. They froze as soon as they touched each other. The part surrounding the place of contact rapidly melted away, but the two pieces continued for a time united by a narrow bridge of ice. The bridge finally melted away, and the pieces were for a moment separated. But it is well known that bodies which water wets, and against which it rises by capillary attraction, moved together when caused to swim upon water. The ice morsels did so, and immediately regelation set in. A new bridge was formed, which in its turn was dissolved, and the pieces closed up as before. Thus a kind of pulsation was kept up by the two pieces of ice. They touched, froze; a bridge was formed and melted, leaving an interval between the pieces. Across this interval they moved, touched, froze, the same process being repeated over and over again. We have here the explanation of the curious fact, that when several large lumps of ice are placed in warm water and allowed to touch each other, regelation is maintained among them as long as they remain undissolved. The final fragments may not be the one-hundredth part of the original ones in size; through the process of closing up just described, they incessantly lock themselves together until they finally disappear. What is true of ice-fragments in a basin of water, is also true of the ice-blocks in the Märgelin See; and probably, if attention were directed to it, the play of regelation might be observed in the icebergs of the ocean. With a certain amount of skill, numerous interesting observations may be made on the regelation of floating fragments of ice.

According to Professor James Thomson's theory, to produce regelation the pieces of ice have to exercise pressure, to draw from the surrounding ice heat necessary for the liquefaction of the compressed part; and then this water must escape and be refrozen. All this requires time. In the foregoing experiments, moreover, the water liquefied by the pressure issued into the surrounding warm water, but notwithstanding this the floating fragments regelated in a moment. It is not necessary that the touching surfaces should be flat; for in this case a film of water might be supposed to exist between them of the temperature 0° C. The surfaces in contact may be convex: they may be virtual points that touch each other, clasped all round by the warm liquid, which is rapidly dissolving them as they approach each other. Still they freeze immediately when they touch.

There are two points urged by Helmholtz—one in favour of the view he has adopted, and the other showing a difficulty associated with the view of Faraday—on which a few words may be said. "I found," says Helmholtz, "the strength and rapidity of the union of the pieces of ice in such complete correspondence with the amount of pressure employed, that I cannot doubt that the pressure is actually the sufficient cause of the union."

But, according to Mr. Faraday's explanation, the strength and quickness of the regelation must also go hand in hand with the magnitude of the pressure employed. Helmholtz rightly dwells upon the fact that the appressed surfaces are usually not perfectly congruent—that they really touch each other in a few points only, the pressure being thus concentrated. Now the effect of pressure exerted on two pieces of ice at a temperature of 0° C. is not only to lessen the thickness of the liquid film between the pieces, but also to flatten out the appressed points, and thus to spread the film over a greater space. On both theories, therefore, the strength and quickness of the regelation ought to correspond to the magnitude of the pressure.

The difficulty referred to above is thus stated by Helmholtz:—"In the explanation given by Faraday, according to with the regelation is caused by a contact action of ice and water, I find a theoretic difficulty. By the freezing of the water a very sensible quantity of heat would be set free; and it does not appear how this is to be disposed of."

On the part of those who accept Faraday's explanation, the answer here must be that the free heat is diffused through the adjacent ice. But against this it will doubtless be urged that ice already at a temperature of 0° C. cannot take up more heat without liquefaction. If this be true under all circumstances, Faraday's explanation must undoubtedly be given up. But the essence of that explanation seems to be that the interior portions of a mass of ice require a higher temperature to dissolve them than that sufficient to cause fusion at the surface. To be perfectly distinct, let a beam of solar heat, or a beam from the electric lamp, be sent through a mass of ice. The substance breaks up into those six-petaled liquid flowers which have been described elsewhere. The flowers expand as the beam continues to act, the energy of the absorbed portion of the beam being almost wholly expended in enlarging the flowers formed during the first few seconds of exposure, and not in forming new flowers. Now the assumption involved in Faraday's theory is, that, before a flower makes its appearance in the interior, the ice there must have been raised to a temperature higher than 0° C., while at the surface the ice fuses at this temperature. When therefore two moist surfaces of ice at the temperature 0° are pressed together, and when, in virtue of the contact action assumed by Faraday, the film of water between them is frozen, the adjacent ice (which is now in the interior, and not at the surface as at first) is in a condition to withdraw by conduction, and without prejudice to its own solidarity, the small amount of heat set free. Once granting the contact action claimed by Faraday, there seems to be no difficulty in disposing of the heat rendered sensible by the freezing of the film.

When the year is advanced, and after the ice imported into London has remained a long time in store, is closely examined, parcels of liquid water will be found in the interior of the mass. I enveloped ice containing such water-parcels in tinfoil, and placed it in a freezing-mixture until the liquid parcels were perfectly congealed. Removing the ice from the freezing-mixture, I placed it, covered by its envelope, in a dark room, and found, after a couple of hours' exposure to a temperature somewhat over 0° C., the frozen parcels again liquid. The heat which was sufficient to fuse this interior ice, passed through the firmer surrounding ice without the slightest visible prejudice to its solidity. But if the freezing temperature of the ice-parcels be 0° C., then the freezing temperature of the mass surrounding them must be higher than 0° C., which is what the explanation of Faraday requires.

In a former paragraph I have attached a description of the precaution used by Professor Helmholtz the query "why?" He states that water freed of its air sinks, without freezing, to a temperature far below 0° C.; while went piece of ice is in the same water it cannot so sink in temperature, but is invariably deposited in the solid form at 0° C. This surely proves ice to possess a special power of solidification over water. It is needless to say that the fact is general—that a crystal of any salt placed in a saturated solution of the salt always provokes crystallization. Applying this fact to the minute film of water enclosed between two appressed surfaces of ice, it seems to me in the highest degree probable that the contact action of Faraday will set in, that the film will freeze and cement the pieces of ice together.

Apart from the present discussion, the following observation is perhaps worth recording. Is well known that ice during a thaw disintegrates so as to form rude prisms whose axes are at right angles to the planes of freezing. I have often observed this action on a large scale, during the winters that I spent as a student on the banks of the Lahn.[25] The manner in which these prisons are in some cases formed is extremely interesting. On close inspection, a kind of cloudiness is observed in the interior of the mass apparently perfect ice. On closer inspection, this cloudiness appears arranged in striae at right angles to the plains of freezing, and when the direction of vision is across these planes the ends of the striae are apparent. The spaces between the striæ are composed of clear unclouded ice. The objects which produce this cloudiness are exceedingly small, but when duly examined they turn out to be piles of minute liquid flowers, whose planes are right angles to the direction of the striae.

JOHN TYNDALL. | Royal Institution, | November 1865.

Since writing the above, I have been favoured with a copy of a Discourse delivered by Professor De la Rive,[26] at the opening of the forty-ninth meeting of the Société Helvétique, which assembled this year at Geneva.[27] From this admirable *résumé* of our present knowledge regarding glaciers I make the following extract, which, together with those from the lecture of Helmholtz, will show sufficiently how the subject is now regarded by competent men:—

"Such, gentlemen," says M. De la Rive, "is a description of the phenomena of glaciers, and it now remains to explain them, to consult observation, and deduce from it the fundamental character of the phenomena. Observation teaches us that gravity is the motive force, and that this force acts upon a solid body—ice—imparting to it a slow and continuous motion. What are we to conclude from this? That ice is a solid which possesses the property of flowing like a viscous body—a conclusion which appears very simple, but which was nevertheless announced for the first time hardly five and twenty years ago by one of the most distinguished philosophers of Scotland, Professor James D. Forbes. This theory, for it truly is a theory, basing itself on facts as numerous as they are well observed, enunciates the principle that ice possesses the characteristic properties which belong to plastic bodies. Although he did not directly prove it, to Professor Forbes belongs not the less the great merit of insisting on the plasticity of ice, before Faraday, in discovering the phenomenon of regelation, enabled Tyndall to prove that the plasticity was real, at least partially.

"The experiment of Faraday is classical in connexion with our subject. It consists, as you know, in this, that if two morsels of ice be brought into contact in water, which may be even warm, they freeze together. Tyndall saw very quickly the application of Faraday's experiment to the theory of glaciers; he comprehended that, since pieces of ice could thus solder themselves together, the substance might be broken, placed in a mould, compressed, and this compelled to take the form of the cavity which contained it. A wooded mould, for example, embraces a spherical cavity; placing in it fragments of ice and squeezing them, we obtain an ice sphere; placing this sphere in a second mould with a lenticular cavity and pressing it, we transform the sphere into a lens. In this way we can impart any form whatever to ice.

"Such is the discovery of Tyndall, which may well be thus named, particularly in view of its consequences. For all these moulds magnified become the borders of the valley in which a glacier flows. Here the action of the hydraulic press which has served for the experiments of the laboratory is replaced by the weight of the masses of snow and ice collected on the summits, and exerting their pressure on the ice which descends into the valley. Supposing, for example, between the spherical mould and the lenticular one, a graduated series of other moulds to exist, each of which differs very little from the one which precedes and from that which follows it, and that a mass of ice could be made to pass through all these moulds in succession, the phenomenon would then be continuous. Instead of rudely breaking, the ice would be compelled to change by insensible degrees from the spherical to the lenticular form. It would thus exhibit of plasticity which might be compared to that of soft wax. But ice is only plastic under *pressure*; it is not plastic under *tension*: and this is the important point which the vague theory of plasticity was unable to explain. Well a viscous body, like a bitumen or honey, may be drawn out in filaments by tension, ice, far from stretching in this way, breaks like glass under this action. These points well established by Tyndall, it became easy for him to explain the mechanism of glaciers, and by the aid of an English geometer, Mr. William Hopkins,[28] to show how the direction of the crevasses of a glacier are the necessary consequences of its motion."

I have quite recently had a mould constructed for me by Mr. Becker*,[29] and yesterday (November 16th) made with it an experiment which, on account of the ease with which it may be performed, will interest all those who care about exhibiting in a striking and instructive manner the effects of regelation. The mould is shown in fig. 7. It consists of two pieces of cast iron, ABC and DFG, slightly wedge-shaped and held together by the iron rectangle RE which is slipped over them. The inner face of ABC is shown in fig. 8. In it is hollowed out a semi-ring MN, with a semicylindrical passage O leading into it. The inner face of DFG is similarly hollowed out, so that when both faces are placed together, as in fig. 7, they enclose a ring 4 inches in external diameter, from M to N, and ¾ of an inch in thickness, with the passage of O, 1 inch in diameter, into which fits the polished iron plug P leading into it. At *q* and *r*, fig. 8, are little pins which, fitting into holes corresponding to them, keep the slabs ABC and DFG from sliding over each other.

Fig. 7.

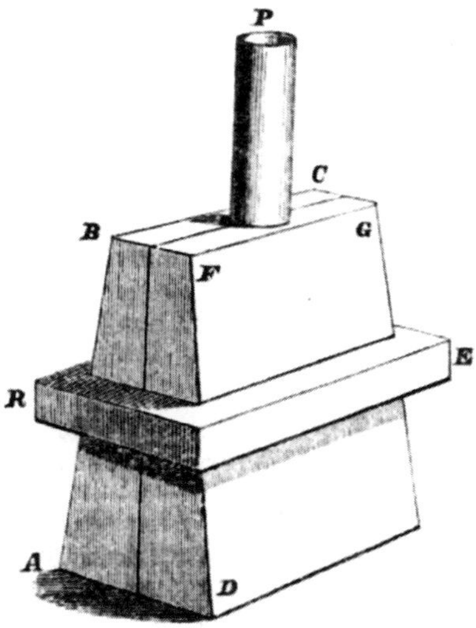

Fig. 8.

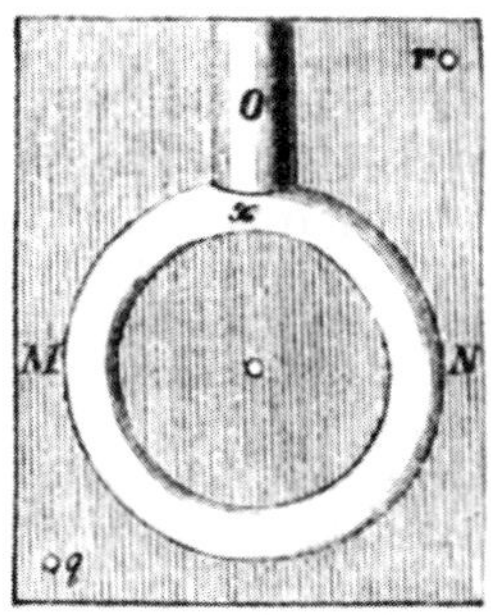

The mould being first cooled by placing it for a short time in a mixture of ice and water, fragments of ice are stuffed into the orifice O and driven down with a hammer by means of the plug P. The bruised and broken ice separates at *x*, one portion going to the right, the other to the left. Driving the ice thus into the mould, piece after piece, it is finally filled. By removing the rectangle R E, the two halves of the mould are then separated, and a perfect ring of ice is found within. Two such rings soldered by regelation at *a* are shown in fig. 9. It would be easy thus to construct a chain of ice. An hydraulic press may of course be employed in this experiment, but it is not necessary; with the hammer and plug beautiful rings of ice are easily obtained by the regelation of the crushed fragments.

I have now to add the description of an experiment which suggested itself to my ingenious friend Mr. Duppa[30] when he saw the ice-rings just referred to,

Fig. 9.

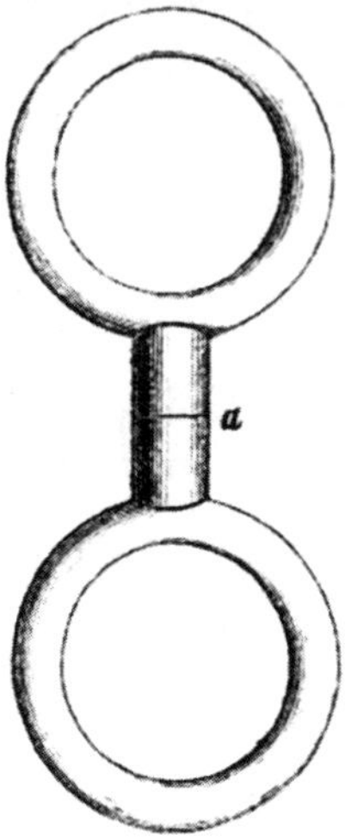

and which was actually executed by him yesterday (the 16th) in the laboratory of the Royal Institution. Pouring a quantity of plaster of paris into a proper vessel, an ice-ring was laid upon the substance, and additional quantity of the cement being then poured over the ring. The plaster "set," enclosing the ring within it: the ring soon melted, leaving its perfect matrix behind. The mould was permitted to dry, and, molten lead being poured into the space previously occupied by the ice, a leaden ring was produced. Now ice can be moulded into any shape: statuettes, vases, flowers, and innumerable other ornaments can be formed from it. These enclosed in cement, in the manner suggested by Mr. Duppa, remain intact sufficiently long to enable the cement to set around them; they afterwards melt and disappear, leaving behind them perfect plaster moulds, from which casts can be taken.—J. T.

J. Tyndall, 'Professor Helmholtz on Ice and Glaciers', *Phil. Mag.*, 30 (1865), pp. 393–407

---

1. *Editors of the Phil. Mag.*: see letter 2245, n. 1.
2. *17 November 1865*: there is no date for the first portion of the letter, but Tyndall stated in his addendum that it was written on 17 November.
3. *brochure entitled "Popular Scientific Lectures"*: H. Helmholtz, *Populäre Wissenschaftliche Vorträge* (Braunschweig: Friedrich Vieweg und Sohn, 1865). His lecture on ice and glaciers, 'Eis und Gletscher', is on pp. 93–133.
4. *Goethe's*: Johann Wolfgang von Goethe (1749–1832), the outstanding German literary figure of the late eighteenth and early nineteenth centuries. His extensive writings included

his theory of colours and a study of the structure and function of plants. During his time in Marburg, Tyndall became familiar with some of Goethe's works.

5. *translation into English*: see letter 2290, n. 6.
6. *my notions and experiments on the formation and motion of glaciers*: see letter 2309, n. 5.
7. *hot wind called Föhn*: or foehn, a warm dry south wind which blows down the valleys on the north side of the Alps (*OED*).
8. *Messrs. Schlagintweit*: Hermann Schlagintweit (1826–82) and Robert Schlagintweit (1833–85), members of a German family of five brothers, all explorers. Adolf Schlagintweit (1829–57) had gone to India with Hermann and Robert in 1854 to study the territory of the East India Company. Adolf remained exploring in Asia when his brothers returned to Europe and he was beheaded in Kashgar when suspected of being a Chinese spy. Two other brothers were Eduard Schlagintweit (1831–66) and Emil Schlagintweit (1835–1904) (*NDB*).
9. *M. Tresca*: Henri Édouard Tresca (1814–85), a French mechanical engineer and professor at the Conservatoire National des Arts et Métiers in Paris ('Obituary', *Minutes of Proceedings of the Institution of Civil Engineers*, vol. 83 (London: Institution of Civil Engineers, 1886), pp. 425–30).
10. *his audiences at Heidelberg and Frankfort*: see letter 2290, n. 5 and n. 6.
11. *paper published by Mr. Huxley and myself*: J. Tyndall and T. H. Huxley, 'On the Structure and Motion of Glaciers', *Phil. Trans.*, 147 (1857), pp. 327–46.
12. *suggested to us by Dr. Hooker*: Tyndall credited Hooker with suggesting the term 'regelation' to describe the freezing together of thawing pieces of ice upon contact. Though not mentioned in the abstract of his Friday Evening Discourse at the RI on 23 January 1857, 'Observations on Glaciers' (*Roy. Inst. Proc.*, 2 (1854–8) pp. 320–7), an account in the *Saturday Review* of 31 January 1857 stated: 'This phenomenon the lecturer attributed, not to viscosity, but to the property which was announced by Dr. Faraday in 1850, and which Dr. Hooker has named "regelation"' ('Dr. Tyndall's Theory of Glaciers', *Saturday Review*, 3 (1857), pp. 102–3, on p. 102). Tyndall noted in his journal on 6 February 1857: 'In the Saturday Review of last Saturday is an article on my Friday evening—The writer is a well informed man a friend to and well up in Forbes: nevertheless he admits the formidable nature of some of the objections; towards the conclusion he reasons very philosophically: Hooker remarked that the conclusion of the article cannot have been written by the man who wrote the commencement' (RI MS JT/2/13c/918).
13. The following footnote appeared here in the original: '* Phil. Trans. vol. cxlvii. p. 327'.
14. *Principal Forbes*: James David Forbes.
15. The following footnote appeared here in the original: '† Proc. Roy. Soc. vol. ix. p. 141; and vol. x. p. 152. Phil Mag. S. 4. vol. xvi. pp. 347 & 544; and vol. xvii. p. 162'.
16. The following footnote appeared here in the original: '‡ Proc. Roy. Soc. Edinb. February 1850'.
17. The following footnote appeared here in the original: '§ Pogg. *Ann*. vol. lxxxi. p. 168'.
18. The following footnote appeared here in the original: '|| Phil. Mag. August 1850'.

19. The following footnote appeared here in the original: '¶ Proc. Roy Soc. vol. viii. P. 455'.
20. *M. Jamin*: Jules Jamin (1818–86), a French physicist and professor at École Polytechnique in Paris from 1852–81 ('Discours prononcés aux obsèques de M. Jamin', *Paris, Comptes Rendus*, 102 (1886), pp. 337–43.
21. The following footnote appeared here in the original: '* *Traité de Physique*, vol. ii. p. 105'.
22. *während dafür ein Theil des gepressten Eises fortschmiltz*: German for the preceding words, 'a portion of the compressed ice continues to be melted'.
23. The following footnote appeared here in the original: '* I have already corrected this slight inadvertence'.
24. *Mr. Becker*: Carl Ludwig Christian Becker (1821–75), a German instrument maker who moved to London (1849), where he became the manager of the scientific instrument maker Elliott Brothers (1858), and later (1873) a partner in the firm ('Obituary', *Monthly Notices of the Royal Astronomical Society*, 36 (1876), pp. 136–7). On Elliott Brothers, see letter 2524, n. 2.
25. *Lahn*: the Lahn River, a tributary of the Rhine River passing through the University of Marburg in Germany.
26. *Discourse delivered by Professor De la Rive*: A. de la Rive, 'Discours Prononcé le 21 Aout 1865 à l'ouverture de la quarante-neuvième session de la Société Helvétique des Sciences Naturalles Réunie a Genève', *Archives des sciences physiques et naturelles*, 25 (1865), pp. 48–75.
27. *forty-ninth meeting . . . at Geneva*: see letter 2270, n. 4.
28. *Mr. William Hopkins*: William Hopkins (1793–1866), an English mathematician and geologist, and renowned private tutor to undergraduate students at Cambridge (*ODNB*).
29. The following footnote appeared here in the original: '* I am continually indebted to this able mechanician for prompt and intelligent aid in the carrying out of my ideas'.
30. *Mr. Duppa*: see letter 2229, n. 4.

To Hermann Helmholtz 27 November 1865 2331

29th Nov. 1865

My dear Helmholtz

I send you by this post a short abstract[1] which will show you that the Vorträge[2] have reached me safely. I read these lectures with extreme pleasure, and more than once cried Bravo! to the sentiments therein uttered.

I expected to see you at Birmingham,[3] but was sorry to learn in Switzerland that your health was not so good as might be wished: I hope you are all right now, and that you will keep your promise to the public, and let them have the remainder of the lectures.

Will you present my kindest regards to Bunsen and Kirchhoff[4]

Believe me | Yours ever truly | John Tyndall

BBAW NL Helmholtz, Nr. 477 BI_6r-v
RI MS JT/1/T/483

1. *short abstract*: very likely an abstract for J. Tyndall, 'Professor Helmholtz on Ice and Glaciers', *Phil. Mag.*, 30 (1865), pp. 393–407 (this is letter 2330 in this volume).
2. *the Vorträge*: lectures (German). See letter 2290, n. 5 and n. 6, and letter 2330.
3. *at Birmingham*: the thirty-fifth annual meeting of the BAAS was held in Birmingham from 6–13 September 1865 (*Brit. Assoc. Rep. 1865*). Helmholtz had accepted an invitation to attend (see letter 2290), but as noted here by Tyndall he was unable to attend.
4. *Kirchhoff*: Gustav Kirchhoff.

From Joseph Plateau 27 November 1865 2332

Gand, 27 N[bre] 1865

Mon Cher Monsieur Tyndall,

J'ai le plaisir de vous adresser deux exemplaires d'un travail que mon fils vient de publier sur la Force Musculaire des Insectes. Mon fils espère que vous voudrez bien accepter l'un de ces exemplaires, et offrir l'autre de sa part à la Société Royale.

J'ai remarqué avec peine que vous n'avez fait encore aucun usage de nos appareils pour la réalisation des figures d'équilibre soit pleines soit laminaires. Quant à ces dernières, je pourrai, j'espère vous offrir dans un *[mois] [ou]* deux, un exemplaire de ma 7ème Série, où vous trouverez des procédés d'une certitude à fort peu près complète pour la préparation d'un bon liquide glycérique.

Agréez, mon cher Monsieur, l'assurance de tous mes sentiments d'affection. | J. Plateau | T.S.V.P.

P.S. Je ne veux pas oublier de vous dire avec quel intérêt j'ai pris connaissance de vos belles expériences sur les rayons calorifiques obscurs provenant du passage à travers une solution d'iode.

Ghent, 27 November 1865

My dear Mr Tyndall,

I have the pleasure of sending you two copies of the work that my son[1] had just published, sur la Force Musculaire des Insectes.[2] My son hopes that you would be kind enough to accept one of those copies, and to offer the other on his behalf to the Royal Society.

I am painfully aware that you have not yet made any use of our apparatuses for the *[implementation]* of the either *[full]* or *[laminar]* figures of equilibrium. As for the latter, I will be able to offer you, I hope in a month or two, a copy of my 7th Series,[3] where you will find almost completely certain methods for the preparation of good glyceric liquid.

Please accept, my dear Sir, the assurance of my best regards and affections | J. Plateau | Please turn over.[4]

P.S. I do not want to forget to tell you with how much interest I read your great experiments on the obscure heat rays resulting from their passing through an iodine solution.[5]

RI MS JT/1/P/97

---

1. *my son*: Félix Auguste Joseph Plateau (1841–1910), a Belgian naturalist who became professor of zoology and comparative anatomy at the University of Ghent in 1870.
2. *sur la Force Musculaire des Insectes*: F. Plateau, 'Sur la force musculaire des insectes', *Bulletins de l'Académie royale des sciences, des lettres et des beaux-arts de Belgique*, 20 (1865), pp. 732–57.
3. *my 7th Series*: J. Plateau, 'Recherches expérimentales et théoriques sur les figures d'équilibre d'une masse liquid sans pesanteur', *Mémoires de l'Académie royale des sciences, des lettres et des beaux-arts de Belgique*, 36 (1867), pp. 1–66. This was translated and published as: J. Plateau, 'Experimental and Theoretical Researches into the Figures of Equilibrium of a Liquid Mass without Weight', *Phil. Mag.*, 33 (1867), pp. 39–47. Plateau published eleven articles in this series, all in the same journal from 1843–68, and later published them altogether as: J. Plateau, *Statique Expérimentale et Théorique des Liquides Soumis aux Seules Forces Moléculaires*, 2 vols (Paris: Gauthier-Villars, 1873).
4. *Please turn over*: in the French version, this was 'T.S.V.P.', for 'tournez s'il vous plait', meaning 'please turn over'.
5. *your great experiments . . . an iodine solution*: see letter 2226, n. 3.

To George Gabriel Stokes [November 1865][1] 2333

Royal Institution of Great Britain | Saturday

My dear Stokes

Many people have experienced difficulty and danger in attempting to make the experiments on Calorescence. In the paper[2] I have described various methods of performing the experiments with ease & safety—Would you think it worthwhile to show them?[3]

I should much rather err in proving the fellows too little than in giving them too much.

Yours very sincerely | John Tyndall.

CUL SC—7656

---

1. *[November 1865]*: the month and year are given based on when Tyndall read his paper to the RS.

2. *paper*: J. Tyndall, 'On Calorescence', *Phil. Trans.*, 156 (1866), pp. 1–24.
3. *to show them*: Tyndall read his paper at the RS on 23 November 1865.

From George Wynne[1] [November 1865][2] 2334

[3]I consider the offer made to you by the University of Oxford[4] a just appreciation of your merits and coming from them an honour, which tho' you declined, you must yet feel gratified by, but I am glad at the same time that you are sticking to London a better field for you in every respect while your health <u>lasts</u> which take care of.

I am not quite in accord with you in your views of what is "Courage". I do not think a man however strong his convictions may be is justified like Mause Headrigg in Old Mortality[5] in lifting up his voice on every occasion like the sparrow on the house top to testify against an Erastian[6] and backsliding Church if the writer in the P.M.G. had concluded his letter by saying, keep silence now John Tyndall, I could imagine the blood of the old reformer with its Irish admixture being roused and justifying <Handwritten Letter Begins> its convictions, but I think reticence is not cowardice, in matters of religious controversy the settlement of which would involve folios not pages[7]—

Frank[8] has told you that I am ordered to Ireland.[9] My going there has a bright and a dark side. The appointment is a higher indeed one of the highest we have got and a better paid one and I shall be a swell,[10] but what do I now care about being a swell—then on the other hand my only brother[11] has just sold his place near Dublin and came to settle near London in a great measure to be near me and like buckets in a well we are passing one another. And again I have just launched my boy Frank into the arena of London and I considered it of great importance to keep him to a certain extent under home influence, and the absence of the power to exercise that influence I regret in my move most of all—

I will manage to see you before I go. I am sorry I cannot ask you to pay me a visit here but I have begun already to dismantle my house. I expect to be relieved in my command here on the 1st Jan[uar]y.

Lucy[12] desires her kindest remembrance and she will also pay you a visit before we leave

Ever most truly y[ou]rs | Geo Wynne

RI MS JT/1/W/1090
RI MS JT/1/TYP/5/1860

1. *George Wynne*: George Wynne (1804–90), an Irish surveyor, engineer, and military officer. He met Tyndall while employed in the Irish Ordnance Survey (1835–40). Wynne later served with distinction in Greece, China, and the West Indies, rising to general by the time he retired in 1877 ('The Late General Wynne', *Times*, 12 July 1890, p. 11).
2. *November [1865]*: the year is given by mention of Tyndall declining the offer of a position at Oxford (see n. 4).
3. This letter is a fragment, with at least the first page missing. The first section of the fragment is based on the LT Typescript only.
4. *offer made to you by the University of Oxford*: Tyndall had been offered the professorship of natural philosophy at Oxford following the death of the previous incumbent Robert Walker (see letters 2320 and 2324).
5. *Mause Headrigg in Old Mortality*: 'Mause Headrigg' is a character in the 1816 novel *Old Mortality* by Walter Scott (1771–1832), an irrepressible woman whose speech is often full of vitriol.
6. *Erastian*: an adherent of the (supposed) doctrines of Erastus; one who maintains the complete subordination of the ecclesiastical to the secular power (*OED*).
7. *the writer in the P.M.G. . . . folios not pages*: Wynne was probably referring to Tyndall's recent letters to the *PMG* regarding prayer and cholera (see letters 2314 and 2318).
8. *Frank*: Frank Wynne (see letter 2235, n. 4).
9. *ordered to Ireland*: George Wynne served as a Commander of the Royal Engineers (C.R.E.) in Ireland from 1866–70.
10. *swell*: a fashionably or stylishly dressed person; hence, a person of good social position, a highly distinguished person (*OED*). Wynne was probably referring to the uniform for his position with the Royal Engineers.
11. *my only brother*: not identified.
12. *Lucy*: Lucy Wynne, the only daughter of George Wynne and Anne Wynne (née Osborne, 1808–64).

To Jane Welsh Carlyle [4 December 1865][1] 2335

Royal Institution | Albemarle Street | Monday evening

Dear M^rs^. Carlyle

I want Mr Carlyle to come with me to see Jefferson act Rip Van Winkle.[2] I am told it is first rate. Will you let him come? If so be good enough to fix the day so as to enable him to let me know when we meet at the Athenaeum[3] on Wednesday.

Could he manage it on Friday evening?[4]

Yours ever | John Tyndall

RI MS JT/1/T/146
RI MS JT/1/TYP/1/189

1. *[4 December 1865]*: LT wrote this date on the letter under 'Monday evening' and included it in her typescript. This is likely the postmark date from the envelope, and the letter was written the day before, Monday, 4 December 1865, and sent the following morning.
2. *Jefferson act Rip Van Winkle*: 'Rip Van Winkle' is a short story by American author Washington Irving (1783–1859), which follows a Dutch-American villager in colonial America named Rip Van Winkle who falls asleep in the Catskill Mountains and wakes up 20 years later, having missed the American Revolution. American actor Joseph Jefferson (1829–1905) performed various dramatizations of the character on the nineteenth-century stage. In 1865, he began performing the character at the Adelphi Theatre in London.
3. *Athenaeum*: see letter 2253, n. 8.
4. *on Friday evening?*: there is no indication in Carlyle's own correspondence that he went with Tyndall to this play.

From Thomas G. Bonney[1] 8 December 1865 2336

My dear Professor Tyndall

I am exceedingly obliged to you for the copy of your valuable article on Prof[essor]: Helmholtz' memoir.[2] I have read it with much interest—One paragraph of it is of peculiar interest to me, that on page 12, in which you speak of the effect of thaw on ice. During the summer I visited three Glaciers near Annecy[3] and there observed the prismatic shadow of which Mr Browne[4] has spoken in his book 'Ice Caves'.[5] The walls (of rock) in places were covered by sheets of transparent ice, some three inches thick, which was marked all over with a delicate network caused by the want of optical contact between the sides of the prisms whose axes were perpendicular (roughly) to the surface—I broke out a large fragment with the axe—and then by the pressure of my hands broke it up into its component prisms. In the enclosed sheet[6] fig 1. 2. are made pictures from memory of the *[prisms]* of the mass and of one very perfect single prism—the impression produced in my mind was that the normal shape of the faces of the prisms was a hexagon: many certainly had more than six sides but then the angles did not look to me sharp and clear—so that had they been crystals I should have said they had been imperfectly crystallized—the surface of the mass reminded me of those septaria stones (where found I forget) in which you have the cracks produced by shrinking of the clay filled with mineral mixture—the temperature of the air in the caves raised from about 1° to 2° centigrade—the air outside being about 9° or 10° The surface of the ice was damp and a little water lay about on the floor in puddles. So that there was a slight thawing going on. I have no doubt that at night there would be frost. I should say that Mr Browne, who was with me, doubted my hexagonal rule. I wish that you or some competent person would go and examine these caves. The excursion can be made in a day from Annecy; the scenery (mid Alpine) is very fine; and there is a wild beauty in the caves that would delight you. I could give you fuller information, if desired—Unfortunately I know nothing about crystallography mineralogy or chemistry—so I contented myself with a general observation of the phenomena of these prisms, so that I might prevail in some more learned man to go and settle the question. I don't care about committing myself to statements which may turn out to be mistaken. Fig. 3[7] is a rough section of a sheet of ice.

Yours very truly, | T. G. Bonney | S. John's College | Cambridge Dec. 8. 1865

RI MS JT/1/B/118

1. *Thomas George Bonney*: Thomas George Bonney (1833–1923), an English geologist, mountaineer, lecturer at the University of Cambridge, member of the Alpine Club, and later president of the Geological Society of London from 1884–6 and Vice President of the RS from 1898–9 (*ODNB*).
2. *your valuable article on Prof[essor]: Helmholtz' memoir*: J. Tyndall, 'Professor Helmholtz on Ice and Glaciers', *Phil. Mag.*, 30 (1865), pp. 393–407 (this is letter 2330 in this volume).
3. *Annecy*: a city in southeastern France.
4. *Mr Browne*: George Forrest Browne (1833–1930), an English bishop, archaeologist, and one of the founding members of the Alpine Club (*ODNB*).
5. *'Ice Caves'*: G. F. Browne, *Ice-Caves of France and Switzerland: A Narrative of Subterranean Exploration* (London: Longmans, Green, and Co., 1865).
6. *enclosed sheet*: enclosed sheet missing.
7. *Fig. 3*: the enclosed sheet in which this figure was given is missing.

To Elizabeth Dawson Steuart 10 December 1865 2337

The Athenaeum Club[1] | 10th. Dec. 1865

My dear Mrs Steuart,

In a note[2] from my sister[3] received a few days ago I was reminded that I had not written to you, the friend of my boyhood, for a long, long time.[4] This is too true. The pressure of my duties here, the concentrated attention which they demand, so direct and urge my thoughts in a special channel as to make me forget those promptings of friendship which would be potent with a less occupied mind. Add to this my intervals of imperfect health, when my heart and its energies are low, and you have the causes of that omission of which my sister's letter reminds me.

Her condition moreover has been a cause of anxiety to me.[5] Its first announcement, in rather incautious terms was a great shock to me. I went to Ireland as soon as I could and made the arrangement which on the face of it seemed best by engaging a regular medical man[6] to look after her safety. I did not expect that he could cure her, but his care was the best means that I could invoke to prevent her from doing injury either to herself or my mother. I was exceedingly disturbed on being applied to some time ago for money by her attendant,[7] and more so when I learned that without my mother's[8] knowledge or consent they had gone to Tramore.[9] I myself am of opinion that going to Tramore at the time might have been beneficial to my sister, but the manner in which it was done caused me extreme annoyance.

I should be indeed grateful for any suggestions regarding the care and treatment of my sister. To the great kindness of Miss Roche[10] I have been already deeply indebted.[11] Surrounded as she is it is difficult for me now to ascertain the real wishes of my mother regarding her daughter. Several weeks ago I received a note[12] bearing my mother's name in which it was affirmed that she was aware of my sister's going to Tramore; but the general diction of the note was so different from any thing that my mother would be likely to suggest that I paid no attention to it.

I am at this moment hard at work preparing for a course of lectures to be commenced on the day after Christmas day.[13] For many years Faraday was in the habit of giving a course of lectures at Christmas to young people.[14] But he is now too feeble for the work and so it has fallen upon me. I am also making an effort to write out the lectures with a view to future publication. Another course[15] at the Royal Institution and my annual course at the School of Mines[16] come afterwards: Add to this my original investigations on which alone my permanent reputation will rest and you have before you the map of a busy man.

Wishing you with all my heart a happy new year
believe me | Yours ever | John Tyndall
Excuse the scratches.

RI MS JT/1/TYP/10/3350–1
LT Typescript Only

1. *The Athenaeum Club*: see letter 2253, n. 8.
2. *note*: letter missing.
3. *my sister*: Emily Tyndall.
4. *I had not written . . . long, long time*: the previous extant letter from Tyndall to Steuart is one dated 8 September 1863 (see letter 2032, *Tyndall Correspondence*, vol. 8).
5. *anxiety to me*: Emily Tyndall was suffering from a mental illness which manifested as an extreme form of religiosity (see letters 2313 and 2316).

6. *medical man*: probably the 'Dr. Phillips' referred to in letters 2313 and 2316.
7. *her attendant*: Abbey Doyle.
8. *my mother's*: Sarah Tyndall.
9. *gone to Tramore*: see letter 2313, n. 3.
10. *Miss Roche*: Annette Roche.
11. *already deeply indebted*: Tyndall arranged Annette Roche, a local friend, to help with managing money, accommodations, and suitable clothing for his mother and sister.
12. *note*: letter missing.
13. *course of lectures . . . day after Christmas day*: the RI's annual juvenile Christmas Lectures began in 1827 and still continue. Tyndall delivered the Christmas Lectures a dozen times over his time with the RI, his first in 1861 (see F. A. J. L. James (ed), *Christmas at the Royal Institution: An Anthology of Lectures* (Hackensack, NJ: World Scientific, 2007)). His third time, in 1865–6, was on 'Sound', and the lectures were given on 26, 28, and 30 December and 2, 4, and 6 January (*Roy. Inst. Proc.*, 4 (1862–6), p. 485; *The Athenaeum*, 16 December 1865, p. 825).
14. *Faraday was in . . . to young people*: over his time with the RI, Michael Faraday delivered the annual juvenile Christmas Lectures nineteen times (ten of those in a row from 1851–60).
15. *Another course*: Tyndall delivered ten lectures 'On Heat' at the RI on Tuesdays and Thursdays from 23 January to 23 February 1866 (*Roy. Inst. Proc.*, 4 (1862–6), pp. 485; *The Athenaeum*, 16 December 1865, p. 825).
16. *annual course at the School of Mines*: a course of thirty-two lectures at the Royal School of Mines on 'Magnetism, Electricity, Sound, Light, and Heat', starting on 1 May 1866, on Mondays through Fridays at 3pm (*The Athenaeum*, 21 April 1866, p. 513). On the School of Mines, see letter 2259, n. 17.

From Elizabeth Dawson Steuart — 15 December 1865 — 2338

Private | Dec. 15—65

My dear John,

It gave me very great pleasure to see a letter from you[1] once more, for although I made every allowance for want of time, I felt it strange that you should be so long without sending a line to one whose friendship and kind feelings, you could not doubt. I should indeed be delighted, if I could in any way suggest or advise, about the management of your Mother[2] and Sister.[3] Being on the spot here, I have many opportunities of observation, which you, living at such a distance cannot command; and as from what I hear, they are by no means satisfied with the present mode carried on, I would candidly give you my opinion, in strict confidence, as I do not wish to interfere with some of the parties,[4] who I am told exercise unlimited authority, to the great annoyance of those who should be made happy and comfortable in every way. Poor Emma seems very well in health, and perfectly rational divested of all the odd fancies which had taken possession of her mind. You are paying a heavy sum yearly, without benefit to her, in fact the mind is beyond the control of medicine, and in case of illness, there can be no difficulty in having advice, giving your Mother the option of calling in whom she pleases, she being the natural guardian of her child. It may be right to retain an attendant for Emma, and the present one[5] is, I believe, attentive, but being a Romanist,[6] is not calculated to sooth or give comfort, where both are needed. I have no doubt a respectable Protestant could be found, if sought for.

I have now written you my ideas, and if you think well of acting upon them, it is easy for you to write and say that "as your Sister's health is now so good, you think she can dispense with the care of a medical attendant." Your money is going for nothing, in that direction, no benefit is derived from it, but an irritating system of espionage adopted, which does more harm than good, and is much complained of. I heard that your Mother was dreadfully hurt and annoyed by her pass book being demanded of her, and I can fully enter into her feelings. I believe she was quite aware of Emma going to Tramore,[7] and it was a great loss to the poor thing to order her home, for the change was doing her good. It pleased and gratified both, that your last remittance was sent directly to your Mother. I have trespassed too long on your time, but my subject is one in which I feel such deep interest you will forgive me.

Above all, please do not allude to me, in any way. You have every right to act as you think best, and I trust the Lord will direct you to the best path.

With every good and kind wish, believe me

my dear John | most sincerely your friend | E. D. Steuart.

RI MS JT/1/TYP/10/3352–52a
LT Typescript Only

---

1. *letter from you*: letter 2337.
2. *Mother*: Sarah Tyndall.
3. *Sister*: Emma Tyndall.
4. *parties*: Abbey Doyle, Annette Roche, and the 'Dr. Phillips' referred to in letters 2313, 2316, and 2337.
5. *present one*: Abbey Doyle.
6. *being a Romanist*: Steuart was just as anti-Catholic as her husband, William Richard Steuart (1798–1852; married in 1820), a local landowner, Captain, and local magistrate High Sherriff of Carlow in 1821, who had employed Tyndall's father John Tyndall (1792–1847). Captain Steuart also was elected for the Office of Poor Law Guardians.
7. *going to Tramore*: see letter 2313, n. 3.

From François-Napoléon-Marie Moigno 16 December 1865 2339

Mon cher Mr Tyndall

L'expérience d'acoustique que j'ai faite dans l'une de mes revues orales du progrès est celle du piano géologique formé de cailloux ou silex convenablement choisis, tenus en suspension sur des cordes, et qui rendent de très beaux sons quand on les frappe avec un autre petit caillou. Peut-être verrez vous à Londres au printemps cet instrument antédiluvien.

En fait d'expériences d'acoustique, il y a toute la collection de Koenig dont vous n'avez je crois qu'une très petite partie. Il a du vous envoyer ces jours-ci son dernier catalogue.

Je viens de quitter mon éditeur Mr Giraud, parce qu'il est indolent et qu'il est impossible d'obtenir de lui aucun compte. Croiroz vous que depuis huit mois, je lui demande en vain tous les jours le compte bien facile à faire de la Chaleur sans pouvoir l'obtenir. J'espère avoir trouvé un éditeur actif, intelligent, habile, très répandu qui donnera une nouvelle impression à mon journal Les Mondes et à mes autres publications. Faites pour moi un peu de propagande au commencement de l'année. Je crains de perdre l'abonnement de Sir Roderik Murchison, obtenez, je vous en prie, qu'on abonne cette année la bibliothèque du musée de géologie pratique Jermyn-Street. Vous verrez que dans le dernier numéro des Mondes j'ai fait en votre faveur une importante réclamation contre Mr Delaunay.

A vous de coeur | l'abbé F. Moigno | 2 rue d'Erfurt. | 16 décembre 1865.

My dear Mr Tyndall,

The acoustic experiment that I have in one of my oral reviews of progress[1] is that of the geological piano,[2] made of suitably chosen pebbles or flint, held suspended on ropes, and which produce beautiful sounds when you hit them with another small pebble. Perhaps you will see in London, at spring, this antediluvian instrument.

On the subject of acoustic experiments, there is an entire Koenig collection,[3] of which you have only a small part. He should have sent you his latest catalogue these days.

I have just left my publisher Mr Giraud,[4] because he is lazy and it is impossible to get anything from him. Could you believe that for the past eight months I have been asking him in vain, every day, about the account of Heat,[5] which is easily made, without managing to obtain it. I hope I found an active, intelligent, skillful and very famous publisher,[6] who will give a new impression to my journal Les Mondes[7] as well as my other publications. Do a little propaganda for me in the beginning of the year. I fear losing Sir Roderick Murchison's subscription, I beg you to please make sure of the subscription of the library of the Museum of Practical Geology Jermyn-street.[8] You will see that in the latest issue of the Mondes I made a substantial claim in your favour against Mr Delaunay.[9]

Yours truly | Father F. Moigno | 2 rue d'Erfurt[10] | December 16th 1865.

RI MS JT/1/M/107

---

1. *one of my oral reviews of progress*: 'Piano géologique de M. BAUDRE', in F.-N.-M. Moigno, *Résumé oral du progrès scientifique et industriel. Conférenece de Juin 1864* (Paris: Êtienne Giraud, 1864), pp. 244–5.
2. *geological piano*: the 'geological piano', later called the 'silex piano' (*silex* being French for *flint*) was an iron-framed 'piano' with flint-stones suspended horizontally at the end of double strings. In order to produce the sound, the stones are struck with a hard instrument. The force of the vibration is not equal upon the entire surface of the stones. If a person wished to play like a virtuoso, it is necessary therefore for him to know the sensitive place of each key. The instruments inventor, Honoré Baudre (dates unknown) of France, exhibited it in London in 1866 ('The Antideluvian Piano', *The Mechanic's Magazine*, 6 April 1866, pp. 214–5).
3. *Koenig collection*: referring to a collection of acoustic instruments created by Karl Rudolph Koenig (1832–1901), a German physicist and instrument maker whose main subject of study was acoustics (*CDSB*). See D. Pantalony, *Altered Sensations: Rudolph Koenig's Acoustical Workshop in Nineteenth-Century Paris* (New York: Spring, 2009).
4. *Mr Giraud*: Étienne Giraud.
5. *<u>Heat</u>*: Giraud published the French translation of Tyndall's *Heat* (1863), which appeared as: J. Tyndall, *La Chaleur considérée come un mode de mouvement*, trans. F. Moigno (Paris: Étienne Giraud, 1864).
6. *publisher*: Jules Rothschild (1838–1900), a Paris-based publisher and naturalist.
7. *Les Mondes*: after leaving the journal *Cosmos*, which he had edited since 1852, Moigno founded the popular science journal *Les Mondes: revue hebdomadaire des sciences* in 1863.
8. *Museum of Practical Geology Jermyn-street*: the Royal School of Mines at Jermyn Street in London, previously called the Museum of Practical Geology and the Government School of Mines and Science Applied to the Arts until 1863, where Tyndall held the Chair of Natural Philosophy between 1859–68.
9. *Mr Delaunay*: Charles-Eugène Delaunay (1816–72), a French astronomer and mathematician best known for his theory of lunar motion. He was director of the Paris Observatory from 1870 until his death by drowning in 1872 (*CDSB*). Moigno was referring to a section of his report on the 11 December 1865 meeting of the Académie des Sciences in his journal *Les Mondes*, in which he shared Delaunay's contribution to the meeting (*Les Mondes*, 9 (1865), pp. 651–7, on pp. 652–3). Moigno quoted from Tyndall's *Heat* (*La Chaleur*; see n. 9) to counter a claim of Delaunay's regarding lunar motion and tides.
10. *2 rue d'Erfurt*: in Paris.

From Joseph Dalton Hooker [18 December 1865][1] 2340

Kew | Monday

Dear old Tyndall

I write principally to let you know, that our Xmas dinner will be at 1.30 and that if you and Hirst, either or both, have no better engagement, there will be Nitrogeneous Elements for both, and no ceremony beyond a hearty welcome.

I was reading your little brochure on Helmholtz[2] with my accustomed gravity and calm enjoyment, when I came upon my name in connection with "Regelation",[3] and my heart gave a bound; it recalled many pleasant days, that present events have relegated almost back to the dark ages of our early friendship. Well! people may ask, "what's in a name"; nothing in many cases, <u>except</u> for the association, and really I mentally squeezed your hands (and did not sing out) for the recollection, and felt myself emotional.

Ever, dear old T. | Aff[ectionate]ly yours | J. D. Hooker.

RI MS JT/1/TYP/8/2564
LT Typescript Only

---

1. *[18 December 1865]*: the day is given as Tyndall joined the Hooker family on Christmas Day, a Monday (as he noted in letter 2343 to Ellen Busk), making this letter written also on a Monday from the week before, 18 December.
2. *little brochure on Helmholtz*: J. Tyndall, 'Professor Helmholtz on Ice and Glaciers', *Phil. Mag.*, 30 (1865), pp. 393–407. This is letter 2330 in this volume.
3. *my name in connection with "Regelation"*: Hooker was referring to Tyndall's recent paper on Helmholtz's views on glacier formation (see n. 2), in which Tyndall wrote: 'It is perhaps worth stating that the term "regelation" was first introduced in a paper published by Mr. Huxley and myself' and that 'it was suggested to us by Dr. Hooker, now Director of the Royal Gardens at Kew' (p. 395). Tyndall's own theory of glacial motion—through the fracture and regelation of ice—is the topic of many letters in, and discussed in the introduction to, *Tyndall Correspondence*, vol. 6. The paper referred to by Tyndall was: J. Tyndall and T. H. Huxley, 'On the Structure and Motion of Glaciers', *Phil. Trans.*, 147 (1857), pp. 327–46; this was also published the following year as: J. Tyndall and T. H. Huxley, 'On the Structure and Motion of Glaciers', *Phil. Mag.*, 15 (1858), pp. 365–88.

To Joseph Dalton Hooker [21 December 1865][1] 2341

Royal Institution. Thursday.

My dear Hooker

I accept your offer of hospitality on Christmas day[2]—Hirst also would gladly come but he is going to Paris to hand over the Copley medal to his friend Chasles.[3]

Wherever and whenever the glaciers are spoken of in future years, our names will be linked together,[4] and there is no association on the face of the earth that I prize more.

Yours affectionately | John Tyndall

RI MS JT/1/TYP/8/2564
LT Typescript Only

1. *[21 December 1865]*: Hooker invited Tyndall to Christmas in a Monday letter that has been dated to [18 December 1865], making this Thursday letter 21 December (see letter 2340). Additionally, Hirst left for Paris on Friday, 22 December 1865 (see letter 2343, n. 3), making this letter written on a Thursday before Hirst left.
2. *offer of hospitality on Christmas day*: see letter 2340.
3. *going to Paris . . . his friend Chasles*: see letter 2343, n. 3, n. 4, and n. 5.
4. *names will be linked together*: see letter 2340.

From Richard Dawes 23 December 1865 2342

Deanery, Hereford, | Dec. 23rd, 1865.

My dear Tyndall,

Many thanks for a copy of your Article on Ice and Glaciers[1] and I need not tell you that anything which comes from you always interests me and Mrs Dawes[2] also. I see you are going to give the Xmas lecture to youngsters at the Institute.[3] I wish I were one of them.

We have had very mild weather here during the last two months, but constant fogs and therefore somewhat gloomy, almost as bad as in London.

Mrs Dawes joins me in kind and affectionate regards, and believe me, my dear Tyndall,

Always sincerely yours | R. Dawes.

RI MS JT/1/TYP/1/327A
LT Typescript Only

1. *your Article on Ice and Glaciers*: J. Tyndall, 'Professor Helmholtz on Ice and Glaciers', *Phil. Mag.*, 30 (1865), pp. 393–407. This is letter 2330 in this volume.
2. *Mrs Dawes*: Mary Helen Dawes.
3. *Xmas lecture to youngsters at the Institute*: see letter 2337, n. 13.

To Ellen Busk[1] 24 December 1865 2343

24th. Dec 1865.

My dear Mrs Busk

I found your cards and was sorry that I had missed yourself. Still their appearance as I came in from the gray drizzle of external nature was like a gleam of blue sky in the gray firmament of my brain. I should like to have wished you audibly a merry christmas and a happy new year; but the strength of this wish is not impaired by the intervention of this sheet of paper. Health and happiness then to you and yours for many a coming year.

I am writing a little book on Sound[2] which I trust will please you. I seem to myself a muddy mortal—so muddy that it often requires long effort to clear the mud away and permit the sunshine to penetrate the mind. But once permitted to enter my light is probably as bright as that of other men: or perhaps it is the contrast with the preceding darkness that makes me think it so. I have been cheered at intervals by pleasant flashes of illumination as I continued to think and write on this subject of sound—so pleased indeed that were sufficient time afforded me to realize my own ideal regarding the work I think it might be made agreeable to the world in general. The illumination has had reference merely to form and order of exposition, but it has been accompanied by a delight almost equal to that which belongs to actual discovery. Knowledge may be in the brain, but it may be there in a chaotic state, and the falling of this chaos into order under the action of the intellect is like the resolution of noise into music.

Hirst with the piety which belongs to his nature has gone to Paris[3] to present the Copley medal[4] to the great geometer Chasles:[5]—a man whose spirit has so over-ridden the infirmities of his flesh that at the age of 72, and in opposition to illness of the gravest character, he has accomplished work so grand as to cast the earlier achievements of his own powerful intellect into the shade.

Tomorrow I have an early dinner with Hooker's boys and girls,[6] while I hope to spend the evening with the boys and girls of Bence Jones.[7] I trust the gods will grant me rest tomorrow night so that I may meet my young friends next day with a cheerful face.

Goodbye | Yours always | John Tyndall

RI MS JT/1/TYP/1/184–5
LT Typescript Only

1. *Ellen Busk*: Ellen Busk (1816–90), the daughter of Jacob Hans Busk (1767–1844). In 1843 she married her cousin George Busk with whom she had four children (*ODNB*).
2. *a little book on Sound*: *Sound* (1867).
3. *Hirst... gone to Paris*: as he noted in his journal, Thomas Hirst left London for Paris on 22 December 1865, and returned on 2 January 1866 (Brock and MacLeod, *Hirst Journals*, pp. 1767–9).
4. *Copley medal*: the Copley Medal, named after Godfrey Copley (1653–1709), is an award given by the RS for outstanding achievements in research in any branch of science. Chasles received the medal in 1865 for 'his historical and original researches in pure geometry'. Hirst presented Chasles with the medal on 24 December 1865 (Brock and MacLeod, *Hirst Journals*, pp. 1768).
5. *Chasles*: Michel Floréal Chasles (1793–1880), a French mathematician who taught at the École Polytechnique (1841–51) and held the chair of geometry at the Sorbonne from 1846 until his death (*CDSB*).
6. *Hooker's boys and girls*: at this time, the children of Joseph Hooker and his wife Frances Harriet Hooker (née Henslow, 1825–74), were: William Henslow Hooker (1853–1942); Harriet Anne Hooker (1854–1945), who would marry the British botanist William Turner Thiselton-Dyer (1843–1928); Charles Paget Hooker (1855–1933); and Brian Harvey Hodgson Hooker (1860–1932). Another daughter, Maria Elizabeth Hooker (1857–1863), died at age 6.
7. *boys and girls of Bence Jones*: the children of Henry Bence Jones and his wife Millicent Bence Jones (née Acheson, 1817–87), were: Millicent Mary Bence Jones (1843–1932), Henry Robert Bence Jones (1844–1912), twins Olivia Mary Bence Jones (1845–1911) and Ralph Noel Bence Jones (1845–1866), Edith Mary ('Edy') Bence Jones (1853–1919), and Archibald ('Archie') Bence Jones (1856–1937). Another daughter, Anabella Mary Bence Jones (1846–1851), died at age 4.

From Julius Robert Mayer 30 December 1865 2344

Hochverehrter Herr Professor!

Endlich gelange ich dazu, wenigstens einen kleinen Theil des Ihnen in Zürich gegebenen Versprechens zu erfüllen und Ihnen probeweise eine kurze Beschreibung einiger meiner Erlebnisse zu schicken. Calamitäten verschiedener Art die inzwischen mich und meinen engeren Familienkreis betroffen, veranlassten eine sonst unverzeihlichen [sic] Verzögerung. Es würde mich sehr freuen, wenn Sie mir mittheilen wollten, auf welche Weise Sie etwa weitere Mittheilungen abgefasst zu sehen wünschten, um, so weit es meine schwachen Kräfte gestatten, Ihren Erwartungen nachzukommen. Ausser dem erwähnten Manuscripte erhalten Sie hiebei [sic] einen Aufsatz „über das Fieber" und in Abschrift einen kleinen Artikel „über die Herzkraft" nebst einer Beurtheilung desselben von Donders; auch habe ich Ihnen Scherzes halber meine Inaugural-Dissertation beigelegt. Von meiner, im Manuscripte erwähnten Frau Schwägerinn, von der ich Ihnen auch in Zürich gesprochen, die ich eine wahre Beschützerinn wissenschaftlicher Bestrebungen nennen kann, lege ich auch eine Photographie bei.—Seit dem Sie sich meiner in London so kräftig angenommen, wird mir die Freude der Anerkennung auch in meinem engeren Vaterlande mehr und mehr zu Theil und ich kann nicht aufhören mit den Gefühlen des höchsten Dankes und der Verehrung Ihrer zu gedenken.—Sollten Sie Herrn Prof. Brayley, an welchen ich zu Anfang dieses Jahres geschrieben, sprechen, so bitte ich, mich demselben zu empfehlen.—

Indem ich schliesslich zum eingetretenen Jahreswechsel meine besten Glückwünsche darbringe und um Erhaltung Ihres Wohlwollens bitte, zeichne ich verehrungsvoll

Ihr | ergebenster | Dr. J. R. Mayer

Heilbronn (Württemberg) | 30 Decemb. 1865.

Most respected Professor!

I am finally able to fulfill at least a small part of the promise that I made to you in Zurich[1] and am sending you tentatively a short description of some of my experiences.[2] Calamities of various kinds that in the meantime have affected me and my closest relatives were responsible for an otherwise inexcusable delay. I would be very glad if you would tell me how you preferred to receive future communications and will try to fulfill your expectations as far as my limited powers allow. Besides the manuscript mentioned, I enclose an essay "on fever"[3] and a copy of a short article "on cardiac force",[4] including a review of the same[5] written by Donders;[6] for amusement's sake, I have also included my inaugural dissertation.[7] I also enclose a photograph of my sister-in-law,[8] who is mentioned in my manuscript, of whom I spoke to you in Zürich and whom I see as a true protectress of scientific endeavours.—Ever since you advocated so actively on my behalf in London,[9] I have also enjoyed more and more recognition in my own, closer fatherland and I cannot but think of you with the greatest feelings of gratitude and admiration.—

Should you speak to Prof. Brayley,[10] to whom I wrote at the beginning of this year, I ask you kindly to recommend me to him.—Finally, in offering my best wishes on the occasion of the turn of the year and praying for your continued good will, I sign respectfully,

Your | most devoted | Dr. J. R. Mayer

Heilbronn (Württemberg) | 30 December 1865.

RI MS JT/1/M/84
RI MS JT/1/M/85
RI MS JT/1/TYP/7/2515–6

1. *in Zürich*: Tyndall had passed through Zurich in August 1865 on his way to a scientific meeting in Geneva (see letter 2270, n. 4).
2. *short description of some of my experiences*: Mayer had enclosed an untitled, 18-page autobiographical sketch, a transcription of which is held at the RI (the original was returned to Mayer's sister-in-law Emma Closs in 1897). See K. L. Caneva, *Robert Mayer and the Conservation of Energy* (Princeton: Princeton University Press, 1993), pp. 9–10, 336, 411.
3. *essay "on fever"*: J. R. Mayer, 'Ueber das Fieber. Ein iatromechanischer Versuch', *Archiv der Heilkunde*, 3 (1862), pp. 385–94.
4. *short article "on cardiac force"*: J. R. Mayer, 'Ueber die Herzkraft', *Archiv für physiologische Heilkunde*, 10 (1851), pp. 512–5.
5. *review of the same*: not identified.
6. *Donders*: Franz Cornelius Donders (1818–89), a Dutch physiologist and pioneer in the field of ophthalmology (P. J. Blok and P. C. Molhuysen (eds), *Nieuw Nederlandsch biografisch woordenboek*, 10 vols (Leiden: Sijthoff, 1911–37), vol. 1 (1911), pp. 727–9).
7. *inaugural dissertation*: J. R. Mayer, *Ueber das Santonin: eine Inaugural-Dissertation welche zur Erlangung der Doctorwürde in der Medicin & Chirurgie unter dem Praesidium von Wilhelm von Rapp, im July 1838 der öffentlichen Prüfung vorlegt Julius Robert Mayer aus Heilbronn* (Heilbronn: Maximilian Müller, 1838).
8. *my sister-in-law*: Emma Closs, the sister of Mayer's wife Wilhelmine Regine Caroline Closs (1816–99; married on 31 May 1842) (K. L. Caneva, *Robert Mayer and the Conservation of Energy* (Princeton: Princeton University Press, 1993), p. 330).
9. *advocated so actively on my behalf in London*: Tyndall was a champion of Mayer's work. See the introduction to *Tyndall Correspondence*, vol. 8; and *Ascent of John Tyndall*, pp. 170–87.
10. *Prof. Brayley*: Edward William Brayley (1801–70), a British writer and lecturer on science and frequent contributor to the *Phil. Mag.*, to which he also contributed editorial assistance from 1823–44 (*ODNB*).

From Henrietta Huxley [31 December 1865][1] 2345

Dear Dr Tyndall

Hal[2] has a ticket of admission to the platform for you—for next Sunday and we want you to come here and go with us.[3]

If I can, you shall have some dinner at five—but if not will you put up with a late tea—alias cold dinner—as one of my maids will be away on a sad errand—to attend her sister's funeral.

I am quite sure that Hal will give a good lecture—you know my faith in him but I am nervous as to the surrounding circumstances next Sunday. The secretary[4] seems to be such an idiot—the circular issued so absurd—that there is really legitimate scope for making the report of the whole affair farcical.

Ever | most truly Yours | Nettie Huxley | If Mr Hirst is back my best regards to him.[5]

RI MS JT/1/TYP/9/2873
LT Typescript Only

1. *[31 December 1865]*: the day is given by relation to a lecture presumably a week later, on 7 January 1866, as Henrietta twice referred to the lecture as occurring on 'next Sunday'.
2. *Hal*: nickname for Thomas Huxley.
3. *go with us*: Henrietta was inviting Tyndall to the first of a short-lived series of Sunday lectures, delivered by her husband Thomas Huxley. A predecessor to the Sunday Lecture Society (SLS) founded in 1869, Sunday Evenings for the People was initiated in 1866 by the National Sunday League (NSL). Huxley's famous lecture, 'On the Desirableness of Improving Natural Knowledge', took place on the evening of 7 January 1866. On Sunday Evenings for the People, see *Ascent of John Tyndall*, pp. 220–21; R. Barton, 'Sunday Lecture Societies: Naturalistic Scientists, Unitarians, and Secularists Unite against Sabbatarian Legislation', in G. Dawson and B. Lightman (eds), *Victorian Scientific Naturalism: Community, Identity, Continuity* (Chicago: University of Chicago Press, 2014), pp. 189–219); and *The X Club*, p. 427–40.
4. *secretary*: R. M. Morrell (b. 1823), a jeweler, skilled workman, and founder and honorary secretary of the NSL (R. Barton, 'Sunday Lecture Societies', p. 193–4).
5. *If Mr Hirst is back my best regards to him*: see letter 2343, n. 3.

1866

From Charles Mackay[1] 1 January 1866 2346

34 Kensington Gardens Square | Jan. 1, 1866.

My dear Sir,

I have just returned to England, after a long sojourn in the United States, and find among a list of names of those whose acquaintance I determined to renew on my return, that of my old friend John Tyndall, I write this from the Reform Club,[2] where I look in almost every day from half past 12 to half past one, or two, and shall be truly glad to see you once again. It is many years since we met, but I have watched your career with interest, and rejoiced in your success.

Believe me | Yours very truly | Cha[rle]s. Mackay

John Tyndall Esq. LL.D.[3]

RI MS JT/1/TYP/2/783
LT Typescript Only

---

1. *Charles Mackay*: Charles Mackay (1814–89), a Scottish poet and journalist, mostly known for his study on crowd psychology in *Extraordinary Popular Delusions and the Madness of Crowds* (London: Robert Bentley, 1841), which covered such subjects as economic bubbles, the Crusades, witch mania, haunted houses, alchemy, and fortune telling, among others (*ODNB*). Mackay's son attended Queenwood College when Tyndall taught there.
2. *Reform Club*: a private gentleman's club in London, founded in 1836, whose members espoused progressive political ideas.
3. *LL.D.*: see letter 2274, n. 2.

To Henry Bence Jones 4 January 1866 2347

Royal Institution, | 4th. Jan. 1866. | 11. P.M.

My dear Friend

Your very anxiety to day to say or do nothing disagreeable to me caused me to think more seriously than I otherwise should have done of the subject of our conversation. It has been my heart's most earnest desire to see the evening of Mr Faraday's life peaceful and happy,[1] and I never to my knowledge committed the slightest act inconsistent with this desire. I have never showed a trace of a wish to interfere in any way with the rule and governance of the Royal Institution, and the fact that for a dozen years I remained in this place[2] without being once invited to the board of managers is to be entirely ascribed to my respect and attachment for him. Under other circumstances I should not have done it. I now desire to continue this line of conduct, and to interfere neither directly nor indirectly with the arrangements of the Institution. This desire will be best carried out by leaving me and my assistants entirely to ourselves, and I would therefore pray that this may be the case.

Regarding the relation which subsists between me and the assistants granted to me I would ask permission to say that if the managers[3] thought it necessary to limit my control over them—which of course they now have a perfect right to do—I should deem it my duty to relinquish my connexion with the Institution. I trust that I have given practical proofs of the sorrow that this would cause me, but a divided allegiance on the part of my assistants I could not accept.

This subject ran through my thoughts during the whole of to days lecture, and did not by any means contribute to its efficiency. I am sorry to think that it will also interfere with my concluding lecture.[4]

If you deem it desirable I should not object to your making this letter known to Miss Barnard.[5] Indeed I should like you to do so, and to accompany it by the assurance that she labours under a grave misapprehension if she imagines that there is the rudiment of a tendency in me to interfere with the arrangements of this Institution. I pray only to be left alone.

Believe me | Yours very truly | John Tyndall

Dr Bence Jones | Hon. Sec. R.I. &c, &c.

RI MS JT/TYP/3/795
LT Typescript Only

1. *evening of Mr Faraday's life peaceful and happy*: at this time Michael Faraday was aged 74 with his health declining. He died on 25 August 1867, an event that would further cement

Tyndall to the RI (the effect of Faraday's death on Tyndall and how it affected his career path is detailed in *Tyndall Correspondence*, vol. 10).

2. *dozen years I remained in this place*: Tyndall was appointed Professor of Natural Philosophy at the RI in 1853.
3. *Managers*: the following were elected Managers of the RI on 1 May 1865: John Peter Boileau (1794–1869), George Busk (1807–86), Warren de la Rue (1815–89), George Everest (1790–1866), Douglas Galton (1822–99), John Peter Gassiot (1797–1887), John Hall Gladstone (1827–1902), Caesar H. Hawkins (1798–1884), Roderick Impey Murchison (1792–1871), Algernon Percy (1810–99), William Pole (1814–1900), Lewis Powell (*c.* 1797–1867), Robert P. Roupell (1798–1886), Joseph William Thrupp (1799–1873), and James Parke (1782–1868). Henry Holland (1788–1873) was elected President, William Spottiswoode (1823–83) was elected Treasurer, and Henry Bence Jones (1813–73) was elected Secretary ('Annual Meeting', *Roy. Inst. Proc.*, 4 (1862–6), p. 434; *ODNB*).
4. *to day's lecture . . . concluding lecture*: Tyndall had been delivering that season's annual juvenile Christmas lectures at the RI, with the last two lectures on 4 and 6 January 1866 (see letter 2337, n. 13).
5. *Miss Barnard*: Jane Barnard.

From Henry Bence Jones 4 January [1866][1] 2348

4 Jan[uar]y 11.30 PM

My dear friend

I must have expressed myself very badly to day & I shall not make it any better by writing.

As far as I possibly can do it I & every one else will let you & your assistant alone, but changes must of necessity come & poor Andersons departure[2] must change arrangements that were under him.

This & nothing else has led to this subject being mentioned now & it was necessary something should be done at once. Had I thought it would have troubled you I would have kept it until Saturday evening. Pray think no more about it but consider that I am sure today to do the best I can for you and everyone in the house.

It is much better I think that your letter[3] did not reach Miss Barnard.[4]

I shall try & see you to morrow but please do not let a thought that I or anyone else had any desire to interfere with you enter your mind.

If I had known that you had any thought about the managers committee I should have spoken about it when I became secretary.[5]

Pray do not let this subject trouble you tomorrow or any day

I am | y[ou]rs ever | H Bence Jones

RI MS JT/1/J/107
RI MS JT/1/TYP/2/727

1. *[1866]*: the year is given as Charles Anderson is recorded as having worked at the RI right up until his death, which is recorded in the RI managers minutes book as occurring on 8 January 1866. The reference to his 'departure' likely refers to him having to leave working at the RI prior to his death several days later. This letter is a reply to letter 2347.
2. *Andersons departure*: Charles Anderson (1790–1866), a laboratory assistant at the RI. He was first employed at the RI in 1827 to assist Faraday in the project (on behalf of the Board of Longitude) of improving optical glass. With the termination of that project in 1830 Anderson became Faraday's laboratory assistant and also assisted him in his RI lectures (*Faraday Correspondence*, vol. 5, p. xlv). On his departure, see n. 1 above.
3. *your letter*: in letter 2347, Tyndall had given permission to Bence Jones to share the letter with Jane Barnard.
4. *Miss Barnard*: Jane Barnard.
5. *when I became secretary*: Bence Jones was elected as secretary of the RI on 3 December 1860.

From Elizabeth Dawson Steuart — 10 January 1866 — 2349

Private | January 10. 66.

My dear John,

Many thanks for the book you so kindly sent me.[1] I know you are very much engaged, at present, and it is with reluctance that I send a few lines to tell you of your poor Mother[2] and Sister.[3] I hear constantly of them, from very respectable people, and they are going on well, but as I believe there are efforts being made to effect their medical attendant[4] being reinstated, and Emma's servant[5] dismissed, I would beg of you <u>not to give credit</u> to any statements that may reach you, from those who are looking to <u>their own interests alone</u>, without making further enquiries, as to the truth of them. I have reason to fear you <u>are</u> or <u>will</u> be worried with letters of complaint, but do <u>not allow</u> them to annoy you, and let matters remain as they are, until you hear further from <u>me</u>, as I have <u>no</u> object, but that of regard and friendship, and never will deceive you. Medical advice did <u>no</u> good to poor Emma, and you may believe <u>me</u> that your money was <u>thrown away</u>: both she and your Mother would be much better if they were not interfered with by disappointed and meddlesome people, whom <u>you</u> do not know, as <u>I</u> do. Emma is quite sensible and collected, and I am sure you may rely on anything she asserts. I <u>know</u> you will not betray me in <u>any</u> way, or mention my name <u>at all</u>, as it would only subject <u>me</u> to annoyance, and I do not wish to bring vindictive tongues upon me.

Whenever you wish to write, I shall be truly glad to hear from you but please direct in different writing to your own, as you will see I am obliged to do.

Ever my dear John | your sincere friend | E. D. Steuart.

Your letter[6] has just reached me, and I add a line to say I am sure the matter written to you about, has been much exaggerated.[7] Your Mother gave Mrs D.[8] leave to go home, and she left her husband and daughter there until her return: She is (I hear from truthful people) most kind and attentive to them and they are both very fond of her, and surely if dismissed it would be very difficult to find a better. Both hate another, who intrudes and annoys them, whose interference now should not be allowed. Your Mother is wonderfully well, for her age and infirmities. If you wish me to go and see her, send me a line which I can show, but the other lady named to you is a relative of ———. I shall make further enquiries and let you know the result. Your Mother entreats of you not to be annoyed about these statements. I hear Mrs D. is quite honest, and does all she can to save expense. The drift of all, is a re-instatement, and dislike to Mrs Doyle. Do not think of it.

RI MS JT/1/TYP/10/3353–53a
LT Typescript Only

1. *book you so kindly sent me*: the book cannot be identified; possibly one of Tyndall's own such as *Heat* (1865) which had been published the year before.
2. *poor Mother*: Sarah Tyndall.
3. *Sister*: Emily Tyndall.
4. *medical attendant*: probably an unidentified man who worked for the 'Dr. Phillips' referred to in other letters (see letter 2313, n. 7, and letter 2383, n. 10).
5. *servant*: Abbey Doyle.
6. *Your letter*: letter missing.
7. *matter written to you about, has been much exaggerated*: letter missing.
8. *Mrs D.*: Abbey Doyle.

From Thomas H. Sanger[1] 10 January 1866 2350

British and Irish Magnetic Telegraph | Company (Limited) |
Dublin Station

John Tyndall Esq. | Royal Institution | Albemarle Street. London.

Sir,

In reply to your favor of the 6th inst,[2] I have known Mr Henry Tyndall[3] for eleven years, during which period, with the exception of a few months, he was in the service of this Company and under my control. So far as Mr

Tyndall's connexion with this Company was concerned, his conduct was above reproach, and he discharged his duties with great ability. He left Queenstown[4] so he told me in consequence of a repugnance to carry out a matrimonial engagement into which he had entered, for the particulars of which I beg to refer you to the Rev. Mr Lombard of Queenstown, who wrote to me on the subject.

I am, Sir, | Your obedient servant | T. H. Sanger.

RI MS JT/2/13c/1304
LT Typescript Only

1. *Thomas H. Sanger*: Thomas H. Sanger (*c.* 1853–82), superintendent of the Dublin station of the British and Irish Magnetic Telegraph Co., at 23 College Green (*Thom's Almanac and Official Directory of the United Kingdom of Great Britain and Ireland: for the Year 1859* (Dublin: Alexander Thom and Sons, 1859), p. 1258; 'Death of Mr. Sanger', *The Electrician*, 9 (21 October 1882), p. 531).
2. *of the 6th inst*: letter missing.
3. *Mr Henry Tyndall*: in his journal entry for 4 February 1866, Tyndall wrote: 'About two years ago I received a letter from a Mr Henry Tyndall who wished me to help him to a situation[.] On enquiry I found that he was the son of a half brother of my father's[.] I could not help him by asking for an improvement of his position as telegraphic operator, and told him frankly that I was too proud to ask favours' (RI MS JT/2/13c/1303). Tyndall continued the journal entry, however, by describing his efforts to help Henry to improve his position, until he received a letter from a woman in Ireland stating that Henry had broken their engagement and taken her money. Henry admitted breaking the promise of marriage but denied taking her money. The woman continued to write and hired a lawyer, but Tyndall ignored her letters (RI MS JT/2/13c/1303–5).
4. *Queenstown*: now Cobh, a seaport town on the south coast of County Cork, Ireland.

To Rudolf Clausius 11 January 1866 2351

11th. January 1866

My dear Clausius

I have often thought since the old year has passed away of writing to wish you and your wife,[1] and the little girls and Johnny[2] "a happy new year". But I have been very busy and therefore put it off. I had this year to give a course of lectures to the young people on Sound.[3] It was Faraday's custom to give those juvenile lectures as they were called, but Faraday will never lecture more, and so the Christmas lectures fall on Frankland[4] and myself. Faraday is like a fine old vine destined I fear soon to fall and pass away as far as his physical frame is concerned. His progress downwards has been exceedingly rapid during the last two years. He seemed to become quite suddenly old, and he is now so feeble as to be hardly able to walk without help—a grand old vine surely, but it is sad to see.[5]

His old assistant Anderson died a day or two ago—a very remarkable man Anderson was. With little intellect, but with a character almost as trustworthy as a law of Nature.[6]

I am now preparing for one of our Friday evening lectures,[7] and it is a difficult piece of work—The subject does not admit of many fine experiments so that its success will depend mainly upon the form in which it is delivered. In these lectures seven or eight hundred people come together, some highly scientific, some not. Lords, Dukes, and Earls, Ladies, Princes and professional men. A curious mixture of humanity which the lecturer must send home contented. I have been working at Radiation and absorption by powdered substances, chemical precipitates for example and find all the experiments that have been hitherto made erroneous. You know the current opinion is that the state of mechanical division in the case of pure powders is the predominant influence—that chemical constitution does not come sensibly into play—this is wholly and entirely wrong.

Did I tell you that I had very nearly gone to Oxford to fill the chair of Natural Philosophy in that university?[8] I was very strongly urged to go by those who had the election in their hands. But when I thought of the antecedents of the old Royal Institution—and when I thought of my friends in London I could not bring myself to desert them. It was an exceedingly nice appointment—they have elected a man named Clifton,[9] an excellent teacher I am told, but he has not yet done any original work. I will now mention a circumstance, not for your sake but for that of your wife. When it was found difficult to get an Englishman of sufficient weight & experience to take the chair, they thought of some foreign man of renown. You & Helmholtz were spoken of, and I firmly believe from what I saw that if you had been able to lecture fluently in English you would have been invited to fill the chair of Natural Philosophy in the University of Oxford.

Mit herzlichen Grüssen[10] | Dein[11] | Tyndall

I am writing out & expanding the lectures on sound—It is a beautiful subject and will make an interesting little book.[12] Vieweg[13] is going to publish 1500 copies of my book on Heat[14]—I suppose Vieweg never thinks of paying the author, though other publishers in Germany do so.

I also send two photographs, the one to Mousson,[15] the other to Escher.[16]

RI MS JT/1/T/194

1. *your wife*: Adelheid Clausius.
2. *little girls and Johnny*: see letter 2270, n. 5.
3. *course of lectures to the young people on Sound*: see letter 2337, n. 13.
4. *Frankland*: Edward Frankland was Professor of Chemistry at the RI from 1863–8.
5. *sad to see*: Faraday died the following year, on 25 August 1867.
6. *His old assistant Anderson . . . as a law of Nature*: for Charles Anderson's death, see letter 2348, n. 1 and n. 2. Although Tyndall found Anderson, Faraday's laboratory assistant of twenty-five years, very useful, he was surprised by his lack of understanding. He wrote in his journal for 6 February 1854, that Anderson had 'mere mechanical knowledge, he never appears to reason upon the experiments' (RI MS JT/2/13b/657).
7. *Friday evening lectures*: Tyndall delivered a Friday Evening Discourse at the RI, 'On Radiation and Absorption with reference to the Colour of Bodies and their State of Aggregation', on 19 January 1866 (*Roy. Inst. Proc.*, 4 (1862–6), pp. 487–92).
8. *nearly gone to Oxford to fill the chair of Natural Philosophy in that university*: late in 1865, Tyndall was approached to become Professor of Natural Philosophy at the University of Oxford but he ultimately declined (see letters 2320 and 2324).
9. *Clifton*: see letter 2320, n. 4.
10. *Mit herzlichen Grüssen*: with kind regards (German).
11. *Dein*: your (German).
12. *interesting little book*: *Sound* (1867).
13. *Vieweg*: Eduard Vieweg.
14. *my book on Heat*: for the German translation of Tyndall's *Heat* (1865), see letter 2296, n. 6.
15. *Mousson*: Johann Rudolf Albert Mousson (1805–90), a Swiss physicist, glaciologist, and malacologist. After the founding of the University of Zurich in 1833 he became a lecturer, then professor of physics there. He held the professorial chair of physics at the École Polytechnique Fédérale and the University of Zurich (1855–78) (*HLS*).
16. *Escher*: Arnold Escher (von der Linth).

To John William Draper 11 January 1866 2352

11th January 1866

My dear Sir

I have long thought of writing to you in acknowledgement of the kind notes with which you have from time to time favoured me;[1] and at length though more tardy than ought to have been the case, I translate intention into act. It was I who sent you the Pall Mall Gazette containing that heretical letter—not too heretical for you, as I might infer from the passages to which you have directed my attention, had I not known your sentiments on such questions long ago.[2]

With regard to your quitting science for history I may say that in my opinion science is a serious loser by your choice. Nor do I think, if you will allow me to say so, that you quite correctly appreciate the estimation in which your scientific labours are held. They certainly are well known and valued immensely by many competent men. My own opinion indeed is that had you held on to science no name on the American continent was more likely than yours to surround itself by a lasting reputation. In fact you did not forsake science to soon to secure this for yourself.[3]

But though you have ceased to pursue science formally I notice in your books the method and spirit of science. Those books thus advance the good cause though in an indirect way.

I trust you will long enjoy health and strength to pursue your labours, and to hold up the example of an independent and courageous mind, in an age when such examples are less numerous than they ought to be.

Would you kindly mention to Professor Youmans[4] that I am working hard at my book on Sound,[5] and that when it is complete I entertain hopes that he will be pleased with it:—it is a beautiful subject.

That letter of mine in the Pall Mall Gazette has brought down upon me an avalanche of religious criticism.[6]

I wish you a happy new year;—the thought just strikes me to ask you to accept a bad photograph of me and to beg of you to have the kindness to hand this second one to Prof. Youmans.

Believe me dear Sir | Most truly yours | John Tyndall

Library of Congress, John William Draper Family Papers

---

1. *kind notes with which you have from time to time favoured me*: letters 2294 and 2326.
2. *such questions long ago*: see letter 2326. In his *A History of the Intellectual Development of Europe*, Draper argued, among many other things, that the 'supernatural' disappears when 'men and nations approach maturity'. Tyndall likely sent Draper a book review of W. E. H. Lecky's *The Rise and Influence of Rationalism in Europe* from the 5 April 1865 issue of the *PMG*.
3. *to secure this for yourself*: Draper discussed leaving science for history in both letters 2294 and 2326.
4. *Professor Youmans*: Edward Youmans.
5. *my book on Sound*: *Sound* (1867).
6. *religious criticism*: Tyndall had written to the *PMG* on the subject of prayer and cholera; see letters 2314 and 2318. See also *Ascent of John Tyndall*, pp. 212–4.

To Thomas Archer Hirst 11 January 1866 2353

11th Jan. 1866

My dear Tom

I am unlucky in almost always missing you when you call upon me. I am very sorry that I missed—your friend if I remember aright (and my memory is sometimes bad) wished to attend my lectures. I send him a ticket for my course in Albemarle Street,[1] which begins on Tuesday the 22nd. Will you kindly hand it to him with my compliments?

The lectures at the School of Mines[2] in Jermyn Street do not commence until the 2nd of April. Should he wish to attend these also he shall find all doors open to him.

A happy new year my dear Tom & believe me always | Affect[ionatel]y yours | John Tyndall

Mss.Ms.Coll.200. American Philosophical Society

1. *my course in Albemarle Street*: see letter 2337, n. 15.
2. *lectures at the School of Mines*: Tyndall was probably referring to the lectures that would actually start in May 1866 (see letter 2337, n. 16). In letter 2409 to Mary Anne Coxe, Tyndall wrote: 'But all through May and June I lecture at the School of Mines'. On the School of Mines, see letter 2259, n. 17.

To Julius Robert Mayer 11 January 1866 2354

Royal Institution | London 11th. Jan. 1866.

My dear friend

I have received and read with extreme interest the memoir of your life.[1] I can quite fancy the effect of excitement upon a sensitive brain, because I have experienced it myself. It is the dull leaden brain that is proof against excitement, not the brain that is able to pierce as you have done the system of physical nature.

I am a plain blunt man who speaks his mind, and therefore you must accept my words as without a trace of flattery when I say that I never turn to your scientific writings without wonder. That you, in a small provincial town, and occupied with the duties of your profession should have shot so far in advance of all other men is to me astonishing. I know no similar case in the history of science.

My dear friend you can well afford to look tranquilly on the men who so grossly misunderstood you. Your name will live and be honoured in science, long after they have been reduced to forgotten dust.[2]

I thank you also for the photograph of your sister-in-law:[3] an exceedingly pleasing countenance. Was the lady in Zurich? If so I shall not forgive you for not introducing me to her. Goodness is imprinted in her countenance and her letter[4] proves her to be as wise as she is good.

I should like much to be informed how you have occupied yourself since 1853, and what is your present occupation. I shall no doubt make use of these documents at some future day.

And now I wish you a happy new year, and many such to come. Yours ever sincerely | John Tyndall

I send you two bad photographs, one for yourself and one for Madame Closs.

RI MS JT/1/T/1066
LT Typescript Only

1. *memoir of your life*: see letter 2344, n. 2.
2. *forgotten dust*: Tyndall was a champion of Mayer's work. See the introduction to *Tyndall Correspondence*, vol. 8; and *Ascent of John Tyndall*, pp. 170–87.
3. *your sister-in-law*: Emma Closs (see letter 2344, n. 8).
4. *her letter*: letter missing.

From Elizabeth Dawson Steuart 12 January [1866][1] 2355

My dear John,

I have made every enquiry in my power into the stories so injudiciously conveyed to you, through one on whose veracity I can depend, Miss Moore,[2] my schoolmistress, and if I ever had any doubts of the motives which dictated the annoying letter sent to you,[3] they would be now quite removed, by the conviction that money in the first place and revenge in the second, were the true reasons of making mountains of molehills—it was very unfair to torment you about such a trifle: the facts of the case, as learned from your Mother[4] were these—On 12th night there was a little party at Mrs Doyle's,[5] her own children, some of the Boughtons and one or two others, amongst them "the servant", brought there, not by Mrs D., but by one of the boys—When time to send them all home your Mother gave Mrs D. leave to go and do so, and just after she got into the house Dr. P.[6] arrived in a great flurry to look for his man.[7] Both your Mother and Emma[8] were in bed, when Mrs D., left them, and her husband was in the kitchen, and daughter of 11 or 12 years old, sitting beside Emma's bed, until Mrs D. returned. This statement made to you was a distorted and exaggerated one, but such is not uncommon, in that quarter: I am credibly informed that Mrs D. is most watchful, kind and attentive, to those in her care, and that they like her very much. If a change were made they might not be suited so well and your pocket would suffer by it. I therefore think it would be very unwise. The interference exercised is not disinterested, for self is uppermost to a very great degree. If you would like to come over and look into matters yourself, I shall be most happy to see you here, if you just send me one line to say when I may expect the pleasure of seeing you. Should you wish me to look after your Mother and Emma, I shall go to them as often as I can, and Miss Moore will see them constantly; both she and I will do all in our power for them. I fear the lady mentioned to you is too much a friend of their late M.D.[9] to see things in their true light. Now my dear John, do not hesitate a moment, in telling me your wishes, for it will be a real pleasure to me to help you in any way. I only want to let it be known that you have asked me to look after those so dear to you, as otherwise, it might be considered officious of me to interfere. If I had you for one hour alone, the day you made that arrangement with ——,[10] I would have tried to prevent it, well knowing how it was likely to turn out. Your poor Mother was put into a very nervous excited state by him,[11] the night he wrote to you, and says she will desire him not to go there until he is sent for—she cannot bear him, and no wonder, for she has had too much annoyance already—His hatred to Mrs Doyle is caused by the suspicion that she was accessory to his dismissal, He was the means of putting her husband out of a situation, leaving a large family starving, some years ago, from revenge also. I have now endeavoured to explain matters and motives, in strict confidence, to you, and will once more request you will make any use of me, you wish. Your Mother is as well as could be expected at her advanced years, and Emma quite sensible, but subject to fits of nervousness, which medical treatment can not remove. We have a first rate Doctor[12] at Bagenalstown,[13] far superior to any in this Co., a nice gentleman most kind and attentive, a general favourite.

Ever most sincerely yours | E. D. S.

RI MS JT/1/TYP/10/3348–9
LT Typescript Only

---

1. *[1866]*: the typescript shows '1865' but LT left a note on it stating that 1866 is more likely.
2. *Miss Moore*: not identified.
3. *letter sent to you*: letter missing; sender not identified, but presumably Tyndall's mother's

medical attendant who grew dissatisfied, the 'man' referred to later in this letter (see n. 7) and in letter 2383.

4. *Mother*: Sarah Tyndall.
5. *Mrs. Doyle*: Abbey Doyle. Her husband and children referred to are not identified.
6. *Dr. P.*: Dr. Phillips (see letter 2313, n. 7).
7. *his man*: not identified, but possibly the same person who wrote the letter noted in n. 3.
8. *Emma*: Emily Tyndall.
9. *M.D.*: not identified.
10. ——: not identified.
11. *him*: not identified, but the same person who wrote the letter noted in n. 3.
12. *first rate Doctor*: James John Trayer (*c.* 1816–77), a doctor in Bagenalstown, County Carlow, Ireland. He was educated at Trinity College, Dublin where he earned a BA (1837) and MB (1839). In 1855 he married Jane Carey (G. D. Burtchaell and T. U. Sadleir (eds), *Alumni Dublinenses* (1593–1860) (Dublin: Alexander Thom & Co., 1935), p. 822; *Thom's Irish Almanac and Official Directory 1870* (Dublin: Alexander Thom & Co., 1870), p. 973). He was a doctor for Tyndall's mother Sarah and his sister Emily, after they split from 'Dr. Phillips' (who is mentioned in letters 2313, 2316, 2337, 2338, 2355, and 2356).
13. *Bagenalstown*: a small town on the River Barrow in County Carlow, Ireland.

From Elizabeth Dawson Steuart 15 January 1866 2356

January 15 66.

My dear John,

I went to see your Mother[1] this day, and I am happy to tell you she is wonderfully well, cheerful and quite clear in intellect. She told me she is entirely satisfied to take the care of Emma[2] on herself, that she does every thing she desires her, and is nearly quite well. In reply to my query as to the present arrangement, she said, in a very decided manner, "yes it <u>is</u> my wish, Dr. Phillips[3] <u>never</u> did anything for Emma, and he shall never come here again." She sent you her love, and seemed pleased to see me. Emma had walked to Bagnalstown,[4] to see a cousin, who is not well, so I did not see her, but I heard from those who do see her constantly, that she is perfectly sensible, and nothing the matter with her, except that at times she is nervous. I hope you have received a letter from me written on the 11th,[5] in which I endeavoured to explain matters to you. Now that I have your directions for going to see your Mother, you need not disguise your writing, my reason for asking it was that I might have been thought interfering. I hope you will send me directions as to any thing you wish done, and make any use you please of me. Miss Payne[6] told me in your Mother's presence, that Mrs Doyle[7] is very kind and attentive. I shall hope to hear from you whenever you can spare a few minutes, and most happy shall I be, if my efforts can in any way contribute to the happiness of those so dear to you.

Ever, my dear John, | sincerely yours | E. D. S.

RI MS JT/1/TYP/10/3356
LT Typescript Only

1. *your mother*: Sarah Tyndall.
2. *Emma*: Emily Tyndall.
3. *Dr. Phillips*: see letter 2313, n. 7.
4. *Bagnalstown*: Bagenalstown (see letter 2355, n. 13).
5. *letter from me written on the 11th*: probably letter 2355, dated 12 January [1866].
6. *Miss Payne*: possibly Maria Payne, a cousin of Tyndall's who lived at Bagenalstown. The family connections have not been positively determined, but she was probably the daughter of Thomas and Mary Payne, and born on 17 November 1818 (Irish Church Records).
7. *Mrs Doyle*: Abbey Doyle.

From Rudolf Clausius 16 January 1866 2357

ZÜRICH 16/1/66

Lieber Tyndall,

Du hast mich durch Deinen lieben Brief sehr erfreut, und ich sage Dir in meiner Frau und meinem Namen den herzlichsten Dank für die freundlichen Glückwünsche zum neuen Jahre, die wir auch unsererseits durch die besten Glückwünsche für Dich und Hirst erwiedern.

Dass Du nach Oxford berufen warst, hatte ich noch nicht gewusst. Ich denke mir, dass Du dort eine sehr angenehme Stellung gehabt haben würdest, die vielleicht weniger aufregend und dafür für Deine Gesundheit zuträglicher gewesen wäre, als deine jetzige Stellung. Aber auf der anderen Seite kann ich mir wohl denken, dass Du eine Anstalt, wo H. Davy und Faraday Deine Vorgänger waren, und die Weltstadt London und Deine vielen Freunde daselbst nicht gern verlassen mochtest. Dass man in Oxford auch an mich gedacht hat, betrachte ich als eine grosse Ehre für mich, bin mir aber ganz bewusst dass ich nicht im Stande gewesen sein würde, Vorlesungen in englischer Sprache vor englischen Zuhörern zu halten.

Vor Kurzem habe ich die Anzeige erhalten, dass das Professorencollegium des Polytechnikums in Wien beim dortigen Ministerium meine Berufung nach Wien beantragt hat. Ich weiss noch nicht, was für Bedingungen man mir von Wien ausstellen wird, halte es aber nicht für unmöglich dass wir hingehen.

Du fragst in deinem Briefe, ob Vieweg bei der deutschen Herausgabe

Deines Buches nicht daran denke, den Autor zu bezahlen. Leider kann ich Dir darüber keine Auskunft geben. Wiedemann, welcher in Braunschweig wohnt und mit Vieweg verkehrt, hat die deutsche Herausgabe mit ihm verabredet; und da Wiedemann, wie ich glaube, in Genf mit Dir über die Sache gesprochen hat, so zweifelte ich nicht daran, dass er alles auf die Herausgabe bezügliche und somit auch die Bedingungen, unter denen sie stattfinden kann, regeln würde. Ich wurde nur gefragt, ob ich eine Bearbeitung oder eine Uebersetzung des Buches zweckmässiger halte, worauf ich schrieb, dass mir eine Uebersetzung zweckmässiger scheine.

Indessen, welches auch die bisherigen Ansichten von Vieweg über die Sache sein mögen, so glaube ich doch, dass man die Sache auch jetzt noch zur Sprache bringen kann. Da das Buch noch nicht herausgegeben ist, so kann Vieweg ja nach Belieben den Preis desselben hoch genug stellen, dass er, auch wenn er dem Autor Honorar giebt, noch gute Geschäfte macht. Ich glaube dass Du in dieser Beziehung ohne Bedenken an ihn oder an Wiedemann schreiben kannst. Solltest Du es aber lieber sehen, wenn ich über die Sache entweder bei Vieweg oder bei Wiedemann anfrage, so bin ich sehr gern dazu bereit, und ich möchte Dich für diesen Fall nur bitten, mir anzugeben, welches Honorar Du, wenn Du befragt wärest, gefordert haben würdest.

Deine Photographie, welche sehr gelungen ist, hat uns ungemein gefreut, und auch Mousson und Escher lassen bestens danken. Escher hat mir versprochen mir eine Photographie von sich zur Gegensendung zuzustellen, und ich werde sie, wenn ich sie rechtzeitig erhalte, einlegen. Mousson muss erst wieder von sich Photographien anfertigen lassen.

Bei der Absendung der Correcturbögen an Hirst, habe ich gesehen dass die hiesigen Postämter nicht darüber unter sich einig sind, ob Correcturbögen nach England als Drucksachen gehen können, oder als Brief frankirt werden müssen. Es würde mir daher lieb sein zu erfahren, ob Hirst in den Fällen, wo ich sie als Drucksachen geschickt habe, in London noch hat eine Nachzahlung machen müssen. Wenn das der Fall wäre, so würde ich sie natürlich künftig als Brief frankiren.

Bei uns geht alles gut. Meine Frau verlangt ausdrücklich von mir, dass ich Dir schreiben soll, der kleine Johnny gedeihe sehr niedlich und sei ein prächtiger Bube. Sie hat mir schon mehrmals ihren Unwillen darüber ausgedrückt, dass ich in meinen Briefen an Dich zu wenig von den Kindern schreibe. Für sie sind die Kinder so sehr Hauptsache, dass ein Brief, der nichts von den Kindern enthält, ihr durchaus unvollständig vorkommt.

Mit herzlichen Grüssen von meiner Frau und mir an Dich und Hirst,

Dein | Clausius.

P.S. Escher beauftragt mich bei Uebersendung des Bildes den herzlichsten Gruss zu bestellen.

ZURICH, 16/1/66

Dear Tyndall,

I was very glad to receive your letter.[1] On behalf of my wife[2] and myself, I want to thank you for your kind wishes for the New Year. We are also replying with our best wishes for you and Hirst.

I did not know that you had been appointed to Oxford.[3] I think that you would have had a very agreeable position there, which would possibly have been less exhausting and therefore more beneficial for your health than the one you have now. On the other hand, I fully understand that you would not want to leave the institute where H. Davy[4] and Faraday were your predecessors, nor would it have been easy to leave the metropolis of London or your many friends. I consider it a great honour that I also was considered at Oxford,[5] but I am perfectly aware that it would be have been impossible for me to give lectures in English to an English-speaking audience.

I have recently been informed that the faculty council of the Technical University in Vienna has made a request to their Ministry regarding my appointment to Vienna. I do not know yet what conditions they will provide for me but I do not consider it implausible to go there with my family.[6]

In your letter you ask whether Vieweg[7] would not consider paying the author for the German edition of your book.[8] Unfortunately, I am not able to give you any information regarding this. Wiedemann, who lives in Braunschweig and associates with Vieweg, has discussed the German edition with him. Since I believe that Wiedemann had discussed the matter with you in Geneva,[9] I never doubted that he would be able to arrange everything related to the publication, including the conditions under which it can be realised. I was merely asked whether I considered a translation or an adaptation more appropriate and I wrote that I found a translation more appropriate.

However, regardless of Vieweg's previous views on this matter, I believe that one can still raise the question. As the book has not been published yet, Vieweg is free to put a price that would be sufficient to pay the author his royalties and still make a nice profit. I believe that you can write to him or to Wiedemann in this regard without hesitation. Should you like me to enquire in this matter with Vieweg or Wiedemann, however, I will gladly do it, but in this case I would like to ask you to indicate the royalty amount that you would have asked for, were you to be asked yourself.

Your photograph, which is very well done, delighted us excessively and Mousson[10] and Escher[11] likewise send their thanks. Escher has promised to send me a photograph of himself in return, and I will enclose it if I receive it in time. Mousson will first have to have photographs made of him.

I noticed when mailing the proof sheets to Hirst[12] that the local post offices are disagreed whether proof can be sent to England as printed matter or stamped as a letter. I would therefore very much like to know whether on the occasions when I sent them as printed matter Hirst had to pay an additional fee in London. Should this be the case, I will certainly send them as stamped letters in the future.

We are all doing fine. My wife insists particularly to write to you that little Johnny[13] is growing into a cute and gorgeous little fellow. She has expressed her disappointment to me on several occasions that I write too little about the children.[14] For her, the children are so much at the centre of everything that she regards a letter to be quite incomplete if it does not contain anything about them.

With best and warmest wishes from my wife and myself to you and Hirst,

Your | Clausius.

P.S. Escher asks me to send you his best wishes upon sending the picture.

RI MS JT/1/TYP/7/2284–6
LT Typescript Only

---

1. *your letter*: see letter 2351.
2. *my wife*: Adelheid Clausius.
3. *appointed to Oxford*: see letter 2320, n. 4.
4. *H. Davy*: Humphry Davy (see letter 2245, n. 19).
5. *I was also considered at Oxford*: in letter 2351, Tyndall informed Clausius that 'When it was found difficult to get an Englishman of sufficient weight & experience to take the chair, they thought of some foreign man of renown. You & Helmholtz were spoken of, and I firmly believe from what I saw that if you had been able to lecture fluently in English you would have been invited to fill the chair of Natural Philosophy in the University of Oxford'. In November 1865, Tyndall had also written to Clausius that during a trip to Oxford he 'was pleased to observe the high opinion of you which is entertained by the scientific men of that university' (letter 2328).
6. *go there with my family*: Clausius decided not to take a position in Vienna (see letters 2387 and 2426).
7. *Vieweg*: Eduard Vieweg.
8. *German edition of your book*: for the German translation of Tyndall's *Heat* (1865), see letter 2296, n. 6.
9. *in Geneva*: Tyndall had attended the 1865 meeting of the Société Helvetique des Sciences Naturelles (see letter 2270, n. 4).
10. *Mousson*: Johann Rudolf Albert Mousson (see letter 2351, n. 15).
11. *Escher*: Arnold Escher (von der Linth).
12. *mailing the proof sheets to Hirst*: see letter 2282, n. 4.
13. *little Johnny*: Rudolf John Clausius.
14. *the children*: see letter 2270, n. 5.

To William Frederick Pollock [19 January 1866][1] 2358

Friday

My dear Pollock,

Can you take charge of Carlyle from his house[2] to the Institution this evening:[3] I earnestly hope you can

Yours ever | John Tyndall

It was Mrs Carlyle's[4] <u>idea</u> to ask you to do this—though it is my <u>act</u>.

RI MS JT/1/TYP/6/2197
LT Typescript Only

1. *[19 January 1866]*: the date is given since Pollock wrote in his journal for 19 January 1866: 'Tyndall lectured at the R. I. on Radiation of Heat, the amount of which depends on chemical constitution, and not on colour. I fetched Carlyle from Chelsea to hear the lecture' (*Remembrances*, vol. 2, p. 140).
2. *his house*: 5 Cheyne Row, Chelsea, the London home of Thomas Carlyle since 1834.
3. *to the Institution this evening*: Tyndall delivered a Friday Evening Discourse at the RI, 'On Radiation and Absorption with reference to the Colour of Bodies and their State of Aggregation', on 19 January 1866 (*Roy. Inst. Proc.*, 4 (1862–6), pp. 487–92).
4. *Mrs Carlyle's*: Jane Welsh Carlyle.

From William Frederick Pollock [19 January 1866][1] 2359

59, Montagu Square, | London, W.[2] | Friday Morning

My dear Tyndall

I shall be proud to assist in promoting the object you have for this evening—armed with the Habeas Corpus of one great Philosopher,[3] I will proceed to the dwelling of the other,[4] & bear my humble part in bringing together Morals & Physics—I will write a note to Carlyle by post to prepare him for the arrival of my triumphal car at about 8—

Juliet,[5] Valentia Donne[6] & Fred[7] will go from home—

Yours ever | W. F. Pollock

RI MS JT/1/P/256
RI MS JT/1/TYP/6/2198

1. *[19 January 1866]*: the date is given since Pollock wrote in his journal for 19 January 1866: 'Tyndall lectured at the R. I. on Radiation of Heat, the amount of which depends on chemical constitution, and not on colour. I fetched Carlyle from Chelsea to hear the lecture' (*Remembrances*, vol. 2, p. 140).
2. *59, Montagu Square, | London, W.*: the Pollocks' residence in Marylebone, London.
3. *Habeas Corpus of one great Philosopher*: here Pollock is metaphorically using 'habeas corpus', a Latin term meaning 'a writ issuing out of a court of justice, or awarded by a judge in vacation, requiring the body of a person to be brought before the judge or into the court for the purpose specified in the writ' (*ODNB*), to refer to Tyndall's request in letter 2358 that Pollock bring Carlyle to his Friday Evening Discourse at the RI (see letter 2358, n. 3).
4. *dwelling of the other*: 5 Cheyne Row, Chelsea, the London home of Thomas Carlyle since 1834.
5. *Juliet*: Juliet Pollock.
6. *Valentia Donne*: a daughter of William Bodham Donne (1807–82), an English journalist and essayist (*ODNB*), and a friend of the Pollocks.
7. *Fred*: Frederick Pollock.

From Jacques-Louis Soret [January 1866][1] 2360

X. *Intelligence and Miscellaneous Articles.* |
ON THE DENSITY OF OZONE. BY M. SORET. |
[(Abstract of a letter to Professor Tyndall.)][2]

I OUGHT to have written to you sooner, but I have been much engaged in trying to finish part of my investigation on the density of ozone,[3] of which I have spoken to you. You are aware of the difficulties of the question, and of the impossibility of determining this density by weighing; for, on the one hand, ozone is never obtained quite pure, and, on the other, bodies like iodide of potassium do not absorb ozone, but merely part of its constituent atoms, without change of volume. You are familiar with the hypothesis which I have suggested, which is analogous to that of Weltzien,[4] adopted by Clausius; it consists in considering ordinary oxygen as consisting of two atoms, OO, and ozone of three OO, O. On this hypothesis the density of ozone ought to be one and a half times as much as that of ordinary oxygen.

I have therefore endeavoured to arrive at this value, and I believe I have succeeded. I first made some attempts by diffusion, but hitherto without any great success; but I expect an apparatus which I have had constructed, and with which I hope to succeed. I then sought for some substance which should really absorb ozone, but without decomposing it as iodide of potassium does. I have found it in oil of turpentine and in oil of cinnamon, which destroy ozone in a mixture of this gas and of oxygen, producing a diminution of volume. It is natural therefore to suppose that these oils absorb the ozone entire: hence the possibility of determining the density, not perhaps with great accuracy, but with a sufficiently close approximation to decide whether it is 1½, 2, or 3 times that of oxygen. It is sufficient to compare the diminution in volume of the gas treated with the oil, with the volume of oxygen absorbed by the iodide of potassium, or, what comes to the same thing, with the increase of volume which it experiences under the influence of heat. I have thus found that the diminution of volume by the oil is the double of the increase of volume by heat; and I thence conclude that *the density of oxygen is once and a half that of oxygen*, and have thus confirmed the hypothesis mentioned above. The following are the results which I have obtained:—

Absorbent.	Diminution in volume.	Expansion by heat.		
		Calculated.	Observed.	Difference.
	cub. centims.	cub. centims.	cub. centims.	cub. centims.
Oil of turpentine … …	6·8	3·4	3·77	+0·37
" " … …	5·7	2·85	3·20	+0·35
" cinnamon … …	5·8	2·90	3·14	+0·24
" turpentine … …	5·6	2·80	3·32	+0·52
" " … …	6·7	3·35	3·30	-0·05
" cinnamon … …	6·9	3·45	3·45	
" " … …	5·7	2·85	2·72	-0·13

The agreement is as close as can be expected in this kind of measurement, especially for the three last experiments, in which I avoided a slight source of error which I had not at first perceived. I worked with flasks of about 230 cubic centims. capacity, and with oxygen ozonized by electrolysis.

Phil Mag, 31 (1866), pp. 82–3

1. *[January 1866]*: the date is given as this letter was published in the January 1866 issue of the *Phil. Mag.*
2. *(Abstract of a letter to Professor Tyndall.)*: this insertion is from the editors of the *Phil. Mag.*
3. *my investigation of the density of ozone*: see M. J.-L. Soret, 'Recherches sur la densité de l'ozone', *Comptes rendus de l'Académie des sciences*, 61 (1865), pp. 941–4; and M. J.-L. Soret, 'Recherches sur la densité de l'ozone (Seconde partie)', *Comptes rendus de l'Académie des sciences*, 64 (1867), pp. 904–8.
4. *Weltzien*: Karl Weltzien (1813–70), a German chemist who was Professor of Chemistry at the Technische Hochschule of Karlsruhe from 1848–69 (*ADB*).

To Henry Bence Jones 3 February 1866 2361

3rd February 1866 | To Dr Bence Jones, Hon[orary]. Secretary of the Royal Institution.

My dear Dr Jones

I have read the letter addressed by Dr V. Arnott[1] to the Managers of the Royal Institution.[2] I have also read the drafts of three replies, one from our president,[3] another from Mr Gassiot[4] and the third from yourself. Finally I have read the article in the Reader newspaper to which Dr. Arnott refers.[5]

Of the justice or injustice of the review, and whether it fairly or unfairly represents the treatise to which it refers, I am not in a position to speak, as I am not acquainted with the work of Dr. Arnott.[6]

With regard to the portion of Dr Arnott's letter which refers to myself it may be briefly stated that I for my part wholly and entirely acquit Dr Arnott of any desire to act otherwise than justly towards me. He may or may not have been sufficiently careful in his references, but I could not think him consciously unjust. To me personally moreover the importance of the particular point referred to is infinitely small, though it attracted the attention of the reviewer.

I think your reply to Dr Arnott embodies the utterance of a sound principle; at all events it is one which I accept for my own personal guidance. But there seems to me a certain abruptness—perhaps straightforwardness would be a better term—in the style of your note, which, if altered, might render it more agreeable to Dr Arnott. For the word "refuse" at the beginning of the third paragraph I would substitute some softer term. There are a few other trifling points which, without weakening the affirmation of your principle, might also I think be somewhat toned down.

On the whole it appears to me that the reply of our president fairly meets the exigencies of the case. It sufficiently affirms the principle of non-intervention, and the kindly feeling which pervades it cannot be otherwise than agreeable to Dr Arnott.

It is hardly necessary to say that Mr W. F. Barrett, assistant in the physical laboratory of the Royal Institution is the writer of the review in question. He had been employed to assist in preparing the scientific portion of the Reader newspaper at a time when an attempt was made to establish sufficient weekly organ for scientific men. To the great loss of all workers in science the attempt proved a failure. The journal passed into other hands, but for a brief interval Mr Barrett continued to assist the new proprietor. It was during this interval that Dr Arnott's book was sent to Mr Barrett, and that the review was written.

It is now more than two years since Mr Barrett was appointed assistant in the physical laboratory of the Institution, and during this period he has discharged the duties and done the work committed to him in an able and conscientious manner. From the Managers, as such, he deserves nothing but praise and encouragement. In first engaging Mr Barrett I was partly actuated by the hope of being able to train up in him a competent experimenter and he has already done original work of which this Institution has no reason to be ashamed. He is entirely devoted to his duties, and I firmly believe it to be the earnestness with which he regards his vocation that caused him to write the review which has given so much dissatisfaction to Dr. Arnott. Rightly or wrongly he believed Dr Arnott's book unsuited to the scientific necessities of the day, and he thought it his bounden duty to give expression to this belief. The subject it will be observed, has two sides, the one involving the question of well-merited deference to Dr Arnott, the other the question of rigid faithfulness to science. Mr Barrett may have too much forgotten the former and suffered himself to be influenced too exclusively by the latter:—he may moreover have wholly erred in his estimate of Dr Arnott's work. But I am persuaded that what he did was done through a sense of duty, and that if shown to be wrong he will make honourable amends to Dr Arnott in the columns of the journal which published his review.

It is perfectly manifest that Mr Barrett has no wish to evade the responsibility of his act, for, though not required to do so, he appended to his article the initials of his name, which led to his identification by Dr Arnott.

I would finally add, that as regards the relation subsisting between me and my assistant I have been always careful to avoid exercising the least control over Mr Barrett, save as regards his position in the Institution. Within these walls I tolerate no neglect of duty, but beyond them Mr Barrett is free from my control or interference. This is the attitude which, in the presence of Mr Barrett's father (now dead) I assumed towards his son. And passing from our present subject to the weightier possibility glanced at in your note I would express the hope that a similar attitude would be assumed in cases of greater gravity by the Managers of the Royal Institution. Demand from your Professors that they and their assistants shall effectually discharge the duties which they have undertaken to perform, but beyond this let their own discretion and gentlemanly feeling be their sole guides in the exercise of their right of free expression.

Believe me Yours v[er]y. faithfully | J. T.

RI MS JT/2/13c/1301–2

1. *Dr V. Arnott*: Neil Arnott (1788–1874), a Scottish physician and inventor (*ODNB*). He received the RS's Rumford Medal in 1854 for 'the successful construction of the smokeless

fire grate lately introduced by him, and for other valuable improvements in the application of heat to the warming and ventilation of apartments'.

2. *Managers of the Royal Institution*: for the Managers of the RI at this time, see letter 2347, n. 3.
3. *our President*: Henry Holland.
4. *Mr Gassiot*: John Peter Gassiot (1797–1877), an English electrician and wealthy wine merchant. He was one of the founding members of the London Electrical Society (1837), and played an important role in reforming the RS during the 1840s. He pursued electrical research in the laboratory he established in his home (*ODNB*).
5. *article in the Reader newspaper to which Dr. Arnott refers*: William Fletcher Barrett (see letter 2245, n. 31), wrote a very critical review of N. Arnott, *Elements of Physics, or Natural Philosophy. Written for General Use, in Non-Technical Language*, 6th edn (London: Longmans, Green, and Co., 1865), which appeared as: W. F. Barrett, 'Dr. Arnott's natural philosophy', *The Reader*, 6 (14 October 1865), pp. 432–3. Arnott's work was then defended by Lector [no first name or initial given], 'Arnott's Physics', 6 (21 October 1865), p. 466, and J. Jeremiah, 'Dr. Arnott's Natural Philosophy', 6 (4 November 1865), p 521. Barrett subsequently responded in 'Dr. Arnott's Natural Philosophy', 6 (11 November 1865), p. 548. On *The Reader*, see letter 2283, n. 3.
6. *work of Dr. Arnott*: some of Arnott's work includes N. Arnott, 'On the construction of Fire-places', *Quarterly Journal of Science*, 6 (1819), pp. 364–7; N. Arnott, 'On Safety Lights for Mines', *Brit. Assoc. Rep. 1837*, pp. 54–6; N. Arnott, 'On the Regulation of Combustion', *Brit. Assoc. Report 1847*, pp. 47–8; N. Arnott, *On the Smokeless Fire-place, Chimney-valves, and Other Means, Old and New of obtaining Healthful warmth and ventilation* (London: Longman, Green, Longman & Roberts, 1855); N. Arnott, *A Survey of Human Progress, from the Savage State to the Highest Civilization Yet Attained* (London: Longman, Green, Longman & Roberts, 1861); and six editions of his *Elements of Physics, or Natural Philosophy* from 1827–65.

To Eliza Spottiswoode[1] 4 February 1866 2362

4th Feb. 1866

My dear Mrs. Spottiswoode

I chanced this morning to turn over the pages of my old journal which has lain neglected for some time. There is a tenderness in reviving the memory of one's friends through these old pages. There was a weekly flashing up & the word "Lamas".[2] and once I dropped on the following entry under the date Sunday 13th. Nov. 1864.[3]—"Called at Spottiswoode's and saw the fine little baby—and the splendid photographs of Egypt. A drizzling miserable day, but I suppose a day of infinite value to the country which is thirsty".

Hugh[4] was then little better than a jelly fish, though out of compliment to you I called the bit of animated pulp "a fine baby". I wish you would explain to me the changes of texture which made the little man what he was yesterday—the optical change interests me most—as manifested in the light of his bonny brown eye[s] when he smiled.

Yours ever | John Tyndall

RI MS JT/1/T/1292
RI MS JT/TYP/3/1244

1. *Eliza Spottiswoode*: Eliza Taylor Spottiswoode (née Arbuthnot, 1837–94), wife of William Spottiswoode (married in 1861).
2. *"Lamas"*: the home of John and Ellen Lubbock at Chislehurst in southeast London before they took up residence at High Elms in 1865 (this was Lubbock's parents' home, very near Charles Darwin's Down House). His father John William Lubbock had died on 21 June 1865, and his mother Harriet Lubbock (née Hotham, 1810–73), probably continued to live at High Elms with her son and daughter-in-law.
3. *entry under the date Sunday 13th. Nov. 1864*: Journal, RI MS JT/2/13c/1265.
4. *Hugh*: William Hugh Spottiswoode (1864–1916).

From Henry T. Coxwell[1] 10 February 1866 2363

Tottenham. N. | 10th Feb—1866—

Dear Sir,

In reply to your kind note of yesterday,[2] I beg to say that I shall be glad if it is in your power to name the month, and the probable week, in which you could undertake your experiment at Munich or Bern.

I generaly make my *[*experiments*]* early in the Spring, and I will make it my business to suit your convenience. I enclose two letters which will afford you a criterion of the amounts I am paid for public ascents, which depend, of course, on the distance from London and the attendant traveling expenses.

All my high ascents for the British Association were made for a sum less by one half than what I have for mere public ascents—the larger portion of them, which have been reported as ascents for Scientific purpose by Mr Glaisher,[3] were made by payment for a seat in the car (viz £5)—

As regards yourself I am really at a loss what sum to name, but if I say—"Fifty Guineas"—(50) and pay for the gas myself, with my own expenses, I will be content—

It is an invariable rule that those who engage us supply the gas and attendance—but in our case we will reverse the order, and I think you will receive the spirit I am disposed to work in, and in which I have previously acted towards the Balloon Committee.[4]

—. I remain | Dear Sir, | Yours truly | Henry Coxwell.

Professor Tyndall | Athenaeum Club.[5]

RI MS JT/1/C/56

1. *Henry T. Coxwell*: Henry Tracey Coxwell (1819–1900), an English aeronaut and balloonist who began working closely with the BAAS in 1862 to investigate the upper atomosphere (*ODNB*). Coxwell mentioned conducting several experiments in his balloon requested by Tyndall and others in his *My Life and Ballon Experiences* (London: W. H. Allen & Co., 1889), p. 102.
2. *note of yesterday*: letter missing. Tyndall biographer Roland Jackson has noted that it does not appear that Tyndall ever did participate in a balloon ascent with Coxwell (*Ascent of John Tyndall*, pp. 194–5).
3. *Mr Glaisher*: James Glaisher.
4. *Balloon Committee*: the Balloon Committee of the BAAS, which according to Jennifer Tucker served 'to legitimate the balloon as a "philosophical instrument" for field work in the new science of the Earth's atmosphere', was active from 1858–67 (J. Tucker, 'Voyages of Discovery on Oceans of Air: Scientific Observation and the Image of Science in an Age of "Balloonacy"', *Osiris*, 11 (1996), pp. 144–76, on p. 147). Tyndall was a member.
5. *Athenaeum Club*: see letter 2253, n. 8.

To the Editors of the *Phil. Mag.*[1] 14 February 1866 2364

XXVII. *On the Black-bulb Thermometer.* | *By* Professor TYNDALL, *F.R.S.* | *To the Editors of the Philosophical Magazine and Journal.*

GENTLEMEN,

In the February Number of the Philosophical Magazine, Mr. Wilson[2] of Rugby has published an interesting letter,[3] in which he describes an observation[4] made by Mr. Glaisher,[5] to the effect that the difference of reading between a black-bulb thermometer exposed to direct sunshine, and a second thermometer shaded from the sun's direct action, diminishes as we ascend in the atmosphere. This observation points to the following conclusions which have been drawn by Mr. Wilson.

1. "If the experiment could be made, the shaded and exposed thermometers outside the earth's atmosphere would show the same very low reading."

2. "All those views of the state of the surface of the moon which represent it as alternately exposed to fiery heat and intense cold will disappear. The temperature on the dark and bright halves of the half moon is the same if there is no atmosphere."

"We learn," continues Mr. Wilson, "one more marvellous function of our atmosphere before unsuspected. We knew that it and the aqueous vapour it bears diffused through it save us from loss of heat. We did not know that but for it, or some part of it, we should have no heat to lose. Hence, too, we learn that conclusions as to the temperatures of other planets are still more open to uncertainty, as being in this way also influenced by their atmospheres."

These are very weighty inferences. They affirm that what is heat in our atmosphere, is not heat in stellar space; that something is emitted by the sun which warms a body surrounded by an atmosphere, but which is powerless to warm the body if not so surrounded. A planetary atmosphere, in short, has a power of transmuting into heat an agency which, prior to its entering the atmosphere, is not heat, and this power of transmutation, if possessed in a very high degree, might raise the most distant planet to a high temperature.

I must frankly confess that I do not consider these conclusions, however fairly drawn, to be representative of natural facts. I am ignorant of the details of Mr. Glaisher's observation, and therefore not in a position to offer any explanation of it. But the following remarks, which are in substance transferred from a paper presented some time ago to the Royal Society, and ordered to be printed in the Philosophical Transactions, may have some bearing upon the question; or if not, they may bring to the notice of meteorologists a possible defect in their observations on solar radiation, which, as far as I know, has been hitherto overlooked:—

1. Solar heat as it reaches us consists partly of visible and partly of invisible rays, a portion of the latter possessing very high calorific power.

2. The ordinary black-bulb thermometer absorbs the visible rays of the sun, but the black glass of the bulb may be highly transparent to the invisible radiation.

3. The invisible calorific rays, especially, augment in power as we ascend in the atmosphere; for it is those rays which suffer most diminution of intensity during their passage through the aqueous vapour of the air.

4. Hence the black-bulb thermometer must, with reference to the total radiation falling upon it, become more and more transparent as we ascend in the atmosphere. It is not an exaggerated estimate that at the limits of our atmosphere 50 per cent. of the solar heat might cross the glass of the bulb, be reflected by the mercury within it, and contribute in no degree to the heating of the thermometer.

If these remarks be correct, or so far as they are correct, the indications of the black-bulb thermometer, as at present constructed, are delusive, and they are especially so at great elevations.

I am, Gentlemen, | Your obedient Servant, | John Tyndall.

Royal Institution. | February 14, 1866.

Phil. Mag., 31 (1866), pp. 191–3

---

1. *Editors of the Phil. Mag.*: see letter 2245, n. 1.
2. *Mr. Wilson*: James Maurice Wilson.
3. *published an interesting letter*: J. M. Wilson, 'Some Remarks on an observation of Mr. Glaisher's,' *Phil. Mag.*, 31 (1866), pp. 104–6.

4. *observation*: Wilson was responding to Glaisher's paper, 'An Account of Meteorological and Physical Observations in Three Balloon Ascents made in the years 1864 and 1865', presented at the thirty-fifth meeting of the BAAS at Birmingham (*Brit. Assoc. Rep. 1865*, pp. 145–91). On Glaisher's observation, according to Wilson: 'That there is always a difference of reading of two thermometers, one with blackened bulb exposed to direct sunshine, and the other with bulb concealed, is well known. The latter makes the temperature of the air surrounding it; the former marks the sum of this and of the direct radiated heat which it receives from the sun. Mr. Glaisher has shown that the difference of reading of the two thermometers, when taken to great heights, instead of increasing or remaining constant, as would seem at first sight to be expected, actually diminishes, and appears as if it would diminish without limit' (p. 104).
5. *Mr. Glaisher*: James Glaisher.

From Louis Bergeret[1] 16 February 1866 2365

16 février 1866

Monsieur,

Vos éclatantes expériences sur les *[*ondulations*]* calorifiques, votre belle analyse de la lumière, vos savantes recherches sur le pouvoir qu'a la vapeur d'eau d'absorber la chaleur rayonnante, votre récent article sur la constitution de l'univers etc, m'ont causé une bien vive satisfaction.

Votre association récente avec MM. Ch. Lyell, Lubbock, Darwin, Huxley, Hodgson, Bowring, etc mérite toutes les sympathies des libres-penseurs. La conférence de Mr Huxley à Saint-Martin's Hall fait reporter complètement l'utilité des sciences naturelles. La philosophie sympathique doit faire place au Positivisme.

Permettez-moi donc, cher Monsieur, de vous offrir un petit livre que je viens d'écrire dans le sens des idées de votre programme de Sermons-Laïques. Vous y verrez que j'ai mis plusieurs fois vos savants travaux à contribution.

Ainsi en même temps que cette lettre, vous recevrez le volume que je vous annonce.

Je vous écris en Français car je ne connais pas l'Anglais. Recevez, cher Monsieur, l'expression de la vive sympathie de votre tout dévoué

Dr Bergeret

Docteur Bergeret | à Chalon-sur-Saône | Saône-et-Loire (France).

February 16 1866

Sir,

Your brilliant experiments on the calorific *[*undulations*]*, your beautiful analysis of light, your scholarly works on the ability of water vapour to absorb radiant heat,[2] your recent article on the constitution of the universe[3] etc., gave me endless satisfaction.

Your recent partnership with Misters Ch. Lyell,[4] Lubbock, Darwin, Huxley, Hodgson,[5] Bowring[6], etc., deserves all the support of the freethinkers.[7] M[r] Huxley's conference at Saint-Martin's Hall[8] completely *[presented]* the usefulness of the natural sciences. Philosophy of sympathy should give way to Positivism.[9]

Allow me then, dear Sir, to offer you a small book[10] that I just wrote in the direction of your secular-sermons program.[11] You will see that I often put forward your scholarly work.

Also, with this letter, you will receive the volume that I have announced to you.

I am writing to you in French because I do not know any English. Please accept, dear Sir, the expression of the deepest support of your devoted

Dr. Bergeret

Doctor Bergeret | in Chalon-sur-Saône | Saône-et-Loire (France).

RI MS JT/1/B/89

1. *Louis Bergeret*: Louis François Étienne Bergeret (1814–93), a French physician and one of the first forensic entomologists (M. Benecke, 'A Brief History of Forensic Entomology', *Forensic Science International*, 120 (2001), pp. 2–14).
2. *your scholarly works on the ability of water vapour to absorb radiant heat*: see letter 2229, n. 2.
3. *article on the constitution of the universe*: J. Tyndall, 'The Constitution of the Universe', *Fortnightly Review*, 3 (1865), pp. 129–44, in which he argued that 'a miracle is strictly defined as an invasion of the law of the conservation of energy . . . Hence arises the skepticism of scientific men when called upon to join in national prayer for changes in the economy of nature' (p. 144). While he did not reject the importance of prayer on the individual, he did suggest that 'if our spiritual authorities could only devise a form in which the heart might express itself without putting the intellect to shame, they might utilise a power which they now waste, and make prayer, instead of a butt to the scorner, the potent inner supplement of noble outward life' (p. 144). This was omitted from the version of the article published as 'The Constitution of Nature' in the sixth edition of Tyndall's *Fragments of Science*, 2 vols (London: Longmans, Green, and Co., 1879), vol. 1, pp. 3–27.
4. *Ch. Lyell*: Charles Lyell.
5. *Hodgson*: William Ballantyne Hodgson (1815–80), a Scottish educationist and economist (*ODNB*).
6. *Bowring*: John Bowring (1792–1872), a British politician, diplomatist, and writer (*ODNB*).
7. *freethinkers*: a person who professes to or is known for independence of thought, *esp.* one who withholds assent to widely held beliefs or ideas; *spec.* a person who refuses to submit his or her reason to the control of authority in matters of religious belief (*OED*).
8. *Mr Huxley's conference at Saint-Martin's Hall*: a reference to Huxley's Lay Sermon 'On the Advisableness of Improving Natural Knowledge', delivered on the evening of 7 January

1866 at St. Martin's Hall in London. This was published as T. H. Huxley, 'On the Advisableness of Improving Natural Knowledge', *Fortnightly Review*, 3 (15 January 1866), pp. 626–37, and a few years later placed first in T. H. Huxley, *Lay Sermons, Addresses, and Reviews* (London: Macmillan and Co., 1870). On this sermon, see J. Paradis, *T. H. Huxley: Man's Place in Nature* (Lincoln: University of Nebraska Press, 1978), pp. 76–81.

9. *Positivism*: Positivism, founded by French philosopher Auguste Comte (1798–1857), is the belief that opposes traditional philosophical approaches in favor of scientific methods across a range of fields. Claims about the world should be based on analyses of phenomena, without speculation about anything beyond those phenomena. Thus positivists argue that only scientific inquiry, not theology or metaphysics, can produce certain knowledge.
10. *small book*: not identified.
11. *secular-sermons program*: Bergeret was referring to Sunday Evenings for the People (see letter 2345, n. 3)., which included lectures by Huxley, Hodgson, and Bowring.

To the Editor of *The Athenaeum*[1] 12 February 1866 2366

THE ROMAN ROAD AND SILBURY HILL.

Royal Institution, Feb. 12, 1866.

Lest the idea that crops obliterate the traces of the Roman road should retard inquiry, I would suggest to Mr. Fergusson[2] that a crop is an extremely delicate re-agent, and in many cases the only one competent to reveal the position of the ancient highway.[3] It was the differential action of the soil on a crop of clover which enabled Sir John Lubbock to follow the track of the Roman road near Silbury Hill; and neither he nor I had, or has, the slightest doubt that the position assigned to the road upon the Ordnance map is the correct one.

Silbury Hill was a splendid landmark. On it, I believe, the Roman engineer planted his flagstaff and worked towards it over miles of country; he would thus save himself considerable trouble in the setting out of his line. He did not, however, run his road up to the base of the hill, but changed his direction before he reached it,—the two branches of the road forming near the hill a very obtuse angle.

JOHN TYNDALL.

J. Tyndall, 'The Roman Road and Silbury Hill', *The Athenaeum*, 17 February 1866, p. 239

1. *Editor of The Athenaeum*: William Hepworth Dixon (1821–79), an English journalist and writer (*ODNB*), was the editor of *The Athenaeum* from 1853–69.
2. *Mr. Fergusson*: James Fergusson (1808–86), a Scottish architectural historian (*ODNB*).

3. *the ancient highway*: Tyndall was responding to a short note by Fergusson on 'Avebury', *The Athenaeum*, 23 December 1865, pp. 888–9, in which he argued that its Neolithic monuments were constructed during the early Medieval period to commemorate the final battle of the legendary King Arthur, and that Arthur's slain warriors had been buried there. John Lubbock responded in 'Secret of the Druidical Stones', *The Athenaeum*, 6 January 1866, pp. 18–9, arguing that Fergusson's views were mistaken and that both he and Tyndall inspected the site the previous year. Fergusson quickly responded in 'Stonehenge and Avebury', *The Athenaeum*, 13 January 1866, pp. 52–3, contending that Lubbock and Tyndall had confused the location of the Roman road, and offered several surveying studies to demonstrate his point. Lubbock, for his part, responded again in 'Secret of the Druidical Stones', *The Athenaeum*, 20 January 1866, pp. 95–6, asserting that Fergusson had misunderstood his point and that Fergusson was still wrong. The disagreement went on for several more numbers. Fergusson later wrote on this topic in detail in his *Rude Stone Monuments in All Countries: Their Age and Uses* (London: John Murray, 1872).

From Henry T. Coxwell[1] 20 Feburary 1866 2367

Tottenham. Feb^{r}[uary]. 20$^{\underline{th}}$. 1866. |
To Professor Tyndall. F.R.S. *[*&c &c &c*]*

Dear Sir,

Whenever you choose to command my services[2] I shall esteem it an honour to place them at your disposal—

I have not mentioned to any one that you meditate a High Ascent, nor shall I do so—

Should your preliminary experiments be sufficiently advanced this Summer I will accomodate, provided you make me acquainted with your wishes in due time to make the necessary arrangements.

The letters I enclosed to you[3] came safely to hand,—and I thank you sincerely for the sentiments you express as to my past labours in behalf of science,

I am, dear Sir, | truly yours | Henry Coxwell.

RI MS JT/1/C/57

---

1. *Henry T. Coxwell*: see letter 2363, n. 1.
2. *command my services*: see letter 2363, esp. n. 2.
3. *letters I enclosed to you*: in letter 2363, Coxwell wrote: 'I enclose two letters which will afford you a criterion of the amounts I am paid for public ascents, which depend, of course, on the distance from London and the attendant traveling expenses'.

From François-Napoléon-Marie Moigno 20 February 1866 2368

Mon cher Mr Tyndall

Il y a bien longtemps que je n'ai rien reçu de vous; et j'en suis très inquiet. On ne m'a même pas envoyé cette année les leçons de Royal Institution. Donnez vos ordres, je vous en prie pour qu'elles me soient fidèlement envoyées.

Vous avez su sans doute, ne fut-ce que par l'interruption dans l'envoi des Mondes, que j'avais été obligé de rompre avec Mr Giraud qui s'est montré complètement incapable et négligeant à l'excès. Il perdait ma propriété et me faisait contracter des dettes au lieu de me donner des bénéfices. Il m'a fait attendre trois ans les comptes de mes Mondes; j'en suis encore à savoir ce qu'il a vendu d'exemplaires de la Chaleur, et ce qui vous revient.

Mon éditeur actuel Mr Rotschild est très capable, très actif, très intelligent, j'espère beaucoup de lui.

J'ai besoin que vous recommandiez mes chers Mondes à nos amis communs d'Angleterre; et je vous rappelle l'abonnement du musée de géologie pratique sur lequel j'ai compté pour remplacer Sir Roderick Murchison. Envoyez moi bientôt quelque jolie nouveauté; qui produise de l'effet. Je me sens plus que jamais, malgré *[tant]* de tribulations plein de courage et d'ardeur; je veux absolument que la nouvelle série de mes Mondes marque une ère nouvelle. Accordez moi votre *[1 word illeg]* patronage.

A vous de coeur | l'abbé F. Moigno | 2 rue d'Erfurt | 20 février 1866.

My dear Mr Tyndall,

It has been a long time since I heard anything from you, and I am rather worried.[1] I have not even received the lectures from the Royal Institution this year. Please give them your directives so that they will be sent to me regularly.

You probably knew, if only by the interruption of the shipment of Les Mondes,[2] that I was forced to break off with M[r] Giraud,[3] who was completely incompetent and excessively negligent. He lost my property and made me incur debt instead of giving me a profit. He made me wait three years for the account of my Les Mondes; and I have yet to know how many copies of Heat[4] he sold and what belongs to you.

My current publisher, M[r] Rotschild,[5] is very capable, very active, very intelligent; and I have high hopes for him.

I need you to recommend my dear Mondes to our mutual friends in England; and I remind you about the subscription for the Museum of Practical Geology, on which I counted to substitute Sir Roderick Murchison.[6]

Send me some nice news soon; that produces its effects. I feel more than ever, despite all the tribulations, full of courage and zeal; I immensely wish for the new series of my Les Mondes to mark a new era. Give me your *[1 word illeg]* patronage.

Yours truly | Father F. Moigno | 2 rue d'Erfurt[7] | 20 February 1866.

RI MS JT/1/M/108

---

1. *I am rather worried*: see Moigno's letter to Tyndall from 16 December 1865 (letter 2339).
2. *Les Mondes*: see letter 2339, n. 7.
3. *Mr Giraud*: Étienne Giraud.
4. *Heat*: see letter 2339, n. 5.
5. *Mr Rotschild*: see letter 2339, n. 6.
6. *subscription for the . . . Sir Roderick Murchison*: see letter 2339.
7. *2 rue d'Erfurt*: in Paris.

From Charles Frédéric Martins[1] 1 March 1866 2369

JARDIN DES PLANTES DE MONTPELLIER. | Montpellier, le 1er mars 66.

Mon cher Monsieur,

Je réunis des matériaux pour un Article de la Revue des deux mondes sur l'état actuel de la question glaciaire. Glaciers actuels et ancienne extension. Je manque de documen sur les glaciers arctiques de l'Amérique du Nord. Pourriez vous me dire dans lequel des voyages entrepris à la Recherche de Franklin je trouverais le plus de détails sur ce sujet, je vous en serais bien obligé.

Je compte retourner cet été dans les Pyrénées pour y refaire les expériences comparatives sur l'échauffement relatif de l'air et et du sol sur une haute montagne et dans la plaine. Je serais bien heureux si le hasard faisait que j'eusse une alternative de Jours secs et des jours humides avec un ciel serein. Si vous avez quelque recommandation à me faire ou quelque bon conseil à me donner je serais bien heureux de les recevoir. Je vous renvoie ma note afin que vous n'ayez pas la peine de la chercher dans les Comptes-rendus.

J'ai lu et relu votre Lecture <u>On radiation</u> c'est excellent et du plus haut intérêt pour moi.

Votre dévoué serviteur, | <u>Ch: Martins</u>

BOTANICAL GARDEN OF MONTPELLIER[2] |
Montpellier, 1 March 66.

My dear Sir,

I am gathering information for an Article in the Review des Deux Mondes on the current state of the glacier question.[3] Today's glaciers and their past extension. I am missing some documentation on the arctic glaciers of North America. I would be much obliged if you could tell me in which of the voyages undertaken in the Search for Franklin[4] I would find the most details on this subject.

I plan to go back to the Pyrenees[5] this summer to rddo the comparative experiments on the relative warming of air and soil on a high mountain and in the plain. I would be very pleased chance gave me both dry days and humid days with clear skies. I would welcome any recommendations or good advice you might have for me. I will send you my remarks directly[6] so that you do not have to take the trouble of looking for it the *[*published*]* reports.

I have read and reread your Lecture On radiation[7] it is excellent and of great interest to me.

Your devoted servant, | Ch. Martins

RI MS JT/1/M/52

1. *Charles Frédéric Martins*: Charles Frédéric Martins (1806–89), a French physician, botanist, and geologist. He held several professorships at the Faculté de Médecine, Paris, the Sorbonne, and Montpellier ('Charles Frédéric Martins', *Annals of Botany*, 3 (1889–91), pp. 480–2).
2. *BOTANICAL GARDEN OF MONTPELLIER*: the oldest botanical garden in France, founded in 1593.
3. *Article in the Review des Deux Mondes on the current state of the glacier question*: C. Martins, 'Les Glaciers actuels et la Période glaciaire', *Revue des Deux Mondes*, 67 (1867), pp. 407–32; Tyndall is cited extensively by Martins.
4. *the Search for Franklin*: a reference to the missing Royal Navy ships HMS *Erebus* and HMS *Terror*, commanded by John Franklin (1786–1847) in an 1845 failed expedition to chart and navigate the Northwest Passage. The crews of both ships perished. An official search began in 1848 and continued well into the second half of the century. The remains of the *Erebus* were discovered in 2014, and the remains of the *Terror* discovered in 2016. See N. K. L. Kaalund, 'What Happened to John Franklin? Danish and British Perspectives from Francis McClintock's Arctic Expedition, 1857–59', *Journal of Victorian Culture*, 25 (2020), pp. 300–14.
5. *Pyrenees*: a mountain range along the border of France and Spain. Martins published on the Pyrenees; see: C. Martins, *La Végétation du Spitzberg comparée à celle des Alpes et des*

Pyrénées (Montpellier: Boehm et Fils, 1865); and C. Martins and E. Collomb, *Essai sur l'Ancien Glacier de la Vallée d'Argelès (Hautes Pyrénées)* (Montpellier: Boehm et Fils, 1868).

6. *send you my remarks directly*: letter missing.
7. *your lecture [on] radiation*: *On Radiation* (see letter 2231, n. 1).

From William Wright[1] 6 March 1866 2370

Springfield Clack,[2] | March 6th 1866. |
Dr. John Tyndall | London, England.

Dear Sir,

Although slow to acknowledge my obligation to you for the notice you would have taken of my friend King[3] had he been so fortunate as to meet with you when in London last summer or his time not so limited as to make a second call, I hope you will not think that I was equally slow to appreciate your kindness for the attention he would have otherwise received from you when your engagements would not interfere.

I suppose that we poor business men have but a faint idea of the close application that you scientific men give to your researches and unable to realize the difficulties which such interruptions might occasion. Still I think for "Auld Lang Syne's" sake I would be excused.

A young friend[4] of mine from Cincinnati has been studying in the Medical Schools of Paris and I learn has passed his examinations very successfully and obtained his Degree of M.D. His father[5] is an old associate of mine in the Rail Road business, and George himself was for some time connected with it. Before he returns to the United States he intends spending a few months in London and I have taken the liberty of giving him an introduction to you which I trust will meet with your approbation as George E. Walton is one of those young men who start out in life without aid save their own individual exertions and perseverance to advance themselves. During our late war[6] he had an appointment in the U.S. army as assistant surgeon and having accumulated means started for Europe to perfect himself in the Medical Schools. He did intend going to Dublin to attend lectures there but I learn from his father that he is so well pleased with his progress in Paris that he has given up going to Dublin except on a short visit and I intend him to call on my friends there. I trust while in London your time will permit of your being able to give him some attention and confer additional favours on me.

I hope it is not too late for me to assure you that the reception of your very kind favour of August 22/65 met with a hearty welcome and often have I since then commenced to address you and as often deferred it for some more convenient season that to me seems never to arrive.[7] Its arrival soon brought out the thoughts of former years and I seemed to live over again. The old familiar scenes passed before me as if it was only yesterday. Many more important events of my life have passed out of my mind and are completely forgotten. The babbling weir[8] at Madling[9] is more vividly displayed before my recollection than the thundering of Niagara's cataract; the old moat more familiar than any of the Earth Works I have seen in the late West and visited during the continuance of the rebellion. I was very glad to find that even with yourself who has so many opportunities to be weaned off the old associations they will come back almost unbidden. I have often thought that I would prefer to meet with you of any one now living and still fear if we should that it might dispel the pleasing fancies which dwell around our long, long ago acquaintance. For our paths in life have been so very opposite of each other that we could not expect to agree as we did in former years. I would be unable to appreciate the man of science and you could see but little to interest you in the struggling trader or old Rail Road employer who has now completed his 21 years with one company an unusual thing in this country. I suppose you would ask, well what have you done for those 21 years? I would have to answer but very little for myself or others except to raise a family of 10 children who may do something. My eldest boy Richard commanded a company in the Co Ohio cavalry in the Potomac Army. He is now in California in the 1st U.S. Cavalry working his way up to a commission in the regular army. Samuel my second son marched with Sherman[10] to the Sea. Both suffered in those horrible prison pens of the South, but not before their adversaries had an opportunity of testing their qualities. Each served 4 years in the U.S. Volunteers. Samuel is now out on the plains beyond Fort Riley with a company surveying the route from Fort Riley to the Rocky Mountains for the Pacific Rail Road that will eventually extend across our whole Continent, and if its present progress continues, 10 years will not elapse before it is completed, from New York on the Atlantic to San Francisco on the Pacific.[11] Willie my third served his 100 days in the Ohio Native Guard and had an opportunity of seeing as hard service for the time as any old campaigner. He is out at Leavenworth city doing business in a hardware store but talks of crossing the Plain the coming Spring and going to the Mines.[12] Kate my eldest daughter is married to Captain McArthur[13] who also did good service with General Sherman on his campaign thro' the South. So that you see if I have not done much myself for my country my children have done their share in trying to preserve its liberty in the field. But I fear we have much more to dread from its opponents in the forum. From the late action of our President and the acclamation his veto met with in the South and its sympathising allies in the North bodes no good to our Republic or the faithful coloured race who help to sustain our cause and were promised that they would be protected from the machinations of their late masters.[14] But to our shame be it said there are some in the Union Ranks that think all has been done that ought to be done for the coloured man, forgetting that altho' the shackles has been removed from his limbs[15] they have not removed from his late masters the power of again shackling both mind and body. But justice must be done to him before this Republic is safe and the sooner done the better for our people. I suppose the disturbance we have had here or that which has lately occurred at our old home by the Fenians has given you but little concern.[16] The latter I suppose would not, if you were acquainted with most of the material of which it is composed although some of the leaders Sweeny[17] for instance was a gallant soldier in our Army and a few others of them bore a conspicuous part in our war. The majority of their dupes were men who when called upon to do their duty to this country were the first to seek British protection to save them from the draft. The noisy blustering man who now wants to go to Ireland or Canada were very careful not to cross the Ohio or Potomac on a campaign to Dixey.[18] Indeed I think if they thought the thing was feasible they would not talk so much about it. I know some little of the spirit which pervades the loyalists in Canada and if a raid is attempted our Fenians will return in a hurry. At home in old Ireland there is a portion of its inhabitants that will as they have successfully done heretofore handle them without gloves. The only fear I have is that they may entangle our Governments before the excitement is over. There is very little encouragement given them by the thinking portion of our community, and it is chiefly looked upon as a swindler's scheme to cheat the poor Irish out of their hard earned dimes, and before long it is expected that they will discover their error of supplying funds to keep in idleness the originators of the plot. Do you ever hear from Leighlin Bridge[19] or know anything of the old inhabitants. Very few from that section of our county has settled out here, and having no correspondence with any one in the County Carlow I am in complete ignorance of how matters are there. I frequently meet with an old school mate of yours and mine William G. Brune,[20] he is settled at Dayton[21] a flourishing town some 20 miles distant from here and is doing a very respectable trade in the clothing and furnishing line. He has acquired a very comfortable competence and raised quite a respectable family. We rarely meet without your name being mentioned in some way. If you can recollect it, Billy considered himself an equal for you and me until we formed an alliance offensive and defensive against him and in one encounter in the old Sand Pit near the school we put him hors de combat[22] and quell'd his ardeur for the strife ever afterwards. When a favourable opportunity presents itself I would take it as a very great favour to hear from you and

believe me ever yours truly | William Wright.

RI MS JT/1/TYP/5/1833–6
LT Typescript Only

1. *William Wright*: William Wright (1818/9–92), from Leighlin Bridge, a close childhood friend of Tyndall's, with whom he subsequently maintained a long-term correspondence. He and his new wife emigrated to America in March 1841, possibly due to opposition from her family. They settled in Cincinnati, Ohio, where William worked in the dry goods and grocery trades. He frequently urged Tyndall to join him in America.
2. *Springfield Clark*: Springfield, Clark County, Ohio.
3. *King*: not identified.
4. *young friend*: George E. Walton (b. 1839), an American doctor who studied in Paris, returning in 1867 to Cincinnati (and later moved to Florida). He wrote on cholera, climate, and published *The Mineral Springs of the United States and Canada* (New York: D. Appleton & Co., 1874) (O. Juettner, *Daniel Drake and His Followers: Historical and Biographical Sketches* (Cincinnati: Harvey Publishing Company, 1909), p. 309).
5. *His father*: not identified.
6. *our late war*: the American Civil War (1861–5).
7. *seems never to arrive*: favor not identified.
8. *weir*: a dam, of which there are various forms, constructed on the reaches of a canal or navigable river, to retain the water and regulate its flow (*OED*).
9. *Madling*: a town in County Kilkenny, Ireland.
10. *Sherman*: William Tecumseh Sherman (1820–91), General in the Union Army during the American Civil War.
11. *from New York on the Atlantic to San Francisco on the Pacific*: construction of the Transatlantic railroad was completed on 10 May 1869 by the driving of a gold spike in Promontory, Utah.
12. *the Mines*: possibly to California for the gold mines.
13. *Captain McArthur*: possibly William M. McArthur (1832–1917), who served as Captain of Co. I, 8th Maine Volunteers, which formed part of the Union forces at the Battle of Port Royal in South Carolina in early November 1861 under General Sherman and Admiral Du Pont. He would increase rank as Major, Lieutenant Colonel, Colonel, and finally Brigadier General (*Annual Report of the Adjutant General of the State of Maine, for the year ending December 31, 1866* (Augusta: Stevens & Sayward, 1867), pp. 499–500).
14. *From the late action of our President . . . of their late masters*: Abraham Lincoln (1809–65), sixteenth President of the United States, presided over the Union victory in the American Civil War, which dominated his presidency. Lincoln pocket vetoed the Wade-Davis Bill of 1864, by Ohio Senator Benjamin F. Wade (1800–78) and Representative Henry Winter Davis (1817–65), which required that fifty percent of a state's white males take a loyalty oath to be readmitted to the Union, in addition to granting voting rights to 'persons held to involuntary labor or servitude'. Congress passed the Bill but Lincoln chose not to sign it because he felt it was too strict and radical.
15. *shackles has been removed from his limbs*: the Thirteenth Amendment to the United States Constitution, which abolished slavery and involuntary servitude (except as a punishment for a crime), was, after passing the United States Senate and House and ratified by 27 of 36 States, officially adopted on 18 December 1865 by Secretary of State William H. Seward (1801–72).

16. *but little concern*: 'Fenians' was an umbrella term for the Fenian Brotherhood and Irish Republican Brotherhood, fraternal organizations dedicated to the establishment of an independent Irish Republic. There were unsuccessful uprisings in early 1867. On 13 December 1867 Fenians blew up a wheelbarrow laden with explosives against a wall of Clerkenwell Prison in London, to attempt (unsuccessfully) to release Ricard O'Sullivan Burke. The explosion killed several people across the street (N. Whelehan, *The Dynamiters: Irish Nationalism and Political Violence in the Wider World, 1867–1900* (Cambridge: Cambridge University Press, 2012), pp. 71–2).
17. *Sweeny*: this name is handwritten in this space in the typescript. Probably Thomas William Sweeny (1820–92), an Irish-American soldier who served in the Mexican-American War and for the Union Army in the American Civil War. Following the Civil War, Sweeny joined the Fenian movement (see n. 16), commanding a failed attack against the British in Canada in June 1866 (R. T. Cornish, 'Sweeny, Thomas William', in D. S. Heidler and J. T. Heidler (eds), *Encyclopedia of the American Civil War: A Political, Social, and Military History* (New York: W. W. Norton & Company, 2002), p. 1912).
18. *Dixie*: the southern United States; the South (*OED*).
19. *Leighlin Bridge*: Leighlinbridge, County Carlow, Ireland, where Tyndall was born and his mother lived.
20. *William G. Brune*: not identified.
21. *Dayton*: Dayton, Ohio.
22. *hors de combat*: out of fight, disabled from fighting (French; *ODNB*).

From John Percy[1] 8 March 1866 2371

1. Gloucester Crescent | Hyde Park. | March 8/66

My dear Tyndall.

Let me entreat you seriously to reflect on the enclosed extract[2] from the sermon of a Reverend Divine,[3] whose treasure is in Heaven, whose thoughts as well as actions are unearthly, and who undoubtedly must be much better acquainted than yourself with all that relates both to the visible and the invisible world. Learn wisdom and humility and no longer allow yourself to be entangled in the meshes of that Devil's net, called "science".[4] How vain and hollow is your philosophy which rests only on that vulgar thing termed evidence of the bodily senses. Who can trust to the observations of the "senses", so individual in their nature, so liable to derangement, and as no one ought to know better than John Tyndall, so intrinsically fallacious? No longer believe your eyes, your ears, your touch, but trust implicitly to the inward teaching of that highly sublimated essence, infinitely less fallible than your pretended luminiferous ether, the soul. Yes, the soul. I hope you possess a soul Ah! My dear Colleague, if that soul of yours were illuminated with the true light how magic would be the transformation! Your electric ray is utter darkness in comparison. Pray, pray for light, that you may never be condemned hereafter to witness the combustion of sulphur without oxygen.[5] The age of miracles is not over; & I may yet live to see you become a disciple of the true faith.

Your sincere well wisher | John Percy

Professor Tyndall.

P.S. Buy up all the numbers of the "Fortnightly Review" containing your infidel paper,[6] & burn them!

RI MS JT/2/10

---

1. *John Percy*: John Percy (1817–89), an English metallurgist and professor at the Royal School of Mines (*ONDB*).
2. *enclosed extract*: not identified.
3. *Sermon of a Reverend Divine*: the sermon and the Reverend are not identified.
4. *called "science"*: Percy, a colleague of Tyndall's at the Royal School of Mines, was presumably being sarcastic in tone with this whole letter and not actually railing against him about religion.
5. *to witness the combustion of Sulphur without oxygen*: meaning Hell.
6. *your infidel paper*: probably J. Tyndall, 'The Constitution of the Universe', *Fortnightly Review*, 3 (1865), pp. 129–44. For some detail on this, see letter 2365, n. 3.

From Arthur Penryn Stanley[1] — 8 March 1866 — 2372

Deanery, Westminster | March 8. 66

Dear Sir,

The kind tones of your note,[2] the other day, encourage me to make a request, which I had intended to make, before I received it, if ever the opportunity should occur, which is now presented to me.

In your article in the Fortnightly Review for Dec 1,[3] you express a hope that a form of Prayer might be desired (on the late disputed subject) in which the heart might express itself without putting the intellect to shame.

I have now the duty of arranging a session for the Abbey on the 23rd of March, the day appointed by the Bishop of London[4] for the Cattle Plague[5] service,—& it would give me sincere pleasure if you could furnish me with any such help for this purpose, as I gather from your words that you think might be possible.[6]

I have not mentioned to any one that I have written to you—& I shall of course perfectly understand, if you find that you cannot comply with my request; as you no doubt will also, if, after all, I fail that I could not make use of your kindness.

But I could not forbear to take advantage, in all sincerity, of the hope that you seem to hold out—in the belief that only by such sort of help shall we be able to do our duty rightly to the Church and to the claims of science.

Yours faithfully | A. P. Stanley

RI MS JT/2/13c/1317–8

---

1. *Arthur Penryn Stanley*: Arthur Penrhyn Stanley (1815–81), an Anglican clergyman. He was educated at Rugby School and Oxford University (1837) where he obtained a fellowship. He accepted a canonry at Canterbury (1851) before becoming professor of ecclesiastical history at Oxford (1856). In 1863 he married Lady Augusta, one of Queen Victoria's women of the bedchamber. They were subsequently frequently guests at Windsor castle. From 1864 to 1881 he was Dean of Westminster, during which time he continued to publish lectures and sermons (*ODNB*).
2. *your note*: letter missing.
3. *your article in the Fortnightly Review for Dec 1*: see letter 2371, n. 6.
4. *Bishop of London*: Archibald Campbell Tait (1811–82), Bishop of London from 1856–68; formerly Dean of Carlisle (1849–56) and later Archbishop of Canterbury (1868–82) (*ODNB*).
5. *Cattle Plague*: see letter 2314, n. 5.
6. *might be possible*: on the issue of prayer and the cattle plagues, see F. M. Turner, 'Rainfall, plagues, and the Prince of Wales: A Chapter in the Conflict of Religion and Science', *Journal of British Studies*, 13 (1974), pp. 46–65. On Tyndall and the prayer debate in general, see R. B. Mullin, 'Science, Miracles, and the Prayer-Gauge Debate', in D. C. Lindberg and R. Numbers (eds), *When Science & Christianity Meet* (Chicago: University of Chicago Press, 2003), pp. 203–24.

To the Editor of the *Spectator*[1] 11 February 1866 2373

PROFESSOR HUXLEY'S DOCTRINE. | (TO THE EDITOR OF THE "SPECTATOR.")

SIR,—In the last number of the *Spectator*,[2] Professor Huxley has paid me the great compliment of stating that were I to tell him to-morrow that I could by a word cause a stone to fall upwards, he would feel bound to suspend his judgment until such time as the matter could be fully investigated. It is not often that I find myself unable to reciprocate the sentiments of my eminent friend. But on the present occasion I feel bound to say that were he to confide to me the statement of his ability to reverse by a word the action of gravity, my judgment regarding him would find most mournful expression in the line:—

> "O what a noble mind is here o'erthrown!"

The evidence of the permanence of natural laws is not that of a generation, but of many generations; and against its accumulated weight the averment of an individual, or even of a group of individuals, though it may claim a theoretic chance of being right, is practically of no account.

The real strength of science is not, I think, manifested either by logic or by authority. Its influence is slower and surer than any that could be brought to bear upon a jury of the present day. It gradually attunes the mind to the methods of nature, and superstition dies, not through any formal demonstration of its error, but by its simple incapacity to live in the presence of scientific truth.

I would add, in conclusion, that the body of that truth, its character and magnitude, the knowledge which it embraces of the heavens and the earth and the fires under the earth, of sensation and thought and life and death, have deepened indefinitely the impression which the mystery of this Universe has produced, from the time of Job[3] to that of Hamilton,[4] on every profoundly religious mind. Under the influence of this impression Zophar[5] asks, "Canst thou by searching find out God?"[6] Under its influence Hamilton affirms that "to think God is as we can think Him to be is blasphemy;"[7] and he states in language as bold as that so impressively uttered in St. Martin's Hall,[8] and which has evoked so much criticism, wise and unwise, fair and unfair, that "*the highest consecration of all true religion must be an altar to the Unknown and Unknowable God.*"—I am Sir, your obedient servant,

JOHN TYNDALL. | *Athenaeum Club,*[9] *February* 11, 1866.[10]

RI MS JT/2/13c/1314–5
J. Tyndall, 'Professor Huxley's Doctrine', *Spectator*, 39 (17 February 1866), p. 188

---

1. *Editor of the Spectator*: Richard Holt Hutton (1826–97) was the editor of the *Spectator* at this time.
2. *last number of the Spectator*: T. H. Huxley, 'Mr. Huxley's Doctrine', *Spectator*, 39 (10 February 1866), p. 158, in which Huxley himself was responding to the piece 'Science and Miracle', *Spectator*, 39 (27 January 1866), p. 93. Hutton had ridiculed Huxley's 'worship of the Unknown and Unknowable', found in 'On the Advisableness of Improving Natural Knowledge', *Fortnightly Review*, 3 (15 January 1866), pp. 626–37. On Tyndall and Huxley's views on science, religion, and morality, see A. Dewitt, *Moral Authority, Men of Science, and the Victorian Novel* (Cambridge: Cambridge University Press, 2013), pp. 21–52.
3. *Job*: of the Book of Job in the Bible, whose faith in God is tested by numerous tragedies.
4. *Hamilton*: William Hamilton (1788–1856), a Scottish philosopher and Professor of Logic and Metaphysics at the University of Edinburgh (J. Veitch, *Memoir of Sir William Hamilton, Bart.* (Edinburgh: William Blackwood and Sons, 1869)).
5. *Zophar*: Job's friend Zophar the Naamathite, who comforts Job during his illness.
6. *"Canst thou by searching find out God?"*: from Job 11:7–9.

7. *"to think God is as we can think Him to be is blasphemy"*: the phrase 'to think God is, as we think Him to be, is blasphemy', comes from W. Hamilton, *Discussions on Philosophy and Literature, Education and University Reform*, 2nd edn (London: Longman, Brown, Green, and Longmans, 1853), p. 15, footnote.
8. *uttered in St. Martin's Hall*: see letter 2365, n. 8.
9. *Athenaeum Club*: see letter 2253, n. 8.
10. *1866*: the editor (see. n. 1) included the following note after Tyndall's letter: 'Mr. Huxley's expression was not "on the altar of the Unknown and Unknowable God," but "on the altar of the Unknown and Unknowable," an important difference, to which in great part our error was due. There was nothing in the lecture even to suggest to us a belief in God, and much reason to have expected some guarding word, if Mr. Huxley had been anxious to prevent misunderstanding. We exceedingly regret our mistake, but it was not unnatural, and our language certainly conveyed no tinge even of reproach, though much of profound respect'.

To Elizabeth Dawson Steuart 10 March [1866][1] 2374

10th March

My dear Mrs Steuart,

I wrote you this morning[2] and I am now going to bother you again. I tried some time ago to get a situation here for the son[3] of Mr John Conwill[4] the schoolmaster; but apparently there is no vacancy. Conwill has desired me to ask you to help him in procuring a situation in the excise. If you could do any thing for him it would be an act of kindness done to a man who well deserves it. I have not his letter[5] now, and the precise form of what he would wish you to do has escaped my memory. But if you can do it it would give me pleasure and make me your debtor if you would lend him a helping hand.

Yours ever | John Tyndall.

RI MS JT/1/TYP/10/3480
LT Typescript Only

1. *[1866]*: the year is given based on a note from LT on the typescript: 'prob[ab]ly 1866 or after (see letters about "Young Conwill" wh[ich]. are dated M[ar]ch. 19–26, 1866)'. Unfortunately, the letters referred to by LT are missing. While we place this letter in 1866, we cannot establish this with absolute certainty.
2. *wrote to you this morning*: letter missing.
3. *son*: not identified.
4. *Mr John Conwill*: John Conwill (1802–90), was Tyndall's schoolmaster at the National School at Ballinabranna from *c.* 1836. He was central in Tyndall's early education

particularly with regard to mathematics (see the first volume of *The Correspondence of John Tyndall*). Although he was a Catholic, Conwill and Tyndall developed a close personal relationship.

5. *his letter*: letter missing.

To John Edward Kempe[1] 13 March 1866 2375

13th. March 1866

Dear Sir

I wish I could promise myself the pleasure of attending your meeting on the 5th of April,[2] but on the 2nd. I intend to hear what Mr Carlyle has to say to the Edinburgh students,[3] & for a few days subsequently I shall be in the Highlands.

Let me thank you for your invitation, the manliness of which I trust I can appreciate.

Yours very faithfully | John Tyndall

I hope your friend Mr. Birks[4] will let the public know his news on "science & prayer"[5]

Wellcome Library MS. 7777/09

---

1. *J. E. Kempe*: John Edward Kempe (1810–1907), Rector of St. James's Church in Piccadilly, London.
2. *your meeting on the 5th of April*: Tyndall had been forwarded a letter from J. E. Kempe, dated 7 March 1866, addressed to a 'Mr. Bowman', that read: 'I believe you know Professor Tyndall. Will you kindly convey to him, with my compliments, an invitation to my next monthly meeting—April 5th—when we shall discuss a subject to which he has given a great deal of consideration—"Science and Prayer". The discussion will be opened by a paper by the Rev. T. Birks—an extremely able man—and I hope to bring together several men who the Professor might find "foemen worth of his steel." But I do not wish the arguments to be all or chiefly on our side, and for this reason I am very anxious to get Professor Tyndall himself to represent the views of which he has been so able an opponent. We begin to muster at 8.30 and I take the chair precisely at 9' (Journal, RI MS JT/2/13c/1321). The Reverend Kempe held these monthly meetings with 'deputies from various Metropolitan parishes... in the Vestry-room of St. James's Church, Piccadilly' (*Spectator*, 31 (16 January 1858), p. 61).
3. *to the Edinburgh students*: in 1865, Thomas Carlyle was elected Lord Rector of the University of Edinburgh by its students. On Easter Monday, 2 April 1866, he delivered his famous inaugural address at the University of Edinburgh's Senate House, which displayed, according to the *Spectator*, Carlyle's characteristic 'vague religion of wondering, awakening intellect'. See 'Mr. Carlyle's Religion', *Spectator*, 39 (7 April 1866), pp. 377–9. Tyndall had

accompanied Carlyle to Edinburgh; they left London on Wednesday, 28 March, stayed at Fryston Hall (an estate home of Richard Monckton Milnes) for two nights, and left York for Edinburgh on Saturday 31 March. Thomas Hirst arrived at Fryston on 29 March, and Thomas Huxley on 30 March. Tyndall returned to London on that Friday, 6 April (Journal, RI MS JT/2/13c/1325–30). For Tyndall's description of the speech to Carlyle's wife, see letter 2400. The speech was published as: T. Carlyle, *Inaugural Address at Edinburgh, April 2nd, 1866* (Edinburgh: Edmonston and Douglas, 1866). Tyndall later wrote of this Edinburgh journey in his 'Personal Recollections of Thomas Carlyle (1890)', first published in the *Fortnightly Review* in January 1890 and published again in J. Tyndall, *New Fragments* (London: Longmans, Green, and Co., 1892), pp. 347–97, on pp. 357–67.

4. *Mr. Birks*: Thomas Rawson Birks (1810–83), an English theologian at the University of Cambridge who was involved in the debate to try to resolve theology and science. In 1872, he became Knightbridge Professor of Moral Philosophy at the University of Cambridge (*ODNB*).
5. *on "science & prayer"*: the subject of Birks' talk at the meeting (see n. 2).

To Arthur Penryn Stanley[1] 13 March [1866][2] 2376

13th. March

Dear Sir

I have waited and thought over the subject of your note[3] but I regret to say without much practical result. With regard to humbling ourselves in the expectation that a single beast the less will die I would say in all frankness that I consider it to be mere paganism. An article appeared some time ago in Macmillan[4] from the pen of my friend Davies[5] where I think he struck the true note in expressing a kind of contempt for prayer offered with a view of obtaining any material good. I hope and think—and pardon me for thus hoping and thinking—that you will not pray as others will for the staying of this plague,[6] but will ask on the contrary for strength of heart and clearness of mind to meet it manfully and fight against it intelligently

Yours very sincerely | John Tyndall

The very Rev. Dean Stanley

RI MS JT/2/13c/1318

1. *Arthur Penryn Stanley*: see letter 2372, n. 1.
2. *[1866]*: the year is given by relation to letter 2372.
3. *your note*: letter 2372.
4. *article appeared some time ago in Macmillan*: J. L. Davies, 'Nature and Prayer', *Macmillan's Magazine*, 13 (1866), pp. 233–40.

5. *Davies*: J. Llewelyn Davies (1826–1916), a liberal clergyman who believed prayer was 'a kind of mechanical prophylactic' (*ODNB*).
6. *plague*: cattle plague (see letter 2314, n. 5).

From Louis Lortet[1] 14 March 1866 2377

Lyon 14 Mars 1866

Monsieur et très honoré professeur,

Depuis quelques temps vos ouvrages obtiennent en France un succès qu'on ne pouvait légitimement espérer chez une nation légère et inconstante comme la nôtre et encore peu habituée aux voyages et à l'étude des grands phénomènes physiques de notre globe. Cependant, Monsieur, la clareté d'exposition et l'élégance d'exposition que vous savez mettre dans vos ouvrages, ont triomphé de tous les obstacles. Votre brochure sur la radiation et vos leçons sur la chaleur sont lues ici par tout le monde avec *[1–2 words illeg]*. Aussi je crois que le moment est favorable pour initier notre public aux sauvages mais attrayants mystères des Alpes, et je vous fais la proposition monsieur, de traduire votre livre sur les Glaciers des Alpes, si toutefois Mr Moigno n'a pas l'intention de le traduire. Fils d'un père qui a été l'ami des Agassiz, Escher, Vogt, Desor etc. j'ai de bonne heure été initié à tous leurs travaux, et depuis mon enfance chaque année j'ai foulé les glaciers et les neiges des hautes cimes. J'espère donc que mieux que beaucoup d'autres je pourrai rendre fidèlement votre pensée ayant par moi-même ma longue expérience des montagnes. Au reste, Monsieur, toutes les épreuves vous seraient soumises. Si votre libraire ne voulait pas se charger d'éditer la traduction je crois que mon libraire de Paris J. Baptiste Baillière 19 rue hautefeuille se chargerait volontiers de ce soin.

En attendant une réponse de votre part et en l'espérant favorable, veuillez Monsieur accepter les témoignages de ma très respectueuse et haute admiration.

Docteur L. Lortet | 69 Avenue de Saxe | Lyon

Lyon, March 14 1866

Sir and very honoured professor,

For some time your books in France have been achieving such success that we could not legitimately expect in a light and fickle nation as ours, and which is still unaccustomed to travels and to the study of major physical phenomena of our globe. However, Sir, the clarity of the explanation and the elegant display that you put in your books, has triumphed over all these obstacles. Your brochure on radiation[2] and your lessons on heat[3] are read here by everyone with *[1–2 words illeg]*. Also, I think that the time is ripe to introduce our audience to the wild but attractive mysteries of the Alps, and I am offering you, Sir, to translate your book on the Glaciers of the Alps,[4] though if Mr Moigno[5] does not intend to translate it. As the son of a father[6] who was friend of Agassiz,[7] Escher,[8] Vogt,[9] Desor,[10] etc., I was initiated into all their work early on, and every year since my childhood I have trodden the glaciers and snows of the high peaks. Thus I hope that better than many others, I could faithfully deliver your thought, having had my own long experience in the mountains. As for the rest, Sir, all the *[drafts]* would be submitted to you. If your bookseller does not wish to take care of the publishing for the translation, I believe that my bookseller in Paris J. Baptiste Baillière 19 rue Hautefeuille will gladly arrange for it.

While waiting for a reply from you and hoping for a positive one, please accept, Sir, the assurances of my deepest admiration.

Dr. L. Lortet | 69 Avenue de Saxe | Lyon

RI MS JT/1/L/33

---

1. *Louis Lortet*: Louis Charles Émile Lortet (1836–1909), a French physician, botanist, zoologist, and mountaineer (C. Gaillard, 'La vie et les travaux de Louis-Charles Lortet', *Archives du Muséum d'histoire naturelle de Lyon*, 11 (1912), pp. 1–31).
2. *brochure on radiation*: J. Tyndall, *Sur la radiation*, trans. F. Moigno (Paris: E. Giruad, 1865). Tyndall had given the Rede lecture at Cambridge in May 1865 (see letter 2231, n. 1).
3. *lessons on heat*: J. Tyndall, *La Chaleur considérée come un mode de mouvement*, trans. F. Moigno (Paris: Étienne Giraud, 1864).
4. *translate your book on the Glaciers of the Alps*: while Lortet translated another of Tyndall's mountaineering books—*Mountaineering in 1861* as *Dans les Montagnes*—it appears that the plan to publish a translation of *Glaciers of the Alps* fell through (see letter 2255, n. 2).
5. *Mr Moigno*: François-Napoléon-Marie Moigno.
6. *father*: Pierre Lortet (1792–1868), a French politician.
7. *Agassiz*: Louis Agassiz (1807–73), a Swiss naturalist who studied medicine, botany, geology, and zoology in Germany and France before moving to the United States in 1847 as professor of zoology and geology at Harvard University, where he founded the Museum of Comparative Zoology. He was a key figure in developing the study of the Earth's history, proposed the theory of ice ages, and made a close study of glaciers in the Alps. In 1840 he published his important *Études sur les Glaciers* in two volumes (Neuchâtel: Gent et Gassmann), which helped establish the scientific study of glaciers (*CDSB*; *HLS*).
8. *Escher*: Arnold Escher (von der Linth).
9. *Vogt*: Karl Vogt (1817–95), a German scientist, philosopher, and politician living in Switzerland (*ADB*).
10. *Desor*: Pierre Jean Édouard Desor (1811–82), was a German-Swiss geologist and mountaineer who studied law at Giessen and Heidelberg. After meeting Louis Agassiz in 1837

Desor became his chief collaborator and close friend, and followed him to America when Agassiz became professor of zoology and geology at Harvard. In the 1840s he made several ascents, including the Jungfrau, with James David Forbes in 1841. In 1852 he returned to Switzerland and was appointed professor of geology at the Academy of Neuchâtel (*CDSB*; *HLS*).

From Jacques-Louis Soret 16 March 1866 2378

Genève 16 Mars 1866

Mon cher Tyndall

Voici longtemps que je veux vous écrire, le temps a passé je ne sais comment. Je voulais d'abord vous dire que l'expérience du regel de la glace m'a très bien réussi et qu'elle a été très appréciée par mes auditeurs. Je l'ai répétée une autre fois à la Société de physique, en faisant un rapport sur votre dernière publication sur le regel et sur celle de Mr Helmholtz. Elle a beaucoup intéressé les assistants y compris le G^d^ Duc de Baden qui était venu à la séance. Vous voyez que l'envoi de vos moules m'a été très utile et je vous en remercie encore.

En second lieu j'ai aussi à vous remercier de votre photographie qui m'a fait le plus grand plaisir, malgré votre air un peu [féroce].

En troisième lieu je suis un peu embarrassé de vous dire que Mme Soret n'a pas reçu les livres dont vous m'annonciez l'envoi dans votre lettre du 5 Décembre. Elle se réjouissait beaucoup de les recevoir et vous était très reconnaissante. Il y aura eu probablement quelque erreur du libraire, et quoique j'aie l'air indiscret de vous parler de cela, je pense qu'il vaut mieux vous en informer tout franchement, pour que vous me disiez si je dois réclamer et auprès de qui.

Si vous avez le temps de me répondre quelques lignes j'aimerais beaucoup avoir des renseignements sur les laboratoires de chimie anglais. Il s'agit d'un jeune homme, un de mes cousins qui étudie la chimie; il a passé déjà six mois à Edimbourg dans le laboratoire de Mr Playfair; mais il trouve qu'il n'y a pas là beaucoup d'activité et que les élèves sont laissés trop à leurs propres forces.—Ses parents désireraient qu'il pût prolonger un peu son séjour dans le G^d^ Bretagne afin qu'il fut tout à fait maître de la langue anglaise avant de l'envoyer ailleurs. Ils m'ont demandé des renseignements que je n'ai pu leur donner. La question serait donc de savoir si l'on peut trouver à Londres ou ailleurs un bon laboratoire d'élèves dans le genre de ceux que l'on trouve en si grande abondance en Allemagne. Comme le jeune homme n'est pas encore très avancé je pense que ce serait surtout la chimie minérale et l'analyse qu'il lui conviendrait d'étudier, plutôt que la chimie organique. Si cela ne vous cause pas de perte de temps une indication de votre part sur ce sujet me serait très précieuse.

J'espère que vous avez été satisfait des appareils qui vous ont été envoyés de notre atelier d'instruments de physique. A ce propos permettez moi d'attirer votre attention sur le fait que nous allons construire des appareils de Mr de la Rive pour l'aurore boréale pour lesquels nous avons quelques demandes, et que si l'un des établissements de Londres voulait en avoir un, ce serait le moment le plus favorable pour le commander. Le prix de cet appareil est de 1000 francs, et pour pouvoir le donner à ce prix il faut que notre atelier puisse en construire plusieurs à la fois, c'est pourquoi nous avons fait dans différentes villes l'offre d'en construire maintenant. Les Institutions royale et polytechniques ont probablement reçu ma circulaire à cet égard. Vous *[connaissez]* l'appareil que vous avez vu à Zurich, *[avant]* que je *[n'ai]* pas de renseignements à vous donner.

Mr de la Rive est bien; vous savez qu'il s'était cassé le bras cet automne, il a eu beaucoup de peine à se bien remettre et il a longtemps souffert; mais il est mieux maintenant.

Madame Soret me charge de ses compliments pour vous et j'y joins l'expression de mon dévouement affectueux.

L. Soret

Geneva 16 March 1866

My dear Tyndall,

I have wanted to write to you for a long time now; I do not know how the time has slipped away. Firstly, I wanted to tell you that I successfully performed the ice <u>refreezing</u> experiment and that my audience liked it very much.[1] I repeated it once more at the Société de Physique,[2] when I reported on your latest publication on <u>refreezing</u> and on that of Mr Helmholtz.[3] It was very interesting to the attendees, including the Grand Duke of Baden,[4] who came to the session. So, you see, your moulds was very useful to me, and I thank you for them once again.

Secondly, I also thank you for your photograph, which I greatly enjoyed despite your <u>somewhat fierce</u> look.

Thirdly, I am a little embarrassed to tell you that Mrs Soret[5] did not receive the books that you mentioned you were sending in your letter of December 5th.[6] She was very excited to receive them and very grateful to you. There was probably some error made by the bookseller, and although it may seem tactless to broach the subject, I thought it better to inform you directly so you could tell me if and to whom I must inquire about it.

If you have the time to write a few extra lines, I would much like to get some information on English chemistry laboratories for a young man, one of my cousins who studies chemistry. He has been at Mr Playfair's[7] laboratory in Edinburgh for the past six months, but he finds that there is not much activity there and the students are left too much to their own devices.—His parents wanted him to stay in Great Britain a little longer so that he could fully master the English language before being sent elsewhere. They asked me for information that I could not provide them. The question is this: could we find, in London or elsewhere, a good student laboratory in the style of those that are found so abundantly in Germany? Since the young man is not yet too advanced, I think that he would be best suited to study inorganic chemistry and analysis rather than organic chemistry. If it would not take too much of your time, any information you might give me on this subject will be extremely valuable.

I hope that you were happy with the apparatuses sent from our physics instrument workshop. Speaking of which, allow me to bring your attention to the fact that we will be constructing Mr de la Rive's[8] apparatus for the aurora borealis,[9] for which we have had some requests; if one of the London establishments would like to have one, this would be the best moment to order it. The price of this device is 1000 francs, and to be able to sell it for this price, our workshop needs to be able to build several of them at the same time. This is why we are currently soliciting orders in many different cities. The Royal and Polytechnic Institutions[10] have probably received my memorandum on the subject. You know the device, which you saw in Zurich before I had any information to give you.

Mr de la Rive is well. You know that he broke his arm this autumn; he had a very difficult recovery and suffered for a long time, but he is better now.

Mrs Soret asks me to send you her compliments, and to them I add this expression of my devoted affection.

L. Soret

RI MS JT/1/S/89

1. *my audience liked it very much*: in his letter of 29 September 1865, Soret wrote to Tyndall that 'This winter I will have to do a public course of about ten lessons on Glaciers' (see letter 2309).
2. *Société de Physique*: Société de Physique et d'Histoire Naturelle de Genève, founded in 1791.
3. *latest publication on refreezing and that of Mr Helmholtz*: J. Tyndall, 'Professor Helmholtz on Ice and Glaciers', *Phil. Mag.*, 30 (1865), pp. 393–407. This is letter 2330 in this volume.
4. *Grand Duke of Baden*: Friedrich Wilhelm Ludwig (1826–1907), Frederick I, Grand Duke of Baden from 1856–1907.
5. *Mrs Soret*: Clémentine Soret (née Odier).
6. *your letter of December 5th*: letter missing.
7. *Mr Playfair's*: Lyon Playfair.
8. *Mr de la Rive's*: Auguste de la Rive.

9. *apparatuses for the aurora borealis*: de la Rive had developed a theory of the cause of Aurora Borealis, and had devised scientific apparatus to demonstrate it experimentally. See A. de la Rive, *A Treatise on Electricity: In Theory and Practice*, 3 vols, trans. C. Walker (London: Longman, Brown, Green, Longmans, & Roberts, 1853–8), vol. 3, pp. 283–309. A description of 'Apparatus designed by Auguste de la Rive, for the demonstration of the Electric Theory of the Aurora Borealis', labeled as the 'property of Messrs. Soret, Perrot, & Sarasin, Geneva', appeared in *Catalogue of the Special Loan Collection of Scientific Apparatus at the South Kensington Museum*, 1st edn (London: George E. Eyre and William Spottiswoode, 1876): 'A large sphere, made of wood, represents the earth. Two iron cylinders represent the two extremities of the terrestrial magnetic axis. They penetrate into two globes filled with rarefied air, which simulate the higher regions of the polar atmosphere. The electric discharge, which takes place in this rarefied air, following rays all around a point situated in the prolongation of the terrestrial axis, turns about this point, and so turns in a different direction at either pole, when the two cylinders are charged by means of a horse-show electro-magnet, in accordance with the observations upon the rotation of the rays of the Aurora Borealis' (p. 232). The apparatus and its use were also discussed in: A. de la Rive, 'Further Researches on the Aurorae Boreales, and the Phenomena which attend them', *Phil. Mag.*, 23 (1862), pp. 546–53.
10. *Royal and Polytechnic Institutions*: the RI and the Royal Polytechnic Institution (founded in 1838, now part of the University of Westminster).

To Richard Monckton Milnes [19 March 1866][1] 2379

Royal Institution of Great Britain | Monday.

My dear Lord.

It will give me great pleasure to join you on the 27th I am just going down to Mr. Carlyle to make out his movements[2]—He possibly may be one of the "congenial spirits".

Most truly yours | John Tyndall

Huxley & myself travel together: if he cannot go till the 28th I sh[oul]d not like to forsake him.

Trinity College Cambridge, Houghton. Add.ms.c.26/78

1. *[19 March 1866]*: the date is given based on when Tyndall went to Edinburgh with Carlyle (see n. 2), with Mondays before 27 March being 19 March and 26 March. If this letter had been written on 26 March, Tyndall most likely would have written 'join you tomorrow', so we place this letter as 19 March.
2. *his movements*: Tyndall accompanied Carlyle to the University of Edinburgh for his speech to the students as Rector on 2 April 1866 (see letter 2375, n. 3).

From Louis Lortet 20 March 1866 2380

Lyon 20 Mars 1866.

Monsieur

Je vous remercie de votre lettre quoique j'ai le malheur d'arriver trop tard. Cependant si M. Felix Foucou ne terminait pas sa traduction ou se décidait à ne pas la publier, je vous prie instamment de ne pas m'oublier, car je serais trop heureux de faire connaître à notre pays votre beau et bon livre.

Veuillez, très honoré Monsieur, accepter le témoignages de ma très haute considération.

Docteur L. Lortet

Lyon 20 March 1866.

Sir

Thank you for your letter,[1] although I had the misfortune of being too late. However, if Mr. Felix Foucou does not finish his translation[2] or if he decides not to publish it, I urge you to not forget me, since I would be more than happy to acquaint our country with your excellent book.

Please, much honoured Sir, accept the expression of my highest regards.

Doctor L. Lortet

RI MS JT/1/L/34

1. *your letter*: letter missing.
2. *his translation*: see letter 2255, n. 2.

From George Gabriel Stokes 20 March 1866 2381

Lensfield Cottage Cambridge. | 20th March 1866

My dear Tyndall,

At the last meeting of the Council R.S. your paper was ordered to be printed in the Philosophical Transactions.[1] I was one of the referees,[2] and I recommended that your attention should be directed to one point, which you should be requested to reconsider, without however at all recommending that this should be made a condition of printing.

The point in question is the argument in favour of the supposed allotropy of Platinum.

The allotropy is supposed to exist in consequence of the very different emissive power of Pt for radiant heat according as it is in the form of

Platinum black or continuous metal. To this must be added the law observed in other cases that bodies of a simple chemical composition absorb non-luminous radiant heat comparatively feebly. Yet this latter circumstance does not really help the argument, for a thin metallic plate, suppose platinum foil, would doubtless be opaque to radiant heat as it is to light of all kinds. So that the argument in favour of allotropy is in fact simply that which I first mentioned.

I will endeavour to put succinctly and clearly the argument I used at the meeting when your paper was read, but which I feel I did not then put clearly.

1. The consideration of a body within a heated envelope shows that powerful reflexion implies not very powerful emission.

Metals in mass are as a general rule powerful reflectors, and therefore in mass they ought not to be powerful emitters.

2. Powerful (or at least strong) reflexion is bound up with two perfectly distinct properties, namely

A. powerful refraction,

B. powerful absorption.

The reflexion bound up with A hardly deserves the epithet "powerful", at least so far as regards transparent substances, for which alone (independently of theoretical considerations) we know the refractive index. Thus for diamond $\mu=2$ *[about]*, which is unusually high. Yet for reflexion at a ⊥ incidence Fresnel's[3] formula $(k\text{-}1 \ / \ \mu+1)^2$ gives $(2-1 \ / \ 2+1)^2 = 1/9 = 11\%$ nearly. But at a ⊥ incidence silver reflects something like 90%.

But it may be said Silver if transparent would be a very powerful reflector; may it not be with powerful refraction that its powerful reflexion is associated?

It would require little short of a treatise to go into the mathematico-optical reasons for connecting the strong reflexion of silver with its powerful absorption, and therefore I will confine myself to purely physical considerations.

I would remark in the first place that platinum, which is nearly twice as dense as silver, is much less reflective. We can hardly doubt, especially as the two substances are similar in their chemical nature, that the nearly double density of platinum would be associated, if the two substances were transparent, with a much higher refractive power.

I might refer to the coloured metals gold & copper, but in fact the nature of metallic reflexion is best elucidated by referring to certain non-metallic substances which exhibit for <u>certain</u> <u>parts</u> <u>only</u> of the spectrum a quasi-metallic reflexion. This reflexion agrees with that of metals not only in its copiousness, but also in its other properties, namely, in not vanishing (or all but) at a certain angle (the polarizing angle) in the case of light polarized ⊥ly to the plane of incidence, and in showing a retardation of phase in light polarized ⊥ly to the plane of incidence relatively to light but || to that plane.

Among these chromatically metallic (in an optical sense) substances, crystallized permanganate of potash is peculiarly instructive. On this subject I would refer you to a paper of mine in the Phil. Mag. Vol 6 (1853) p. 393 especially p. 397 et seq.[4] There can be no possible doubt here that metallic reflexion is bound up with intense absorption, and not with high refraction.

(3) In the case of chromatically metallic (optically) substances, the light reflected by virtue of the metallicity is distinguishable by its colour. In the case of carthamine (safflower red) for instance it is yellowish green, in the case of indigo copper-colour and so forth. Now observation shows that the colour belonging to quasi-metallic reflexion is only seen when the substance is in the state of a mass continuous or nearly so. Thus indigo in fine powder is blue, which is the transmission tint; on burnishing the copper colour appears. Carthamine in fine suspension or in a dyed ribbon is pink, the greenish-golden reflexion is observed on a "pink saucer" and so forth.

Seeing then that in the case of the chromatically metallic bodies the colour due to metallicity disappears when the substance is in a state of extremely fine division, we may infer that in the case of the true metals the metallic reflexion (that which is bound up with their intense opacity) would disappear when the substance was in a state of extreme division. The colour and quantity of reflected light would then be determined by the residue left after the absorption intervening between the multiplied reflexions which take place in a heap of powder; and since all metals are intensely opaque in powder, they look black or all but.

I think it probable that the intense absorption exhibited by metals is due to the mode in which the molecules are built up in the mass, and not to the character of the individual molecules; and it is not unlikely that there may theoretically exist a state (realized perhaps in metallic vapours) in which they would be transparent for radiant heat, and which has as much right to be called an allotropic state as diamond and charcoal to be considered allotropic forms of carbon; but if so I feel persuaded that platinum foil and platinum black are both one and the other analogues of charcoal-carbon.

It would be proper that you should notice the experiments[5] of Balfour Stewart[6] which were mentioned when the paper was read, and you will doubtless be desirous of doing so.

Your paper is lying at Burlington House. If there are any slight changes to make in consequences of what I have written, they may as well be made before it goes to press. You can get the paper back for the purpose if you like.

Yours very truly | G. G. Stokes.

RI MS JT/1/S/241
RI MS JT/1/TYP/4/1400–2

1. *in Philosophical Transactions*: J. Tyndall, 'Sixth Memoir on Radiation and Absorption.—Influence of Colour and Mechanical Condition on Radiant Heat', *Phil. Trans.*, 156 (1866), pp. 83–96.
2. *one of the referees*: despite their radically different upbringing, temperament, and religious orientation, Tyndall and Stokes maintained a strong friendship and correspondence. Moreover, as editor of the *Phil. Trans.*, Stokes also shaped both Tyndall's papers and ideas about scientific theories. See, M. Baldwin, 'Tyndall and Stokes: Correspondence, Referee Reports and the Physical Sciences in Victorian Britain', in B. Lightman and M. S. Reidy (eds), *The Age of Scientific Naturalism* (London: Pickering & Chatto, 2014), pp. 171–86.
3. *Fresnel*: Augustin-Jean Fresnel (1788–1827), a French civil engineer and natural philosopher whose work helped to establish the wave theory of light. He invented the stepped lens which bears his name for use in lighthouses (*CDSB*).
4. *397 et seq*: G. G. Stokes, 'On the Metallic Reflexion exhibited by certain Non-metallic Substances', *Phil. Mag.*, 6 (1853), pp. 393–403.
5. *experiments*: probably B. Stewart, 'An Account of some Experiments on Radiant Heat, involving an extension of Prevost's Theory of Exchanges', *Transactions of the Royal Society of Edinburgh*, 22 (1858), pp. 1–20.
6. *Balfour Stewart*: Balfour Stewart (1828–87), a Scottish physicist, meteorologist, devoted churchman, director of Kew Observatory (1859–71), and later professor of natural philosophy at Owens College, Manchester. Stewart was co-author with P. G. Tait of *The Unseen Universe* (London: Macmillan and Co., 1875) (*ODNB*).

From Lyon Playfair 21 March 1866 2382

Private | 21 March /66

My dear Tyndall

Is it true that you are coming to Edinburgh to be present at Carlyles inauguration as Rector on the 3^{d}.[1] If so I shall propose to the Senators that you be made L.L.D[2] on the occasion.

I have spoken to Brewster[3] who quite approves.

From controversial feeling there may be a little objection to the proposal on the part of one or two members[4]—but I hope not, and we hope to carry the proposition unanimously—Let me know by return that I may take the necessary formal steps.

If Huxley[5] & Ramsay[6] come, prior invitation will fill my beds but not the board. So I hope if you sleep out of the house that you will make the bed a sleeping berth & my home your home—

Yours Sincerely | Lyon Playfair

RI MS JT/2/10/396

1. *on the 3d*: Tyndall accompanied Carlyle to University of Edinburgh for his speech to the students as Rector on 2 April 1866 (see letter 2375, n. 3).
2. *L.L.D*: Doctor of Law. Just as he received an honorary LL.D (Doctor of Law) from the University of Cambridge in 1865, Tyndall also received the same from the University of Edinburgh in 1866. Thomas Huxley also received the degree, but Carlyle declined (see letter 2391).
3. *Brewster*: David Brewster.
4. *one or two members*: not identified.
5. *Huxley*: Thomas Huxley accepted the honorary degree (L. Huxley (ed), *Life and Letters of Thomas Henry Huxley*, 2 vols (London: Macmillan and Co., 1900), vol. 2, p. 473).
6. *Ramsay*: Andrew Crombie Ramsay (1814–91), a Scottish geologist who held the chair of geology at the School of Mines from 1851–72, and in 1872 succeeded Roderick Murchison as director of the Geological Survey. Ramsay argued that glaciers formed some lake basins, and had accompanied Tyndall to the Alps in 1858 (*ODNB*). Ramsay accepted the degree (A. Geikie, *Memoir of Sir Andrew Crombie Ramsay* (London: Macmillan and Co., 1895), p. 285).

From Elizabeth Dawson Steuart 21 March 1866 2383

March 21. 66.

My dear John,

Your three letters[1] reached me all together yesterday evening, and I lose no time in trying to ease your mind. I should have written to you sooner, but heard you were going to Scotland for 3 or 4 weeks,[2] and I waited until you returned to London. This and illness, must plead my apology. With respect to the two letters enclosed to me, the one from your Mother[3] was written by one of the Misses Warr,[4] the other by Mrs Doyle.[5] I sent one in whom I place great confidence,[6] over to your Mother to make enquiries respecting the management of her money, and from her own lips it came that "she keeps her money herself, pays all her bills, and will swear that Mrs Doyle is honest." These are her own words, and I believe them to be perfectly true. The things bought of Lynch were for herself and Emma,[7] winter clothing which could not be done without. I have seen them on the latter. As to the money borrowed by your Mother of Miss Payne[8] I know nothing except that she did so, and am very sorry for your sake, that such is the fact. Now as regards Emma, I cannot see what is to be gained by a "consultation", her health is good except for a cold just now, and she has an excellent appetite, and as to her mind, a great deal of her apparent state of it, is caused by temper and a very wayward disposition, engendered in her from her birth, by injudicious indulgence, and even yet, she is never controlled or contradicted in any way. She can be perfectly calm whenever she pleases, and the reverse at will—She is worrying your poor Mother, and every one, about going to Tramore,[9] and I do think if you would write in a kind manner, but a determined one, saying that whenever the season is good enough, you will send her there, it may have a good effect. I may as well tell you the truth of your Mother's feelings towards her late medical attendant.[10] She hates him, and says so openly, and declares he shall never attend her or Emma again. Such is her antipathy to him that I verily believe if forced upon her, her life might be sacrificed. He has never gone near her since the day his dismissal from you[11] arrived, or made the smallest enquiry for her, and this proves the source of all his attention heretofore, the love of gain. The great annoyance your poor Mother suffered from himself and wife[12] while he had authority over them, was not to be told and he has purposely poisoned your mind against Mrs Doyle because he blames her for losing the salary you gave him, for he is very vindictive. You would find it very difficult to replace Mrs Doyle, for few would tolerate the whims and humours of poor Emma, or go through all her attendant has to encounter. Whenever a physician is required, a first rate one can be had in Bagnalstown,[13] a gentleman of the highest standing, skill, and disposition. But I assure you, at present there is nothing to consult about, and your money will be thrown away. I have now told you all I can in reply to your queries, and will feel it at all times a pleasure to do anything in my power for yourself or those naturally so dear to you; I shall go over myself as soon as possible to see them, but I very seldom can get out further than the garden, my health is so broken and uncertain. I would give you a hint about the way some spend what you send to them (not your own family), but you might think me officious and interfering.

With kindest regards, | believe me, | very sincerely yours | E. D. Steuart.

RI MS JT/1/TYP/10/3357–8
LT Typescript Only

---

1. *Your three letters*: letters missing.
2. *going to Scotland for 3 or 4 weeks*: Tyndall accompanied Carlyle to the University of Edinburgh for his speech to the students as Rector on 2 April 1866 (see letter 2375, n. 3), leaving on 28 March and returning to London on 6 April, so not quite a three or four week trip.
3. *Mother*: Sarah Tyndall.
4. *one of the Misses Warr*: not identified. LT suggested it could be 'Wall'.
5. *Mrs Doyle*: Abbey Doyle.
6. *one in whom I place great confidence*: possibly the 'Miss Moore" referred to in letter 2355.
7. *Emma*: Emily Tyndall.
8. *Miss Payne*: see letter 2356, n. 6.

9. *Tramore*: a seaside and resort town in County Waterford, Ireland. Emily had gone to Tramore in 1865 (see letters 2313, 2337, and 2338).
10. *late medical attendant*: not identified, but presumably the 'man' referred to in letter 2355.
11. *his dismissal from you*: presumably in a letter, which is missing.
12. *his wife*: not identified.
13. *first rate one can be had in Bagnalstown*: James John Trayer (see letter 2355, n. 12).

To Lyon Playfair 22 March [1866][1] 2384

Royal Institution of Great Britain | 22nd. March

My dear Playfair

It is quite true that I intend to be present at Carlyle's inauguration:[2] I join the grand old fellow on Wednesday, go down first to Lord Houghtons[3] & hope to accompany him & Huxley to Edinburgh on the Saturday.

I stay with my old friend Lady Coxe.[4]

My dear Playfair your cordial kindly letter[5] is I swear to you more pleasant to me than any degree[6] that the great men of Edinburgh could confer upon me. In this matter I must leave myself entirely in your hands, but I should certainly abhor being the cause of any dissension.[7]

I could wish that the "controversy" had never occurred, or that my blood had run less hotly through what I have written.[8] But I am not Christian enough if smitten on the right cheek to turn the other also.[9] I hit my antagonist—I trust not below the belt—and bear him no malice afterwards.

Believe me dear Playfair | Yours ever truly | John Tyndall

Give my kindest regards to your Veteran President[10]

RI MS JT/2/10/396
RI MS JT/2/13c/1321–2

---

1. *[1866]*: the year is given by reference to Tyndall accompanying Carlyle to the University of Edinburgh (see n. 2).
2. *Carlyle's inauguration*: Tyndall accompanied Carlyle to the University of Edinburgh for his speech to the students as Rector on 2 April 1866 (see letter 2375, n. 3).
3. *Lord Houghtons*: Richard Monckton Milnes.
4. *Lady Coxe*: Mary Anne Coxe.
5. *your cordially kind letter*: letter 2382.
6. *degree*: LL.D (Doctor of Law). See letter 2382, n. 2.
7. *cause of any dissension*: in letter 2382, Playfair wrote that because of 'controversial feeling there may be a little objection to the proposal on the part of one or two members'.
8. *I have written*: Tyndall was referring to his recent letters to the *PMG* regarding prayer and

cholera (see letters 2314 and 2318), as well as an article in the *Fortnightly Review* (see letter 2371, n. 6).

9. *smitten on the right cheek to turn the other also*: a reference to Jesus' 'Sermon on the Mount', as recorded in the Gospel of Matthew: 'You have heard that it was said, "An eye for an eye and a tooth for a tooth". But I say to you, Do not resist an evildoer. But if anyone strikes you on the right cheek, turn the other also; and if anyone wants to sue you and take your coat, give your cloak as well' (Matthew 5:38–40).
10. *your Veteran President*: probably referring to David Brewster, who was President of the Royal Society of Edinburgh from 1864 until his death in 1868, and also Principal and Vice-Chancellor of the University of Edinburgh from 1859–68.

From Johann Conrad Bohn 23 March 1866 2385

Giessen d. 23 März 1866

Hochverehrter Herr Tyndall

Ihre Arbeiten über Calorescenz, insbesondere die erste, in welcher durch die Wärme einer fast ausschliesslich dunkle Strahlen aussendenden Wasserstoffflamme, Platindraht veranlasst wurde leuchtende Strahlen auszusenden, haben mich angeregt einen Versuch anzustellen, der Ihrem ersten nahe verwandt ist, aber, wie mir scheint, in noch schlagenderer Weise die negative Fluorescenz darthut. Ich hing an einem Baumwoll-oder Leinen-Faden in dem Kochraume eines gewöhnlichen eisernen und dann auch eines sogenannten Porzellan Stubenofens ein Stück Flussspath und insbesondere ein Stück von der Varietät Chlorophan auf. Dieses Mineral wurde aus geringer Entfernung bestrahlt durch die Strahlen, welche die mässig erwärmten,—jedenfalls weit vom Glühen entfernten 6 eisernen oder 2 geschwärzten metallnen und 4 glasirten thönernen Platten (die Wände des Kochraumes) aussendeten. Nach 2 Minuten leuchtete der Chlorophan, eine andere Varietät nach 2½ Minuten und die Intensität der leuchtenden Ausstrahlung steigerte sich rasch zu einem recht lebhaften und schönen Glanze.—Die Erwärmung des an dem schlecht leitenden (sich nicht durch die Hitze bräunenden) Faden befestigten Minerals, geschah allerdings zum Theil auch durch Berührung mit der heissen Luft in der Ofenröhre. Allein einmal war diese Luft nicht so heiss, dass nicht meine Hand in ihr, während sie gleichzeitig noch durch strahlende Wärme afficirt wurde, längere Zeit verweilen konnte ohne dass ich Beschwerde empfand; ferner kann die so wenig dichte Luft, die dazu noch eine so geringe specifische W40 Min.chten des Flussspathes wesentlich ducine merkliche Erwichte Luft, die dazu noch eine so geringe specifische Wärme besitzt, in der kurzen Zeit nicht wohl eine merkliche Erwärmung des Flussspathes bewirken. Dass die Erwärmung, und dadurch das Leuchten des Flussspathes wesentlich durch die dunklen, von den Ofenwänden ausgehenden Wärmestrahlen bewirkt wurde, bewies sehr deutlich folgender Versuch. Ich brachte ein Stück Flussspath in eine dünnwandige Glasröhre und hing diese im Ofenraume auf, daneben aber an einem Faden frei in der Luft schwebend ein anderes Stück Flussspath, das von demselben grösseren Kristall abgeschlagen worden war. Die Luft hatte zu den Kristallen in der Glasröhre möglichst freien Zutritt. Nun, während der frei aufgehängte Flussspath schon nach 2½ Minuten merklich leuchtete, konnte der erste Schimmer an dem in der Glasröhre befindlichen Stücke erst nach Verlauf von 18 Minuten wahrgenommen werden. Die Glasröhre selbst war sehr früh so heiss geworden, dass ihre Berührung mit der Hand Schmerz erregte. Der Versuch wurde in einem sehr dunklen Zimmer angestellt und die Glasröhre mittelst eines Holzstäbchens in Schwingungen versetzt, weil ein Lichtschimmer eher wahrgenommen wird, wenn er von einem bewegten Gegenstand, als wenn er von einem ruhenden ausgeht.—Da die Glaswände der Röhre die dunklen Wärmestrahlen fast vollständig vom Flussspath abhielten, so wurde dieser nur durch Berührung mit der heissen Luft und durch Berührung mit der allmählig sich stark erhitzenden Glasröhre erwärmt. Da das Leuchten des Flussspathes in der Glasröhre so sehr viel später eintrat, so folgt, dass die hauptsächliche Wirkung bei Erregung des Leuchtens den dunklen Wärmestrahlen zuzuschreiben ist. Wir haben hier also wirklich eine, durch den Flussspath vermittelte, Umwandlung dunkler, wenig brechbarer Strahlen in leuchtende, stärker brechbare Strahlen,—was Sie negative Fluorescenz genannt haben.—Es wäre vielleicht nicht uninteressant den Versuch dahin abzuändern, dass der Chlorophan in einem Behälter von Steinsalz aufgehängt würde, ich vermuthe, dass er darin, da die dunklen Wärmestrahlen nur sehr unbedeutend vom Steinsalze absorbirt werden, fast sobald zum Leuchten kommt, als in der freien Luft. Mir fehlt es an passendem Material um den Versuch anstellen zu können; Ihnen würde es leicht sein den Versuch zu machen, wenn Sie ihm einiges Interesse abgewinnen könnten.

Uebrigens besteht doch ein wesentlicher Unterschied zwischen dem Leuchtendwerden des Platindrahtes in Ihrem, dem Leuchtendwerden des Flussspaths in meinem Versuche, gegen die gewöhnlichen Fluorescenzerscheinungen. Aesculin, Chinin, Uranglas und die übrigen hierhergehörig Substanzen alle, fluoresciren augenblicklich sowie sie von geeignetem Lichte getroffen werden. Das Leuchtendwerden durch dunkle Wärmestrahlen tritt aber erst dann merkbar auf, <wenn> die Strahlung eine gewisse Zeit hindurch gewährt hat, wenn eine merkliche Quantität absorbirter Wärme in dem bestrahlten Körper sich angesammelt hat. Ich vermuthe, dass dieses selbst in den mächtigen Brennpunkten dunkler Strahlen, welche Sie in so ingenieuser Weise hergestellt haben, der Fall ist.—Liess ich fein vertheilten Flussspath auf die heisse Ofenplatte fallen, so trat das Leuchten allerdings scheinbar sofort ein. Man könnte nun sagen die Aussendung leuchtender Strahlen beginne allerdings im Moment selbst, in welchem die erregenden dunklen Strahlen den Körper treffen, nur sei in der ersten Zeit die Intensität so gering, dass das Leuchten noch nicht wahrgenommen werde. Allein eine dementsprechende Zunahme der Intensität ist eben bei der Fluorescenz im eigentlichen Sinne nicht zu bemerken; die Körper fluoresciren scheinbar im ersten Augenblicke so stark als in den späteren. Was den Flussspath angeht, so besteht noch ein anderer sehr bemerkenswerter Unterschied seiner durch erwärmende Bestrahlung hervorgebrachten Phosphorescenz gegen seine durch Bestrahlung mit ultravioletten oder sehr brechbaren Strahlen hervorgebrachte Fluorescenz. Die Phosphorescenz nimmt eine Zeit lang an Intensität zu, dann aber auch wieder ab. Ein Stück Flussspath das über eine halbe Stunde bei meinen Versuchen von dem Ofenraume bestrahlt war, erlosch fast gänzlich; Stücke die in directer Berührung mit der heissen Ofenplatte waren, erloschen noch früher. Bei der Fluorescenz findet sich nichts ähnliches. Nebenbei bemerke ich, dass eine Angabe Becquerel's ein stark erwärmter Flussspath phosphorescire erst dann wieder, wenn er vorher einer Bestrahlung durch Licht erfahren habe, nicht richtig oder mindestens nicht allgemein ist. Ich habe an demselben Abend, beständig im dunklen Zimmer verweilend, dasselbe Stück Flussspath sechs mal hintereinander phosphoresciren lassen;—die letzten Male wurde allerdings das Leuchten nicht so lebhaft;—wahrscheinlich hatte ich die Erhitzung vorher zu weit getrieben.

Wenn Sie diese Mittheilung der Ehre einer Veröffentlichung im Philosoph. Magazine werth erachten, so würde es mich sehr freuen, dieselbe dort erscheinen zu sehen. Für diesen Fall würde ich Sie bitten mir gefälligst einige Separatabdrucke von dem Verleger ausbitten zu wollen.

In Ihrer durch Francis mir zugekommenen Abhandlung über Calorescenz (aus Philos. Trans f. 1866), für welche ich Ihnen noch bestens zu danken und zu sagen habe, dass ich sehr grosse Freude an den interessanten Versuchen hatte, wird Bezug genommen auf Figuren, die mir nicht zugekommen sind. Die bildliche Darstellung der Apparate könnte wohl ohne die Deutlichkeit zu beeinträchtigen aus der Uebersetz. (Anhang zu Cap. XII) wegbleiben; müsste aber auch die andre Figur fehlen, so wäre eine kleine Aenderung im Texte erforderlich. Am angenehmsten wäre es wenn ich in den Besitz einer Copie der Figuren gelangen könnte.

Sie haben mir seiner Zeit von Herrn Vieweg als Verleger meiner Uebersetzung Ihres Buches on Heat abgerathen. Hätte ich Ihrem Rathe Folge gegeben, so hätte ich mir einigen Aerger erspart und das Buch wäre früher erschienen. Ich habe mit H. Vieweg Streit bekommen. Er liess den Satz der Bogen unterbrechen, ohne mich im geringsten zu benachrichtigen, endlich nahm er sich heraus an meiner Arbeit Aenderungen zu machen oder machen lassen und muthete mir zu jene sehr ungeschickten, unmotivirten und zum Theil geradezu Unsinn hervorbringenden Aenderungen zu genehmigen. Der Ausgang der Storie war eine Auflösung unseres Vertrages, wobei Vieweg mir eine Geldentschädigung für meine verlorene Zeit gewährte. Seit gestern besitze ich wieder mein Manuscript und bin freier Herr darüber. Ich habe sofort mit Theod. Fischer in Cassel, den Sie mir seiner Zeit als Verleger angelegentlich empfohlen, Unterhandlungen angefangen und dieselben sind *[1 word illeg]* zu Ende gekommen. Herr Fischer will mit thunlichster Beschleunigung meine Uebersetzung in bester Ausstattung erscheinen lassen und hat mir in sehr angenehmer Weise viel günstigere Honorarbedingungen gewährt als Vieweg. Da Sie selbst mir Fischer anriethen, so glaubte ich Ihre Einwilligung im Voraus als sicher annehmen zu dürfen. Ich bitte Sie nachträglich darum, dass Sie Herrn Theod. Fischer das gewähren, was Sie H. Vieweg erlaubten.—Um keine Verzögerung der Herausgabe eintreten zu lassen bemüht sich H. Fischer die Clichés der Figuren zu erhalten. Er ersucht mich Sie zu bitten Ihren englischen Verleger zu fragen unter welchen Bedingungen die Clichés zur Verwend. in der deutschen Uebersetzung abgegeben würden, meint auch Herr Dr Debus würde, wenn er in London sei, aus Gefälligkeit für Fischer gerne die etwa nöthigen Gänge thun. Den grösseren Theil der Figuren hat Vieweg, und zwar sehr hübsch schon stechen lassen; vielleicht verkauft er die Clichés an Fischer, jedenfalls wäre es aber gut zu wissen, dass dieselben auch in England zu haben wären und man nicht von Herrn Vieweg abhängt. Verzeihen Sie mir, dass ich Sie auch noch mit dieser Bitte belästige. Verzeihen Sie mir überhaupt diesen ungebührend langen Brief.—In einigen Tagen erwarte ich von Berlin Abdrucke einer kleinen Abhandlung, von der ich Ihnen schon schrieb, die Correcturen habe ich schon gehabt. Ich werde mich beeilen Ihnen einen Abdruck zu schicken und hoffe dass Sie bei dem Lesen dieser rein theoretischen Studie über die Absorption der Strahlen, will not find, that I was mounting the high horse of *[1 word illeg]* and so on . . . Nehmen Sie freundlich und nachsichtig auf, was ich Ihnen biete und genehmigen Sie die erneute Versicherung der ganz ausgezeichneten Verehrung, mit welcher ich verbleibe

Ihr ergebenster | D Bohn

Dear Mr Tyndall,

Your works on calorescence,[1] particularly the first one, in which the heat of a hydrogen flame emitting almost exclusively dark radiation induces a platinum filament to glow brightly, motivated me to attempt an experiment similar to yours, but which, it appears to me, shows the negative fluorescence in an even more striking manner. I suspended a piece of fluorspar, more specifically a piece of the chlorophane variety, from a cotton or linen thread in the burning chamber of a conventional iron stove and then also in that of a so-called parlour stove made of porcelain. The mineral was exposed to the radiation emanating from different plates at a short distance—6 made of iron, or 2 of blackened metal and 4 of glazed clay respectively (the walls of the burning chamber)—that were only moderately heated and, in any case, far from glowing. The chlorophane glowed after 2 minutes, a different type after 2½ minutes, and the intensity of the glowing radiation quickly increased to a very vivid and beautiful luster. Certainly, the mineral, hanging from the poorly conducting thread (which did not turn brown in the heat), was also partially heated through contact with the hot air in the stovepipe. However, that air was not so hot that I could not keep my hand—even though still affected by the radiating heat—inside it for some time without feeling discomfort. Furthermore, the air should not be able to cause marked warming of the fluorspar in such a short period, given its low density and low specific heat. That the fluorspar warmed up and glowed primarily as a result of the dark heat *radiation* emanating from the stove walls was very clearly proven in the following experiment. I inserted a piece of fluorspar into a thin-walled glass tube and suspended the latter in the burning chamber. Next to it, but suspended freely in the air by a thread, I hung another piece of fluorspar that had been cut from the same larger crystal. The air's access to the crystals in the glass tube was as free as possible. However, while the free-hanging fluorspar glowed noticeably after 2½ minutes already, the first shimmer of the piece contained in the glass tube was only detectable after 18 minutes. The glass tube had early on become so hot, that it caused pain upon touch. The experiment was done in a very dark room and the glass tube was made to oscillate with a small wooden stick, since a gleam of light is more easily perceived in a moving object than in a static one. Given that the walls of the glass tube prevented almost all of the dark heat radiation from reaching the fluorspar, it follows that it warmed up only due to contact with the hot air and the glass tube, which gradually became very hot. As the fluorspar in the glass tube started to glow so much later, it follows that the main influence causing the glowing should be attributed to dark heat *radiation*. Here, then, we have an actual transformation—facilitated by the fluorspar—of dark, barely refracting radiation into a glowing and more strongly refracting one, which you had called negative fluorescence. It would probably not be uninteresting to modify the experiment by suspending the chlorophane within a container made of rock salt. As rock salt only minimally absorbs dark heat radiation, I suppose that chlorophane so placed would start to glow almost as quickly as in the open air. I do not have suitable materials to be able to perform the experiment; it would be easier for you to do it, should you find it of any interest.

Incidentally, there is a significant difference between inducing the glowing of the platinum filament in your experiment, the fluorspar in mine, and the common phenomena of fluorescence. Aesculin, quinine, uranium glass, and the rest of substances that belong here, all fluoresce *immediately* as soon as they are exposed to adequate light. But glowing generated by dark heat radiation noticeably only occurs when the radiation has persisted *over a certain period*, when a considerable amount of absorbed heat has accumulated in the irradiated object. I suspect that this is the case also in the powerful focal points of dark radiation that you have created with such ingenuity. It seems that the glow appeared immediately, however, when I let finely dispersed fluorspar fall onto the hot stove plate. One could now say that the emission of luminous rays begins at the very moment in which the dark radiation hits the object, but that its intensity is initially so negligible that the luminescence cannot yet be perceived. Except that a corresponding increase in intensity actually cannot be noticed in the case of fluorescence per se; apparently, bodies fluoresce as intensely in the first moments as later on. Concerning the fluorspar, there is another very noticeable difference between the phosphorescence produced by heating radiation and the fluorescence produced through irradiation with ultraviolet or strongly refracting radiation. The phosphorescence increases for a while, but decreases subsequently. A piece of fluorspar that was irradiated for more than half an hour by the stove stopped glowing almost entirely; pieces that were in direct contact with the hot stove plate stopped glowing even earlier. It is different with fluorescence. Incidentally, I note that a claim made by Becquerel that fluorspar that has been intensely heated will only phosphoresce again if previously irradiated with light, is not correct, or at least not generally so. I was able to make the same piece of fluorspar phosphoresce six times in a row on the same night, while remaining in a dark room throughout. Admittedly, the glow was not as lively on the last occasions; I had probably taken the previous heating too far.

Should you consider this communication worthy of being published in the Philosophical Magazine, I would be very glad to see it appear there.[2] In that case, I would kindly ask you to solicit the publisher for some separate prints.

Your article on calorescence[3] (from Philos. Trans. 1866), which you sent me via Francis[4] and for which I still want to thank you and tell you that I greatly enjoyed the interesting experiments, includes references to figures that I did not receive. The images of apparatuses could be left out entirely from the translation (appendix to Chap. XII) without compromising the clarity. Should the other figure have to be left out, however, it would be necessary to modify the text slightly. It would be most convenient if I could obtain a copy of the figures.

You previously advised me against publishing my translation of your book on Heat with Mr Vieweg.[5] Had I followed your advice, I would have been spared quite a bit of *[annoyance]* and the book would have appeared earlier. I got into an argument with Mr Vieweg. He interrupted the setting of the sheets without informing me in the least, and in the end he was presumptuous enough to make—or allow—changes to my work and expected me to approve changes that were clumsy, unmotivated, and that in some cases nonsensical.[6] The outcome of this story was that we terminated our contract and that Vieweg granted me a monetary compensation for my lost time. Since yesterday, I am again in possession of my manuscript and free to decide what to do with it. I immediately began negotiations with Theod. Fischer,[7] in Cassel whom you once warmly recommended to me, and the negotiations came *[to a] [1 word illeg]* end. Mr Fischer is willing to publish my translation as soon as possible and in the best possible edition. He has also kindly offered me much more favourable royalties than Vieweg. As it was you who recommended Fischer, I assumed that you would approve my decision without hesitation. I therefore ask you, albeit belated, to allow Mr Theod. Fischer what you had allowed Mr Vieweg to do. To avoid delaying the publication, Mr Fischer is trying to obtain the print blocks for the figures. He requests that I ask you to inquire with your English publisher under what conditions he would hand over the print blocks for the figures so that they can be used in the German translation. He also thinks that Dr Debus would gladly run the necessary errands when he is in London. Vieweg has already let the largest part of the figures be engraved, actually very nicely. He may be interested in selling the print blocks to Fischer, but it would be good to know that they could also be obtained from England and that we are not dependent on Mr Vieweg. Please excuse me for troubling you with this request. Excuse me altogether for writing this unreasonably long letter. From Berlin I expect to receive in a few days copies of a short article about which I already wrote to you.[8] I have already had the corrections. I will hasten to send you a copy and hope that, when reading this purely theoretical study on the absorption of radiation, you will <u>not</u> find, that I was mounting the high horse of *[1 word illeg]* and so on . . . Please accept with kindness and indulgence what I am offering and allow me to express the renewed assurance of my admiration, with which I remain,

Your humble | J Bohn.

RI MS JT/1/B/110

---

1. *Your works on calorescence*: J. Tyndall, 'On the History of Negative Fluorescence', *Phil. Mag.*, 29 (1865), pp. 44–55; J. Tyndall, 'On the History of Calorescence', *Phil. Mag.*, 29 (1865), pp. 218–31 (this is letter 2245 in this volume); J. Tyndall, 'On Calorescence', *Phil. Trans.*, 156 (1866), pp. 1–24; and J. Tyndall, 'Sixth Memoir on Radiation and Absorption.—Influence of Colour and Mechanical Condition on Radiant Heat', *Phil. Trans.*, 155 (1866), pp. 83–96.
2. *see it appear there*: Bohn's notes never appeared in the *Phil. Mag.*
3. *on calorescence*: see letter 2333, n. 2.
4. *Francis*: William Francis.
5. *You previously advised . . . with Mr. Vieweg*: see letter 2296, n. 6.
6. *nonsensical*: a different account is given by Rudolf Clausius to Tyndall in a letter dated 24 March 1866 (see letter 2387).
7. *Theod. Fischer*: Theodore Fischer.
8. *wrote to you*: article not identified. Previous letters from Bohn to Tyndall in this volume do not mention a short article, and it may be from a letter that is missing.

To Elizabeth Dawson Steuart 23 March [1866][1] 2386

23rd. March

My dear Mrs Steuart,

I cannot sufficiently thank you for your most useful letters,[2] neither can I conceal from you my annoyance to find these old debts arising when I had every reason to believe none such existed. In self defence I fear I shall have to become more indifferent as to how matters proceed in Leighlin[3]—simply sending the proper amount of money and giving myself no further anxiety. These annoyances damage, not only my purse but my intellect, for they disqualify me from pursuing the work of my life which demands the entire concentration of my powers.

At the present moment, for example, I am quite unfit to engage in the hard study which ought to be the sole object of my attention.

I can assure you that I find myself very heavily weighted. It was only the day before yesterday that I lost 12 pounds in sending a son[4] of Caleb Tyndall's[5] to America. Caleb Tyndall himself borrowed 30 pounds from me some time ago but has broken his most solemn promises of re-payment. Another gentleman—who had no other claim upon me than the fact of his being kind to my mother—has borrowed £60 from me; but his promises have been so often broken that I do not expect ever to receive a farthing of the money. Then come the people in Bagnalstown[6]—In fact instead of being permitted to pursue an arduous career with my whole strength I find myself like a weighted horse dragging the burdens of others up the steep hill which it has been my fate to climb.

I shall be very thankful to you for any hints that you may give me. On Wednesday next I accompany Mr Carlyle down to Lord Houghton's[7] whence we proceed to Edinburgh and the North.[8]

Yours most sincerely | John Tyndall.

RI MS JT/1/TYP/10/3361
LT Typescript Only

---

1. *[1866]*: the year is given by reference to Tyndall accompanying Thomas Carlyle to the University of Edinburgh (see n. 8).
2. *your most useful letters*: recent letters from Steuart to Tyndall are 2355, 2356, and 2383, but there are likely some that are missing.
3. *Leighlin*: Leighlinbridge, County Carlow, Ireland, where Tyndall was born and his mother lived.
4. *son*: John Tyndall, a son of Caleb Tyndall. In 1867, Tyndall wrote to Steuart that he

'received a begging letter from the blockhead that I sent to America' (letter 2717, *Tyndall Correspondence*, vol. 10); and 'Caleb Tyndall's son has again written to me, but I have thrown his letter in the fire. He wants the means of carrying him home to Ireland. If his sister thinks it right to devote a portion of the twenty pounds to this purpose she may do so. But I can have no direct communication whatever with this very troublesome and very stupid person' (letter 2634, *Tyndall Correspondence*, vol. 10).

5. *Caleb Tyndall*: Caleb Tyndall (*c.* 1798–1879), an Irish farmer and uncle of John Tyndall. In 1841 he achieved notoriety by shooting Mary McAssey while facing a pro-Repeal mob. At his trial he was cleared of malicious wounding. He subsequently served as High Constable for the Barony of Idrone West and Overseer of the Poor.
6. *Bagnalstown*: Bagenalstown, a small town on the River Barrow in Country Carlow, Ireland.
7. *Lord Houghton's*: Richard Monckton Milnes (see letter 2379).
8. *proceed to Edinburgh and the North*: Tyndall accompanied Thomas Carlyle for his speech to the students as Rector on 2 April 1866, leaving London on 28 March (see letter 2375, n. 3).

From Rudolf Clausius 24 March 1866 2387

ZÜRICH, 24/3/66

Lieber Tyndall,

Auf Deine Anfrage ob Vieweg nicht daran denke, Dir ein Honorar bei der Herausgabe der Übersetzung Deines Buches zu geben, schrieb ich Dir, dass ich nichts davon wisse, dass ich aber, wenn Du es wünschest und wenn Du es mir schreiben wollest, was Du gefordert haben würdest, wenn Du gefragt wärst (damit ich für einen etwaigen Vorschlag an Vieweg einen Anhaltspunct hätte) ich gern an Vieweg oder Wiedemann schreiben würde. Da Du darauf nicht geantwortet hast, so nahm ich an, Du wünschest nicht, dass ich von meiner Seite etwas in der Sache thue. Als nun aber die Zeit herannahte, wo Wiedemann von Braunschweig abreist, und ich lange Zeit gar nichts über die Herausgabe der Übersetzung gehört hatte, hielt ich es für gut, einmal bei Wiedemann über die Sache anzufragen. Darauf habe ich nun eine Antwort erhalten, die mich sehr überrascht und betrübt hat. Ich halte es für das Beste, dass Du genau weisst, wie die Sache steht, und schicke Dir daher Wiedemann's Brief mit. Du wirst ihn hoffentlich lesen können, obwohl er schlecht geschrieben ist, da Wiedemann in grosser Eile war, und, wie er sagt, statt eines Tisches eine Kiste benutzen musste.

Aus diesem Briefe habe ich zuerst erfahren, weshalb gerade Bohn die Übersetzung übertragen ist, nämlich weil Vieweg mit Liebig zusammengetroffen ist, und mit dem die Sache verabredet hat. Bohn ist ein Günstling von

Liebig und hat die Professur in Giessen, wie ich glaube, vorzugsweise durch Liebig's Verwendung erhalten.

Wie Du aus dem Briefe ersehen wirst, hat Bohn die Übersetzung so schlecht gemacht, dass Vieweg sie nicht herausgeben will. Eine andere Übersetzung kann Vieweg nicht machen lassen, weil er nicht weiss, ob er es thun darf, indem dieses davon abhängt ob und in welcher Form Du Bohn das alleinige Übersetzungsrecht gegeben hast. Vielleicht kannst Du die Sache wieder in ein richtiges Geleise bringen. Wenn ich irgendwie dazu behülflich sein kann so bin ich gern dazu bereit. Auch Wiedemann wird, wie er schreibt, gern thun, was er kann. Für den Fall dass Du an ihn schreiben wolltest, will ich Dir seine Adresse, die er etwas undeutlich geschrieben hat, hier noch einmal schreiben: bis Donnerstag nach Ostern in Berlin bei Geheimraetin Mitscherlich, Albrecht's Strasse, 12a; später in Carlsruhe (Polytechnicum). Am besten würde es mir scheinen, wenn Du Dich mit Vieweg selbst in Verbindung setztest. Soweit ich mit ihm zu thun gehabt habe, habe ich gefunden dass sich gut mit ihm verhandeln lässt, und er kann natürlich am besten sagen auf welche Weise die Sache etwa wieder in Gang zu bringen ist. Ich wünsche von Herzen, dass sie noch zu Deiner Zufriedenheit geordnet wird.

Die französische Übersetzung meines Buches ist jetzt vollendet und E. Lacroix in Paris hat die Herausgabe übernommen. Folie hat mir die Übersetzung zur Durchsicht geschickt, und ich finde sie, abgesehen von kleinen Stellen, die ich leicht corrigiren kann, sehr gut. Von der englischen Herausgabe habe ich wieder seit längerer Zeit keinen Bogen erhalten. Könnte die Druckerei nicht veranlasst werden, etwas schneller zu drucken? Hirst schrieb mir früher einmal dass er auch wünschte, dass die Sache schneller vorwärts ginge.

Die Frage, ob wir nach Wien gehen, ist noch immer nicht entschieden. Sollte jetzt ein Krieg zwischen Oesterreich und Preussen ausbrechen, so würde sich daran die Sache zerschlagen, denn dann könnte ich als Preusse keine Anstellung in Oesterreich annehmen.

Ich habe Dir und Hirst noch nicht mitgetheilt dass meine Frau unsere Familie wieder mit einem Töchterchen vermehrt hat. Mutter und Kind befinden sich wohl, und ebenso auch die ganze übrige Familie. Mit herzlichen Grüssen von meiner Frau und mir an Dich und Hirst,

Dein | Clausius

ZURICH, 24/3/66

Dear Tyndall,

You asked me if Vieweg[1] was thinking about paying you an honorarium for the translation of your book[2] and I answered that I did not know anything about it but that upon your wish I would gladly write to Vieweg or Wiedemann and ask them, provided that you told me how much you would have demanded if so asked (so that I would have an indication for a potential proposal to Vieweg). Since you did not answer, I assumed that you did not want me to get involved. Since Wiedemann's departure from Braunschweig was approaching and I had not heard about the publication of the translation in a long time, I deemed it best to ask him about this matter. I have now received his answer, which very much surprised and saddened me. I deem it best that you know exactly how things stand and am therefore enclosing Wiedemann's letter. I hope you will be able to read it, although it is badly written as Wiedemann was in a rush and, as he says, had to use a box instead of a desk.

First of all, I have learned from this letter for what reason Bohn[3] had been assigned to do the translation, namely because Vieweg has met with Liebig[4] and discussed the matter with him. Bohn is Liebig's favourite and, as I believe, received his professorship at Giessen preferentially through Liebig's intercession.

As you will see in the letter, Bohn's translation is so poorly done that Vieweg does not want to publish it. Vieweg cannot order another translation, for he does not know if he is allowed to do so, as it depends on if and in which form you had given Bohn exclusive translation rights. Perhaps you can bring this back on the right track. If I can be of assistance in any way, I am prepared to do so gladly. Wiedemann, as he wrote, will likewise be glad to do whatever he can. In case you wanted to write to him, I am sending you again his address, which he did not write clearly enough: until Thursday after Easter he will stay in Berlin at: Geheimraetin Mitscherlich, Albrecht's Strasse, 12a[5] and after that in Karlsruhe (Polytechnic Institute).[6] It seems to me that it would be the best if you contacted Vieweg directly.[7] As far as I have had dealings with him, I found that it is possible to negotiate with him and he is obviously the one who can tell best how to put things back on track. I wish with my whole heart that everything be resolved to your satisfaction.

The French translation of my book[8] has now been completed, and E. Lacroix from Paris has taken over the publication.[9] Folie[10] sent me the translation for review and I find it very good, aside from some little things that I can easily correct. Of the English translation[11] I have again not received any proof sheets for a long time. Could the print shop not be convinced to work faster? Hirst wrote to me previously that he would also wish that the matter progressed faster.

We have still not decided whether or not we will go to Vienna.[12] If a war between Austria and Prussia should now break out,[13] the matter would be dashed by the fact that it would be impossible for me as a Prussian to accept a position in Austria.

I have not yet told you or Hirst that my wife[14] has again expanded our family with a little daughter.[15] Mother and child are both well, as is the rest of the family. With best wishes to you and to Hirst from my wife and myself.

Your | Clausius

RI MS JT/1/TYP/7/2287–8
LT Typescript Only

1. *Vieweg*: Eduard Vieweg.
2. *translation of your book*: the German translation of *Heat* (1865); see letter 2296, n. 6.
3. *Bohn*: Johann Conrad Bohn.
4. *Liebig*: Justus Liebig (see letter 2304, n. 5).
5. *Geheimraetin Mitscherlich, Albrecht's Strasse, 12a*: the residence of the widow of German chemist Eilhard Mitscherlich (1794–1863). Mitscherlich held an advisory position to the government in public health, and had the title *Geheimrat*, with his widow having the title *Geheimraetin*. Wiedemann was married, since 1851, to Mitscherlich's daughter Clara Wiedemann (née Mitscherlich, 1827–1914).
6. *Karlsruhe (Polytechnic Institute)*: the Polytechnic school in Karlsrule was founded in 1825, and later became the University of Karlsruhe and now Karlsruhe Institute of Technology.
7. *contacted Vieweg directly*: there are no extant letters between Eduard Vieweg and Tyndall concerning the translation of *Heat* (1865).
8. *my book*: see letter 2282, n. 4.
9. *E. Lacroix from Paris has taken over the publication*: see letter 2288, n 7.
10. *Folie*: François-Jacques-Philippe Folie.
11. *English translation*: see letter 2282, n. 4.
12. *go to Vienna*: in January, Clausius, then a professor at the Federal Polytechnic School in Zurich, wrote to Tyndall that he had been offered a position in Vienna (see letter 2357). In May he shared his decision to stay in Zurich (see letter 2426).
13. *war between Austria and Prussia should now break out*: the Austro-Prussian War broke out on 14 June 1866.
14. *my wife*: Adelheid Clausius.
15. *little daughter*: the Clausius's fourth child, Hedwig (1866–1946). For their first three children, see letter 2270, n. 5.

To Elizabeth Dawson Steuart [24 March 1866][1] 2388

Royal Institution | Saturday.

My dear Mrs Steuart,

I thought it was understood between us that Emma[2] was to go to Tramore[3] if you thought it advisable. I am quite willing that she should go.

I am rather bewildered by the curious entanglements to which you refer, and but poorly able to see my way through them.

Mrs Doyle[4] told me that her wages were £4.18.0, but that out of this £1- had been paid to her which reduced it to the sum mentioned in my memorandum.

Of Mrs Doyle's conduct in this whole affair no doubt rests upon my mind. It was shortsighted conduct on her part, for had she behaved aright I should not have forgotten her.

The house on the Green I confess does not please me. I thought its aspect rather miserable when I passed it. Perhaps for the present it would be better to leave them where they are.

My head is very bad to-day; perhaps when it improves I may be able to see my way more clearly

Yours ever sincerely | John Tyndall.

I think with regard to Mrs Doyle's wages &c., my mother[5] ought to be consulted. The whole matter irritates me to an extent I cannot describe.

RI MS JT/1/TYP/10/3360
LT Typescript Only

---

1. *[24 March 1866]*: the date is given by relation to letter 2383, dated 21 March 1866 (a Wednesday), in which Steuart wrote of Tyndall's sister Emily: 'She is worrying your poor Mother, and every one, about going to Tramore, and I do think if you would write in a kind manner, but a determined one, saying that whenever the season is good enough, you will send her there, it may have a good effect'. This letter, written on a Saturday, is then from 24 March 1866. Letter 2394 also appears to be a reply from Tyndall to 2383.
2. *Emma*: Emily Tyndall.
3. *Tramore*: see letter 2383, n. 9.
4. *Mrs Doyle*: Abbey Doyle.
5. *my mother*: Sarah Tyndall.

To Thomas Carlyle 26 March 1866 2389

26th. March 1866

My dear General.

Your Lieutenant[1] will wait upon you on Wednesday morning at 10 ½ A.M. He will then take you, bag & baggage, to King's Cross Station whence we start for Knottingley at noon.[2]

Please have all packed by half past ten.

I am sorry to find that I shall not have the pleasure of taking care of Mrs Carlyle.[3]

Yours ever | John Tyndall

RI MS JT/1/T/148
RI MS JT/1/TYP/1

1. *Lieutenant*: Tyndall commonly referred to himself as Carlyle's 'Lieutenant', with Carlyle as 'General', in an allusion straight out of Carlyle's *On Heroes, Hero-worship, and the Heroic in History* (London: James Fraser, 1841).
2. *He will then take you . . . for Knottingley at noon*: Tyndall accompanied Carlyle to the University of Edinburgh for his speech to the students as Rector on 2 April 1866 (see letter 2375, n. 3).
3. *Mrs Carlyle*: Jane Welsh Carlyle.

To David Brewster [before 27 March 1866] 2390

The Athenaeum[2]

My dear Sir David

I do not know whether I am betraying confidence in saying that Lady Coxe[3] forwarded your note to me.[4] If so I know she will forgive me. I had also been informed by Dr Playfair[5] of your kindly sentiments towards me in regard to the LL.D.[6]

I trust that you will with equal kindness listen to a few words from me upon this latter subject. I prize as much as anybody the good opinion of the eminent body over which you preside. Again the occasion would give the honour a special value in my eyes; for from my youth up I have honoured Carlyle, and there is certainly no earthly prince or potentate in whose company I would rather find myself. But with all this I confess to a certain degree of uneasiness regarding the acceptance of this honour. You say there was some difficulty but it was removed. My warmest thanks are due to those who have removed it but I believe an honour of the kind loses more than half its value when there is the least doubt as to the propriety of its being conferred. I have no fear that you will misinterpret me when I say that I shall visit Edinburgh with a freer heart, if I enter it as I first intended, see my friends "doctored" and join in the applause which is so <*1 word missing*> their due. Were the honour offered to me as a mark of amity I should accept it in the spirit of amity. But if overruling the opposite sentiment you have resolved to grant me the distinction simply because you believe that on scientific grounds I merit it, then I honour you for that resolve, and I cannot express this feeling better than by saying that I could not accept a distinction at the hands of such noble-minded men in which their hearts could not join.

RI MS JT/2/13c/1324–5
LT Typescript Only

1. *[before 27 March 1866]*: this letter clearly precedes that from Brewster to Tyndall dated 27 March 1866 (letter 2391).

2. *The Athenaeum*: see letter 2253, n. 8.
3. *Lady Coxe*: Mary Anne Coxe.
4. *forwarded your note to me*: Brewster's letter to Mary Anne Coxe that she then forwarded to Tyndall is missing.
5. *Dr Playfair*: Lyon Playfair.
6. *your kindly sentiments towards me in regard to the LL.D*: in letter 2382, Playfair informed Tyndall that Brewster 'quite approves' of Tyndall receiving a Doctor of Law degree from the University of Edinburgh.

From David Brewster 27 March 1866 2391

My dear Prof[esso]r Tyndall,

I am very sorry that I should have referred, so vaguely, to "the difficulty"[1] which existed, and was removed, about the degrees which the Senatus[2] was anxious to confer upon Mr Carlyle, yourself and Mr Huxley.[3] The difficulty, & it really was no difficulty, was that Mr Carlyle declined to take the degree which appeared to some of my Colleagues as putting an end to the scheme of conferring degrees at Mr Carlyles Installation[4] No objection to any of the proposed Graduates was ever made; and if any had been made it certainly would not have been to you. I am, in a great hurry,

Ever Most Truly Y[ou]rs | D Brewster

College | March 27th 1866

RI MS JT/2/10/399

1. *"the difficulty"*: Brewster was referring to his letter to Mary Anne Coxe, which she then forwarded to Tyndall (see letter 2390).
2. *Senatus*: Latin for 'the Senate'; here referring to the Senate of the University of Edinburgh.
3. *Mr Huxley*: Thomas Huxley accepted the honorary degree (L. Huxley (ed), *Life and Letters of Thomas Henry Huxley*, 2 vols (London: Macmillan and Co., 1900), vol. 2, p. 473).
4. *Mr Carlyles Installation*: see letter 2375, n. 3.

To Rudolf Clausius 27 March [1866][1] 2392

27th March

My dear Clausius

I am just starting for Edinburgh[2] but must send you a line before I go. Thank you a thousand times for your useful letter.[3] Bohn[4] wrote to me yesterday and to my astonishment told me that he had broken with Vieweg.[5] At the recommendations of Debus I had at the beginning mentioned Fischer of

Cassel[6] to him, but once having agreed with Vieweg I gave him no authority to break the agreement, and was exceedingly surprised to find that he did so without consulting me.

He wrote to me twice last autumn[7] asking permission to translate the book, and I thinking that he would do it in a first-rate manner at once gave him the permission. I should never have done so were I not under the impression that he would do it in a first-rate manner.

He has gone to Fischer & made an agreement with him & he informs me that he receives from Fisher much better pay than from Vieweg. He asks my consent to their arrangement.[8]

I will write to him[9] by this post saying that I cannot decide without further information whether I can comply with his requests. I must also write a line to Wiedemann.[10]

I think I ought not to permit a bad translation of the book to appear. Supposing I write to Bohn[11] giving Wiedermann authority over the translation—do you not think it would be well.

Give my love to the children[12] & kindest regards to your wife.[13]

I will speak to Hirst who has prob[ably] quitted London[14] about your book—I shall see him on Thursday.

Yours ever | <u>John Tyndall</u>

RI MS JT/1/T/228

---

1. *[1866]*: the year is given by mention of going to Edinburgh and discussion of the German translation of *Heat* (1865).
2. *starting for Edinburgh*: Tyndall accompanied Thomas Carlyle to the University of Edinburgh for his speech to the students as Rector on 2 April 1866, leaving London on 28 March (see letter 2375, n. 3).
3. *your useful letter*: letter 2387.
4. *Bohn*: Johann Conrad Bohn, who had been working on the German translation of *Heat* (1865); see letter 2296, n. 6.
5. *broken with Vieweg*: Eduard Vieweg. See letter 2385.
6. *Fischer of Cassel*: Theodore Fischer.
7. *wrote to me twice last autumn*: see letters 2296 and 2302.
8. *their arrangement*: see letter 2385.
9. *write to him*: letter missing. For Bohn's letters after this, see letters 2396, 2406, and 2407.
10. *write a line to Wiedemann*: letter missing. For Wiedemann's possible reply, see letter 2410.
11. *write to Bohn*: see n. 9.
12. *children*: see letter 2270, n. 5, and letter 2387, n. 15.
13. *your wife*: Adelheid Clausius.
14. *quitted London*: Hirst also went to Edinburgh for Carlyle's speech (see letter 2375, n. 3).

From Lyon Playfair 27 March 1866 2393

27th March 1866.

My dear Tyndall

Brewster[1] has shown me your letter.[2] The "difficulty" to which he alluded was that Carlyle refused to be doctored and had nothing to do with you personally.

To morrow the Senatus[3] meet and I intend to push the motion that Lord Houghton[4] and you three members "of the Jermyn Street Gang"[5] should be doctored.[6] I have secured Tait's[7] silence, although I wanted him to go a step further and second your nomination. Therefore I expect both your's and Huxley's and Ramsay's degrees to pass unanimously, if the Senatus grant degrees at all, in the face of Carlyle's refusal.

I think that it will decide to grant them, but this question is the only difficulty I expect to have. A few Holy men shuddered at a Huxley and a Tyndall, but to do them justice immediately shook off their shivers and recollected that they were <*1 word missing*> and not Theologians. One of them made a joke and said that to these heterodox men the degree of D---D[8] would be better than LL.D[9]

Yours sincerely | Lyon Playfair.

RI MS JT/2/13c/1324
LT Typescript Only

1. *Brewster*: David Brewster.
2. *your letter*: letter 2390.
3. *Senatus*: Latin for 'the Senate'; here referring to the Senate of the University of Edinburgh.
4. *Lord Houghton*: Richard Monckton Milnes.
5. *three members of the "Jermyn Street Gang"*: referring to Tyndall, Thomas Huxley, and Andrew Ramsay (see letter 2382, n. 6), whom were all employed by the Royal School of Mines, located at Jermyn Street in the St. James's area of Westminster, London. Huxley himself used the phrase in an 1890 letter: 'I am the last of the original Jermyn Street gang left in the school now' (L. Huxley (ed), *Life and Letters of Thomas Henry Huxley*, 2 vols (London: Macmillan and Co., 1900), vol. 2, p. 260). On the School of Mines, see letter 2259, n. 17.
6. *should be doctored*: Tyndall, Huxley, and Ramsay all accepted the degrees bestowed on them, while Carlyle declined (see letter 2382, n. 5 and n. 6).
7. *Tait's*: Peter Guthrie Tait (1831–1901), a Scottish mathematician and natural philosopher, who was senior wrangler in the mathematics tripos at Cambridge University in 1852. He was professor of mathematics at Queen's College, Belfast (1854–60), and held the chair

of natural philosophy at the University of Edinburgh from 1860 until shortly before his death (*ODNB*). During the 1860s Tait forged a friendship with William Thomson and began writing their *Treatise on Natural Philosophy* (1867). During the 1860s Tait and Thomson had been in a dispute with Tyndall over his claims about Julius Mayer's priority over Joule in developing the mechanical equivalent of heat. This heated dispute spilled over into popular journals before they resolved their differences at the Dundee meeting of the BAAS in 1867, where, Tyndall wrote they 'seemed like brothers together'. Their reconciliation did not last.

8. *D---D*: damned.
9. *LL.D.*: Doctor of Law degree.

To Elizabeth Dawson Steuart [28 March 1866][1] 2394

Wednesday.

My dear Mrs Steuart,

I hope Dr. Trayer[2] will consider himself at liberty to adopt as generous a treatment as he may deem necessary in Emma's[3] case. My friend Bence Jones[4] has done me wonderful service by cutting off beer, and recommending mutton and dilute pale brandy to my attention. I think mutton and wholesome claret (to secure goodness I would not care about the expense) would be admirable things for Emma. At all events I hope that he will give directions for the supply of whatever he deems needful to her health and comfort. Tea and coffee I think can hardly be good for her.

If he is of opinion that a trip to Tramore[5] would do her good I am quite willing that she should go.

Believe me always | Yours most truly | John Tyndall.

If you would kindly let me know when Dr. Trayer's account amounts to £5 I would send a cheque.

RI MS JT/1/TYP/10/3359
LT Typescript Only

---

1. *[28 March 1866]*: the date is given by relation to letter 2383, dated 21 March 1866 (a Wednesday), in which Steuart wrote of Tyndall's sister Emily: 'She is worrying your poor Mother, and every one, about going to Tramore, and I do think if you would write in a kind manner, but a determined one, saying that whenever the season is good enough, you will send her there, it may have a good effect'. This letter, written on also on a Wednesday, is then from a week later, 28 March 1866. Letter 2388 is also a reply from Tyndall to 2383.
2. *Dr. Trayer*: James John Trayer (see letter 2355, n. 12).

3. *Emma's*: Emily Tyndall.
4. *Bence Jones*: Henry Bence Jones.
5. *Tramore*: see letter 2383, n. 9.

To George Gabriel Stokes [30 March 1866][1] 2395

Fryston Hall | Yorkshire[2] Good Friday

My dear Stokes

I had no thought of "coercion": but I thought & think that any point on which you have thought, and regarding which you have arrived at an opinion opposite to mine, demands from me very careful consideration before committing myself to print upon the subject—

For the present I think it will be better to withold that passage.[3]

My spare hours last week were spent in thinking over an apparatus for the examination of aqueous vapour & of mercury vapour at the boiling temperature. Years ago Miller[4] & myself had some conversation on the subject, and your question will stimulate me to go more resolutely *[after]* it.

If the "ruby" be an oxide I should think it must be a very low one.

Yours very truly | John Tyndall

CUL SC—7656

1. *[30 March 1866]*: the date is given as Tyndall stayed at Fryston Hall on 29 and 30 March 1866 while traveling to Edinburgh with Thomas Carlyle (see letter 2375, n. 3). In 1866, Good Friday fell on 30 March.
2. *Fryston Hall | Yorkshire*: see letter 2375, n. 3.
3. *that passage*: probably referring to a removed passage from one of two papers Tyndall published in the *Phil. Trans.* in early 1866: J. Tyndall, 'On Calorescence', *Phil. Trans.*, 156 (1866), pp. 1–24; and J. Tyndall, 'Sixth Memoir on Radiation and Absorption.—Influence of Colour and Mechanical Condition on Radiant Heat', *Phil. Trans.*, 155 (1866), pp. 83–96. As the letter from Stokes discussing this is missing, we cannot determine for certain which paper.
4. *Miller*: probably William Allen Miller (see letter 2513, n. 13).

From Johann Conrad Bohn 29–30 March 1866 2396

Hochverehrter Herr Tyndall!

Ich habe Unrecht gethan mit H. Vieweg neue Verabredung (auf Aufhebung des Vertrages) zu treffen, ehe ich Sie davon benachrichtigt und Ihre

Meinung darüber erbeten hatte. Zur Entschuldigung, wenn auch nicht zur Rechtfertigung meiner Unterlassung mag dienen, dass Sie mir früher energisch von Vieweg abriethen und erst in Ihrem 4[n]. Briefe vom 21 Sept. nachträglich Ihre Zustimmung gaben dass das Buch bei Vieweg erscheinen soll,—dass Sie mir Theod. Fischer als Verleger empfahlen—ich hiernach als wahrscheinlich annehmen dürfte, Sie würden, nachdem Sie mich persönlich zur Uebersetzung Ihres Buches autorisirt hatten, die Uebertragung des Verlags von Vieweg an Fischer genehmigen, sobald ich Sie, wie es geschehen, darum bitten werde.

Ferner führe ich zu meiner Entschuldigung an, dass mich Vieweg sehr gekränkt und zu einer sofortigen Antwort gereizt hat;—dass nicht ich beabsichtigte den Vertrag aufzulösen, trotz dem ernsthaften Grund zur Klage, sondern der Vorschlag von Vieweg ausging—endlich dass ich wünschte das Erscheinen des Buches nicht länger zu verzögern, als durchaus unvermeidlich schien. Dies sind meine Entschuldigungen,—eine völlige Rechtfertigung gaben Sie nicht,—Sie haben Grund sich zu beschweren, das erkenne ich willig an. Indem ich die Bitte wiederhole Sie mögen gütigst Ihre Einwilligung dazu *[1 word missing]*, dass Fischer das Buch verlege, spreche ich Ihnen mein Bedauern über meine Voreiligkeit aus und bitte Sie dieselbe zu verzeihen.

Ihren Wunsch „to learn something more about the old engagement" verstehe ich dahin, dass Sie über die Gründe der Auflösung des Vertrags mit Vieweg vollständig unterrichtet sein wollen. Ich beeile mich dem zu entsprechen; sollten Sie auch unterdessen bereits nach Schottland abgereist sein, so wird Sie dann mein Brief alsbald bei Ihrer Rückkehr nach London und zu den Geschäften treffen.

Hr. Vieweg hat plötzlich, und ohne irgend eine Nachricht zu geben, den Satz des Werkes unterbrochen. Nach 3 Wochen frug ich nach der Ursache dieser Unterbrechung bei ihm an und erhielt zur Antwort V. habe bei der Lektüre des Buches die Form nicht fliessend genug gefunden und bemerkt, dass von mir manche englische Worte und Ausdrücke nicht richtig übersetzt worden seien. Er habe sich desshalb veranlasst gesehen meine Arbeit einer genauen Revision unterwerfen zu lassen, schicke mir das Ergebniss dieser und ersuche mich die Abänderungen zu genehmigen. Ausserdem müsse er mir p. Bogen 1*[1/2]* Thaler Kosten in Rechnung bringen.

Über eine Arbeit, die, wie meine Uebersetzung, für die Öffentlichkeit bestimmt ist, hat Jedermann das Recht mir seine Meinung zu sagen und mehr noch als jeder Andere der Verleger. Ist ein Tadel gegen die Form meiner Uebersetzung oder ein noch schlimmerer gegen mein Verständniss des Originales gerechtfertigt, so kann ich es nur mit Dank aufnehmen, wenn ich darauf aufmerksam gemacht werde und dadurch Gelegenheit fände Verbesserungen anzubringen. Allein ein Vorwurf wie der von V. gemachte wird zur Beleidigung wenn er ungegründet ist, wenn er von einem Manne gemacht wird, welcher nur beweisst, dass er nicht competent ist über die Sache zu urtheilen, wenn es in so ungehöriger Form vorkommt, wie diess geschehen ist. Dass sich V. erlaubte, die ihm wünschenswerth erscheinenden Aenderungen, ohne mein Vorwissen, von einem Andern, mir Unbekannten, kurzweg machen zu lassen, war eine Impertinenz. Er hat nicht das geringste Recht auch nur die kleinste Aenderung an einem ihm zum Drucke anvertrauten Manuscript zu treffen;—kein Mann, der etwas auf seine Ehre und seinen Namen hält, wird sich eine solche Anmassung gefallen lassen. Das Entscheidende in der Sache ist aber, dass die an meiner Arbeit gemachten Ausstellungen ungegründet sind, dass die angeblichen Verbesserungen etwas anderes hervorbrachten, als das Original sagt, dass in vielen Fällen Ihre Worte in dieser abgeänderten Uebersetzung zu einem Unsinne verkehrt worden sind, in manchen Fällen ist die Aenderung gleichgültig an und für sich betrachtet, allein von der Treue abweichend, welche die erste Pflicht des Uebersetzers gegen den Verfasser des Originals ist. Es würde Sie, so setze ich voraus, ermüden, wollte ich, wie ich es Vieweg gegenüber gethan habe, alle einzelnen Aenderungen besprechen; ich beschränke mich einstweilen auf einige wenige. Und zuvor, damit nicht der Anschein eintreten kann, als wähle ich tendenziös dies, werde ich die im allerersten Abschnith des Buches gemachten, besprechen und bitte um Ihre gütige Geduld. Ich bemerkte dass ich, wenn Sie es wünschen, bereit bin die ganze weitläufige Antikritik und Vertheidigung meiner Fassung mitzutheilen.

§1. The aspects . . . Ich übersetzte: Der Anblick der Natur erregt im Menschen den Geist der Untersuchung. View. lässt ändern: Die Anschauung der Natur erweckt im Menschen den Forschungstrieb. Aspect ist Anblick—hätten Sie Anschauung sagen wollen, so würden Sie das Wort contemplation gebraucht haben. Hätten Sie vom Triebe zur Forschung oder Untersuchung reden wollen, so hätten Sie instinct, impulse . . . oder dergleichen gesagt.

Den folgenden Satz übersetze ich: „Wie das Auge zum Sehen, das Ohr zum Hören bestimmt sind, so ist der menschliche Geist zum Verstehen der Erscheinungen der körperlichen Welt bestimmt." Viewegs Fassung lautet: „Wie das Auge zum Sehen, das Ohr zum Hören geschaffen ist, so ist der menschliche Geist befähigt die Erscheinungen der körperlichen Welt zu verstehen". Es ist ersichtlich dass die letztere, ungetreue Uebersetzung Sie eine Plattheit sagen lässt. Sie bilden die Analogie

<table>
<tr><td>Auge zum Sehen
Ohr zum Hören</td><td>} made</td><td rowspan="2">} gebildet</td><td rowspan="2">und der Analogie wegen
machen die unbestrittene
zwei ersten Sätze den dritten</td></tr>
<tr><td colspan="2">Geist zum Verstehen formed</td></tr>
</table>

Satz wahrscheinlich. Wenn man aber schriebe

<table>
<tr><td>Auge zum Sehen
Ohr zum Hören</td><td>} made</td><td rowspan="2">} geschafft</td><td rowspan="2">So hört die Analogie auf; der
Zusammenhang zwischen den
in den 2 ersten Sätzen enthaltenen</td></tr>
<tr><td colspan="2">Geist zum Verstehen befähigt</td></tr>
</table>

Urtheilen und dem Urtheil des 3ⁿ. Satzes ist ganz genommen und die 2 ersten Aussagen sind dadurch unnöthig und trivial geworden. Wenn von der Befähigung des Geistes zum Verstehen die Rede gewesen wäre so würde die richtige Zusammenstellung der Sätze gewesen sein

Eye can (oder is able, is apt, capable, fit) for seeing
Ear " (" " " " " " ") for hearing
Mind " (" " " " " " ") for understanding

Die nächste Aenderung ist eine Folge der eben erwähnten und ich bespreche sie desshalb nicht besonders.

Very characteristical—übersetze ich: „Sehr bezeichnend, Vieweg: sehr charakteristisch, an Stelle des vortrefflichen deutschen Wortes, setzt er also in dem deutschen Buche ein zweideutiges Fremdwort.

Ihre Worte: „growth" of sc. übersetze ich „Wachsthum der Wissensch., V. aber lässt dafür „Entwicklung" setzen. Von der Entwicklung sprechen Sie bald nachher, wo es heisst expansion and development. Diese Ausdrücke übersetze ich mit „Ausbreitung und Entwicklung", allein V. der schon die Entwicklung verbraucht hat, setzt dafür, ganz ungetreu: „lebhaftere Entfaltung"

„All its facts are fruitful" habe ich übersetzt: „Jeder Zuwachs ist fruchtbar", aber V. „alle ihre Erfolge tragen neue Früchte". Facts kann in der That hier nicht wohl wörtlich übersetzt werden, aber nicht die Erfolge (success) sind fruchtbar, sondern die Vermehrung unserer Kenntniss, der Zuwachs ist es. Statt „fruchtbar sein" zu sagen „neue Früchte tragen" ist, an und für sich betrachtet, ganz einerlei, aber es ist lächerlich und anmassend wenn meine vorgefundene, dem Originale be fruitful genau entsprechende Fassung durch eine andere ersetzt werden soll.

Die Worte, welche unmittelbar auf all its facts are fruitful folgen, übersetze ich „indem von jeder neuen Entdeckung sofort weitere Forschungen ihren Ausgang nehmen"—Vieweg: „da jede neue Entdeckung sofort der Ausgangspunkt für weitere Forschungen wird." Gegen diese Aenderung habe ich nichts zu erinnern, als dass sie gänzlich überflüssig ist und keine Verbesserung erzeugt.

Ich übersetze: kein hervorragenderes Beispiel des Wachsens. Vieweg kein schöneres Beispiel dieser Entwicklung.—Wegen Entwicklung für growth habe ich meine Bemerkung schon gemacht. Das Wort noble, welches Sie anwenden, bedeutet niemals schön, noblemen sind z.B. keine schönen Männer, sondern (der Wortbedeutung nach) ausgezeichnete hervorragende Männer.

Ihre bestimmte Zeitangabe twenty two years, habe ich, da unterdessen Zeit verflossen ist umgewandelt in „seit dem Jahre 1842", weil ich glaube, Sie haben den Zeitpunkt des Erscheinens der Abhandlungen von *[Mayn]* und von *[Junk]* im Sinne,—Vieweg setzt „innerhalb der letzten 20 Jahre".

Ich halte hier ein, indem der erste Abschnitt erörtert ist. Sie sehen daraus, welcher Art die Vieweg'schen Aenderungen sind und haben damit einiges Material um sich ein Urtheil zu bilden. Wie gesagt, es steht Ihnen, wenn Sie es verlangen die ganze umfangreiche Schrift zu Gebote, die ich an Vieweg zur Vertheidigung meiner Lesart sandte—freilich müsste ich sie neu machen, da ich keine Abschrift davon besitze.

Wollte ich unter den gemachten Aenderungen willkürlich auswählen, so könnte ich Sie zu einem viel ungünstigeren Urtheile über Vieweg's Aenderungen bewegen, ich könnte meine Behauptung, dass geradezu Unsinn producirt worden ist, beweisen.

Ich antwortete an V. dass ich bereit sei eine erneute Revision der bereit gedruckt Bogn zu machen, um jene sprachliche Ausfeilungen vorzunehmen, welche bei grösseren Anforderungen an die Form möglich und nützlich seien, dass ich aber die Verpfuschung des Textes, wie sie durch seinen Gehilfen bewerkstelligt worden, nicht billigen könnte. Zugleich forderte ich ihn auf, mir die Stellen zu bezeichnen, wo ich seiner Behauptung gemäss, unrichtig übersetzt habe. Da dieselben, wie er schrieb, alle bezeichnet seien, so habe er ja nur einen Umschlag um die Bogen zu machen und an mich zu adressir. V. blieb diesen geforderten Nachweis für seine Behauptung schuldig. Er antwortete er finde sich nicht veranlasst auf den Inhalt meines Schreibens einzugehen; wie die Sachen liegen sei die Fortsetzung eines weiteren Verkehrs zwischen uns unangenehm, nicht *[gentlemanlike]* (ipsissima verba), er schlage mir vor unsere Verbindung fördersamst zu lösen; die materiellen Verluste *[trafen]* ihn und er wolle *[eintragen]*. In meiner Antwort constatirte ich, dass V. meine Behauptungen gegen ihn und meine Anklage nicht bekämpft habe, sagte, auch ich empfinde kein Vergnügen daran mit ihm ferner in Verkehr zu bleiben, ich sei bereit den Vertrag zu lösen unter der Bedingung dass bis zu einem gewissen Termine mein Manuscript in meinen Händen sei und, da ich nicht in der Lage sei die aus der Aufhebung, mir unverschuldet entstandenen Nachtheile zu tragen, so verlangte ich als weitere Bedingung eine Geldentschädigung, die ich äusserst mässig zu 100 Gulden (8 1/3 lb Sterling) ansetzte.—Ich erhielt mein Manuscript & 100 *[Gl]* und damit war die Sache erledigt.

Sie haben hiermit eine möglichst kurz gefasste Schilderung des Vorgangs. Wünschen Sie noch mehr Detail zu Ihrer Information, so halte ich es für meine Pflicht Ihnen dieselben zukommen zu lassen.

Sie schreiben mir: „I had agreed with an excellent English scholar to look over the proof before they finally went to press and I did not think that you would in the least object to this.—Sie haben ganz richtig vermuthet. Nicht nur würde ich nicht das Geringste gegen eine solche Anordnung einzuwenden haben, sondern dieselbe im Gegentheile mit dem grössten Vergnügen begrüssen, da ich voraussetze, dass jener „scholar" mit Ihnen in persönlicher Berührung steht und über allenfallsige Zweifel mit Ihnen Rücksprache nehmen kann. Es bleiben, bei einem so umfangreichen Werke hin und wieder Zweifel über den genauen Sinn der in einer fremden Sprache gefassten Werke. Ueber die wichtigsten dieser Zweifel, habe ich seiner Zeit mit Ihnen correspondirt und Sie haben die Güte gehabt mit ganz besonderer Liebenswürdigkeit darauf einzugehen und mich zu belehren, wofür ich Ihnen wiederholt danke. Aber es bleiben natürlich noch Stellen genug, über welche eine Rücksprache mit dem Autor sehr wünschenswerth wäre, und nur um nicht gar zu unbescheiden zu sein, habe ich Sie damit nicht belästigt. Kann so etwas mündlich abgemacht werden, so wird es sicherlich von grösserem Vortheil für das Buch sein und ich habe, wie Sie, keinen anderen Wunsch als eine möglichst gute und getreue Wiedergabe Ihrer Arbeit zu Tage zu fördern.

Ich bin sehr gerne bereit, eine Copie der Bogen, mit den von mir für nöthig erachteten Aenderungen in der ersten Revision nach England zu schicken. Dort könnten dann die nothwendig oder nützlich befundenen Aenderungen eingezeichnet werden und der so uebersetzte Bogen an mich zurück gesendet werden. Ich hätte dann Gelegenheit bei der 2^{n} Revision die gemachten Correcturen einzusetzen, und es entstünde hieraus, nicht einmal ein Zeitverlust. Wenn Sie also jetzt noch Jemanden finden, der sich die Mühe nehmen will die Bogen durchzulesen, so würde ich auf dieses Arrangement mit grösstem Vergnügen und mit Dank für die Gefälligkeit des Betreffenden eingehen.

Mir ist aus verschiedenen Gründen sehr viel daran gelegen die Uebersetzung möglichst bald erscheinen zu sehen. Die Mühe und die Arbeit habe ich daran nicht gespart und es ist wohl verzeihlich, wenn mein Wunsch für diese einen Lohn zu finden auch mit unter den Gründen auftritt, die mich wünschen machen, das Buch bald vollendet zu sehen. Desshalb erlaube ich mir auch, so wenig bescheiden es sein mag, nochmals die ergebenste Bitte auszusprechen, Sie *[mögen]* mir gefälligst bald eine gütige Entscheidung geben.

Indem ich wünsche, dass Ihnen Ihr Osterausflug nach Schottland recht genussreich sein möge und Sie bitte mir die Fortdauer Ihrer freundlichen Gesinnung auch fernerhin nicht zu versagen, verbleibe ich wie immer mit ganz vorzüglicher Verehrung

Ihr ergebenster | D. Bohn

Giessen, in der Nacht vom 29 auf 30 März 1866.

Dear Mr Tyndall!

I did wrong in making new arrangements with Mr Vieweg[1] (for the termination of the contract) before informing you and asking for your opinion. In my defence, if not as justification for this omission, I may plead that you had emphatically advised me earlier against Vieweg and gave your approval to publish the book[2] with Vieweg only in retrospect, in your letter from Sept. 21.[3] Since you had recommended Theod. Fischer[4] as a publisher and authorised me personally to translate your book, I consequently assumed that you would likely approve transferring the publication from Vieweg to Fischer as soon as I asked you, as I have done.

I also mention in my defence that Vieweg offended me gravely and obliged me to take immediate action; that it was not I who intended to dissolve the agreement, despite having serious reasons for complaint, and that the proposal had come from Vieweg; finally, that I did not wish to postpone the publication of the book any further than what appeared inevitable. These are my excuses, they did not give a complete justification—you have reason to protest, which I willingly acknowledge. In repeating my plea that you may give your approval that Fischer publish the book, I express my regret about my hastiness and ask you to forgive me for the same.

I understand your wish "to learn something more about the old engagement" as a request to be fully informed about the reasons for terminating the agreement with Vieweg. I hasten to comply with this request; should you already have left for Scotland,[5] you will find my letter upon your return to London and to your affairs.

Mr Vieweg interrupted the setting of the book suddenly and without any notification. After 3 weeks, I asked him about the reasons for this interruption and received the answer that upon reading the book, V. had found that its form lacked fluency and that I had not properly translated some English words and expressions. He had therefore seen himself obligated to subject my work to a detailed revision, would send me the results of the same and ask me to accept the changes. Additionally, that he had to charge me 1*[1/2]* Thaler per sheet for his costs.

Anyone, and above all the publisher, has the right to tell me their opinions about a work, which, like my translation, is destined for the public. If criticism against the form of my translation or, even more seriously, against my understanding of the original is justified, I can only be grateful if I am made aware of it and am thereby given the opportunity to make corrections. However, a reproach such as V.'s turns into offence if it is unfounded; if it is made by a man who only demonstrates that he is not competent to judge on the topic; if it is presented in such an outrageous way as it was. It was impertinent to authorise someone else, a person unknown to me, to modify what V. wished to see changed and without notifying me beforehand. He is not the least bit entitled to make even the smallest change to a manuscript entrusted to him for printing; no man valuing his honour and name will tolerate such presumption. But the most decisive factor here is that the objections to my work are unfounded, that the alleged improvements produced something that differs from the original, that in several cases the modified translation turns your words into nonsense; in some cases the changes are only trivial, a mere deviation from the faithfulness that is the foremost duty of a translator to the author of the original text. I suppose it would be exhausting to you if I discussed all the individual changes, as I have done with Vieweg; for the time being, I will restrict myself to a few ones. Above all, I will discuss the changes made in the very first passage of the book, so as to avoid the suspicion that I am tendentiously selecting passages. I ask for your patience and note that, should you wish me to do so, I am prepared to communicate the full extent of my refutation and the defence of my version.

§1. The aspect . . . I translated: "Der Anblick der Natur erregt im Menschen den Geist der Untersuchung". View. had it changed to: "Die Anschauung der Natur erweckt im Menschen den Forschungstrieb". Aspect is 'Anblick'—had you wanted to say 'Anschauung', you would have used the word contemplation. Had you really wanted to discuss 'Triebe zur Forschung oder Untersuchung', you would have said instinct, impulse . . . or something similar.

I translated the following sentence as: "Wie das Auge zum Sehen, das Ohr zum Hören bestimmt sind, so ist der menschliche Geist zum Verstehen der Erscheinungen der körperlichen Welt bestimmt". Vieweg's version says: "Wie das Auge zum Sehen, das Ohr zum Hören geschaffen ist, so ist der menschliche Geist befähigt die Erscheinungen der körperlichen Welt zu verstehen". It is evident that the latter, unfaithful, translation puts a platitude in your mouth. You make the analogy

Eye for seeing } made } gebildet — and following the analogy the first two sentences, which are undisputed, make the third sentence
Ear for hearing
Mind for understanding formed

sound possible. If one wrote instead

Eye for seeing } made } geschafft — then the analogy ends; the relation between the claims presented in the first 2 sentences and the claim
Ear for hearing
Mind for understanding befähigt

in the 3rd sentence disappears and renders the first 2 statements unnecessary and trivial. Had you spoken about the 'Befähigung des Geistes zum Verstehen', then the correct composition of the sentences would have been

Eye can (or is able, is apt, capable, fit) for seeing
Ear " (" " " " " " ") for hearing
Mind " (" " " " " " ") for understanding

The next change is a consequence of the one just described and I will therefore not specially discuss it.

Very characteristical—I translate as: "Sehr bezeichnend", Vieweg: "sehr charakteristisch"; instead of using the excellent German word, he introduces an ambiguous foreign one into the German book.

I translate your words: "growth" of sc. as "Wachstum der Wissensch.", but V. had it set as 'Entwicklung'. You talk about 'Entwicklung' soon afterwards, when you write expansion and development. I translate these expressions as "Ausbreitung und Entwicklung", but as V. has already used up 'Entwicklung', he therefore sets, entirely unfaithfully: "lebhaftere Entfaltung".

"All its facts are fruitful" I translated: "Jeder Zuwachs ist fruchtbar", V. on the other hand, "alle ihre Erfolge tragen neue Früchte". Facts cannot actually be translated literally in this case, but it is not success (Erfolge) that is fruitful, but the increase in our knowledge, it is the growth. Writing "fruchtbar sein" instead of "neue Früchte tragen" makes no real difference, but it is ridiculous and arrogant that my version, be fruitful, which corresponds exactly to the original, had to be substituted with a different one.

I translate the words that immediately follow after all its facts are fruitful as "indem von jeder neuen Entdeckung sofort weitere Forschungen ihren Ausgang nehmen"—Vieweg: "da jede neue Entdeckung sofort der Ausgangspunkt für weitere Forschungen wird". I have nothing to object against this change other than that it is completely superfluous and creates no improvement.

I translate: "kein hervorragenderes Beispiel des Wachsens". Vieweg "kein schöneres Beispiel dieser Entwicklung".—I have already commented on using Entwicklung for growth. The word noble, which you use, never means beautiful: noblemen, for example, are not beautiful men, but (following the meaning of the word) outstanding excellent men.

I changed your specific indication of time, twenty two years, to "seit dem Jahre 1842", for time has passed since then and I believe you had in mind the time of publication of the articles of *[Mayn]* and *[Junk]*,—Vieweg writes "innerhalb der letzten 20 Jahre".

I stop here, having discussed the first passage. You can already tell the type of changes that Vieweg has made and you now have some material to form your own opinion. As I mentioned before, should you request it, the complete document that I sent to Vieweg in defence of my interpretation is at your disposal. Of course, I would have to prepare it anew since I am not in possession of a copy.

If I were to make a deliberate selection from among the changes, it would likely give you a far less favourable impression of Vieweg's changes, and would prove my allegation that downright nonsense has been produced.

I replied to V. that I was willing to revise those sheets that were already printed again, in order to undertake those idiomatic refinements that greater formal requirements made possible and useful, but that I could not tolerate the bungling of the text contrived by his assistant. At the same time, I asked him to specify the passages where according to him I translated incorrectly. He wrote that since these passages were all identified, he would only have to enclose the sheets in an envelope and address them to me. V. failed to send me the requested evidence for his allegations. He answered that he did not feel compelled to go into the contents of my correspondence; that as things stood, he felt that the continuation of our contact would be uncomfortable and not *[gentlemanlike]* (his very words), and he suggested that it would be more beneficial if we ended our relationship; that he had incurred material losses which he wanted to *[recover]*. In my answer, I noted that V. had not disputed my asseverations and my accusations against him; I said that it was not pleasant for me either to maintain further contact with him, that I was willing to dissolve the agreement on the condition that my manuscript was returned to my hands by a certain date and that, given that I was not in a position to assume any burden due to an annulment that was not my fault, I demanded as further condition a monetary compensation that I set very moderately at 100 guilders (8 1/3 lb Sterling). I received my manuscript and 100 *[Gl]* and the affair was settled.

I have given you the shortest possible description of the happenings. Should you wish more details for your information, I would consider it as my obligation to provide you with the same.

You wrote to me:[6] "I had agreed with an excellent English scholar to look over the proofs before they finally went to press and & did not think that you would in the least object to this". You thought quite correctly. Not only would I not object in the least to such an arrangement: quite to the contrary, I would welcome it with the greatest pleasure, for I suppose that said "scholar" keeps in personal contact with you and will be able to consult with you in case of any doubt. In such an extensive endeavour, there are always some doubts about the exact meaning of works composed in a foreign language. I had corresponded with you at the time about the more important of these doubts and you always had the goodness to address them and to instruct me with especial kindness, for which I thank you again. But of course, there remain enough passages for which it would be desirable to consult with the author, and only in order not to appear too presumptuous did I not bother you with them. If the like could be done verbally, it would certainly be of great benefit for the book and like you, I have no other wish than to bring a good and faithful rendition of your work to life.

I will gladly send to England a copy of the sheets with the changes that I deem necessary for the first revision. The changes that are considered necessary or useful could be marked therein and the resulting translated sheets can be returned to me. I would then have the chance to incorporate the corrections in the 2nd revision and this would not even cause any delays. Therefore, if you still found someone who would be willing to proofread the sheets, I would accept this arrangement with the greatest pleasure and gratitude for that person's helpfulness.

I have various reasons to hope that the translation will be published soon. I did not spare any effort and labour and it is probably pardonable that my wish to see these compensated figures among the reasons for which I wish to see the book completed soon. I therefore allow myself, as immodest it may be, to reiterate my request that you may soon please to give me your kind decision.

Wishing that your Easter journey to Scotland be pleasurable and hoping that you will not deny me the continuation of your kindness towards me, I remain with the greatest admiration

Your humble | D. Bohn

Giessen, in the night from 29 to 30 March 1866.

RI MS JT/1/B/111

1. *Mr Vieweg*: Eduard Vieweg.
2. *publish the book*: the German translation of *Heat* (1865); see letter 2296, n. 6.
3. *your letter from from Sept. 21*: letter missing.
4. *Theod. Fischer*: Theodore Fischer.
5. *left for Scotland*: Tyndall accompanied Thomas Carlyle to the University of Edinburgh for his speech to the students as Rector on 2 April 1866, leaving London on 28 March (see letter 2375, n. 3).
6. *You wrote to me*: letter missing, though some of its contents can be inferred from letter 2392.

From Charles Mackay[1] 31 March 1866 2397

Reform Club[2] | March 31st 1866.

My dear Professor Tyndall,

My name it appears is up for Ballot at the Athenaeum but without a seconder.[3] How many years I have been on the Book, I cannot tell but in the interval, my seconder has shuffled off this mortal coil[4]—and left me without a living sponsor. Will you undertake the task of recommending me? I am assured by many friends that I could not have a better man. If you will—will you kindly see to John Browning[5] my proposer: and arrange matters? If I am for political or other reasons, likely to be black-balled, I would rather withdraw; though I am not aware that any reasons exist, political, social or personal for my exclusion. I think Lord Houghton[6] would help with his influence:—also Sir James Clark, Dr. Pacy,[7] Dr. <*1 word missing*>, Mr Foley,[8] Mr Robert Chambers,[9] Mr Theodore Martin,[10] Mr Duncan Macgregor,[11] Sir E. Bulwer Lytton[12] and many other members with whom I am acquainted more or less intimately

Excuse my troubling you in preference to any of these gentlemen, which I do <*1–2 words missing*>, and

Believe me | Yours truly | Chas. Mackay

Professor Tyndall.

RI MS JT/1/TYP/2/784
LT Typescript Only

1. *Charles Mackay*: Charles Mackay (see letter 2346, n. 1).
2. *Reform Club*: see letter 2346, n. 2.
3. *Ballot at the Athenaeum but without a seconder*: to become a member of the Athenaeum Club (see letter 2253, n. 8), ordinary members were elected by ballot. Mackay had been

entered in the Candidates Book for the Athenaeum on 22 October 1855: 'Charles Mackay LLD Author of Studies from the Antique; Sketches from Nature; The Salamandrine; Ballad and Songs &c &c'. His nomination went to the ballot on 18 February 1867, with Sir John Bowring as his proposer and Tyndall as his seconder. There were 25 signatures of support on his ballot sheet, including Charles Dickens, Wilkie Collins, and John Murray. However, in the ballot he received 80 yes votes and 25 no votes, and so was not elected; eight no votes would have been sufficient to exclude him (Jennie De Protani, Archivist, the Athenaeum Club, personal communication).

4. *mortal coil*: a Shakespearean phrase to refer to the troubles of daily life, strife, and suffering of the world, used in Hamlet's famous 'To be, or not to be' speech.
5. *John Browning*: a mis-transcription of John Bowring (1792–1872), an English political economist and writer (*ODNB*).
6. *Lord Houghton*: Richard Monckton Milnes.
7. *Dr. Pacy*: there does not appear to have been a member of the club named Pacy. This is possibly a mis-transcription of John Percy, who while a metallurgist and professor at the Royal School of Mines (and colleague of Tyndall's), was a trained physician (*ODNB*; see letter 2371).
8. *Mr Foley*: probably John Henry Foley (1818–74), an Irish sculptor (*ODNB*). Among his numerous pieces is a statue of Michael Faraday that resides at the RI.
9. *Mr Robert Chambers*: Robert Chambers (1802–71), a Scottish publisher and journalist, who anonymously authored the evolutionary *Vestiges of the Natural History of Creation* (London: John Spriggs Morss Churchill, 1844) (*ODNB*).
10. *Mr Theodore Martin*: Theodore Martin (1816–1909), a Scottish lawyer, poet, biographer, and translator, best known for his *Life of the Prince Consort* (1874–80), the writing of which was entrusted to him by Queen Victoria (*ODNB*).
11. *Mr Duncan Macgregor*: Duncan Macgregor (1787–1881), of the Scottish clan Macgregor, served during the Napoleonic Wars, in Bengal, and in Canada, retiring in 1858 as Lieutenant-General. He was awarded a civil Knight Commander of the Order of the Bath (K. Stewart-Murray (ed), *A Military History of Perthshire, 1660–1902* (Perth: R. A. & J. Hay, 1908), pp. 511–4).
12. *Sir E. Bulwer Lytton*: Edward Bulwer-Lytton (1803–73), an English writer and politician , the author of *Paul Clifford* (1830) and *The Poems and Ballads of Schiller* (1844) (*ODNB*).

To Jane Welsh Carlyle 1 April [1866][1] 2398

Kinellan | Edinburgh[2] | 1st. April

Dear Mrs. Carlyle.

We arrived safely in Edinburgh yesterday evening[3] when I delivered Mr Carlyle into the hands of his brother.[4]

He had one night's profound sleep at Fryston,[5] which seemed quite to set him up. The night previous to our leaving he did not sleep so well, but yesterday we managed to get a carriage to ourselves and by a skilful arrangement of the cushions he was enabled to rest, and really to enjoy the journey. He smoked at intervals, and I administered a little brandy & water occasionally. He was thus enabled to have some sleep.

Yours ever truly | John Tyndall

RI MS JT/1/T/144
RI MS JT/1/TYP/1

---

1. *[1866]*: the year is given based on when Tyndall went to Edinburgh with Thomas Carlyle (see n. 3).
2. *Kinellan | Edinburgh*: Tyndall used paper printed with 'Kinellan | Edinburgh', the estate of Mary Anne Coxe and her husband James Coxe (1811–78).
3. *We arrived safely in Edinburgh yesterday evening*: Tyndall accompanied Carlyle to the University of Edinburgh for his speech to the students as Rector on 2 April 1866 (see letter 2375, n. 3).
4. *brother*: John Aitken Carlyle (1801–79).
5. *Fryston*: Fryston Hall (see letter 2375, n. 3).

To Jane Welsh Carlyle 2 April 186[6] 2399

THE ELECTRIC AND INTERNATIONAL
TELEGRAPH COMPANY.

The following Message has been received at Knightsbridge *Station,*
2 April 186
FROM John Tyndall
TO Mrs Carlyle | No 5 Cheyne Row | Chelsea[1] | L[on]d[o]n

A perfect triumph.[2]

RI MS JT/1/T/143
RI MS JT/1/TYP/1

---

1. *No 5 Cheyne Row | Chelsea*: the Carlyles' London address since 1834.
2. *a perfect triumph*: Tyndall's telegram from Edinburgh was to inform Jane Welsh Carlyle that her husband Thomas Carlyle's speech as Rector to the students of the University of Edinburgh, on 2 April 1866, was a success (see letters 2375, n. 3 and 2400).

To Jane Welsh Carlyle [2 April 1866][1] 2400

Kinellan | Edinburgh[2] | Easter Monday.

My dear Mrs Carlyle

I have time to do little more than confirm my telegram.[3] He[4] spoke without faltering for an hour and a half unassisted by notes of any kind. His audience was breathlessly attentive, except at intervals when it burst into irrepressive applause. He ended by quoting with exceeding power that brave and beautiful poem of Goethe's[5]

"Choose well your choice is | Brief and yet endless."[6]

and here the audience rose as one man and burst into a universal roar of acclamation. When he quitted the house with his friends he was followed into the street and when I last saw him there was a crowd at his heels battering their hats together in the open air, and cheering the Hero.

Now are you content?

Yours ever | John Tyndall

RI MS JT/1/TYP/1/190
LT Typescript Only

1. *[2 April 1866]*: the date is given by reference to Thomas Carlyle's speech to the students at the University of Edinburgh on 2 April 1866 (see letter 2375, n. 3).
2. *Kinellan | Edinburgh*: although the handwritten letter is missing, Tyndall had used paper printed with 'Kinellan | Edinburgh', the estate of Mary Anne Coxe and her husband James Coxe (1811–78).
3. *my telegram*: letter 2399.
4. *He*: Thomas Carlyle.
5. *Goethe's*: see letter 2330, n. 4.
6. *"Choose well your choice is | Brief and yet endless"*: quoted from Goethe's poem 'The Mason Lodge' (1827), which Carlyle himself had translated.

To Heinrich Debus 2 April 1866 2401

Kinellan | Edinburgh[1] | 2nd April | 1866

My dear Heinrich.

I do not quite understand the movements of Prof. Bohn.[2] Before making so important a change he ought I think to have written to me.[3] I had a letter from him[4] a few days ago on the subject which surprised me much. The fact is (and this must be private between you & me) I fear Bohn is not quite up in the English language.

I have had a letter[5] from Wiedemann, who undertook to look over the proofs, and he finds the translation exceedingly defective. Now I had no other reason for giving the thing to Bohn than the belief that he would make a first rate translation. It is a very difficult position for me to be in. But as I derive no pecuniary profit whatever from the translation I think the least that I can demand is that it should be well done. I cannot therefore for the present give my consent to the transfer of the book. I have written to Bohn[6] saying that I require further information before I can come to a decision.[7]

Yours ever | John Tyndall

RI MS JT/1/T/310

1. *Kinellan | Edinburgh*: Tyndall used paper printed with 'Kinellan | Edinburgh', the estate of Mary Anne Coxe and her husband James Coxe (1811–78).
2. *Prof. Bohn*: Johann Conrad Bohn.
3. *Before making so important a change he ought I think to have written to me*: see letter 2296, n. 6.
4. *letter from him*: letter 2396.
5. *letter*: letter missing.
6. *written to Bohn*: letter missing.
7. *come to a decision*: as letters later in this volume show, Tyndall eventually did not go through with this translation by Bohn (see letter 2296, n. 6).

From Jane Carlyle 3 April [1866][1] 2402

[5][2] Cheyne Row | Chelsea | 3 April

My dear M[r] Tyndall

The Scotch Housemaid was perfectly right, when she pronounced you "The very nicest gentleman that came to this House! First rate charming, in her opinion"! God bless you! Yes!—I never saw anybody so "charming" in all my life! Even Mr. C[3] is charmed! He writes to me;

"Tyndall's conduct to me has been Loyalty's own self; no loving Son could have more faithfully watched a decrepit Father: in fact I shall not forget it."—

Nor shall I forget it dear kind friend! Thank you, thank you, for all you have done to relieve my anxieties.

I was horribly afraid yesterday, till the Telegram[4] came; I had got it in my head that he was to drop down dead at the outset! I should certainly have done so in his place.

The relief your two cleverly chosen words sent thro' me from head to foot could only express itself in a violent fit of crying! But I didn't begin to cry till I had read the words to a gaping audience—comprising every body in the house, Cook, Housemaid, young Lady Cousin—who all fell to dancing and clapping their hands, and exclaiming in their several ways—Jessie the housemaid way was—"Eh!—Hear to that! Hurrah!" But above her rose the voice of the Cook, crying, "Goodness gracious! Some Brandy for the Mistress!—Don't you see she is going off!!! and a table spoonful of brandy was administered and was, I dare say, useful.

The Telegram was in my hand at three minutes after six! and I took it in my hand to John Forster's[5] birthday dinner, where was Dickens[6] and Wilkie Collins,[7] and again there was hurrahing and again on my part a tendency to "go off" (suppressed) "fine fellow that Tyndall!" said Forster, a dear warm-hearted fellow—you can see that from the way he throws himself into the thing"!

I had several letters from persons who had been present, along with yours,[8] but none of them put the scene before the eyes of my heart as yours did. Since the letters, an eye witness has been here Mr Macmillan[9] but dreadfully inarticulate with the best intentions.

Yours affectionately | Jane Carlyle

RI MS JT/2/10

---

1. *[1866]*: the year is given by reference to Tyndall's telegram to Jane Welsh Carlyle (see n. 4).
2. *[5]*: the Carlyles lived at 5 Cheyne Row in Chelsea, London since 1834. Jane's signature appears on the first page of the letter, written vertically at the top and crossing over the address. While "Cheyne Row' is unmistakable, the expected '5' before it does not appear to be '5', and the text is thus unidentified.
3. *Mr. C*: Thomas Carlyle.
4. *Telegram*: Tyndall had sent Jane a very brief telegram following her husband Thomas Carlyle's speech before the students of the University of Edinburgh on 2 April 1866 (see letter 2375, n. 3), stating that it was 'A perfect triumph'. This telegram is letter 2399.
5. *John Forster*: John Forster (1812–76), an English biographer and literary critic (*ODNB*).
6. *Dickens*: Charles Dickens (1812–70), the famed English writer and critic (*ODNB*).
7. *Wilkie Collins*: Wilkie Collins (1824–89), an English writer (*ODNB*).
8. *yours*: letter 2400.
9. *Mr Macmillan*: presumably Alexander Macmillan (1818–96), one of the co-founders of the publishing house Macmillan and Co. (*ODNB*).

From Rudolf Clausius 3 April 1866 2403

ZÜRICH, 3/4/66

Lieber Tyndall,

Du fragst mich in Deinem lieben Briefe, ob ich es nicht auch für gut halte, wenn Du Wiedemann Vollmacht über die Übersetzung giebst.

Es ist schwer, in dieser Sache einen guten Rath zu geben, weil sie etwas verwickelt geworden ist. An sich würde es gewiss gut sein wenn Wiedemann die Sache ganz in der Hand hätte. Ich fürchte nur, dass Bohn, welcher selbst Professor an einer Universität ist, sich dem Urtheil eines anderen, ihm im Range beinahe gleichstehenden Professors nicht wird unterwerfen wollen. Vielleicht wäre es am besten, wenn Du selbst Dir die Übersetzung kommen liessest. Du verstehst und sprichst so gut Deutsch dass Du gewiss ganz gut beurtheilen kannst, ob die Übersetzung brauchbar ist, oder ob sie ganz verworfen werden muss. Wenn die Übersetzung sich noch durch Änderungen brauchbar machen lassen sollte, so kannst Du wohl auch selbst diese Änderungen angeben. Deinem Urtheile wird sich Bohn gewiss gern unterwerfen. Auch Folie, welcher die französische Übersetzung meines Buches gemacht hat, hat sie mir zur Durchsicht zugesandt, und ich habe darin, obwohl ich sie im Allgemeinen sehr gut fand, doch an vielen Stellen Änderungen gemacht, wo ich sie nicht ganz richtig oder nicht ganz genau fand.

Ich will hoffen, dass sich auf irgend eine Weise die Sache noch wieder in ein gutes Geleise bringen lässt.

Mit herzlichen Grüssen an Dich und Hirst,

Dein | Clausius

P.S. Solltest Du schon an Bohn und Wiedemann in dem Sinne geschrieben haben, dass Du dem Letztern Vollmacht gegeben hast, so würde es wohl am Besten sein, die Sache nun einfach ihren Gang gehen zu lassen.

ZURICH, 3/4/66

Dear Tyndall,

You ask me in your kind letter[1] if I agreed that it would be a good idea if you give Wiedemann full powers over the translation.[2]

It is difficult to give good advice in this matter, as it has become quite entangled. In itself, it would be certainly good if Wiedemann had the whole matter in his hands. I only fear that Bohn,[3] who is a professor at a university himself, will not be willing to submit to the decision of another professor who is in a position almost equal to his. It would probably be best if you arranged for the translation to be sent to yourself. You understand and speak German so well that you can certainly judge if the translation is usable, or if it has to be rejected entirely. If the translation could be made usable by introducing some changes, you could indicate those changes yourself. Surely, Bohn will be glad to submit to your judgment. Folie,[4] who made the French translation of my book,[5] also sent me the translation for revision and although I found it very good in general, I changed several passages where I did not find the translation to be entirely correct or precise.

I hope that this issue can yet be brought to a good ending in some way.

With kind regards to you and Hirst,

Your | Clausius

P.S. Should you already have written to Bohn and Wiedemann that you have given full powers to the latter, then it would probably be best to simply let things go their way.

RI MS JT/1/TYP/6/2289
LT Typescript Only

1. *your kind letter*: see letter 2392.
2. *the translation*: the German translation of *Heat* (1865); see letter 2296, n. 6.
3. *Bohn*: Johann Conrad Bohn.
4. *Folie*: François-Jacques-Philippe Folie.
5. *French translation of my book*: see letter 2288, n. 6 and n. 7.

To unidentified[1] 4 April 1866 2404

It will take many years to blot the kindness of his friend from the memory of John Tyndall.

Kinellan 4th. April 1866

RI MS JT/1/T/249
RI MS JT/1/TYP/1/289

1. *unidentified*: this letter is probably to Mary Anne Coxe, who Tyndall had stayed with (at her estate Kinellan) while he was in Edinburgh for Thomas Carlyle's speech at the University of Edinburgh (see letter 2375, n. 3).

From Rudolf Clausius 8 April 1866 2405

ZÜRICH, 8/4/66.

Lieber Tyndall,

Soeben erhalte ich einen Brief von Vieweg über dem Buch, welchen ich beilege. Zunächst will ich einen auf mich bezüglichen Punct berichtigen. Er sagt, ich habe angefragt, ob er Dir nicht ein kleines Honorar zahlen könne. Das Wort "klein" kommt aber in meiner Anfrage nicht vor.

Was nun Dein Buch anbetrifft, so wirst Du aus dem Briefe ersehen, dass er schon bedeutende Kosten gehabt hat, indem er Bohn 100 Guldern Abstandgeld zahlen musste, 120 Thaler für Holzschnitte ausgegeben hat, und 7 Bögen hat vergeblich setzen lassen. Er arbeitet sich nun, eine andere Uebersetzung machen zu lassen, und Dir, wenn Du diese Uebersetzung als die *[richtige]* bezeichnen und ein eigenes Vorwort dazu schreiben willst, £10 zu zahlen.

Obwohl diese Summe gering ist, so würde ich doch im Interesse der Sache, damit Dein schönes Buch in guter deutscher Uebersetzung erscheint, sehr wünschen, dass Du darauf eingehen könntest. Was das geringe Honorar betrifft, so kann man dabei auch in Anschlag bringen, dass Vieweg in der Sache schon anderwertige Kosten gehabt hat. Die Hauptschwierigkeit wird wohl für Dich sein, mit Bohn auseinanderzukommen. Aber da Vieweg schreibt: "eine so schlechte ja stümperhafte Uebersetzung" so ist wohl anzunehmen dass die Uebersetzung wirklich unbrauchbar ist, und das es am besten wäre, sie gerade zu verwerfen. Solltest Du nach meinem früheren Rathe Bohn aufgefordert haben, Dir die Uebersetzung zu schicken, so könntest Du am besten darüber urtheilen, ob sie sich durch einige Verbesserungen in eine gute verwandeln lässt, oder sie so schlecht ist, dass es besser ist, eine neue machen zu lassen. Wenn Du das letztere findest, so kannst du ihm ohne Bedenken schreiben, dass Du zur Veröffentlichung, einer solchen Uebersetzung Deine Einwilligung nicht geben kannst, und ihn ersuchen musst, Dir die ihm gegebene Ermächtigung zurückzustellen.

Du wirst, nachdem Du die Sachlage genau kennst, gewiss selbst am besten sehen, was dabei zu thun ist, und ich wünsche von Herzen, dass sich alles zu Deiner Zufriedenheit gestaltet.

Mit herzlichen Grüssen von meiner Frau und mir an Dich und Hirst,

Dein | Clausius.

ZURICH, 8/4/66.

Dear Tyndall,

I just received a letter about the book from Vieweg,[1] which you will find enclosed.[2] First of all, I want to correct one point that has to do with me. He says that I had asked him whether he should not pay you a small honorarium. The word "small" was not part of my inquiry, however.

Regarding your book, you will read in the letter that he has already invested a significant amount of money: he had to pay 100 guilders to Bohn[3] as a compensation; 120 Thalers for xylography, and 7 sheets he has had typeset in vain. He is working on arranging a new translation, and if you are willing to approve the translation as the correct one and write your own foreword, he will pay you £10.

I understand that this a small amount, but I sincerely hope that you will agree to it for the sake of publishing a good German translation of your fine book. Regarding the small honorarium, you have to take into account that Vieweg has already had other expenses in this matter. The main difficulty for you will be how to manage with Bohn. However, since Vieweg writes: "such a bad, unprofessional translation", it can be assumed that the translation is completely unusable, and the best would be to reject it altogether. If you decided to follow my previous advice and asked Bohn to send you the translation, you will be able to determine whether it can be improved through some modifications, or if it is so bad that it is better to have a new one. If you decide for the latter, you can write to him without concern that you cannot give your permission to publish such a translation, and that you must ask him to return the authorisation he was given.

As soon as you know the situation exactly, you will certainly be in the best position to see what should be done, and I wish with my whole heart that everything works out to your satisfaction.

With best wishes from my wife[4] and myself to you and to Hirst,

Your | Clausius.

RI MS JT/1/TYP/7/2290–1
LT Typescript Only

1. *Vieweg*: Eduard Vieweg.
2. *find enclosed*: enclosure missing. Clausius was referring to the German translation of *Heat* (1865); see letter 2296, n. 6.
3. *Bohn*: Johann Conrad Bohn.
4. *my wife*: Adelheid Clausius.

From Johann Conrad Bohn 9 April 1866 2406

Giessen d. 9 April 1866

Hochverehrter Herr Tyndall

Am 6ⁿ. April schrieb Vieweg an Th. Fischer in Cassel: „Unsere Rechte an eine deutsche Uebersetzung des gedachten Werkes (Tyndall, Heat . . .) werden wir übrigens demnächst zur Ausführung bringen."

Da Herr Vieweg keinerlei Anspruch mehr auf meine Uebersetzung Ihres Buches zu machen erklärt hat, so muss er seine vermeintlichen Rechte auf die Herausgabe der deutschen Uebersetzung anderswoher leiten.

Mittelst Schreiben vom 5ⁿ. Sept. 1865 waren Sie so gütig mir die Autorisation zur Uebersetzung zu ertheilen. Nachdem Sie mir von Vieweg als Verleger abgerathen, hingegen Th. Fischer als solchen empfohlen hatten, gaben Sie auf mein Ersuchen erst am 21ⁿ. Sept. Ihre Zustimmung dazu, dass das Buch in Viewegs Verlag gehe. Als Sie später so freundlich waren mir einen Brief von

Quandt und Händel in Leipzig, welche sich bei Ihnen um die Autorisation zur Herausgabe einer deutschen Uebersetzung beworben hatten, mitzutheilen, schrieben Sie mir, Sie hätten jenen Herrn geantwortet, dass die Uebersetzung in meiner Hand sei und ich Ihre Ermächtigung besässe.

Ich kann demnach kein Recht Viewegs auf Herausgabe einer deutschen Uebersetzung Ihres Buches erkennen und nicht verstehen, dass derselbe sich anders als irrthümlich im Besitze der Concession glauben mag.

Seit Empfang Ihres letzten Briefes vom 27^{n}. März habe ich natürlich in der Angelegenheit nichts thun dürfen, als Ihnen am 29/30 März zu schreiben und Sie um Ihre gütige ausdrückliche Genehmigung zu bitten, dass das Recht des Verlags des Werkes an Th. Fischer ertheilt werde. Sie würden mich zu ausserordentlichem Dank verpflichten, wenn Sie dieser Bitte willfahren und zugleich meiner und meines eventuellen Verlegers Beruhigung erklären wollten, dass Niemand anders autorisirt sei eine deutsche Uebersetzung erscheinen zu lassen. Sie allein können natürlich über diese Frage entscheiden.

Ich bezweifele nicht, dass Vieweg versucht haben wird meiner Arbeit bei Ihnen einen schlechten Credit zu verschaffen,—hat er es doch H. Fischer gegenüber gethan!—Wenn Sie die Anordnung treffen wollen, von welcher Sie in Ihrem letzten Briefe sprachen und zu deren Befolgung ich mit dem grössten Vergnügen meine Bereitwilligkeit erklärt habe, so werden Sie im Stande sein, vor dem Erscheinen des Buches sich ein Urtheil über meine Arbeit zu schaffen und alle Veränderungen an derselben, welche Sie wünschenswerth finden sollten, rechtzeitig zu bewerkstelligen.—Ich habe nicht die Ehre näher von Ihnen gekannt zu sein und kann daher nicht verlangen, dass Sie eine Anschuldigung wie die von Vieweg gemachte, von vornherein als ungegründet zurückweisen sollten. Desshalb ist es mir ein ganz besonders erfreuliches Zusammentreffen, dass Sie schon früher die Absicht hatten, die Bogen vor ihrer endgültigen Drucklegung der Durchsicht einer Ihres Vertrauens geniessenden Persönlichkeit zu unterwerfen.

Mich der Hoffnung auf eine baldige, gütige Antwort hingebend und wünschend dass Ihre Osterreise nach Schottland Sie recht befriedigt habe, zeichne ich wie immer mit ganz vorzüglicher Verehrung

Ihr ergebenst | C Bohn

Giessen 9 April 1866

Dear Mr Tyndall

On April 6th, Vieweg[1] wrote to Th. Fischer[2] in Cassel: "Incidentally, we will soon exercise our rights to produce a German translation of the intended work[3] (Tyndall, Heat . . .)."

As Mr Vieweg has declared to make no further claims to my translation of your book, he must be deriving his alleged rights to publish the German translation from somewhere else.

In your letter from Sept. 5th 1865,[4] you were so kind as to give me the authorisation for the translation. After having advised me against using Vieweg as publisher and having recommended Th. Fischer instead, you consented only on Sept. 21st[5] to my request that the book be published by Vieweg. When you later kindly forwarded me a letter from Quandt und Händel in Leipzig,[6] petitioning you for the authorisation to publish a translation, you wrote to me[7] that you had answered these gentlemen that the translation was in my hands and that I had your authorisation.

I can therefore not recognize any right Vieweg's to publish a German translation of your book, nor can I understand that he can believe himself—other than erroneously—in possession of this concession.

Since receiving your letter from March 27th,[8] I have naturally not been able to do anything in this matter other than to write to you on the 29/30th of March[9] to ask for your kind, express permission to grant the publishing rights to Th. Fischer. You would make me greatly indebted if you satisfied this request and at the same time clarified, to my and my potential publisher's reassurance, that no one else is authorised to publish a German translation. Naturally, only you can decide on this matter.

I do not doubt that Vieweg will have tried to discredit my work before you—he has done so before Mr Fischer! If you wanted to make the arrangement you mentioned in your last letter,[10] and with which I would comply with the greatest pleasure, you would be able to judge my work for yourself, <u>before</u> the publication of the book, and make all the corrections that you should find desirable in a timely fashion.[11]—I do not have the honour of being better known to you and can therefore not expect that you will reject an accusation such as that made by Vieweg as unfounded from the outset. For this reason, it is an especially gratifying conjuncture that you already had the intention to subject the sheets to the inspection of a person whom you trust, before sending a final version to print.

Indulging myself to hope for an early, gracious reply and wishing that your Easter journey to Scotland[12] was quite satisfactory, I sign, as always, with great respect,

Your obedient servant | C Bohn

RI MS JT/1/B/113

---

1. *Vieweg*: Eduard Vieweg.
2. *Th. Fischer*: Theodore Fischer.
3. *German translation of the intended work*: of *Heat* (1865); see letter 2296, n. 6.
4. *your letter from Sept. 5th 1865*: letter missing.
5. *on Sept. 21st*: letter missing.
6. *Quandt und Händel in Leipzig*: German publishers.

7. *you wrote to me*: letter missing.
8. *your letter from March 27th*: letter missing.
9. *write to you on the 29/30th of March*: letter 2396.
10. *your last letter*: probably the missing letter referred to in n. 8.
11. *timely fashion*: Bohn's next letter to Tyndall (2407) indicates his desire to move forward with this translation, but in the end Tyndall would not go through with it (see letter 2296, n. 6).
12. *Easter journey to Scotland*: Tyndall accompanied Thomas Carlyle to the University of Edinburgh for his speech to the students as Rector on 2 April 1866, leaving London on 28 March (see letter 2375, n. 3).

From Johann Conrad Bohn 12 April 1866 2407

Giessen 12 April 1866

Hochverehrter Herr Tyndall!

Soeben erhalte ich Ihren Brief, worin Sie den Wunsch aussprechen das Manuscript meiner Uebersetzung einzusehen und ich sende umgehend einen Theil desselben Ihnen zu, indem, wie ich denke, der Zweck auch durch einen Theil erreicht wird und das Porto sonst gar zu hoch würde. Uebrigens steht Ihnen auf Verlangen auch das Ganze zur Verfügung.—Ich fürchte Sie werden grosse Mühe haben die Blätter zu lesen, namentlich da dieselben von den Druckern, die sie bereits in Händen hatten, sehr beschmutzt wurden. Wenn Sie die Güte haben wollen später die Druckbogen vor der 2n. Revision durchzusehen und eventuell zu annotiren, so werden Sie weit weniger Unbequemlichkeit haben.

Für die in Ihrem vorigen Briefe formell ausgesprochene Ermächtigung die Herausgabe an Herrn Th. Fischer zu übertragen, habe ich Ihnen herzlich zu danken. Ihren Brief sendete ich sofort an H. Fischer mit der Aufforderung an diesen, Vorschläge zu machen, in welcher Weise Sie ein Honorar für die Ertheilung der Erlaubniss zur Herausgabe einer Uebersetzung empfangen sollen. Die Antwort des Verlegers wollte ich abwarten, ehe ich Ihnen wieder schrieb.

Gleich nachdem der Vertrag mit Vieweg aufgelöst war, hatte ich mit Fischer folgendes verabredet, was ich zu Ihrer gefälligen Kenntnissnahme hiermit bringe:

1., Die Ausstattung des Buches wird die von (der deutschen Uebersetzung) von Hoffmanns Elements of modern chemistry sein.

2., Die erste Auflage wird von 1000 oder 1200 Exemplaren sein.

3., Dafür wird ein Honorar von 16 Thaler p Bogen gezahlt.

4., Für spätere Auflagen ist die Normalstärke von 1600 Exempl. festgestellt, doch können die Auflagen auch stärker sein.

5., Für die späteren Auflagen wird ein Honorar von 1 (ein) Thal p. Bogen für je 100 Exemplar gezahlt.

6., Jedoch bleiben von der zweiten Auflage so viele Exemplare ohne Honorar, als bei der ersten Auflage an der Zahl 1600 fehlen.

Gestern schrieb ich nun an H. Fischer, ich fände es, angesichts Ihres Briefes, billig, da er nun weitere Ausgaben (an Sie) habe, dass er nicht mehr an seine Vorschläge hinsichtlich der späteren Auflage gebunden sei. Bei meiner Unkenntniss der einschlagenden Verhältnisse, müsse ich aber ihn bitten neue Vorschläge zu machen. Ferner schrieb ich ihm, er solle sich nun wegen der Clichés direct an Ihr [sic] englischen Verleger wenden.

Ihnen nochmals für die im vorigen Briefe ausgedrückte Gewährung meiner Bitte dankend und mich Ihnen bestens empfehlend, verbleibe ich, wie immer, mit der ausgezeichnetsten Verehrung

Ihr ergebenster | J. C Bohn

Giessen 12 April 1866

Dear Mr Tyndall!

I have just received your letter[1] in which you state the wish to see the manuscript of my translation.[2] I am sending you a part of it immediately and I think that the objective can be reached with only a part, as the postage would be too expensive otherwise. However, the whole of it is available on your request.—I fear that you will have great difficulty reading the pages, because the printers who had them in their hands made them very dirty. If you were so kind to later look through and perhaps annotate the printed sheets before the 2nd revision, you would be significantly less inconvenienced.

My heartfelt thanks for the authorisation, formally expressed in your previous letter,[3] to transfer the publication to Mr Th. Fischer.[4] I sent your letter immediately to Mr Fischer with the request to propose ways in which you can receive royalties for giving the authorisation to publish a translation. I wanted to wait for an answer from the publisher before writing to you again.

Soon after terminating the agreement with Vieweg,[5] I agreed with Fischer on the following, which I would like to bring to your kind attention:

1., The features of the book will be same as those in (the German translation of) Hoffman's Elements of modern chemistry.[6]

2., The first printing will be 1000 or 1200 copies.

3., The royalties for this will be 16 Thaler p. sheet.

4., Further editions will normally be 1600 copies, but it is possible to have a higher print-run for these.

5., The royalties for further editions will be 1 (one) Thaler p. sheet for every 100 copies.

6., However, there may be as many copies of the second edition without royalty as are missing in the first edition to complete 1600.

Considering your letter, I wrote to Mr Fischer yesterday, telling him that since he now has additional expenditures (to you), I would find it adequate if he should not be bound to his proposals concerning the later edition. Given that I am unfamiliar with the relevant conditions, I had to ask him to make new proposals. Furthermore, I wrote to him that he should contact your English publisher regarding the print blocks.

I convey my respects to you, thanking you again for conceding to my petition in the previous letter and remain, as always, in greatest admiration

Your humble | J. C Bohn

RI MS JT/1/B/112

---

1. *your lettter*: letter missing, but presumably Tyndall's reply to Bohn's letter of 9 April 1866 (letter 2406).
2. *my translation*: the German translation of *Heat* (1865); see letter 2296, n. 6.
3. *your previous letter*: letter missing, but presumably Tyndall's reply to Bohn's letter of 23 March 1866 (letter 2385), to which Bohn replied on 30 March 1866 (letter 2396). Tyndall mentioned this missing letter to Rudolf Clausius in letter 2392.
4. *Mr Th. Fischer*: Theodore Fischer.
5. *Vieweg*: Eduard Vieweg.
6. *Hoffman's Elements of modern chemistry*: August Wilhelm Hofmann (1818–92), a German chemist who was the first director of the Royal College of Chemistry in London from 1845–65 and then moved to Berlin, published in 1865 *Introduction to Modern Chemistry, Experimental and Theoretic. Embodying Twelve Lectures delivered in the Royal College of Chemistry, London* (London: Walton and Maberley). The German translation apppeared in 1866 as: A. W. Hofmann, *Einleitung in die moderne Chemie. Nach einer Reihe von vorträgen gehalten in dem Royal college of chemistry zu London* (Braunschweig, F. Vieweg und Sohn).

From Jane Kirk Brewster[1] 12 April [1866][2] 2408

Thursday, the 12th, April

It was so nice of you, in the midst of all the bustle of return to that great Babylon[3]—writing to Willie himself[4]—(though I took possession of the letter all the same. <u>I have every one you have ever written to me</u>

I dont think I have ever told you Sir David's[5] poem upon Miss Phoebe Lyon

Phoebe! | Ye be | Hebe—| I be | <u>D. B.</u>[6]

He justly says it is the most conprehensive in the English language—Naming and describing a lady—also a distinguished philosopher—all in 5 lines!!!

Here is his wording of Deaths Births and Marriages:

Hatches—Matches (or as the case may be—Catches) and dispatches![7]

Think of my having got from him this morning—"More Worlds than one—with the Author's kind regards".[8] Of course I shall not understand a word of it, but I am determined to read it all through—for his sake—and yours!

Well, I don't profess to be up in your works, albeit they are in this house—for I had the conviction they would be Greek and Hebrew to me—but I think I could tell you every single time I have seen yourself.

I have a petition this morning from a lady[9] for the signature of Mr Carlisle.[10] Should you ever happen to stumble on a note of his, from which you would not mind abstracting the name, may be you would kindly let me have it—but do not take any trouble, for I shall hold out no hope to the applicant.

I may perhaps find it in my heart to give her this signature of your note to Willie[11]

Here endeth the 3rd lesson!

RI MS JT/1/TYP/1/280?
LT Typescript Only

---

1. *Jane Kirk Brewster*: Jane Kirk Brewster (née Purnell, b. 1827), second wife of David Brewster. Brewster's first wife Juliet Macpherson died in 1850, and he remarried on 27 March 1857 at the age of 75 to Jane, daughter of Thomas Purnell of Scarborough, who was 46 years his younger. They had met on 17 November 1856 while Brewster was traveling to Cannes on account of his health. In January 1857 Brewster followed her to Nice, where they were married. Brewster and Lady Jane Kirk Brewster had one daughter, Constance Marion, who was born 27 January 1861. See M. M. Gordon, *The Home Life of Sir David Brewster*, 2nd edn (Edinburgh: Edmonston and Douglas, 1870), p. 270.
2. *[1866]*: the year is given by reference to Tyndall's trip to Edinburgh (see n. 3).
3. *return to that great Babylon*: referring to Tyndall's return to London from Edinburgh, where he had accompanied Thomas Carlyle—referred to as 'Carlisle' later in this letter—to the University of Edinburgh for his speech to the students as Rector on 2 April 1866 (see letter 2375, n. 3). Tyndall certainly saw the Brewsters while in Edinburgh, as he noted in his journal that David Brewster was guest for dinner at Mary Anne Coxe's estate Kinnelan, with whom he stayed while in Edinburgh (RI MS JT/2/13c/1329). London was often referred to as 'Babylon'. British politician Benjamin Disraeli (1804–1881) called London 'a modern Babylon' in his novel *Tancred: or, the New Crusade*, 3 vols (London: Henry Colburn, 1847), vol. 3, p. 77.
4. *writing to Willie himself*: 'Willie' is not identified. The Brewsters had one daughter (see n. 1), so this does not appear to be a son. The referred to letter is missing.
5. *Sir David's*: David Brewster.
6. *Phoebe! | Ye be | Hebe—| I be | D. B.*: in his *A Century of Scottish Life: Memorials and*

Recollections of Historical and Remarkable Persons, 2nd edn (London: Charles Griffin & Co., 1872), the Scottish minister Charles Rogers (1825–90) wrote of David Brewster: 'By the gentler sex he was beloved. When he was bordering on eighty, Miss Phoebe Lyon, a charming young lady, begged that he would contribute some lines to her album. In vain did the philosopher protest that verse-making was not his *forte*. The lady would admit of no excuse; so Sir David snatched a pen and wrote thus:—"Phoebe, | Y' be | Hebe. | D. B."' (p. 51).

7. *Hatches—Matches (or as the case may be—Catches) and dispatches!*: this phrasing—'Hatches, Matches, and Dispatches'—was used in journalism as a heading for birth, marriage, and death announcements in newspapers (E. Latham, *A Dictionary of Names, Nicknames and Surnames of Persons, Places and Things* (London: George Routledge & Sons, 1904). It is uncertain if the phrase is indeed original with Brewster or if his wife was unaware he was only repeating the phrase.
8. *"More Worlds than one—with the Author's kind regards"*: D. Brewster, *More Worlds than One: The Creed of the Philosopher and the Hope of the Christian* (London: John Murray, 1854). This book was Brewster's response to William Whewell's *Of the Plurality of Worlds: An Essay* (London: John W. Parker and Son, 1853), in which Whewell argued that the earth may not be the only planet in the universe inhabited by life.
9. *lady*: not identified.
10. *Mr Carlisle*: Thomas Carlyle.
11. *your note to Willie*: see n. 4.

To Mary Anne Coxe 16 April 1866 2409

Royal Institution 16th April 1866

My dear Lady Coxe

I have been waiting for a little breathing time to write you a long letter; but the respite in this Babylon is not likely to come.[1] You are happy in the peace of Kinellan,[2] far away from this turmoil physical and intellectual. Depend upon it you have given me the best reason possible to return to your hospitable house, for assuredly it would be impossible to render man more thoroughly comfortable than you have rendered me. I appreciate your magnanimity—and I feel grateful to you among other things for the generous liberty you granted me. In fact you made me feel your house my home, and if I could I would close with your invitation and return to it in blooming May.

But all through May and June I lecture at the School of Mines.[3] Mine is the loss that I cannot exchange Jermyn Street[4] for Kinellan.

Thank you exceedingly for your beautiful present.

Anent the brandy, I assure you I simply obeyed orders: but I must say that had I known your champagne to be so awfully good as it really is I should certainly have finished the pint you set before me on the evening of my arrival.

Will you kindly remember me to your excellent husband[5] and no less excellent brother.[6] It is worth living for to have a pair of such hearty manly friends, not to speak at all of the friendship of womankind, which the psalmist says is as fleeting as the morning dew. You remember the chapter and verse do you not?

I am going to run down to the Isle of Wight[7] for a few days—starting from here on Thursday morning.

And now good bye and may the good angels guard you.

Yours ever and always | John Tyndall.

RI MS JT/1/TYP/1/289
LT Typescript Only

---

1. *respite in this Babylon is not likely to come*: see letter 2408, n. 3.
2. *Kinellan*: an estate of Mary Anne Coxe and her husband James Coxe (see n. 5) in Edinburgh, where Tyndall stayed during his trip accompanying Thomas Carlyle for his speech to the students at the University of Edinburgh on 2 April 1866 (see letter 2375, n. 3)
3. *I lecture at the School of Mines*: see letter 2337, n. 16.
4. *Jermyn Street*: location of the Royal School of Mines.
5. *your excellent husband*: James Coxe (1811–78), a Scottish physician who married Mary Anne Coxe in 1841. From 1857 he served as one of the commissioners of the General Board of Lunacy for Scotland, and was knighted in 1863 for his services ('Obituary of Eminent Persons', *Lancet*, 111 (18 May 1878), p. 471).
6. *excellent brother*: William Fullerton Cumming (1804–92), a physician and brother of Mary Anne Coxe. He was employed by the East India Company after having obtain his MD from the University of Edinburgh (1827) but returned to Scotland in 1836 due to poor health ('Obituary', *Lancet*, 140 (6 August 1892), pp. 349–50).
7. *run down to the Isle of Wight*: no other letter in April 1866 refers to this trip to the Isle of Wight. In his journal, Tyndall wrote: 'Huxley and myself returned together on the Friday. The week of dissipation did not quite agree with us. I went with DuBois for three days to the Isle of Wight. Saw Tennyson, roamed the downs, climbed down and up the cliffs of Scratchels bay, and returned much benefitted' (RI MS JT/2/13c/1330). He also referred to the trip in his 'Personal Recollections of Thomas Carlyle (1890)', first published in the *Fortnightly Review* in January 1890 and published again in J. Tyndall, *New Fragments* (London: Longmans, Green, and Co., 1892), pp. 347–97, writing that following a visit to Jane Welsh Carlyle after his return to London, 'I went to the Isle of Wight, which was my usual refuge when tired, made Freshwater Gate my headquarters, and was refreshed as I had often been before by the broad-blown, brotherly voice of Tennyson. Two walks in the island have always had a special charm for me; one along "the ridge of a noble down" which stretches from Freshwater Gate to the Needles; the other along the spine of the island from Freshwater Gate to Carisbrook, past ancient Barrows, with the Solent on the one side and the ocean on the other. From Carisbrook it was an easy walk to Cowes, whence steamers plied to Southampton' (pp. 367–8).

From Gustav Wiedemann 19 April 1866 2410

Carlsruhe (Polytechnicum) | 19ten, April 1866

Mein lieber Tyndall!

Endlich kann ich Ihnen nach dem Wechsel meines Wohnsitzes einige Worte als Antwort auf Ihre letzten freundlichen Zeilen senden. Die fatale Verzögerung der Uebersetzung Ihres Werkes durch die Unfähigkeit von Bohn hat hoffentlich jetzt ein Ende erreicht. Nicht Bohn hat mit Vieweg wegen des „miserablen *[1 word illeg]*" *[1 word illeg]* gebrochen, sondern Vieweg mit Bohn. Er hat 7–8 Bogen Schriftsatz geopfert, und Bohn auf sein Verlangen auch eine Abstandssumme, ich glaube von 100 Gulden, gegeben, um nicht die schlechte Uebersetzung herausgeben zu müssen. Ich möchte Ihnen auch nicht empfehlen, Bohn zu gestatten, wo anders seine Uebersetzung erscheinen zu lassen; Ihnen wäre nicht damit gedankt & ich meine, wir wären auch Vieweg diese Rücksicht schuldig, der schon einen Theil der erforderlichen Holzschnitte fertig hat.

Nun komme ich eben mit einem anderen Vorschlag, bei deßen Mittheilung ich freilich Ihre Discretion erbitten möchte.—Frau Helmholtz und meine Frau haben große Lust, sich in der Uebersetzung ihres Werkes zu theilen und wir Männer haben ebenfalls schon das Nöthige besprochen. Helmholtz & ich, wir würden selbstverständlich die Uebersetzung der Damen genau durcharbeiten und sie dann unter unserer eigenen Verantwortung erscheinen lassen. Die Damen zu nennen, wäre wohl nicht angemessen. Ich hoffe, lieber Freund, der Vorschlag wird ihnen genehm sein. Beide Damen beherrschen, wie Sie wissen, die beiden erforderlichen Sprachen vollkommen und haben schon durch ihre Umgebung eine physikalische Richtung. Uns Männern würde es eine große Freude sein, wenn sie neben dem täglichen Einerlei der Wirthschaft im Hause durch die Uebersetzung Ihres Werkes an unserem wissenschaftlichen Treiben, wenn auch im Verborgenen, Theil nehmen könnten.

Sobald Sie nun Ihre Zustimmung zu unserem Plan geben, schreibe ich an Vieweg & die Uebersetzung soll alsbald erscheinen; die Damen haben schon des Versuchs halber an derselben zu arbeiten begonnen.

Sie sehen, lieber Tyndall, daß die Huldigungen, welche Ihnen die Damen in Genf spendeten, sich auch nach Deutschland übertragen. Aber fürchten Sie unsere Eifersucht nicht, wir Männer wollen ja doch bei der Bearbeitung die letzte Oberaufsicht für uns behalten.

Von mir kann ich Ihnen auch wenig berichten, ich bin noch allein, da meine Frau noch bei ihrem Bruder in Schlesien zum Besuch ist. Meine Räume sind glänzend, die Sammlung sehr reich, doch macht mir das Ordnen viele Mühe. *[Doch]* das sind Aeußerlichkeiten, die sehr angenehm sind, aber zu den eigentlichen wissenschaftlichen Leistungen nichts beitragen.

Daß Clausius einen sehr schönen Ruf nach Wien erhalten hat, haben Sie wohl schon erfahren. Wir wünschen für uns und ihn, daß er den Ruf ablehnt. Das Leben in Wien sagt den Protestanten und Preußen, wie man behauptet, sehr wenig zu & und Clausius würde uns, wenn er hinginge, sehr fern gerückt.

Bitte, antworten Sie recht bald mit einigen freundlichen Zeilen.

Freundschaftlichst Ihr | aufrichtig ergebener | G. Wiedemann.

Carlsruhe (Polytechnic Institute) | 19th, April 1866

My dear Tyndall

After my change of residence, I can write to you a few words in response to your last friendly lines.[1]

The fatal delay of the translation of your work[2] through the incompetence of Bohn[3] has now hopefully come to an end. Not Bohn has broken off with Vieweg[4] because of the "miserable *[1 word illeg]*", *[1 word illeg]* but Vieweg with Bohn. He has sacrificed 7–8 typeset sheets, and upon Bohn's demand also given him compensation, I think of 100 guilders, in order not to have to publish the bad translation. I also do not want to recommend you to allow Bohn to publish his translation elsewhere; it would not be to your benefit & I think we also owe this consideration to Vieweg, who has already finished a part of the necessary woodcuts.

I come with a different suggestion, for which I naturally ask for your discretion.—<u>Mrs Helmholtz</u>[5] and <u>my wife</u>[6] are very desirous to share the translation of your work and we men have also already discussed the wherewithal. Helmholtz & I, would naturally revise the ladies' translation very carefully and then publish it under our own responsibility. It would probably not be appropriate to mention the ladies.[7] I hope, dear friend, the suggestion will be agreeable to you. As you know, both ladies have complete mastery of the two languages necessary and through their environment already have a leaning towards physics. It would be a great pleasure for us men if besides the every-day monotony of domestic economy they could participate, albeit in secret, in our scientific endeavours through the translation of your work.

As soon as you give your approval to our plan, I will write to Vieweg & and the translation shall soon be published; the ladies have already started working on it by way of experiment.

You see, dear Tyndall, that the homage the ladies gave you in Geneva[8] is also transferred to Germany. But do not fear our jealousy, we men still want to keep the final supervision of the work for ourselves.

About myself I can only tell you only very little, I am still alone, as my wife is still visiting her brother in Silesia. My rooms are splendid, the collection is extensive, but putting everything in order takes much effort. */But/* these are superficialities that are very pleasant, but do not contribute anything to actual scientific performances.

You have probably heard that Clausius has received a very nice call to Vienna.[9] For us and for him, we hope that he declines the call. Life in Vienna, as it is said, is little suited for a protestant and a Prussian & Clausius, if he went there, would be very far removed from us.

Please, respond quite soon with a few friendly lines.

Most amicably your | sincerely devoted | G. Wiedemann.

RI MS JT/1/W/42

---

1. *last friendly lines*: letter missing.
2. *translation of your work*: the German translation of *Heat* (1865); see letter 2296, n. 6.
3. *Bohn*: Johann Conrad Bohn.
4. *Vieweg*: Eduard Vieweg.
5. *Mrs Helmholtz*: Anna Helmholtz.
6. *my wife:* Clara Wiedemann.
7. *mention the ladies*: both Anna Helmholtz and Clara Wiedemann were experienced translators. Despite this, the intention was that their translation of Tyndall's *Heat* (1865) would appear under their husbands' names.
8. *in Geneva*: probably when Tyndall was in Geneva in August 1865 for the meeting of the Société Helvetique des Sciences Naturelles (see letter 2270, n. 4).
9. *Clausius has received a very nice call to Vienna*: see letter 2387, n. 12.

From Rudolf Clausius 20 April 1866 2411

ZÜRICH, 20/4/66.

Lieber Tyndall,

Vieweg hat wieder an mich einen Brief in Bezug auf die Uebersetzung Deines Buches gerichtet. Ich weiss eigentlich nicht, wie er darauf kommt, gerade durch mich die Correspondenz zu führen, da ich an den früheren Arrangements nicht betheiligt war; indessen, da er einmal den Brief an mich geschickt hat, so übersende ich ihn dir einliegend.

Was die Rechtsfrage anbetrifft die in den Worten liegt „der Herr Tyndall mir sowohl, wie Hrn Bohn die Ermächtigung zur Uebersetzung gegeben hatte", so kann ich darüber gar nichts sagen, da ich gar nichts weiss, was über die Sache mit Vieweg verhandelt ist. Du wirst darüber leicht entscheiden können. Nur will ich erwähnen dass die Honorarangelegenheit welche Vieweg auch wieder berührt, durchaus nichts mit der Rechtsfrage zu thun hat. Ich habe nur zu einer Zeit, wo ich glaubte, dass die Herausgabe der Uebersetzung in vollem Gange sei, Wiedemann gefragt ob er nicht veranlassen könne, dass Vieweg Dir ein Honorar zahle, worauf Wiedemann mir sofort schrieb, dass die Herausgabe der Uebersetzung überhaupt unterbrochen sei. Diesen

Punct kannst Du also als ganz ausserhalb der Sache stehen betrachten. Wie Du aus Vieweg's Briefe ersehen wirst, verlangt Vieweg nicht ein ausschliessliches Uebersetzungsrecht, sondern wünscht nur ermächtigt zu werden, neben der etwa erscheinenden anderen Uebersetzung, für sich auch eine Uebersetzung herauszugeben.

Indem ich diese ganze Frage in wieweit es Dir zulässig und mit Deinen mit Bohn getroffenen Verabredungen vereinbar und nach Bohn's Benehmen in der Sache angemessen scheinen mag, Viewegs Wunsch zu erfüllen, ganz Deiner eigenen Entscheidung überlasse, will ich nur einen Punct erwähnen. Vieweg gilt in Deutschland als der beste Herausgeber physikalischer Werke, und wenn Du ihm die Herausgabe der Uebersetzung gestattest, so glaube ich dass Du darauf rechnen kannst, dass eine gute und schön ausgestattete Uebersetzung erscheint.

Uebrigens glaube ich, wenn es Dir auch augenblicklich lästig sein mag, in den Streit zwischen Vieweg und Bohn hineingezogen zu werden, so kann es Dir doch auf der anderen Seite nur erfreulich sein, dass sich, nachdem eine Uebersetzung als mangelhaft befunden ist, sofort ein Mann wie Vieweg bereit erklärt für eine neue bessere Uebersetzung zu sorgen. Zu einem solchen Erfolge deines schönen Buches kann man Dir gewiss gratuliren. Auch die Bemerkung von Vieweg ist gewiss richtig, dass es für die Verbreitung Deines Buches in Deutschland nicht schädlich sondern vielmehr nützlich sein würde, wenn statt einer Uebersetzung zwei erscheinen.

Indessen, wie gesagt, die Entscheidung über die Sache überlasse ich ganz Dir. Ich denke mir, dass es einen grossen Einfluss haben wird, wie Du selbst beim Durchlesen der Uebersetzung von Bohn findest. Wie Du am Schlusse von Viewegs Briefe sehen wirst, lässt er Dich bitten ihm <u>direct</u> zu antworten.

Mit herzlichen Grüssen | Dein | Clausius.

ZURICH, 20/4/66

Dear Tyndall,

Vieweg[1] has sent me another letter regarding the translation of your book.[2] I do not really understand why he decided to correspond with you through me, since I was not part of the earlier agreements; however, since he already sent his letter to me, I am forwarding it to you.[3]

Regarding the legal question contained in the following words: "as Mr Tyndall had given <u>me,</u> as well as Mr. Bohn,[4] his authorisation for translation", I cannot say anything pertinent since I do not know anything about what has been agreed upon with Vieweg. You can easily make the decision yourself. I would just like to mention that the issue of the honorarium, which Vieweg again touches upon, has absolutely nothing to do with the legal question. At a time when I believed that publication of the translation was well underway, I asked Wiedemann whether he could not induce Vieweg to pay you an honorarium. Wiedemann immediately replied that the publication of the translation had in fact been suspended.[5] You can consider this point as quite independent of the main issue. As you will see in Vieweg's letter, he does not demand exclusive translation rights, but instead would like to have your authorisation to publish a translation, in addition to the other one that may potentially be published.

I leave it to you to decide in how far you find it justifiable to fulfill Vieweg's wish, considering your arrangements with Bohn and given Bohn's behaviour in this affair. I want to point out one thing, however. Vieweg is considered in Germany as the best publisher in the field of physics. Therefore, if you give him permission to publish the translation, I believe you can expect it to be a good and beautifully finished translation.

Beside that, I believe that even if at present it is annoying for you to be involved in the conflict between Vieweg and Bohn, you can on the other hand only be pleased that after one translation was found deficient, a person like Vieweg immediately announced his intention to arrange a new, better one.[6] One can certainly congratulate you on this success of your fine book. I certainly also find Vieweg's observation correct that for the distribution of your book in Germany, it would not be damaging, but rather beneficial, if not one, but two translations were published.

In the meantime, as already mentioned, I leave the decision in this matter entirely to you. I think your own impressions after reading Bohn's translation will have great influence what you decide. As you will see at the end of Vieweg's letter, he asks you to reply to him directly.[7]

With best wishes | Your | Clausius.

RI MS JT/1/TYP/7/2292–3
LT Typescript Only

1. *Vieweg*: Eduard Vieweg.
2. *translation of your book*: the German translation of *Heat* (1865); see letter 2296, n. 6.
3. *forwarding it to you*: enclosure missing.
4. *Mr. Bohn*: Johann Conrad Bohn.
5. *Wiedemann immediately replied . . . fact been suspended*: Wiedemann had reported this to Tyndall also (see letter 2410).
6. *better one*: previous correspondence describes the deficiencies of Bohn's attempted translation.
7. *to him directly*: if Tyndall replied to Vieweg, the letter is missing.

From George Biddell Airy 21 April 1866 2412

Royal Observatory, Greenwich, | London, SE | 1866 April 21

My dear Sir,

With my best thanks I have to acknowledge receipt of your paper "On Caloresence",[1] which I have run through—I need not say superficially—but with great interest. Beginning at the motto, I congratulate you on having attained that happy quotation from Lucretius.[2] {I find that one of my sons, an admirer of Lucretius, had written in the margin opposite another passage, V. 855–877, "Darwinism".}[3] Then the measures of the radiation appear unexceptionable, and the proportion of invisible to visible astounding. I am edified by the notion that black rays of heat have a great admiration for black coats: it seems to be well supported.

Do you remember, at the Vicechancellor's[4] dinner at Cambridge,[5] my asking you what glass would admit solar-heat into a hothouse and prevent plant-heat from escaping? I am confident that (with the cooperation of a good man of business) a lucrative patent might be based on this. I do not despair of seeing the "Tyndall Lights" adopted in every hothouse in the kingdom.[6]

I am, my dear Sir, | Yours ever truly | G B Airy

Professor Tyndall

RI MS JT/1/A/39

RI MS JT/1/TYP/1/29

---

1. *"On Caloresence"*: see letter 2333, n. 3.
2. *from Lucretius*: Tyndall opened his paper 'On Calorescence' by quoting from Lucretius's *De Rerum Natura* ("On the Nature of Things"): 'Forsitan et roseâ sol altè lampade lucens | Possideat multum cæcis fervoribus ignem | Circum se, nullo qui sit fulgore notatus, | Æstiferum ut tantum radiorum exaugeat ictum', which he believed 'reads like divination'.
3. *"Darwinism"*: Airy's son, Hubert Airy (1838–1903), was an English physician and a pioneer in the study of migraines, which his father himself suffered. In the second half of Book V of *De Rerum Natura*, Lucretius envisioned how life first emerged on the earth and how humans developed from nomadic hunters to city-dwellers. This prehistory excludes divine agency of any kind. The fertile young earth naturally sprouted with life, 'of face and members most grotesque; | Monsters half-man, half-woman, not from each | Distant, yet neither total; shapes unsound, | Footless, and handless, void of mouth or eye, | Or from misjunction, maim'd, of limb with limb; | To act all impotent, or flee from harm, | Or nurture take their loathsome days t' uncouth . . . ' (Lucretius, *On the Nature of Things. A Philosophical Poem, in Six Books*, trans. J. S. Watson (London: Henry G. Bohn, 1851), pp. 435–6). Most of these organisms perished, but a minority survived. Among the

Victorians, this account won admiration as it seemed to anticipate a Darwinian view of nature, providing an early account of design without teleology. See F. M. Turner, 'Lucretius Among the Victorians', *Victorian Studies*, 16 (1973), pp. 329–48.

4. *Vicechancellor's*: Henry Wilkinson Cookson.
5. *at Cambridge*: Tyndall delivered the annual Rede Lecture in the Senate House at the University of Cambridge ('On Radiation') on 16 May 1865, later published as *On Radiation*. The dinner with the Vice Chancellor was on 15 May 1865 (Journal, RI MS JT/2/13c/1289).
6. *a lucrative patent . . . in the kingdom*: referring to Tyndall's demonstration of greenhouse gases in 1859 (R. Jackson, N. K. L. Kaalund, and D. A. Finnegan (eds), *The Correspondence of John Tyndall, Volume 7: The Correspondence, March 1859–May 1862* (Pittsburgh: University of Pittsburgh Press, 2019). Tyndall's response to Airy about a patent for glass that 'would admit solar-heat into a hothouse and prevent plant-heat from escaping' is not known, but Tyndall in general throughout his career focused on conducting and promoting pure science without regard of its practical value, and did not care for patents (see *Ascent of John Tyndall*, p. 313). On Tyndall and pure versus practical science, see M. D. Barton, 'The "efficient defender of a fellow-scientific man": John Tyndall, Darwin, and Preaching Pure Science in Nineteenth-Century America' (MA thesis, Montana State University, 2010).

From Harry W. Cook[1] 21 April 1866 2413

Leghorn[2]—April 19.

Sir.

Will you permit me to present to your judgment two philosophical instruments which I have invented and which several of my scientific friends here and in Paris assure me to be both new and of considerable interest. I had thought of describing them to the Academy of Science at Paris[3] but should infinitely prefer (if you would not mind the trouble of examining them) that they should be laid before you not only as a chief of the Royal Institution but as the person best able to decide upon their merits—The first is an instrument for comparing the value of static and dynamic forces—this instrument I have described to several gentlemen at different times in London—the other is for measuring the velocity of the force called gravity—.The former is extremely simple—a plaque of metal of the form indicated in the drawing pivoting at M and kept in equilibrium by the counterweight C acts as a balance to compare any perpendicular pressure applied at P with a horizontal force applied at R—the perpendicular force will be an ordinary weight to receive which there is at P a small table supported on a pin or single central leg the horizontal force at R will be the blow of a pendulum B to receive which there is a hemispherical bosse—the centre of this hemisphere and the pin P are precisely equidistant from M the point at which the plaque pivots so that the machine may act as an impartial balance between a pressure on the table and a pressure against the bosse, when the instrument is in equilibrium the counterweight C almost touches but is not absolutely in metallic contact with E another ball of metal fixed rigidly above it and the two balls are in connection with the opposite poles of a galvanic battery so that the faintest pressure on P will instantly result in the passage of a current which passage of current will cease as soon as the pressure on P is removed or counteracted if then the force exerted on R by a blow of the pendulum be as great as the force of a weight placed upon the table P the current will cease.—Let an electro-magnetic arrangement be introduced into the system and let the magnet of this system be charged with the greatest weight it can support and then the cessation of the current will be nolified by the fall of this weight.—

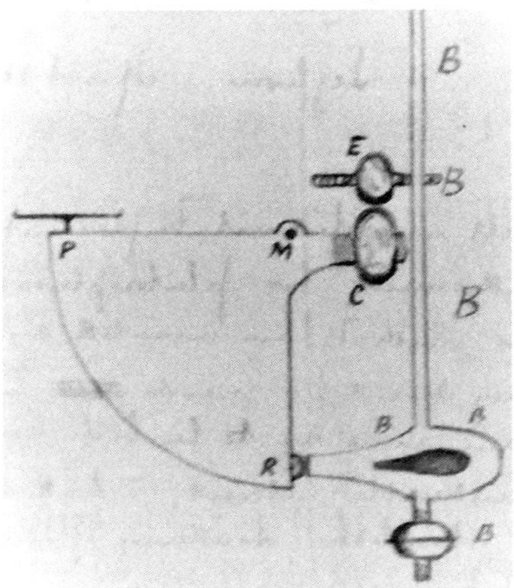

The form of the pendulum should be such that when it hangs in equilibrium it should touch but not press against R and its centre of gravity should be on a level with the centre of R—its weight & its velocity when dropped from some point should be known, let its weight be P and its velocity V then the problem would be to discover the smallest weight which would prevent the blow from breaking the contact CL and the largest weight under which the contact would be broken. the mean of these two weights would represent very closely the absolute static value of the momentum PV—three or four experiments with various velocities and weights would establish some formulae which would reduce the powers to a matter of calculation and would not fail I think to be of considerable use—the error arising from the inertia of the parts of the instrument might be connecting by attaching blocks of metal that without disturbing its equilibrium should double the weight of the plaque of metal and repeating the experiment under thus altered condition—

By the aid of this instrument and another of which I propose now to give a description I have obtained too I fancy a means of determining not perhaps accurately but very approximatively the velocity of the force of gravity

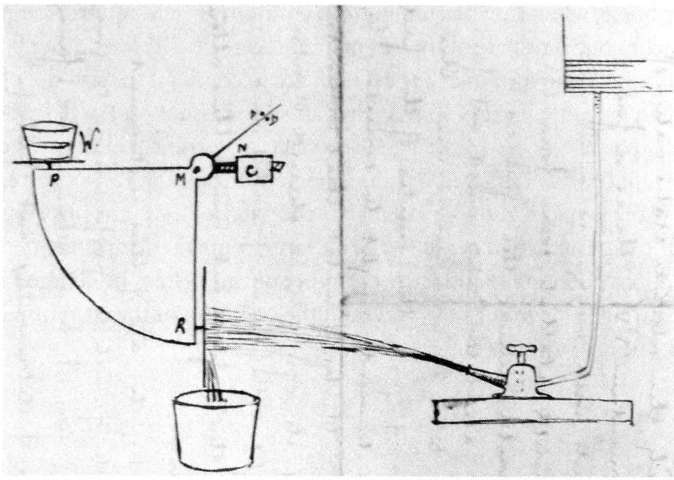

N.B. it would perhaps be better that the water were applied at once as:—

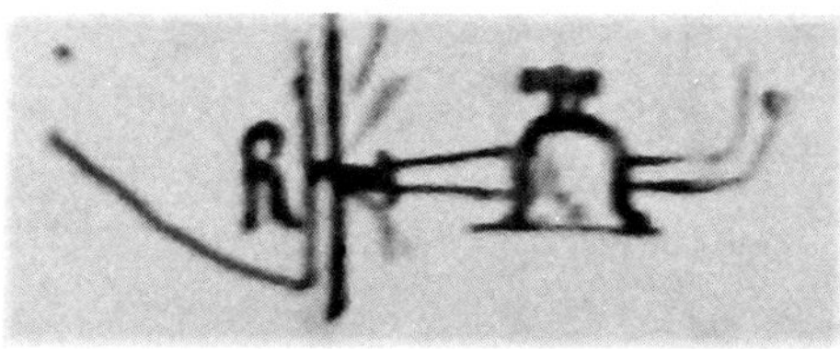

and a separate calculation of the amount of water used made afterwards.

In this purpose a balance similar in form to that employed in the other instrument is provided on the one hand as before with a table to support a weight but the bosse which in the last case received the blow is replaced by a second table rather large and facing horizontally a needle N which vibrates between the two points D-D serves to show when the machine is in equilibrium

Against the lower table R a jet of water is directed from a nozzle O so as to infringe upon it in a horizontal direction and with a calculable velocity V, the upper table is now loaded until the needle vibrating equally between the two points shows that equilibrium is re-established—call this weight W—

Let now the water which during a minute or some determinate time T has inpinged upon the table R be collected and carefully weighed and let its weight be expressed by P—Now in the time T a force P.V. has been applied to R—by means of the other instrument we obtain W' the static value of this momentum, we can therefore say that in the time T a force W' has been applied to R. which as the machine has been always in equilibrium involves that the weight W has in the time T exerted a force W' in the time

T' occupied for the exertion of gravity between it & the centre of the earth it would exert its own ordinary static weight W. therefore as W':W::T:T' in other words the time occupied by the weight in exerting itself is W·T/W'—this result contains of course errors derived from various sources but it would if obtained put an end to the present uncertainty as to whether gravity requires a second or a century for its exertion.—if you would not mind the trouble of letting me know what you think of these ideas I should be very much obliged to you, the more so that I have many other things in my head of the probable value of which I shall judge from the verdict you pass upon these the simpler results of my cogitations—With a thousand apologies for having occupied your time should these instruments be valueless.

I Remain | Sir Your obedient Servant | Harry Whiteside Cook | Villa Regini—Livorno[4]—| April the 21st/66.

RI MS JT/1/C/45

1. *Harry W. Cook*: Harry Whiteside Cook (dates unknown), a British inventor from, at different times, Manchester; Rossall School in Lancanshire; Angel Court, London; Ovington Square, Middlesex; and Niederdorf, Tyrol (Austria).
2. *Leghorn*: 'Leghorn' is the English name for Livorno, a coastal city in western Italy.
3. *Academy of Science at Paris*: l'Académie des sciences, or French Academy of Sciences, founded in 1666.
4. *Villa Regini—Livorno*: a coastal city in western Italy.

From Geraldine Jewsbury[1] 22 April 1866 2414

5 Cheyne Row. Chelsea[2] | S.W | April 22. 1866. Sunday

dear Dr Tyndall

Mrs. Carlyle[3] is dead she died yesterday at four o'clock quite suddenly[4]—in the Brougham[5]—she was taken to St. George's Hospital[6]—poor Mr Carlyle is to come home tomorrow & what will become of him—poor man poor man—

Your telegram about the Inaugural Lecture[7] did her so much good—you don't know how grateful she was for it—it is such a comfort to think how pleased she was to get it after the horrible anxiety she had been in. Mr Forster,[8] Mr Blunt[9] and Mr Froude,[10] have done everything possible and necessary.

She was brought home last night—I am only here till he comes home—Dr Quain[11] says it was paralysis—I don't know—but the end is the same by whatever way it came—I do not realize it yet—but there is the rest of my life for that—

Yours very truly | G. E. Jewsbury

RI MS JT/2/10/396
RI MS JT/2/13c/1331

---

1. *Geraldine Jewsbury*: Geraldine Endsor Jewsbury (1812–80), a novelist and book reviewer, friend of the Carlyles (*ODNB*).
2. *5 Cheyne Row. Chelsea*: the Carlyles' London address since 1834.
3. *M*[rs]. *Carlyle*: Jane Welsh Carlyle.
4. *she died yesterday at four o'clock quite suddenly*: on his return to London from the Isle of Wight (see letter 2409, n. 7), Tyndall learned from the *Times* that Jane Welsh Carlyle had suddenly passed away on 21 April 1866: 'At Southampton on my return I picked up a copy of the Times, and there among other things met the announcement "Sudden death of Mrs Carlyle." I turned from it with a kind of bewildered stupor, and read mechanically one of the leading articles' (RI MS JT/2/13c/1330). In his 'Personal Recollections of Thomas Carlyle (1890)', first published in the *Fortnightly Review* in January 1890 and published again in J. Tyndall, *New Fragments* (London: Longmans, Green, and Co., 1892), pp. 347–97, Tyndall wrote: 'I sped to London, and on my writing-table found a note from Miss Jewsbury. Carlyle had arrived in Chelsea. "For Heaven's sake," said my correspondent, "come and see the old man! he is utterly heart-broken." In a few pathetic words Leslie Stephen has told the story of her death: "Mrs. Carlyle had asked some friends to tea on Saturday, April 21. She had gone out for a drive with a little dog; she let it out for a run, and presently she was found sitting with folded arms in the carriage, dead"' (p. 368). The Carlyles' physician Richard Quain (see n. 12) recorded the cause of death as 'spinal disease' (R. Ashton, *Thomas & Jane Caryle: Portrait of a Marriage* (London: Catto & Windus, 2002), p. 444).
5. *Brougham*: a one-horse closed carriage, with two or four wheels, for two or four persons (*OED*).
6. *St. George's Hospital*: founded in in 1733, St. George's Hospital was a large public hospital located on Hyde Park Corner in London.
7. *your telegram about the Inaugural Lecture*: letter 2399.
8. *M*[r] *Forster*: probably John Forster (1812–76), an English biographer and literary critic (*ODNB*), who is mentioned in letter 2402.
9. *M*[r] *Blunt*: the Reverend Abel Gerald Wilson Blunt (1827–1902), who in 1860 was appointed rector of St. Luke's, Chelsea, the Carlyles' parish (R. Blunt and J. M. Craig, *Memoirs of Gerald Blunt of Chelsea, His Family and Forebears* (London, 1911)).
10. *M*[r] *Froude*: James Anthony Froude (1818–1894), an English historian and family friend of the Carlyles (*ODNB*).
11. *D*[r] *Quain*: Richard Quain (1816–98), the Carlyles' physician (*ODNB*).

From Heinrich Debus 25 April 1866 2415

Wednesday | April 25. 1866.

My dear John,

Excuse me for not writing to you about Bohns[1] translation[2] at an earlier date.[3] The fact is I happen to be very busy just now with a paper for the journal of the chemical Society[4]

a) Bohn does not appear to be quite conversant with the idioms of the English language. His translation is not always correct.

b) He has translated as much as possible verbatim;—this wont do. Everywhere you feel that you read a translation.

c) His German is very crooked and wants elegance and fluency

d) No patching will do but the whole must be rewritten

This holds good for the first 10 or 12 sheets. If you wish it I will read the whole manuscript and see whether or not he gets better towards the end of the book. I have made some pencil marks in different places. It is probable that I shall be in town in about a fortnight.[5] Let me know whether I shall return the manuscript by post or bring it with me to town. If you desire me to read the whole manuscript then I must ask you to grant me some more time. My judgment may be too severe and before you act on it it will perhaps be well to give a few sheets to Max Müller[6] of Oxford and ask him what he thinks of the style.

When I knew Bohn as a student at Marburg[7] he never appeared to me particularly bright. Take the first few words: "Wärme betrachtet als Bewegung"[8] This is not only very indefinite but according to the sense of the words wrong. It ought to be:

Die Wärme | dargestellt als | Bewegungserscheinung | von John Tyndall.[9]

Viewegs[10] letter to Clausius is badly written too and he wants somebody to look after his composition. I dont like the spirit of the letter. There is a contradiction in it. First he says that you have given him as well as Bohn the right to translate, and a little further on he remarks that he will stick to his money offer if you will give him the right to translate, by return of post.

Please write to Vieweg in English and if he replies in German then let me see his replies. Clausius is evidently in favour of Vieweg. Two translations wont do in Germany. Vieweg is anxious to get the translation into his hands, and if he does, I think he will take good care to then shut up any publisher who should venture to publish any other translation. It appears to me that you gave permission to <u>Vieweg and Bohn</u>, both considered as a whole, but if Vieweg and Bohn cannot manage to produce a translation <u>together</u>, then I think their right falls to the ground and neither of them can claim the right of translation alone. I say, so matters appear to me, you of course will know all about this point. Wiedemanns offer is by far the best thing I have yet heard about the matter. The Ladies[11] I take it, will find it rather more difficult to produce a German translation comparable to the original, than they do imagine. I assure you it is not easy to turn your elegant English into German of similar character. It requires facility and experience in writing. But then their husbands[12] will have to look after them and help where the Ladies *[wont]* do. And I should say better translators than Helmholtz and Wiedemann you could not have. I think I should accept Wiedemanns offer if I could in an honorable way separate from Bohn.

Always yours | H Debus

RI MS JT/1/D/36

1. *Bohns*: Johann Conrad Bohn.
2. *translation*: the German translation of *Heat* (1865); see letter 2296, n. 6.
3. *at an earlier date*: Tyndall must have sent to Debus, who was German, pages of Bohn's translation to find out its merit.
4. *paper for the journal of the chemical Society*: H. Debus, 'On the Constitution of some Carbon-compounds', *Journal of the Chemical Society*, 4 (1866), pp. 17–30, 256–89.
5. *in town in about a fortnight*: Debus taught chemistry at Queenwood College near Stockbridge, Hampshire, England, southwest of London, until 1867 when he began teaching at Clifton College, Bristol.
6. *Max Müller*: Friedrich Max Müller (1823–1900), a German-born philologist and Orientalist and often considered one of the founders of *Religionswissenschaft*, the academic study of religions (*ODNB*).
7. *Marburg*: Debus studied at Marburg in Germany from 1845–8 where he met Tyndall, Edward Frankland, and Thomas Hirst.
8. *"Wärme betrachtet als Bewegung"*: Heat considered as motion (German).
9. *Die Wärme | dargestellt als | Bewegungserscheinung | von John Tyndall*: Heat | represented as | an effect of motion | by John Tyndall (German).
10. *Viewegs*: Eduard Vieweg.
11. *Ladies*: Anna Helmholtz and Clara Weidemann (see letter 2296, n. 6).
12. *their husbands*: Hermann Helmholtz and Gustav Weidemann.

To Rudolf Clausius 28 April [1866][1] 2416

28th April

My dear Clausius

Hofmann[2] came to me a few days ago at Vieweg's[3] request and read a letter containing a long and very strong statement of his relations to Bohn[4] & to the translation of my book.[5] He is exceedingly anxious to undertake the publication of the work, and Wiedemann informs me[6] that Helmholtz & himself would look after the translation and bring the book out under their authority.

I received Bohn's manuscript and submitted it to some friends here[7]—I cannot conclude—otherwise than that it is a bad translation and unfit to be published. Vieweg Hofmann assures me is a very cultivated man, and he through curiosity took up a sheet of Bohn's translation after it had been printed. He was astonished to find that even reading the matter three times over did not enable him to understand it. Not only was the meaning unclear but the style was also exceedingly bad. He turned to the original and found matters altogether different. The translation had in fact utterly ruined the book.

I am astonished to find that Bohn himself is unconscious of these defects. I have written to him[8] in the most civil manner possible advising him to come to some arrangement with Vieweg, and to leave the translation & publication entirely in Viewegs hands. I also told Hofmann that Vieweg might transfer those ten pounds which he offered to me to Bohn. Bohn I believe wants money, and these additional ten pounds may possibly satisfy him.

And now I must tell you why I wanted the sketch of your researches. Three foreign members were to be chosen for the Royal Society.[9] A considerable number were proposed, and out of them five were selected. You were one of the five. Well it came to a vote—there are many in the Council (I am not now a member) who always vote for older men, and hence Gustav Rose[10] was chosen, Donders of Utrecht[11] had also several personal friends on the Council, and he was chosen. You and Rieman[12] also received a large number of votes, and you were both equal. You were then voted for a second time, and the consideration that Rieman is dying at Pisa and would probably never have another chance of being elected carried the point in his favour. But you were exceedingly well supported and we must hope for better fortune next time.

Kindest regards to Mrs. Clausius[13] & love to the children.[14]

Ever yours | John Tyndall

RI MS JT/1/T/232

1. *[1866]*: the year is given by content, concerns over Bohn's translation of Tyndall's *Heat* (1865).
2. *Hofmann*: August Wilhelm Hofmann (see letter 2407, n. 6).
3. *Vieweg's*: Eduard Vieweg.
4. *Bohn*: Johann Conrad Bohn.
5. *translation of my book*: the German translation of Tyndall's *Heat* (1865); see letter 2296, n. 6.
6. *Wiedemann informs me*: see letter 2410.

7. *some friends here*: Heinrich Debus was one such friend (see letters 2415).
8. *I have written to him*: letter missing.
9. *Three foreign members were to be chosen for the Royal Society*: for the notice of the three elected foreign members, see *Roy. Soc. Proc.*, 15 (1867), p. 189.
10. *Gustav Rose*: Gustav Rose (1798–1873), a German mineralogist and President of the German Geological Society from 1863–73 (*CDSB*), for which the rose-colored mineral roselite is named.
11. *Donders of Utrecht*: Franz Cornelius Donders (1818–89), a Dutch ophthalmologist and professor of physiology at Utrecht University (*CDSB*).
12. *Rieman*: George Friedrich Bernhard Riemann (1826–66), a German mathematician and professor at the University of Göttingen most known for laying the foundations of the mathematics of general relativity. He was the son of a Protestant minister and dedicated Christian, who saw his life as a mathematician as serving God. He fled Göttingen in 1866 when the armies of Hanover and Prussia clashed, and died of tuberculosis in Italy (*CDSB*).
13. *Mrs Clausius*: Adelheid Clausius.
14. *the children*: see letter 2270, n. 5, and letter 2387, n. 15.

From Johann Conrad Bohn 29 April 1866 2417

Giessen 29 April 1866.

Sehr verehrter Herr Tyndall!

Ihr Brief, der mir anräth meine Uebersetzung Ihres Buches nicht erscheinen zu lassen, hat mich auf das Tiefste betroffen und geschmerzt. Dass ich eine beträchtliche Zeit auf die Herstellung der Uebersetzung gewendet habe und nun keinen Lohn dafür finden soll, ist unter meinen Verhältnissen sehr empfindlich, aber es ist noch nicht das schlimmste. Wenn die Uebersetzung nicht von mir, sondern von einem Anderen erscheint, nachdem viele Personen, an deren guter Meinung mir alles gelegen sein muss, darum wissen, dass ich eine solche gemacht habe, so gibt das meiner Reputation einen Stoss, von dem sie sich nie wieder erholen wird, um so mehr, da ich nicht die Gelegenheit haben würde jenen Personen meine Leistung sehen zu lassen, damit sie ein selbständiges Urtheil darüber sich bilden können. Sie versuchen am Schlusse Ihres Briefes mich darüber zu trösten, indem Sie die Ansicht aussprechen ein Misserfolg als Uebersetzer thue der wissenschaftlichen Reputation eines Mannes keinen Eintrag; allein ich kann diese Meinung nicht theilen. Man wird sagen, dass wenn ich nicht einmal im Stand gewesen sei die Arbeit eines Andern gut zu reproduziren, noch weniger zu erwarten stehe, dass meine eigene Schöpfungen etwas werth seien. Für meine ganze Laufbahn wird diess eine unübersteigliche Barriere sein,—man wird mir dieses stets entgegenhalten, wenn ich versuche vorwärts zu kommen.

Ich hielt bis heute meine Arbeit nicht für schlecht, sondern glaubte Ihre

Gedanken getreu wiedergegeben zu haben. Ihre und Ihrer Freunde abweichende Meinung muss diesen Glauben erschüttern. Allein nicht so *[weit]*, dass ich verzweifeln sollte, der Aufgabe genügen zu können eine gute Wiedergabe Ihres Buches in deutscher Sprache zu machen. Ich fühle und weiss, dass ich von den behandelten Gegenstnden solide behandelten Gegenst Buches in Deutscher Sprache zu machen. Ich fe und Ihrer Freänden solide Kenntnisse habe; Sie selbst haben, als ich über einzelne Punkte schrieb, die *[mir]* Zweifel erregten, diess in früheren Briefen anerkannt. Ihr Ausspruch war also mehr als höfliche Vermuthung, da Sie Material vorliegen hatten um sich über meine speciellen Kenntnisse zu unterrichten. Ich weiss ferner, dass ich meine Muttersprache verstehe und in derselben mich deutlich auszudrücken vermag. Diess aber sind die Erfordernisse für die Herstellung einer guten Uebersetzung. Zur Erklärung der Thatsache, dass meine Arbeit weder Ihren Beifall, noch den Ihrer Freunde fand, kann also nur der Umstand dienen, dass ich nicht die nöthige Sorgfalt und Mühe angewendet habe. Ich glaubte diess gethan zu haben; allein da ich sehr rasch arbeitete und Sie nun meine Arbeit tadelnswerth finden, so muss ich hierbei mich geirrt haben. Ich wiederhole, dass es für mich eine Lebensfrage sein kann die Uebersetzung von meiner Hand erscheinen zu sehen und ich werde kein Opfer scheuen diess durchzusetzen. Ich bin bereit mein M.S. ganz umzuarbeiten und die grösste Sorgfalt auf Herstellung eines guten und richtigen Textes zu verwenden. Ich bin bereit mir jede Controle der Arbeit vor ihrer Veröffentlichung gefallen zu lassen. Ich bin bereit mit dem geringsten Lohne für meine Mühen vorlieb zu nehmen und den ganzen Rest des Buchhändlerhonorars Ihnen für diese Controle, welche Sie entweder selbst vornehmen oder von einem Anderen, zu welchem Sie mehr Vertrauen als zu mir haben, ausführen lassen können, *[zur]* Verfügung zu stellen. Herr Fischer hat mir 16 Thl p. Bogen geboten und auf meine Mittheilung hin, dass Sie nun, auf Rath Ihrer Freunde ebenfalls Honorar beanspruchen zu sollen glauben, hat derselbe vorgeschlagen, bei der 2$^{\underline{n}}$ Auflage Ihnen 20–30 Louis d'or zu zu stellen, welche zur Hälfte von mir, zur Hälfte von ihm zu tragen seien. Vielleicht wird schon für die erste Auflage eine ähnliche Bedingung zu erzielen sein, jedenfalls ist es möglich, da ich, wie schon bemerkt, *[mit]* einem Minimum von Honorar mich zu begnügen bereit bin, den Geldverlust von Ihnen abzuwenden, den Sie dadurch erlitten zu haben schrieben, dass Sie auf die Möglichkeit eine *[Summe]* für das Uebersetzungsrecht zu erhalten verzichtet haben, um mich als Uebersetzer zu erhalten.

Wenn ich meine speciellen wissenschaftlichen Kenntnisse und meine sprachliche Fertigkeit noch so gering anschlage, so kann ich doch nicht bezweifeln, dass durch Ihre direct oder indirect ausgeführte Durchsicht und Correctur meiner Arbeit ein allen Anforderungen genügendes Buch zu Stande kommen wird und muss. Es würde für mich nichts demüthigendes haben, wenn auf dem Titel der Uebersetzung zu lesen wäre, dass dieselbe unter Beihilfe und Aufsicht des Verfassers zu Stande gekommen sei,—wie das häufig bei derartigen Werken der Fall ist.

Lassen Sie mir die Hoffnung zu glauben, dass bei erneutem Ueberlegen Sie meinen Vorschlag für annehmbar halten oder einen andern zu machen sich bereit finden, der das Erscheinen der Uebersetzung unter meinem Namen möglich macht und mir nicht den Schimpf anthut, dass meine Arbeit ungenügend und ich nicht für befähigt gefunden worden sei, von einem Werke meiner Wissenschaft eine Uebersetzung zu liefern. Bei dem hohen Werthe den ich auf diese Angelegenheit lege und legen muss, werden Sie mir vielleicht verzeihen und es nicht indiscret finden, wenn ich mir erlaube Sie daran zu erinnern, dass Sie zu wiederholten Malen, und noch am 9^n April mir in der förmlichsten Weise gütigst die unbedingte Ermächtigung zur Herausgabe der Uebersetzung ertheilten. Dass ich auf diese Zusage hin bedeutende Opfer an Zeit gebracht habe und mich und meinen Ruf so fest engagirte, dass die verderblichste Compromittirung entstünde, wenn ich nicht im Stande sein sollte das Begonnene auszuführen. Und endlich darf ich Sie wohl auch noch bitten sich nicht der Erwägung zu verschliessen, dass Ihr gegenwärtiges Urtheil über mein M.S. und ebenso das Ihrer Freunde kein ganz vorurtheilsfreies sein kann. Durch das Auflösen des <Vertrages> <*several lines missing*> daran setzen wird das Unternehmen in seiner Hand zu haben; dass er in seinem Misswollen gegen mich, mit dem er im Streite lebt, ein Opfer nicht scheut und in der Wahl seiner Mittel, gelind gesagt, äusserst rücksichtslos gewesen ist und noch sein wird.

Indem ich Ihnen nochmals die Eingangs dieses Briefes erwähnten Umstände dringend an's Herz lege, und Sie bitte keinen für mich gerade zu verderblichen Entschluss endgültig zu fassen, verbleibe ich mit der ausgezeichnetsten Hochachtung

Ihr ergebenster | J C Bohn

Giessen 29 April 1866.

Dear Herr Tyndall!

Your letter,[1] which advises me not to let my translation of your book[2] appear, has most deeply affected and hurt me. That I have spent considerable time in producing the translation and should now not obtain any reward for it is in my circumstances very serious, but it is still not the worst. If the translation does not appear by me, but rather by someone else, after many people whose good opinion means everything to me know that I had done one, then that will give my reputation a blow from which it will never recover, and all the more so as I would not have the opportunity to let those persons see my work, so that they are able to form an independent judgment of it for themselves. You try, at the end of your letter, to console me about this by expressing the view that one failure as a translator does not do harm to a man's scientific reputation; but I cannot share this opinion. People will say that if I was not even able to reproduce someone else's work well, it would stand even less to be expected that my own creations were worth something. This will be an insurmountable barrier for my entire career;—people will always hold this against me whenever I try to improve my position.

Until today, I did not think my work was poor but instead believed that I had faithfully reproduced your thoughts. Your and your friends' divergent opinion could only shatter this belief. But not so *[far]* that I should despair of being able to satisfy the task of doing a good reproduction of your book in the German language. I feel and know that I have a sound knowledge of the subjects dealt with; you yourself have, when I wrote about individual points that were causing *[me]* doubts, acknowledged this in earlier letters.[3] Your remark was thus more than a polite assumption, as you had material to inform yourself about my special knowledge. I know furthermore that I understand my mother tongue and can express myself clearly in it. These, though, are the requirements for producing a good translation. Explaining the fact that my work met with neither your approval nor that of your friends can thus only serve the circumstance that I did not apply the necessary care and effort. I <u>believed</u> that I had done this; but as I was working very quickly and you now find my work blameworthy, then I must have been mistaken in this connection. I reiterate that it can be a vital matter for me to see the translation appear by my hand and I shall shun no sacrifice to achieve this. I am prepared to revise my M.S. entirely and to take the greatest care in producing a good and correct text. I am prepared to put up with any monitoring of the work before its publication. I am prepared to make do wiuth the most meagre wage for my efforts and to put at your disposal the entire remaining bookseller's fee for this monitoring, which you can either undertake yourself or leave to be carried out by another person in whom you you have more trust than in me. Herr Fischer[4] has offered me 16 Thalers p. sheet, and on my information that you now beleieve, acting on your friends' advice, you should also be entitled to a fee, he has proposed posting you 20–30 Louis d'or with the 2nd printing, which would have to be borne in half by me and in half by him. A similar term will perhaps have to be obtained for the first printing, at any rate it is possible as I, as noted already, am prepared to be content *[with]* a minimal fee to deflect the monetary loss from you that you wrote you had incurred by renouncing the possibility of receiving a *[sum]* for the translation rights in order to keep me as translator.

If I were to still eliminate my special scientific knowledge and my linguistic skill so modestly, then I cannot doubt that through the inspection and correction of my work that you carry out directly or indirectly, a book satisfying all requirement will and must come about. It would have nothing humiliating for me if it were to be read on the title page of the translation that it was come about with the assistance and under the supervision of the author,—as this is frequently the case with works of this sort.

Leave me the hope of believing that with renewed consideration you will think that my proposal is acceptable or that you will be prepared to make a different one that makes possible the appearance of the translation under my name and spares me of the affront of my work having been found unsatisfactory and I not qualified to supply a translation of a work in my science. Given the great importance that I do and must attach to this matter, you will perhaps forgive me and not find it indiscreet if I permit myself to remind you that on repeated occassions, and most formally on 9th April, you kindly granted me the unconditional authorisation to publish the translation. That following this promise I made significant sacrifices in time and commited myself and my reputation so firmly that the most disastrous embarrassment would arise if I should not be able to carry out what I have begun. And finally, may I possibly still ask you also not to close your mind to considering that your present judgement of my M.S. and that of you friends as well cannot be one entirely free of prejudice. By terminating the <contract> <*several lines missing*> will place in it to have the business in his hands; that in his ill-will towards me, with whom he lives in dispute, he does not shun one sacrifice and in the choice of his means has been, and will continue to be, to put it mildly, extrememly inconsiderate.[5]

In urgently entrusting the circumstances mentioned at the beginning of this letter to you again, and asking you not to make a final decision that will be positively disastrous for me, I remain with the most excellent esteem

Your very devoted | J C Bohn

RI MS JT/1/B/114

1. *Your letter*: letter missing.
2. *my translation of your book*: the German translation of *Heat* (1865); see letter 2296, n. 6.
3. *you yourself have . . . acknowledged this in earlier letters*: Tyndall's letters to Bohn are missing.
4. *Herr Fischer*: Theodore Fischer.
5. *extremeley inconsiderate*: it is not clear to whom Bohn was referring to here since the relevant lines seem to be missing, but he is probably referring to Eduard Vieweg.

From Mary Anne Coxe [April 1866][1] 2418

A Dream | Time and Eternity

I dream't—and as I dream't
My dream appeared too true—
The Earth shook—the rocks were rent
A man came forth to view.

An aged Man—his silver locks
Fluttered in the wind—
And as he lifted up his eyes
I saw that he was blind.
With reverence I approached him
Uncovering my head—
"Old Man! How art Thou yet alive
"When all the world are dead?
"I am the last of Mortals—
"And yet thou art not dead—
"What is thy name, Old man"?
"My name is Time"—he said!
"Before the Summer shall have fled
"I in my grave shall be;
"I shall be numbered with the dead
"For I am old you see".
"And wilt thou really die old man—
"I thought thou'dst live for ever—
"Surely Time wont pass away
"No—Surely—Never"!
"Young man thou art mistaken,
"Hast thou not read it so—
"When Heaven and Earth shall pass away
"Then Time shall also go".
"But then—I asked—"When thou art dead"
Who in thy place shall be?
The old man slowly raised his head
And said—"Eternity"!

The Jew and the Robber

"Thou son of an unbelieving Jew
Come give me thy gold for I hear t'is true
That thou hoardest it up in an iron chest
With the sparkling jewels thou lovest best!

Come—tarry not—quickly! I cannot wait
The day will soon dawn—it is getting late—
Come give me the key of thine iron chest
I will take the jewels that I love best!
"I'm a poor old man—the Jew replied—
Thy friends that told thee have surely lied—
I have no chest, and, I have no gold
My steps are feeble and I am old—

Oh! leave me alone in peace I pray"
But a twinkle appeared in his eyes of grey.

He had told a lie—and the robber knew it—
And he swore with an oath the Jew should rue it
He drew a knife, from his girdle so strong—
And he passed his fingers the blade along
"Do you see this knife old Jew—he said
One stroke with its blade and you are dead"
To a sense of his danger the Jew awoke—
He begged for mercy—he wished no stroke—
"Then give the key of the chest to me
Once for all it is better for thee".
"I am a poor old man the Jew replied
And I must confess that I have lied—
Here is the key of the Iron Chest
But take not the jewels that I love best".
"I'll take what I like—the robber replied
And hark'ye old Jew! if you move aside
From the place where I set you—to any part
This knife shall enter your hardened heart".
The robber he wended his way up stairs—
He opened the chest! what beautiful wares!
What splendid jewels of gold—what bars!
Again and again he blessed his stars!
x x x x x x x x x x x x x x x x
An hour hath passed, and the robber hath fled
Not a sound is heard in the chamber o'er head
With feeble steps the Jew enters the room
He lifts the lid—falls down in a swoon—!

x x x x x x x x x x x x x x x x

The Jew's Complaint

Oh! holy Moses, and Abraham's son—
He hath taken my gold and I am undone—
Oh! Jacob and Esau!—Oh! Peter and Paul!
He has taken my gold—my jewels and my all!

x x x x x x x x x x x x x x x x x

Jew's Complaint continued

He envoked all the Saints in the Universe
—When they heeded him not, he began to curse—
He cursed them from Peter right down to Paul
He cursed the patriarchs—saints and all.

Moral

Now the moral of my tale is this—
Never hoard your gold—
Or like the Jew before us
You may be robbed when old—
Never imitate the Miser
And deny yourself all pleasures.
But be free and open hearted
And never hoard your treasures!

Hope

There is hope for the weary and wayward ones
In trouble or oppressed—
They have but to come to Jesus
And He will give them rest.
There is Hope for the shipwrecked mariner
Tossed on the stormy wave
For God who is omnipotent
Hath He not power to save?
There is Hope for the wicked sinner
Hardened in every vice—
For God can melt the hardest hearts
As the Sun's rays melt the ice
There is Hope for the lowly Christian
Who follows his Lord in love—
The griefs and troubles of this world o'er
He shall have his reward above!

My dear Dr Tyndall

I have copied you out these lines[2] of Charley's.[3] There is really some imagination and poetic feeling in the first—and the "Old Jew" will I am sure give you a good laugh! I only trust myself with this scrap of paper, because I don't want you to consider me a bore—and—please dont think it necessary to "answer" this. I am so glad we have once more met "in the body"[4] and I can now realize you in your den.[5] How I wish you were in Edinburgh!

Ever believe me | very much Yours | M. A. C.

RI MS JT/1/TYP/1/292–3
LT Typescript Only

1. *[April 1866]*: the date is given as April 1866 as Tyndall had gone in late March and early April 1866 to Edinburgh with Thomas Carlyle and stayed with Coxe at her estate Kinellan (see n. 4).
2. *lines*: only one of the poems Coxe shared—'Hope'—seems to have been published (C. L. B. Cumming, *Hymns for the Sabbath: and Other Hymns* (London: Jarrold and Sons, 1863), p. 24). All the others are likely unpublished.
3. *Charley's*: Charles Lennox Bruce Cumming (1844–1912), a nephew and ward of Mary Anne Coxe, after the death of his father George Vanbrugh Cumming in 1856. Cumming attended Queenwood College along with his brothers where Tyndall taught them. He joined the Indian Civil Service in 1867 retiring in 1888 (*The India List and India Office List for 1905* (London: Harrison and Sons, 1905), p. 474). His brother Alick had died the year before (see letter 2252).
4. *once more met "in the body"*: Tyndall had stayed with Coxe at her estate Kinellan while in Edinburgh for Thomas Carlyle's speech at the University of Edinburgh (see letter 2375, n. 3).
5. *in your den*: probably his rooms at the RI in London.

From Georg Balthasar von Neumayer[1] 4 May 1866 2419

To Professor Tyndall, F.R.S., &c. |
Frankenthal, Pfalz a. Rhein, | May 4, 1866.

MY DEAR SIR,

You will perhaps remember that some two years ago I sent you some original observations[2] made at the Flagstaff Observatory, Melbourne,[3] bearing on the absorption of heat by aqueous vapour, and its relation to terrestrial radiation. At the time when I did myself the pleasure to call upon you at London, you stated that great pressure of other matters did not admit of your undertaking the task of discussing those observations; and in consequence of this I had the extensive labour of classifying and condensing so large a number of figures done in my bureau of computation, established with the object of publishing discussions on the magnetic and meteorological observations made by me during the years 1858–63.[4] In the annexed Tables, four in number, you receive the results of my labours in this direction, which I feel great pleasure in putting before you.

In order that you may be able to form an opinion as to the manner in which the observations were made, I subjoin the following remarks.

(1) The observations were made at all times of the day and the night whenever the sky was clear, *i.e.* the zenith perfectly free from clouds.

(2) The tension of aqueous vapour and the degree of relative humidity

were observed by August's psychrometer,[5] continually checked, however, by Döbereiner-Regnault's aspirator hygrometer.[6] The reductions were effected by Regnault's[7] Tables,[8] the thermometers being Kew standards.[9]

(3) The radiation-instrument consisted of a spirit minimum-thermometer of Kew, the bulb of which was placed and carefully adjusted in the focus of a parabolic reflector nicely polished and silvered, 6·4 inches wide and 2·4 inches deep*.[10] This reflector was put in a box filled with cotton and placed in a little house, keeping out the rays of the sun, but in such a manner that the zenith over the instrument remained perfectly free for a space of some 38°. The focus of the reflector was about 1½ feet above the surrounding ground (see introduction to 'Results,' page 4.)

(4) The psychrometer was placed close to this radiation-apparatus; and in order to have a control, a thermometer was placed within the little house (see annexed figure), the same being frequently compared with the Kew standard.

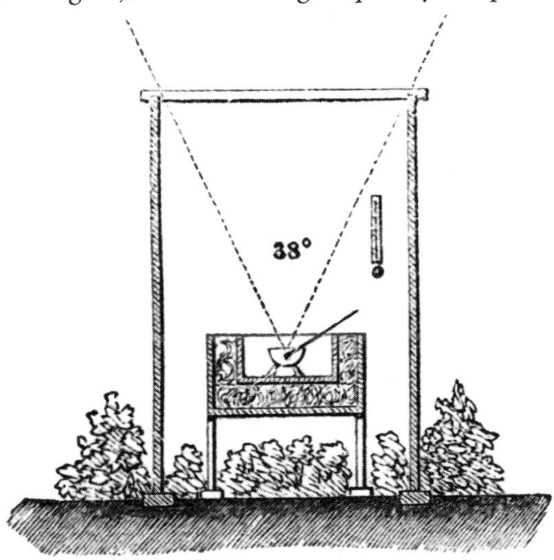

(5) Table I. contains in its first column the number of single simultaneous observations, the second column shows the mean pressure of aqueous vapour, the next the mean temperature of the air, and the last the mean difference between temperature of air and radiation, obtained by the instruments described. The results have been arranged according to the value of the aqueous tension.

(6) Table II. shows the relation between the degree of relative humidity of air and the radiation-difference. For every single set of observations the relative humidity was computed.

(7) It remains yet to be mentioned that the elaborate task of classifying and reducing the 4376 single sets of observations has been twice executed, both computations being quite independent of each other.

(8) With a view to show that the decrease in radiation with an increase in humidity, as evident from Table II., is not due to the decrease in temperature, as may be suspected when glancing over Table II., two more Tables were constructed illustrative of the influence the temperature of the air alone exerts on the radiation-instrument (Tables III. and IV.). For, although fully aware of the principles generally adopted with regard to the influence of the temperature of the air on radiation, and also of the results of your investigations (§ XII. p. 406), to the effect that the temperature of the surrounding air does not sensibly influence the difference of temperature between the radiating body and the temperature of the air, I thought one might be inclined to suspect undue radiation influencing the reading of the parabolic thermometer, and I took for this reason the trouble of classifying the observations merely according to temperature of the air.

The results at which I arrive may be summed up in the following lines:—

(*a*) The absolute quantity of aqueous vapour in the air is in itself alone not sufficient as a criterion for the degree of radiation; or if it should be otherwise, the instrument used for the observations is not sufficiently delicate to show the degree of influence.

(*b*) The absolute quantity of aqueous vapour, together with a certain temperature, *i.e.* the relative humidity of air, greatly influence terrestrial radiation, in such a manner that, the greater the degree of relative humidity, the less radiation is noticeable (see Table II.). With respect to the latter remark, I cannot refrain from adding a few words. The law is even noticeable, though less distinctly expressed, in the lower per cents. of humidity in extreme cases of dryness—although in all these cases a thin veil of cirrostratus clouds covers the sky, so delicate as frequently not to be perceptible by the eye. It is further to be remarked that, during hot winds, when the air is extremely dry, quantities of dust are floating in the air, seriously interfering with radiation, although the sky is apparently quite clear. All these cases have been included in deriving the above results; and considering all, we must be surprised that, notwithstanding all this, an increase of radiation of between 15 and 30 per cent is perceptible.**[11]

It is scarcely necessary for me to dwell on the apparent connexion between the increase of temperature of the air and the amount of radiation shown by Table II., such being fully accounted for by the relation between temperature and relative humidity. Tables III. and IV. sufficiently prove that the radiation-instrument was not influenced by the surrounding air in any undue manner so as to interfere with the recognition of the general law.

I hastened to put these facts before you, in order you might be able to convince yourself of the reliability of the observations and the method of discussion; and further, because I was convinced you would find it important to receive proofs that by the aid of meteorological instruments generally in use we are able to recognize a function of aqueous vapour in our atmosphere which hitherto had only been demonstrated by your own important experiments (as far as my knowledge on this point goes).

You would oblige me exceedingly by giving me your ideas on the various points raised in this letter. According to your opinion, I purpose either turning the results to account by publishing them in my next publication, or hand them over to you with no other condition but that I shall be supplied with whatever you shall write upon this special subject.

Requesting you kindly to excuse the liberty I have taken in thus troubling you,

I remain, my dear Sir, | Yours very truly, | G. NEUMAYER.

TABLE I.

No of obs.	Mean			No of obs.	Mean			No of obs.	Mean		
	Pressure of vapour.	Temp of air.	Temp of air minus temp radiation.		Pressure of vapour.	Temp of air.	Temp of air minus temp radiation.		Pressure of vapour.	Temp of air.	Temp of air minus temp radiation.
	inch.	°	°		inch.	°	°		inch.	°	°
4	0·1712	32·1	1·35	29	0·2703	55·1	3·56	45	0·3483	64·8	4·03
43	·1929	37·7	2·73	44	·2704	50·1	3·82	44	·3485	56·3	3·37
10	·1956	46·9	3·38	44	·2869	54·7	3·74	48	·3493	65·0	3·17
42	·2100	39·9	2·62	41	·2878	56·4	3·73	45	·3497	66·9	3·93
34	·2114	45·3	2·88	43	·2879	54·0	4·31	49	·3677	58·7	3·24
43	·2120	46·4	2·88	41	·2884	56·8	4·00	45	·3680	70·2	3·55
44	·2133	49·5	3·52	43	·2886	52·7	3·39	44	·3687	64·5	3·72
45	·2293	43·2	2·98	43	·2888	52·5	3·85	48	·3687	63·4	3·23
40	·2302	43·6	3·34	45	·2892	59·6	3·56	24	·3692	63·0	3·69
40	·2304	46·2	3·10	40	·2894	57·2	4·00	44	·3692	64·4	4·34
43	·2307	45·4	3·23	45	·2899	50·7	3·13	43	·3701	60·5	3·52
43	·2307	51·3	4·60	40	·2900	56·5	3·91	44	·3876	66·3	4·50
44	·2313	52·9	2·69	49	·2915	53·3	3·97	49	·3888	61·8	3·60
29	·2322	52·6	3·51	45	·2940	61·9	3·70	43	·3894	63·2	3·76
44	·2484	47·3	3·37	44	·3076	53·0	3·21	45	·3900	69·2	3·77
44	·2493	46·5	2·81	40	·3080	61·1	4·58	19	·3909	59·0	3·32
43	·2494	47·1	3·32	40	·3088	57·5	3·38	44	·4081	65·9	3·84
41	·2496	52·6	3·41	43	·3091	57·6	3·95	49	·4087	63·5	3·60
44	·2497	48·2	2·32	43	·3093	58·1	4·10	43	·4092	62·3	4·00
43	·2497	49·4	3·60	44	·3095	54·2	3·83	3	·4100	64·4	3·07
43	·2502	54.9	3·60	24	·3095	55·9	4·18	48	·4280	68·1	4·24
45	·2504	46.3	4·52	44	·3096	55·0	3·95	9	·4290	64·0	3·29
40	·2510	59.8	4·12	45	·3097	64·4	3·47	49	·4295	67·1	4·13
40	·2512	48.2	3·09	44	·3111	57·7	3·82	32	·4360	67·5	3·56
6	·2533	54.9	3·10	4	·3232	55·1	3·30	35	·4476	66·7	3·77
41	·2523	52.1	3·53	44	·3280	57·7	4·03	48	·4497	69·2	4·25
43	·2683	48.5	3·51	48	·3286	55·5	3·82	48	·4692	70·6	3·66
45	·2687	52.7	3·71	43	·3287	59·1	3·80	26	·4698	71·7	3·84
44	·2687	50.2	3·60	41	·3290	59·4	3·90	3	·4886	68·7	2·86
40	·2688	56.3	4·05	44	·3291	55·2	3·90	48	·4893	73·6	3·04
44	·2689	50.2	3·58	45	·3294	59·3	3·64	48	·5096	73·1	3·47
41	·2691	56.4	3·74	40	·3298	66·7	4·33	27	·5284	76·8	3·58
44	·2692	47.7	3·58	40	·3320	68·8	3·68	12	·5460	77·0	4·55
43	·2694	50.3	3·98	43	·3468	62·1	3·85	12	·5715	79·4	3·50
43	·2695	51.8	2·55	49	·3472	56·3	3·48	19	·5908	77·9	3·72
40	·2696	64.8	4·10	44	·3478	60·0	4·23	13	·6061	72·5	2·58
43	·2699	53.2	3·54	34	·3481	61·0	4·41	4	·6292	78·3	2·23

TABLE II.

Groups.	Relative humidity.	Number of observations.	Temperature of air.	Temperature of air minus radiation.
	percent.		°	°
1	22·417	190	84·83	4·79
2	40·220	406	71·43	4·69
3	47·479	244	68·03	4·43
4	52·455	286	66·46	4·37
5	57·521	322	61·99	3·92
6	62·580	301	59·70	3·70
7	67·441	322	58·01	3·71
8	72·485	357	55·12	3·52
9	77·427	438	52·67	3·48
10	82·598	458	51·43	3·36
11	87·397	426	49·63	3·21
12	92·348	403	46·26	3·11
13	96·643	223	43·33	2·55

TABLE III.

Number of		Mean pressure of vapour.	Mean temperature of air.	Temperature of air minus radiation.
Observations.	Groups.			
		inch.	°	°
220	7	0·2009	42·54	2·77
284	7	·2307	47·89	3·35
259	6	·2493	48·52	3·14
215	6	·2514	52·70	3·57
257	6	·2687	52·38	3·70
213	7	·2698	53·30	3·59
385	9	·2885	54·95	3·74
258	6	·3000	57·22	3·79
243	6	·3000	57·53	3·91
268	7	·3253	57·10	3·64
295	7	·3401	62·03	3·95
276	6	·3552	63·65	3·55
296	7	·3746	63·40	3·80
384	11	·4117	64·93	3·69
343	13	·5229	73·5	3·46

In this Table are represented the means of the groups, irrespective of the number of single observations.

TABLE IV.

Number of observations.	Temperature of air.	Pressure of vapour.	Temperature of air minus radiation.
	°		°
251	41·46	0·2108	2·96
747	48·30	0·2502	3·39
881	53·70	0·2799	3·58
919	58·01	0·3163	3·73
740	63·40	0·3628	3·76
522	68·15	0·4081	3·86
135	72·72	0·5187	3·23
74	77·88	0·5732	3·52

N.B. I need scarcely mention that the values arrived at only possess a relative worth with respect to the instruments used. An apparatus measuring terrestrial radiation absolutely and in a manner *which renders it practical*, is, I think, yet to be contrived.

Phil. Mag., 31 (1866), pp. 510–15

1. This letter was forwarded to the *Phil. Mag.* by Tyndall, and his note with it is letter 2442.
2. *I sent you some original observations*: this was probably in manuscript form, as they were not published until 1867 (see n. 4). This letter from Neumayer is the only one in Tyndall's extant correspondence.
3. *Flagstaff Observatory, Melbourne*: Neumayer established the Flagstaff Observatory at Flagstaff Hill in central Melbourne in 1858, and meteorological observations were recorded there until the establishment of the Melbourne Observatory in 1863 (R. Gillespie, 'Georg Neumayer and Melbourne Observatory: an institutional legacy', *Proceedings of the Royal Society of Victoria*, 123 (2011), pp. 19–26).
4. *magnetic and meteorological observations made by me during the years 1858–63*: Neumayer indeed published this, as: G. Neumayer, *Discussion of the Meteorological and Magnetical Observations made at the Flagstaff Observatory, Melbourne, during the years 1858–1863* (Manheim: J. Schneider, 1867). Neumayer discussed Tyndall's investigation on aqueous vapour and terrestrial radiation (pp. 113–4), and followed with this very letter (pp. 114–8), although with 'slight alterations' and only two of the four tables. Of his letter to Tyndall, he wrote: 'Professor Tyndall wrote me on the 10th of May a letter, acknowledging the receipt of my communication, from which I may glean the following passage: "I regard your letter of extreme importance. As regards the science of meteorology I think

it is hardly possible to overrate the value of your results . . . . . . Thinking that you would not object to the publication of your letter I have sent it to the editor of the Philosophical Magazine'" (p. 114). The letter from Tyndall to Neumayer is missing.

5. *August's psychrometer*: an instrument for ascertaining the relative humidity of the air by noting the difference in the readings of two thermometers, one of which has a wet bulb that is cooler as a result of evaporation (*OED*); this design by Ernst Ferdinand August (1795–1870), a German physicist (*ADB*).
6. *Döbereiner-Regnault's aspirator hygrometer*: a scientific instrument for measuring the humidity of the air or a gas (*OED*); this design by the German chemist Johann Wolfgang Döbereiner (1780–1849) (*ADB*) and Henri Victor Regnault.
7. *Regnault's*: Henri Victor Regnault.
8. *Tables*: probably H. V. Regnault, *Relation des expériences pour déterminer les principales lois et les données numériques qui entrent dans le calcul des machines à vapeur*, 3 vols (Paris: Didot, 1847–70). Neumayer would be referencing the first (1847) and second (1862) volumes, as the third was not published until 1870.
9. *Kew standards*: standard thermometers 'were manufactured by London instrument makers and then tested at Kew' (L. T. Macdonald, *Kew Observatory and the Evolution of Victorian Science, 1840–1910* (Pittsburgh: University of Pittsburgh Press, 2018), pp. 71–2).
10. The following footnote appeared here in the original: '* In strict conformity with the instructions of the Royal Society'.
11. The following footnote appeared here in the original: '** I have shown that the smoke of the London air is feeble in comparison with its aqueous vapour.—J.T.'.

From Gustav Wiedemann 5 May 1866 2420

Carlsruhe, Hirschstraße 54, 5 Mai 1866

Mein lieber Tyndall!

Als Sie in Ihrem Briefe vom 27st. März die freundliche Absicht aussprachen, mich mit der endgültigen Ordnung der Angelegenheit betreffend die Uebersetzung ihres Buches betrauen zu wollen, glaubte ich nicht, daß Sie Sich von Neuem mit Bohn einigermaßen gebunden hätten. Da Vieweg Bohn eine Abstandssumme gegeben, nur um die Uebersetzung nicht erscheinen lassen zu müßen, hätte Bohn schon damals zu der Ueberzeugung kommen sollen, daß dieselbe ihm nicht zur Ehre gereichte. (Ich selbst habe die Uebersetzung erst später gesehen, so viel ich weiß, hat Prof. Knapp zuerst das ungünstige Urtheil des Correctors bestätigt.) Ich kann mir indeß nicht denken, daß Bohn jetzt noch auf die Uebersetzung bestehen wird. Er müßte nach Ihrem letzten Briefe dazu doch allzu viel Muth besitzen. Jedenfalls werden Sie doch demnächst Nachrichten von Bohn erhalten und dann sind Sie wohl so freundlich, mich es wissen zu lassen. Aber Vieweg habe ich die Sachlage auch mitgetheilt; es wird aber schwer für ihn sein, selbst etwas zu thun. Eine Antwort habe ich von ihm auch nicht erhalten.—

Vielleicht wäre es besser gewesen, mit den Damen nicht eher zu sprechen, ehe die Sache mit Bohn nicht völlig geklärt war. Ich glaubte aber, Sie waren von ihm ganz frei & hätten sich noch nicht wieder gebunden. Für die Damen ist die Uebersetzung nun eine etwas bedenklichere Aufgabe, da Bohn natürlich mit Argusaugen jeden Mangel bewachen wird. Jetzt meine ich, Sie sollen sich dadurch nicht abschrecken lassen; Helmholtz & ich, wir übernehmen doch die Verantwortung & außer uns & Vieweg braucht Niemand zu wissen, wer der eigentliche Bearbeiter ist. Ich hoffe also *[nun]*, die Angelegenheit wird sich auf eine uns Allen erfreuliche Weise ordnen.—Mir thut es nur sehr leid, daß Sie auch davon so viel Aerger und Unannehmlichkeiten haben. Wenn ich auch nicht gerade Bohn zum Uebersetzen von vornherein gewählt hätte, so glaubte ich doch nicht, daß Vieweg bei ihm so schlecht fahren würde. Freilich lauten auch die Urtheile seiner näheren Bekannten über ihn, wie ich sie jetzt eingeholt habe, nichts weniger als besonders günstig. Ich selbst habe ihn *[nur]* ganz flüchtig auf der Naturforscherversammlung in Gießen gesehen; doch hat er nur daselbst nach keiner Richtung einen besonderen Eindruck hinterlassen.

Sobald Sie Nachrichten haben, lieber Tyndall, schreiben Sie wohl einige Worte; wie ich von Helmholtz höre, möchte Vieweg bei glücklicher Erledigung der Sache gern recht bald die ersten 12 Bogen der Uebersetzung erscheinen lassen.—Was ich noch irgend thun kann, soll gewiß geschehen. Sie wißen, wie gern ich Ihnen zu Diensten bin.—

In aufrichtiger Freundschaft & Ergebenheit | Ihr | G. Wiedemann.

Carlsruhe, Hirschstraße 54, 5 May 1866

My dear Tyndall!

When, in your letter from the 27th of March,[1] you expressed your kind intention to entrust me with the final arrangement of the matter concerning the translation of your book,[2] I did not think that you had once more quite bound yourself to Bohn.[3] Since Vieweg[4] had given Bohn a compensation payment just so he did not have to publish the translation, Bohn should already at the time reached the conclusion that the same would not be to his credit. (I myself have seen the translation only later; as far as I know, it was Professor Knapp[5] who first confirmed the proof-reader's opinion.) I therefore cannot imagine that Bohn will still insist on the translation. After your last letter, he would have to possess altogether too much courage to do so. In any case, you should receive news from Bohn shortly[6] and then you will please be so kind as to let me know. However, I have also informed Vieweg of the situation; but it will be difficult for him, to do something himself. I also have not received a response from him.—

Perhaps it would have been better to not speak to the ladies[7] ere the matter with Bohn was not completely clarified. I did believe, however, that you were entirely free from him and had not yet bound yourself again. The translation is now a somewhat more apprehensive task for the ladies, since Bohn will naturally watch with eagle eyes for any fault. I now think that you should not let this scare you; Helmholtz & I, we will take on the responsibility & except for us & Vieweg, nobody needs to know who the actual person responsible is. Therefore, I *[now]* hope that the matter will sort itself in a manner that will be gratifying to all of us.—I am only sorry that you have so much trouble and inconvenience because of it. Even if I would not necessarily a priori have chosen Bohn for the translation, I did not think that Vieweg would have such bad experiences with him. The opinions of his closer acquaintances, which I have now obtained, are naturally no less than especially favourable. I myself have only seen him very briefly at the meeting of natural scientists in Gießen; however, he has not left a particular impression in any respect there.

As soon as you have any news, dear Tyndall, do write me a few words;[8] as I have heard from Helmholtz, Vieweg would gladly like to publish the first 12 sheets of the translation quite soon if the matter is brought to a happy completion.—Whatever I can do shall certainly happen. You know how gladly I am at your service.—

In sincere friendship and devotion | Your | G. Wiedemann.

RI MS JT/1/W/43

---

1. *your letter from the 27th of March*: letter missing.
2. *translation of your book*: the German translation of *Heat* (1865); see letter 2296, n. 6.
3. *Bohn*: Johann Conrad Bohn.
4. *Vieweg*: Eduard Vieweg.
5. *Professor Knapp*: possibly Jacob Hermann Knapp (1832–1911), a German-American ophthalmologist who from 1860–68 was professor of ophthalmology at Heidelberg (*NDB*, *ANB*).
6. *receive news from Bohn shortly*: the next letter from Bohn is 2437.
7. *ladies*: Clara Wiedemann and Anna Helmholtz (see letter 2296, n. 6).
8. *do write me a few words*: if Tyndall wrote to Wiedemann, the letter is missing.

From George Gabriel Stokes 7 May 1866 2421

Cambridge 7th May 1866

My dear Tyndall,

I return you your paper[1] for the sake of some queries.

P 2. red cross. Is not what you mean rather cheese than milk? Is not "milk" of sulphur properly a watery mixture containing S in fine division? If I am right I would suggest for "The milk" to read "dried milk"

P 7 cross. Is not the term "spongy platinum" by the usage of chemists appropriated to that obtained by igniting the platinum chloride of ammonium, which is rather grey than black, and as to radiation & absorption would probably stand intermediate between the coherent metal and Pt black? If I am right (which Frankland would tell you in a moment) would it not be better to speak of black platinum?

P 8 current for cement escaped you

P 11 Two red crosses

P 12 Two red crosses

Foot note. It struck me that this might be read in an uncomplimentary sense, as much as to say the only experiment of his which was good for anything.

<P> 13 x

In connexion with platinum black I was thinking of suggesting to you an experiment on silver, but I don't think I did.

If to an ammoniacal solution of a salt of silver you add a protosalt of iron with at least enough tartaric acid to prevent precipitation, rendered alkaline by ammonia, the silver is precipitated in the metallic state as a fine black powder, which (at least with dilute solutions) remains long in suspension, but after it has once fallen falls more readily, and may be washed by decantation. I have little doubt that this black silver would exhibit a (for silver) copious radiation and absorption.

Yours very truly | G. G. Stokes

RI MS JT/1/S/242
RI MS JT/1/TYP/4/1403

---

1. *your paper*: J. Tyndall, 'Sixth Memoir on Radiation and Absorption—Influence of Colour and Mechanical Condition on Radiant Heat', *Phil. Trans.*, 156 (1866), pp. 83–96.

To George Gabriel Stokes 8 May [1866][1] 2422

8th May

My dear Stokes

Many thanks for the corrections,[2] all of which I have adopted except that relating to the Milk of sulphur.

This I am informed is the name universally applied to the powdered substance (obtained by precipitation) It is a misnomer, and if you think "dried milk" a better term pray put it in.[3]

Thanks especially for the change regarding Stewart.[4] I should certainly not like to make the impression that the experiment referred to was his only good one.

I hope to attack the question referred to towards the end of your note when I cast my lectures at Jermyn Street[5] off hands, and have completed a little book on sound[6] which I am now writing—I should not forget any of your suggestions.

Yours very truly | John Tyndall

RI MS JT/1/T/1420
RI MS JT/1/TYP/4/1480

1. *[1866]*: the year is given by relation to letter 2421.
2. *corrections*: see letter 2421.
3. *pray put it in*: the term 'Milk of Sulphur' was retained in Tyndall's published paper (see letter 2421, n. 1).
4. *Stewart*: the footnote on p. 94 of the published paper read: 'I think the important experiment first executed by MR. BALFOUR STEWART, of rock-salt radiating through rock-salt, is by itself sufficient to demonstrate in the most unequivocal manner that this substance is not equally pervious to all kinds of rays'. Stokes had written in letter 2421: 'It struck me that this might be read in an uncomplimentary sense, as much as to say the only experiment of his which was good for anything'. For Balfour Stewart, see letter 2381, n. 6.
5. *my lectures at Jermyn Street*: see letter 2337, n. 16.
6. *a little book on sound*: *Sound* (1867).

From Michael Faraday 9 May 1866 2423

Hampton Court Green,[1] | S.W. | 9 May, 1866.

My dear Tyndall,

I have received a letter this morning from Mr. Farrer[2] of the Board of Trade, and in it he tells me of his interview with you, and of your ever kind readiness to assist me.[3] I expect I shall be at home on Friday, and shall then hope to see you; but in the meantime I must send a few lines to say that I accept, as frankly as you offer, your help to me, in giving an opinion on these lighthouse apparatus.

I am, my dear friend, | Yours very truly, | M. Faraday | Dr. Tyndall, | &c. &c. &c.[4]

Faraday Correspondence, 6:4580
RI MS JT/1/TYP/12/4160

1. *Hampton Court Green*: Faraday's house in London since 1858.
2. *Mr. Farrer*: Thomas Henry Farrer (1819–99), an English civil servant and statistician who was permanent secretary of the Board of Trade from 1867 (*ODNB*).
3. *your ever kind readiness to assist me*: while Tyndall accepted this offer to assist an aging Faraday in lighthouse matters, he would in July 1867 officially take over for Faraday as Scientific Adviser to the Board of Trade (a British government department concerned with making policy and instituting regulations regarding economic activity in Britain and its empire) and Trinity House (the lighthouse authority for England and Wales, established by Royal Charter in 1514) (letters concerning this are in *Tyndall Correspondence*, vol. 10). In this role, Tyndall investigated and gave recommendations to Trinity House on matters of lighthouse illumination, and later, sound signaling. On Tyndall and his role with the Board of Trade and Trinity House, see *Ascent of John Tyndall*, pp. 282–5, 316–7, 345–6, 389, 406–13, 420–3, 433–4, and 437; S. Courtney, '"A very diadem of light": exhibitions in Victorian London, the Parliamentary light and the shaping of the Trinity House lighthouses', *British Journal for the History of Science*, 50 (2017), pp. 249–65; and R. M. MacLeod, 'Science and Government in Victorian England: Lighthouse Illumination and the Board of Trade, 1866–1886', *Isis*, 60 (1969), pp. 4–38.
4. Frank A. J. L. James, the editor of the Faraday correspondence, noted that this letter was written in 'Miss [Jane] Barnard's handwriting but signed and addressed by Faraday' (*Faraday Correspondence*, 6, p. 479).

To Georg Balthasar von Neumayer 10 May [1866][1] 2424

I regard your letter[2] of extreme importance. As regards the science of meteorology I think it is hardly possible to overrate the value of your results . . . . . . Thinking that you would not object to the publication of your letter I have sent it to the editor of the Philosophical Magazine.[3]

G. Neumayer, *Discussion of the Meteorological and Magnetical Observations made at the Flagstaff Observatory, Melbourne, during the years 1858–1863* (Manheim: J. Schneider, 1867), p. 114

1. *[1866]*: the year is given by reference to Neumayer's letter to Tyndall of 5 May 1866. The rest of this letter from Tyndall to Neumayer is missing.
2. *your letter*: letter 2419.
3. *I have sent it to the editor of the Philosophical Magazine*: see letter 2442.

From Heinrich Gustav Magnus 12 May 1866 2425

Berlin 12. Mai. 1866

Theurer Freund!

Ich kann meine Note „über den Einfluss der Absorption der Wärme auf die Bildung des Thaues" nicht in die Welt gehn lassen ohne ein Paar Zeilen an Sie zu richten. Sie wissen daß es öfter mein Schicksal gewesen ist, nicht nur gleichzeitig mit bedeutenden Physikern denselben Gegenstand zu bearbeiten, denn das wird einem jeden begegnen der überhaupt thätig in der Wissenschaft ist, sondern daß auch bisweilen noch Differenzen blieben, die auszugleichen waren, was nicht gerade sehr angenehm ist. Ich würde deßhalb die Angelegenheit mit dem Wasserdampf ganz haben ruhen lassen wenn nicht von verschiedenen Seiten immer wieder darauf zurückgekommen würde, so daß ich weniger in meinem Interesse als im Interesse der Sache es für nöthig erachtete den Unterschied zwischen eigentlichen, vollkommen durchsichtigen und den nebelförmigen Dämpfen bestimmter hervorzuheben. Ich hoffe dies in einer Weise gethan zu haben daß weder Sie noch D^r^. Frankland Ursache haben sich zu beschweren. Seien Sie überzeugt daß ich vor Ihrem Talent und Ihren Leistung sowie vor denen von D^r^. Frankland die größte Achtung hege und lassen Sie keine Spur von Trübung in unsere Freundschaft treten die mir werth und lieb ist und die ich erhalten zu sehn aufrichtig wünsche.

In alter Freundschaft | Ihr | G Magnus.

Berlin 12. May. 1866

Dear friend!

I cannot release into the world my note "on the influence of heat absorption on the formation of dew"[1] without directing some lines to you first. You know that it has frequently been my fate, not only to be working on the same topic simultaneously with prominent physicists—for this is something that happens to anyone who is active in science—but that there were at times differences that had to be reconciled, which is not particularly pleasant. I would have simply ignored the discussion about water vapour, if one were not brought back to it from different sides again and again. More for the sake of the topic itself than to satisfy my own interests I therefore considered it necessary to stress more clearly the difference between actual, entirely transparent and foggy vapours. I hope that I did this in such a way that neither you nor Dr Frankland will have reason for complaint.[2]

Rest assured that I have the highest regard for your talent and your achievements, as much as for those of Dr Frankland, and do not allow any clouds to darken our valuable friendship, which I sincerely wish to see maintained.

In old friendship | Yours | G Magnus.

RI MS JT/1/M/31

1. *my note*: G. Magnus, 'Über den Einfluss der Absorption der Wärme auf die Bildung des Thaues', *Poggendorff's Annal.*, 127 (1866), pp. 613–24. This was translated and published as: G. Magnus, 'Prof. Magnus on the Influence of the Absoprtion of Heat on the Formation of Dew', *Phil. Mag.*, 32 (1866), pp. 111–18.
2. *reason for complaint*: for Tyndall's response to Magnus' paper, see: J. Tyndall, 'Prof. Tyndall's Remarks on the Paper of Professor Magnus', *Phil. Mag.*, 32 (1866), pp. 118–20.

From Rudolf Clausius 14 May 1866 2426

ZÜRICH, 14/5/66.

Lieber Tyndall,

Wenn sich nach Deiner letzten Verhandlung mit Bohn die Sache endlich so gestaltet, dass Vieweg die Besorgung der Uebersetzung Deines Buches ganz in die Hand bekommt, so werde ich sehr froh sein. Ich habe zwar nichts von Bohn's Uebersetzung gelesen, aber nach den vielen ungünstigen Urtheilen kann ich nicht zweifeln dass sie wirklich zu schlecht ausgefallen ist; während auf der andern Seite, wenn Wiedemann und Helmholtz die neue Uebersetzung controlliren, gewiss etwas Gutes zu erwarten ist.

Deine Mittheilungen aus der Royal Society haben mich sehr erfreut. Ich betrachte es als eine ausserordentliche Ehre für mich, dass ich mit in der Wahl gewesen bin; auf der andern Seite kann ich auch aufrichtig sagen, dass es mir selbst angenehm gewesen ist, dass bei einer Entscheidung zwischen Riemann und mir, Riemann den Vorzug erhalten hat. Erstens ist Riemann ein so ausgezeichneter Gelehrter dass ich unter keinen Umständen Zurücksetzung darin sehen würde, ihm nachzustehen, und zweitens ist der Umstand, dass er krank ist, und die Früchte seiner herrlichen Arbeiten nur wenig und vielleicht nur noch kurze Zeit geniessen kann, ein sehr wesentlicher. Wenn ich gewählt wäre und hätte zugleich erfahren, dass Riemann deshalb hätte übergangen werden müssen, so würde ich keine rechte Freude an meiner Wahl gehabt haben. Ich danke Dir herzlich für die Freundlichkeit mit welcher Du für meine Wahl gewirkt hast.

Ich glaube, ich habe Dir schon mitgetheilt, dass ich von Wien aus angefragt bin, ob ich eine Professur am dortigen Polytechnikum übernehmen wolle. Ich konnte mich damals, weil die Sache noch zu unbestimmt war, nicht mit Sicherheit entscheiden. Einige Zeit später habe ich dann einen bestimmten Ruf mit guten Anerbietungen erhalten. Man hat mir aber bei der Gelegenheit hier soviel Freundlichkeit erwiesen und hat soviel zur Verbesserung meiner hiesigen Stellung gethan, dass ich mich endlich zum Hierbleiben entschlossen habe. Gegenwärtig, wo sich die Lage von Oesterreich durch Kriegsdrohungen von allen Seiten so verschlimmert hat, habe ich allen Grund mit dem gefassten Entschlusse zufrieden zu sein.

Mit herzlichen Grüssen von meiner Frau und mir, und der Bitte, auch Hirst bestens zu grüssen,

Dein | Clausius.

ZURICH, 14/5/66

Dear Tyndall,

I will be glad to hear if after your latest negotiation with Bohn,[1] Vieweg[2] will finally receive the rights to translate your book.[3] Although I have not read any of Bohn's translation, judging by the many unfavorable reviews I cannot doubt that it really has turned out to be too poor; on the other hand, if Wiedemann and Helmholtz oversee the new translation, we can certainly expect something good.

Your reports from the Royal Society made me very glad. I see it as an extraordinary honour for me to be considered for the election;[4] on the other hand, I can say in all honesty that I am glad that in a decision between Riemann and me, Riemann was given precedence.[5] First of all, he is such a brilliant academic that under no circumstances would I consider it as an insult to be ranked below him; and second, we have to consider the circumstance that he is ill and will perhaps not be able to fully enjoy the fruits of his excellent work, or only for a short time. If I had been elected and at the same time found out that Riemann had to be passed over because of it, I would not have derived any real pleasure from my election. I want to sincerely thank you for the kindness towards me that you showed during the election.

I think I have already informed you that I was asked if I wanted to take a professorship at the Technical University in Vienna. Since this was still too uncertain at the time, I could not make a firm decision. Some time later I received a concrete offer with good conditions. However, given this opportunity, in my present position they have shown so much kindness and done so much to improve my current situation that I finally decided to stay here.[6] Moreover, the worsening situation in Austria due to threats of war from all sides, has given yet another reason to be satisfied with my decision.[7]

With best wishes from my wife[8] and myself and the request to send Hirst the best wishes,

Your | Clausius.

RI MS JT/1/TYP/7/2294–5
LT Typescript Only

---

1. *Bohn*: Johann Conrad Bohn.
2. *Vieweg*: Eduard Vieweg.
3. *to translate your book*: the German translation of *Heat* (1865); see letter 2296, n. 6.
4. *to be considered for the election*: see letter 2416.
5. *Riemann was given precedence*: Bernhard Riemann (see letter 2416, n. 12) had been elected a foreign member of the RS. He died only a few months later.
6. *to stay here*: in January, Clausius, then a professor at the Federal Polytechnic School in Zurich, wrote to Tyndall that he had been offered a position in Vienna (see letter 2357).
7. *the worsening situation in Austria*: the Austro-Prussian War broke out on 14 June 1866.
8. *my wife*: Adelheid Clausius.

From Félix Foucou [15 May 1866][1] 2427

[Mardi]

Mon cher Monsieur Tyndall,

Je vous transmets la lettre que je viens de recevoir de M. Hetzel, en réponse de la communication de votre lettre dernière. D'après cela, je vois que M. Hetzel craint que, même en dehors de vous, un autre éditeur ne vienne le supplanter. Dans son opinion, ni vous ni personne ne peut lui assurer la propriété de la traduction de votre livre, et c'est pour cela qu'il veut vous lier pour les modifications que vous ferez à votre ouvrage.

Le cas me parait embarrassant. Vous ne pouvez évidement pas vous lier de la sorte, et M. Hetzel de son côté ne peut pas s'exposer au danger qu'il redoute, si la première édition ne représente pas pour lui un bénéfice convenable. Aussi ne puis-je que vous dire que je suis prêt à faire ce que vous jugerez utile à l'intérêt du livre.

Les complications *[politiques]* me font rester encore un peu en Europe, vous pouvez donc m'écrire. Voulez-vous que, à tout événement, je vous envoie le manuscrit révisé ?

Cordiales amitiés | F. Foucou

[Enclosure]

Sur quoi pouvons nous dour disputer ?

Je respecte extrêmement le droit *[moral]* de l'auteur sur son livre—si une fois la 1ère édition écoulée M. Tyndall veut changer son livre, le transformer, le modifier, le supprimer—je le veux bien—mais s'il veut le donner à un autre cela ne me parait plus du tout acceptable—pourquoi ma situation serait elle pire parceque *[j'aurais été]* plus délicat.

Je désire que M. Tyndall soit persuadé de mon envie de bien être agréable—mais il est évident qu'il ignore complètem[m]ent que les législations ne font pas aux traductions de conditions d'aucun comme aux oeuvres originales.

Tout à vous | J. Hetzel

[Tuesday]

My dear Mr Tyndall,

I am sending you the letter that I just received from Mr. Hetzel[2] in response to your last letter.[3] You will see that Mr. Hetzel is concened that another editor, even someone aside from you, might supplant him. In his opinion, neither you nor anyone else can guarantee him sole translation right to your book,[4] and this is why he wants to obligate you for any future modifications to your work.

This matter seems awkward to me. Obviously, you cannot legally bind yourself in such a fashion, and Mr. Hetzel, for his part, cannot expose himself to the danger that he fears if the first edition does not bring him adequate returns. I can only tell you that I am ready to do whatever you think necessary in the interest of the book?

Political complications have forced me to stay in Europe a little longer,[5] so you can write to me. Would you like me to send you the revised manuscript if anything happens?

Warmest regards | F. Foucou

[Enclosure]

Then what can we argue about?

I have the deepest respect for the author's *[moral]* right to his book—if, once the 1st edition is sold out, Mr. Tyndall wanted to change his book, transform it, modify it, pull it out of circulation—I would not mind that.—But if he wanted to give it to another, I would no longer consider that acceptable—why should I, for *[having been more tactful]*, find myself in a worse situation?

I hope that Mr. Tyndall will be convinced of my desire for a friendly working relationship—but it is clear that he is completely unaware that current legislation does not give translations conditions anything at all like those for original works.

Truly yours | J. Hetzel

RI MS JT/1/H/71

1. *[15 May 1866]*: the date is given by relation to letter 2431, dated 19 May 1866, a Saturday, making the Tuesday before 15 May.
2. *Mr. Hetzel*: Pierre-Jules Hetzel.
3. *your last letter*: letter missing.
4. *translation right to your book*: for a two-volume French edition of *Glaciers of the Alps* which fell through (see letter 2255, n. 2, and letter 2441).

5. *stay in Europe a little longer*: Foucou was planning a trip to the United States (see letter 2431, n. 3).

From George Hector Tyndale 15 May 1866 2428

Philadelphia May 15th 1866

My dear John

I have not written for a long time, but I am indebted to you for two letters[1] and some papers, with your pamphlet on Radiations,[2] for which I am much indebted to you, all received since my last to you. I wrote you last, I think, a very long letter in the summer of 1863,[3] when in the Service and in Virginia. It was in answer to one in which you spoke of having been to Holland and of some people of our name there and in Ireland.[4] I suppose you never received that letter as it was unacknowledged.[5]

My history since then is soon told. After Chickamauga where our forces[6] were worsted in September 1863, in Tennessee, the 11th & 12th Corps were ordered under Hooker,[7] from Virginia to Chattanooga (about 1000 miles) which place we relieved by opening the Tennessee River and thus giving the beleaguered troops food and safety. We then fought the battles of Wauhatchie, Chattanooga, Lookout Mountain, & Mission[ary] Ridge and then marched to Knoxville which place we relieved from a close and tenacious siege. This campaign to Knoxville, a march, there and return, of 250 miles made in Mid Winter, with ice from one to two inches thick, many of our men, hundreds yes thousands of them entirely barefooted and even the officers so in many cases, this campaign broke me down (I was suffering all the time from wounds and disease contracted in the service) and I became again partially paralysed. In May 1864 I returned home on leave and finding that I grew no better and not wishing to stand in the way of promotion of other Officers who were in the field and had been more fortunate than myself (there being a limitation to the number of General Officers) I resigned in August 1864—just two days before Mrs. Tyndale's[8] Brother[9] (whom I had reared) was killed in action—None more gallant, generous and self-sacrificing than he. In rallying his regiment, broken by a Rebel Division, he leaped upon a breast-work and fell literally torn,—riddled with balls.

After my resignation, not worrying so much, as it is my nature to do when having a duty and for many months idle, I very slowly recovered, until now I am nearly or quite well, except an occasional slow action of the nerves and muscles of the legs.

Mrs. Tyndale is well and sends her regards to you and kindest respects to "Tom" Hirst—long may he flourish as one in whose shade, or Sunshine rather, it is pleasant to be. Mrs. Tyndale was very glad to receive your kind sympathy for her Brother's death.[10]

As you are doubtless aware, our Grand Armies are disbanded and only the small regular force remains—But though contemporaneous History does not see it, those armies have fought the battles of the Lord and the hopes of Mankind are more largely realized than we now know. Like all other Americans I felt bitterly the cold apathy or worse of the English and other men, but that has gone now, and I feel proud of my countrymen and grateful to God who gave them the grandest opportunity the Sun ever shone upon—and the courage to impress it.

For, dear John, this has been no war for races, or territory, or forms or for self aggrandizement—but it has been a religious, not bigoted, martyrdom of a people for their faith in Humanity and in the just providence of God. These men are practical and have no excessive belief in the Inapplicable or, as they think, the Useless—and so they believe most earnestly in the application of God's attributes to his Works and through them to human governments. Nearly half a million of them are dead[11]—died in or because of the Service.

A friend of mine Mr Peter J. Lesley[12] is about to visit Europe and I will give him a large photograph of myself, which Mrs Tyndale had taken, in my uniform, or rather undress uniform, for the officers here rarely or never wear any other. This photograph I ask your acceptance of. It is sent un-mounted to save space in carriage. It was taken when I was suffering—still I think it like.

I will also take the liberty to give Mr. Lesley a letter of introduction to you, and I ask your kind attentions for him. He is one of the most distinguished Geologists of the Country, connected with the most important Institutions. He has visited Europe before, is a member of some European Societies, is an invalid and now goes out for his health. He is a most estimable and amiable Gentleman, whom it is a pleasure to know. If you can aid him in making acquaintances, in making his stay pleasant, I shall feel a personal indebtedness to you, for I esteem him very much, as do all his Countrymen. If you can make him acquainted with Hirst do so. By the way I am much obliged to Hirst for his kind mention of me in your biographical memoir.[13] My regards to him always.

At present I am doing nothing, have speculated in Oil lands (Petroleum lands) but have succeeded only in losing some money!

When you write home please make my kindest respects to your Mother[14] and your Sister.[15]

Affectionately your cousin | Hector Tyndale

RI MS JT/1/T/58
RI MS JT/1/TYP/5/1664–6

1. *two letters*: for one of these letters, see letter 2095 (14 May 1864), *Tyndall Correspondence*, vol. 8. The other is missing.
2. *your pamphlet on Radiations*: *On Radiation*.
3. *I wrote you last, I think, a very long letter in the summer of 1863*: see letter 2026 (6 August 1863), *Tyndall Correspondence*, vol. 8.
4. *one in which you spoke of having been to Holland and of some people of our name there and in Ireland*: see letter 2019 (9 July 1863), *Tyndall Correspondence*, vol. 8.
5. *I suppose you never received that letter as it was unacknowledged*: the letter was indeed received and is held at the RI.
6. *our forces*: Tyndale was a Union General in the American Civil War.
7. *Hooker*: Joseph Hooker (1814–79), an American Civil War general (*ANB*).
8. *Mrs. Tyndale's*: Julia Tyndale (née Nowlen, 1823–97), was the wife of Hector Tyndale. They married in 1842.
9. *Brother*: not identified.
10. *your kind sympathy for her Brother's death*: if this was through a letter, it is missing.
11. *half a million of them are dead*: current estimates put the Union and Confederate death toll at 750,000 (J. D. Hacker, 'A Census-Based Count of the Civil War Dead', *Civil War History*, 57 (2011), pp. 307–48).
12. *Mr Peter J. Lesley*: Joseph Peter Lesley (1819–1903), an American geologist and secretary and librarian of the American Philosophical Society from 1858–85 (*CDSB*, *ANB*). Lesley was selected as one of ten commissioners representing the United States at the International Exposition (or Exposition Universelle) in Paris in 1867.
13. *his kind mention of me in your biographical memoir*: Hirst contributed a brief life of Tyndall to L. Reeve (ed), *Portraits of Men of Eminence in Literature, Science, and Art, with Biographical Memoirs. The Photographs from Life, by Ernst Edwards, BA* (London: Lovell Reeve & Co., 1864), vol 2, pp. 25–32. In a footnote on pp. 26–7 about family of Tyndall's that had emigrated to the United States, Hirst wrote: 'One of these is at the present moment a distinguished officer in the Federal army. This is the Hector Tyndale who, some years ago, to use the words of Mr. Wendell Phillips, went down to Harper's Ferry with his life in his right hand to receive the dead body of John Brown, and deliver it over to his widow. We may add, too, that by a curious coincidence, Major Tyndale afterwards held possession of, and commanded at, Harper's Ferry. At Antietam nearly half his men were cut to pieces; he himself was carried, apparently dead, from the battle-field, and for his gallant behaviour on the occasion was made Brigadier-General'. Hirst noted in his journal entry of 22 March 1864 that 'On Saturday I commenced writing a sketch of Tyndall's life to be published in the series of lives with Portraits by Edwards. I had not the time to give it the finish I desired for I had but the two days to finish it. Nevertheless I warmed towards the work as I wrote' (Brock and MacLeod, *Hirst Journals*, p. 1665).
14. *your Mother*: Sarah Tyndall.
15. *your Sister*: Emily Tyndall.

To James Croll 19 May 1866 2429

19th May 1866

My dear Sir

Almost all the luminous heat, and a vast flux of nonluminous heat, manages to get from the sun to the earth. So that a screen which prevents this from falling on a properly absorbent thermometer must I think make some difference whether close to the earth or at great elevations. I do not in the least understand Mr. Glaisher's[1] observation,[2] but I was rather astonished to see the conclusions which he drew from it, and which were put in a more definite form[3] by Mr. Wilson.[4] I have the greatest of respect for facts, lest this alleged fact is, as you say, so opposed to all our other knowledge that it would require far more evidence than now exists to cause my mind even to entertain the possibility of it. Probably at <Handwritten Letter Ends>[5] some future day we may hear more about it from Mr. Glaisher, but until then all attempts to account for it would in all probability be mere waste of time.—Yours very truly, | JOHN TYNDALL.

P.S.—Instead of the shaded thermometer, one with a silvered bulb might be compared with the black bulb. The difference would assuredly make itself felt, though here we should have some loss by radiation from the black bulb.

BL Add MS 41077, ff. 33–33a

J. C. Irons (ed), *Autobiographical Sketch of James Croll with Memoir of his Life and Work* (London: Edward Stanford, 1896), pp. 148–9

1. *Mr. Glaisher's*: James Glaisher.
2. *observation*: see letter 2364, n. 4. It is not clear to which particular observation Tyndall was referring, but Glaisher frequently drew unfavourable comparisons between balloon observations and those made by Alpine mountaineers. See I. M. McCabe, 'Second best as a researcher, second none as a populariser? The atmospheric science of John Tyndall FRS (1820–1893)' (PhD Dissertation, University College London, 2012), pp. 235–6.
3. *more definite form*: see letter 2364, n. 3.
4. *Mr. Wilson*: James Maurice Wilson.
5. We were missing the third page of the handwritten letter and unable to obtain it from the BL, and thus the remainder of this letter, from 'some future day', is taken from Croll's autobiography.

To George Gabriel Stokes 19 May 1866 2430

19th May 1866

My dear Stokes

I have read Mr. Wilde's[1] paper[2] and have no hesitation in recommending it for publication in the Philosophical Transactions.

I have seen the machine & can testify to the accuracy of Mr. Wilde's statements regarding its performance.

There are some trivial verbal alterations desirable which will occur to you when reading the paper.

In §.27 a result is apparently derived paradoxical which must follow from theory—I have marked the place in pencil.

It would be desirable to have Table III preceded by a series of measurements with the Tangent compass, similar to those in Table I.

Yours faithfully | John Tyndall

Prof. Stokes D.C.L. | Sec. R.S. | &c &c &c

1866 | Wilde | by Tyndall[3]

RS RR/6/297 Tyndall 19 May 1866

---

1. *Mr. Wilde's*: Henry Wilde (1833–1919), an English electrical engineer and inventor. Wilde independently invented the dynamo-electric machine from Werner von Siemens (1816–92), and this resulted in a legal priority dispute (*ODNB*).
2. *paper*: H. Wilde, 'Experimental Researches in Magnetism and Electricity', *Phil. Trans.*, 157 (1867), pp. 89–107. This paper was received by the RS on 26 March 1866, and communicated by Michael Faraday on 26 April 1866.
3. *1866 | Wilde | by Tyndall*: this is written in red pen on the reverse of the letter, probably by Stokes.

From Félix Foucou 19 May 1866 2431

Paris—19 mai 1866

Mon cher Monsieur Tyndall,

Je n'ai remis qu'aujourd'hui le manuscrit complet à M. Hetzel, sauf les 2 ou 3 pages d'avertissement que je rédigerai lorsque les deux volumes seront imprimés.

Conformément à votre désir, les épreuves vous seront envoyées toutes à la fois, un volume après l'autre. Du reste, M. Hetzel me dit que le 1er volume seul sera imprimé pendant mon voyage. Le second sera prêt au mois d'août, et l'ouvrage entier paraîtra aussitôt.

Comme j'ai retardé mon voyage de 14 jours, je ne serai en Angleterre, de retour des États-Unis, que vers la fin de juillet. J'espère vous voir à cette époque, soit à Londres soit à Paris. En vous rendant en Suisse, pendant mon absence, vous passerez sans doute par Paris : dans ce cas, si vous en avez le temps, allez voir M. Hetzel, éditeur, rue Jacob 18.

Je vous envoie le traité que j'ai signé hier avec lui. Cette copie vous appartient. Si le texte vous convient, vous n'avez qu'à écrire à M. Hetzel quelque chose comme ceci:

«J'ai reçu copie d'une convention relative à la traduction et à la publication en français, de mon livre The Glaciers of the Alps, et j'approuve tous les termes de cette convention.»

Enfin, comme j'ai stipulé que vous seriez remboursé de 5£ pour les clichés, il est convenu avec M. Hetzel que vous n'avez qu'à *[1 word illeg]* sur lui pour cette somme; *[*à*]* *[*présentation*]* il paiera votre chèque. Je n'ai pas fait d'objection à la manière dont cette partie de l'acte a été rédigée; mais comme je n'ai déboursé que 10 Shillings pour le transport des clichés, et que vous avez déboursé $4^{£}$ 10^{S} pour les clichés eux-mêmes, j'ai préféré que la somme entière vous fût payée, sauf plus tard à régulariser ensemble le compte minime de 10 Shillings.

Je laisse ici un ami très compétent qui corrigera les épreuves du 1er volume pendant mon absence. Je serai de retour lors de l'impression du second.

Mille bons compliments | F. Foucou

Excusez-moi, je vous prie, du retard que j'ai fait subir à l'impression de votre livre : des circonstances indépendantes de ma volonté m'ont empêchées de faire ce que j'aurais voulu.

Paris—19 May 1866

My dear Mr Tyndall,

It is only today that I gave the complete manuscript to Mr. Hetzel, except for the 2 or 3 pages of prelude that I will write once the two volumes are printed.[1]

In accordance with your wishes, the drafts will be sent to you all together, one volume after another. As for the rest, Mr. Hetzel[2] told me that only the 1st volume will be printed during my trip.[3] The second will be ready in the month of August, and the entire work will appear right after.

Since I have postponed my trip by 14 days, when I return from the United States, I will only be able to come to England by the end of July. I hope to see you at that time, whether it is in London or in Paris. While going to Switzerland in my absence, you will probably pass by Paris: in that case, if you have the time, go visit Mr. Hetzel, editor, 18 Jacob St.

I am forwarding you the treaty that I signed with him yesterday. That copy belongs to you. If the text is satisfactory for you, you can simply write to Mr. Hetzel something like this:

"I received the copy of an agreement regarding the French translation and publication of my book The Glaciers of the Alps, and I approve all of the terms of this agreement."

Finally, as I stipulated that you should be reimbursed by 5£ for the pictures, it was agreed with Mr. Hetzel that you only have to *[1 word illeg]* on him for that sum; *[at the presentation]* he will pay your cheque. I did not object to the manner in which this part of the agreement was written; but as I did not spend more than 10 Shillings for the transport of the pictures, and since you spent 4£ 10S for the pictures themselves, I preferred that the entire sum would be paid to you, except to arrange later the minimal account of 10 Shillings.

I am leaving a very competent friend[4] here, who will correct the drafts of the 1st volume during my absence. I will be back during the printing of the second one.

A thousand warm wishes | F. Foucou

I beg you to forgive me for the delay that I caused in the printing of your book: circumstances independent of my will prevented me from doing what I would have liked.

RI MS JT/1/F/38

1. *the two volumes are printed*: for a two-volume French edition of *Glaciers of the Alps* which fell through (see letter 2255, n. 2, and letters 2427 and 2441).
2. *Mr. Hetzel*: Pierre-Jules Hetzel.
3. *my trip*: Foucou held a particular interest in petroleum deposits, for which he traveled to the United States and then moved there in 1870, where he died.
4. *competent friend*: not identified.

To Mr. Milner[1] 20 June [1866][2] 2432

20th June

Dear Sir,

Would you kindly convey to your brother[3] my best thanks for his offer of hospitality. For some considerable time however, I have been pledged to a friend[4] in Nottingham[5] in the event of my going there. I thank you also for the friendly way in which your brother's invitation has been conveyed. Yours very truly | John Tyndall | Mr. Milner Esq[ui]re

Deutsches Museum HS 1961–2

1. *Mr. Milner*: not identified.
2. *[1866]*: the year is given by reference to going to Nottingham (see n. 4).

3. *your brother*: not identified.
4. *friend*: not identified.
5. *Nottingham*: the BAAS held its thirty-sixth annual meeting at Nottingham from 22–30 August (*Brit. Assoc. Rep. 1866*).

From William Francis 21 May 1866 2433

Printing Office | Red Lion Court, | Fleet Street, E. C. | May 21/66

My dear Tyndall,

Your note with the enclosed[1] was forwarded to me at Swanage[2] where I had been spending the last two weeks; and I should have replied to it at once, but wished before writing to you to consult a friend respecting the latter part of Dr. Wackerbarth's[3] letter[4] I know no one who would undertake the publication of a translation of Thalen's[5] spectral analysis[6] and I can scarcely imagine that any English bookseller would feel inclined to make the venture.

With respect to the first part of his letter, as Mr Bahr's[7] statements have not appeared in the Phil. Mag. I do not see how any notice can be taken of the subject. If however you are of a different opinion and will draw up a short paragraph for publication I will most willingly insert it.[8] I think however Dr. Wackerbarth's observations had better be addressed to the Editor of the Annalen der Chemie und Pharmacie in which Mr. Bahr's paper appeared.[9]

Many thanks for Neumayer's[10] paper[11] which is very interesting. . . . .
Francis

RI MS JT/1/TYP/12/3983
LT Typescript Only

1. *Your note with the enclosed*: letter missing.
2. *Swanage*: a coastal town in southern England.
3. *Dr. Wackerbarth's*: Athanasius Francis Didrik Wackerbarth (1813–84), a London-born German writer who was naturalized as a Swedish citizen in 1853. He then turned to astronomy. He was professor of mathematics at Uppsala University (*Monthly Notices of the Royal Astronomical Society*, 45 (1885), pp. 200–03).
4. *letter*: it is not clear if this refers to a letter to Tyndall from Wackerbarth (which if so, it is missing), or to Francis from Wackerbarth.
5. *Thalen's*: Tobias Robert Thalén (1827–1905), a Swedish physicist who collaborated with Anders Ångström (1814–74) in his researches on spectral analysis, for which he was awarded the RS's Rumford Medal in 1884 ('Prof. T. R. Thalén', *Nature*, 72 (1905), p. 403).

6. *spectral analysis*: R. Thalén, *Spektralanalys Exposé och Historik* (Upsala, 1866).
7. *Mr Bahr's*: Johann Friedrich Bahr (1805–75), a Swedish chemist who studied at Uppsala University and worked with Robert Bunsen in his laboratory in Heidelberg ('Bahr, Johann Fredrik', *Nordisk familjebok*, 20 vols (Stockholm: Gernandts boktryckeri-aktiebolag, 1876–99), vol. 1 (1876), pp. 1445–6).
8. *I will most willingly insert it*: it does not appear that Tyndall took up Francis's suggestion since there is no such note in the *Phil. Mag.*
9. *in which Mr. Bahr's paper appeared*: probably J. F. Bahr, 'Umkehrung der Absoprtions-streifen im Erbinspectrum', *Annalen der Chemie und Pharmacie*, 135 (1865), p. 376.
10. *Neumayer's*: Georg Balthasar von Neumayer.
11. *paper*: see letter 2419.

From Charles Lyell 21 May 1866 2434

53 Harley Street | May 21st/66

Dear Tyndall

I have just heard from Greenwich[1] in answer to those queries which I told you I had sent.[2] Before I tell you the result let me ask you in case I should forget it whether if there was a sea of some depth at the pole & consequently no blanket of snow to prevent the earth from radiating its heat through the cold & dry atmosphere, would there still be a great loss of heat during a long winter in aphelion.[3] In other words, suppose extreme excentricity & not only a much greater distance of the earth from the sun in winter but a month's longer winter than there is at present even at the South Pole, would the loss by radiation be very great from a sea of the average depth of the ocean? The answer to this question would I suppose not be affected by the probable fact that warm currents would be constantly flowing towards the poles & cold currents from the poles. What I want to know is whether in your researches on heat & radiation you have been able to form any estimate as to the relative degree of cooling by radiation of a body of deep salt water as compared to a surface of dry land.

At present there are points where the Arctic Sea is open for a part of the year as far north as lat. 830; but it may be said that here the cumulative effect of snow & ice is checked by the earth happening to be nearest the sun in winter. In the southern hemisphere open sea has not been found so near the pole by five or six degrees.

The answer from Greenwich in regard to Leverrier's[4] formula[5] is that we may depend upon it as such an approximation even if we were to go back four millions of years as would be sufficient for our geological calculations: & secondly that the recent important correction as to the distance of the earth from the sun which lessens the distance by 4 ½ millions of miles does not affect that formula

believe me | ever most truly yours | Cha[rles] Lyell
Prof. Tyndall

RI MS JT/1/L/52
RI MS JT/1/TYP/3/851

1. *Greenwich*: Royal Observatory, Greenwich.
2. *queries which I told you I had sent*: the queries were presumably in reference to Le Verrier's formula.
3. *let me ask you . . . long winter in aphelion*: Lyell had previously asked Tyndall about variations in the earth's orbit affecting climate (see letter 2243, esp. n. 2).
4. *Leverrier's*: Urbain Jean Joseph Le Verrier (1811–77), a French astronomer and co-discoverer of the planet Neptune (*CDSB*).
5. *formula*: Le Verrier's method for calculating the eccentricity of the Earth's orbit around the Sun in the past. See J. Gapaillard, 'By How Much Did Le Verrier Err on the Position of Neptune?', *Journal for the History of Astronomy*, 46 (2015), pp. 48–65.

To William Stanley Jevons[1] 23 May 1866 2435

23rd. May | 1866

Dear Sir,

The proposal you refer to was not mine, but was uttered, I think more in the way of a theoretic possibility than with any idea of its being practical by my epistolary in the columns of the Reader newspaper.[2] I attach no practical importance whatever to that suggestion.

I do not think there is any means of preserving our manufacturing supremacy after our coal fields have been exhausted. We are sure for example to be beaten by America.

I do not believe that the earth's crust, so far as we have explored it, or are likely to explore it, furnishes any substitute for coal.

The source of power is in all cases chemical attraction, and for the body of the earth's crust this attraction has been satisfied long ago & the heat produced by it wasted in space.

I console myself with the idea that though old England will not be in a position to support her present teeming population, the residue left to us may possibly be of a nobler quality than that which now exists.

We have our winds streams and tides; and we have above all the glorious sun. Noble men and women may be manufactured out of these materials; but their number must be limited.[3]

I am very sincerely | John Tyndall

John Rylands Library, University of Manchester, Jevons Archive

---

1. *William Stanley Jevons*: William Stanley Jevons (1835–82), an English economist and logician (*ODNB*). In 1866, Jevons published *The Coal Question: An Inquiry Concerning the Progress of the Nation, and the Probable Exhaustion of Our Coal Mines* (London: Macmillan and Co.).
2. *the Reader newspaper*: probably J. Tyndall, 'Vitality', *The Reader*, 29 October 1864, pp. 545–56 (this was later published in J. Tyndall, *Fragments of Science for Unscientific People: A Series of Detached Essays, Lectures, and Reviews* (London: Longmans, Green, and Co., 1871), pp. 436–44). In this short article Tyndall discussed the philosophical concept of vitality—'vital force, power, or principle as possessed or manifested by living things' (*ODNB*)—as originating from solar energy: 'In short, it is considered as proved that all the energy which we derive from plants and animals is drawn from the sun' (p. 436). He described 'creative' energy as opposed to 'mechanical' energy, noting that 'all the power which we obtain from the muscles of men and animals, as much as that which we develop by the combustion of wood or coal, has been produced at the sun's expense' (p. 437). On *The Reader* as a whole, see letter 2283, n. 3.
3. *number must be limited*: the reply from Jevons is missing, but see letter 2444 for Tyndall's continued discussion.

From Heinrich Gustav Magnus 23 May 1866 2436

Berlin 23 Mai 1866.

Verehrter Freund!

Ihre freundlichen Zeilen vom 17 d.M. erhielt ich gleichzeitig mit den Abhandlungen, die Sie so gut waren mir zu senden und für die ich Ihnen meinen verbindlichen Dank ausspreche. Sie sagen in Ihrem Briefe daß es Ihnen angenehm gewesen wäre, wenn Sie nichts mehr mit der Controverse zwischen uns zu thun gehabt hätten. Seien Sie überzeugt daß ich mich in derselben Lage befinde, und daß es mir ganz besonders schmerzlich ist gerade mit Ihnen öffentlich eine solche Controverse führen zu müssen.—Sie erwähnen ferner daß von mir nicht ein einziger von den vielen „powerful proofs" berührt sei, die Sie für die Wirkung der Wasserdämpfe geliefert hätten. Ich kann Sie versichern daß ich alle die Versuche, die Sie in dieser Beziehung veröffentlicht haben, nicht nur kenne, sondern auch alle oft wiederholt oder zu wiederholen versucht habe und daß ich daher wohl weiß welche Beweiskraft jedem derselben innewohnt. Ich habe es aber bisher vermieden auf eine Discussion dieser Versuche einzugehn, und werde es ferner vermeiden, es sei denn daß ich zum Gegentheil gedrängt werde. Denn, wie sehr man sich auch in Acht nehmen möge, so wird eine solche öffentliche Discussion von Versuchen doch leicht persönlich, und dergleichen möchte ich, bei den freundschaftlichen

Beziehungen welche uns verbinden, so lange als irgend möglich vermeiden. Seien Sie überzeugt, daß meine alte Freundschaft Ihnen unverändert bewahrt ist.

Ihr | G Magnus

Berlin 23 May 1866.

Dear friend!

I received your friendly lines from the 17th of this month[1] together with the articles that you were so kind to send me and I would like to express my grateful thanks. You say in your letter that you would have preferred not having to deal with the controversy between us any more.[2] You can be sure that I am in the same situation and that it is especially painful for me to participate in a public controversy like this particularly with you.—You mention further that I did not touch on any of the "powerful proofs" that you delivered in relation to the effect of water vapours. I can assure that I not only know but have repeated or attempted to repeat all the experiments that you published in this regard, and that I know well the evidential value inherent in each. But so far I have avoided discussing these experiments and will continue avoiding it, unless I am forced to do otherwise. For, no matter how careful one tries to be, such public discussions of experiments easily become personal and this is something that I would like to prevent as much as long as I can, given the friendly relations that unite us. Rest assured that my old friendship to you remains unchanged.

Your | G Magnus.

RI MS JT/1/M/32

1. *your friendly lines from the 17th of this month*: Tyndall's letter in reply to Magnus's previous letter (2425) is missing.
2. *deal with the controversy between us any more*: in his paper on dew (see letter 2425, n. 1), Magnus shared his view that transparent aqueous vapour was not a good absorber of heat. Tyndall disagreed, of course, and published a response to Magnus's paper in the *Phil. Mag.* (see letter 2425, n. 2).

From Johann Conrad Bohn 25 May 1866 2437

Giessen d. 25 Mai 1866.

Hochgeehrter Herr Tyndall!

Ihren letzten Brief habe ich, in vergeblicher Erwartung der darin angekündigten Mittheilung Viewegs bis heute zu beantworten gezögert.

Gegen die von Ihnen getroffene Entscheidung hinsichtlich der Uebersetzung Ihres Buches, steht mir kein Weg der Berufung offen, den ich zu betreten wünsche. Mein Bedauern darüber dass dem so ist, habe ich schon früher ausgesprochen.

Es bleibt mir nur noch übrig Sie um gefällige Rücksendung des noch in Ihren Händen befindlichen Theils meines Manuscriptes zu bitten. Für Sie hat dieses Manuscript keinen Werth, wohl aber für mich.

Genehmigen Sie die Versicherung der ausgezeichneten Hochachtung Ihres ergebenen Dieners | J. C Bohn

Giessen 25 May 1866.

Most honourable Mr Tyndall!

To this day I have delayed answering your last letter[1] in futile anticipation of the communication by Vieweg[2] that had been announced therein.

There is no way I see it feasible to appeal against your decision regarding the translation of your book.[3] I have already expressed my regret on this matter on an earlier occasion.[4]

The only thing left for me to do is to ask you to kindly return that part of the manuscript that is still in your possession. This manuscript has no value to you, but it does to me.[5]

Let me assure you of the highest regard
From your obedient servant | J. C Bohn

RI MS JT/1/B/115

1. *your last letter*: letter missing.
2. *Vieweg*: Eduard Vieweg.
3. *translation of your book*: the German translation of *Heat* (1865); see letter 2296, n. 6.
4. *on an earlier occasion*: see letter 2417.
5. *it does to me*: see letter 2446.

To John Herschel 26 May 1866 2438

26th May 1866

My dear Sir John,

The thought has occurred to me that you might like to hear from an impartial listener some account of the lecture last night.[1] I think it may without the slightest exaggeration be described as entirely successful. The delivery was good, the experiments were well devised and the explanations of a somewhat entangled subject eminently clear and satisfactory. Mr. Herschel[2] has certainly made a great step forward since he last lectured at the Royal Institution[3]

Yours most truly | John Tyndall

RS HS 17/400
RI MS JT/1/TYP/2/539

1. *lecture last night*: Herschel's second son, Alexander Herschel (see n. 2), delivered a Friday Evening Discourse at the RI on 25 May 1866, on 'The Shooting-stars of the Years 1865–66, and on the Probability of the Cosmical Theory of their Origin' (*Roy. Inst. Proc.*, 4 (1862–6), pp. 645–50).
2. *M*[r]*. Herschel*: Alexander Stewart Herschel (1836–1907), an English astronomer and son of John Herschel. He was professor of natural philosophy at Anderson's College, Glasgow (1866–71), and professor of physics and experimental philosophy at the University of Durham College of Science, Newcastle (1871–86). He is known primarily for his work on meteor spectroscopy and for identifying the link between the orbits of comets and the occurrence of meteor showers (*ODNB*).
3. *last lectured at the Royal Institution*: Herschel had previously delivered a Friday Evening Discourse at the RI on 24 April 1866, 'On Luminous Meteors' (*Roy. Inst. Proc.*, 4 (1862–6), pp. 87–97).

From John Herschel 28 May 1866 2439

Collingwood[1] | May 28/ 66

My dear Sir

Many thanks to you for your note[2] of the day before yesterday. I am much gratified with that expression of your opinion as to the improvement in my son's[3] style of Lecturing[4]—The fact is he warms to that subject—for he has taken it in hand with an immense energy & perseverance, and I suppose by this time is in as full possession of that subject as any person now living—and as things hang together, he has now taken to Meteorology with almost equal devotion. I wish I could see him established as a Prof. of Physics[5] in some considerable Institution or one of the many Universities which are daily rising into eminence throughout the country. His devotion to scientific objects is such that I am convinced he will find his happiness & utility in no line of life that will detach him from them.

Once more thanking you for your letter I remain

Dear Sir | Yours very truly | J F W Herschel

RI MS JT/1/H/106
RS HS 24/157
RI MS JT/1/TYP/2/540

1. *Collingwood*: Herschel's home near Hawkhurst in Kent.

2. *your note*: letter 2438.
3. *my son's*: Alexander Herschel (see letter 2438, n. 2).
4. *style of Lecturing*: Alexander Herschel delivered a Friday Evening Discourse at the RI on 25 May 1866 (see letter 2438, n. 1).
5. *see him established as a Prof. of Physics*: Alexander Herschel was appointed professor of natural philosophy at Anderson's College, Glasgow later in 1866.

From Hermann Helmholtz 29 May 1866 2440

Heidelberg. 29 Mai 1866.

Verehrter Freund

Ich hatte eigentlich wenig Lust Bemerkungen gegen Ihre, mir freundlichst mitgetheilten Einwände, die Gletschertheorie betreffend zu veröffentlichen, da ich die Sache noch nicht für spruchreif halte, und die Thatsachen dem wissenschaftlichen Publicum vorliegen. Da ich aber von verschiedenen Seiten über meine schließliche Ansicht der Sache befragt wurde, habe ich mich endlich entschlossen bei Gelegenheit der Veröffentlichung unserer beiden Aufsätze in der Revue des Cours littéraires einige Worte hinzuzufügen, von denen ich Ihnen umstehend Mittheilung mache: Sollten Sie eine Veröffentlichung davon im Philosophical Magazine passend finden, so bitte ich Sie eine solche zu veranlassen.

Wir haben Sie wegen der unangenehmen Erfahrungen sehr bedauert, die Sie bei Gelegenheit der Übersetzung Ihres schönen Buches „On heat" gemacht haben. Ich hatte im vergangenen Winter, ohne von Bohn's Unternehmen etwas zu wissen, an Vieweg geschrieben, und ihm vorgeschlagen die Übersetzung unter meiner Aufsicht von meiner Frau ausführen zu lassen, die schon einige englische Bücher der Mrs. Gaskell übersetzt hat, und mehr Gewandtheit in der Bildung des deutschen Ausdrucks hat, als ich selbst haben würde. Vieweg schrieb mir damals, daß Professor Bohn die Arbeit schon angefangen habe, und erst neuerdings hörte ich durch Vieweg und Wiedemann, daß sich Bohn als ein sehr schlechter Übersetzer erwiesen habe, und daß Sie geneigt wären, wenn sich die Sache mit Bohn erledigen ließe, uns die Übersetzung zu überlassen.

Mit vielem Dank für die zugesandten Abhandlungen und in treuer Freundschaft

Ihr | H. Helmholtz

Heidelberg. 29 May 1866.

Dear friend

I really did not want to publish my remarks on your objections to the theory of glaciers which you kindly sent me, as I do not consider the matter ready for publication nor are the facts before the scientific public. But as I have been asked from various quarters for my final opinion on the matter, I have decided with the publication of our two essays in the Revue des Cours littéraires to add a few words, which I am sending you: if you consider it appropriate to publish them in the Philosophical Magazine, please have it done.[1]

We were very sorry about the unpleasant experience you had with the translation of your fine book "On heat".[2] Last winter, without knowing about Bohn's[3] undertaking, I wrote to Vieweg[4] and proposed letting my wife do the translation under my supervision.[5] She has already translated some of Mrs Gaskell's books[6] and has more skill getting the German expression than I would have. Vieweg wrote at that time that Professor Bohn had begun the work and only recently did I hear from Vieweg and Wiedemann that Bohn had proved to be a very bad translator and that you were inclined to let us do the translation, if you managed to cancel the deal with Bohn.

With gratitude for the essays you sent and in faithful friendship

Your | H. Helmholtz

RI MS JT/H/43

---

1. *please have it done*: Helmholtz erred in giving the title journal as *Revue des Cours littéraires*; it is *Revue des Cours Scientifiques*. See H. Helmholtz, 'La Glace et les Glaciers', *Revue des Cours Scientifiques*, 3 (1866), pp. 433–47; and J. Tyndall, 'Au Sujet de la Leçon de M. Helmholtz sur la Glace et les Glaciers', *Revue des Cours Scientifiques*, 3 (1866), pp. 447–51. The enclosure to this letter, Helmholtz's 'few words', is missing, but they were indeed forwarded to the journal by Tyndall and published as H. Helmholtz, 'On the Regelation of Ice', *Phil. Mag.*, 32 (1866), pp. 22–3. On p. 22 is Tyndall's letter to the editors of the *Phil. Mag.* about Hemholtz's note; this is letter 2456 in this volume.
2. *the unpleasant experience you had with the translation of your fine book "On heat"*: the German translation of *Heat* (1865); see letter 2296, n. 6.
3. *Bohn's*: Johann Conrad Bohn.
4. *Vieweg*: Eduard Vieweg.
5. *letting my wife do the translation under my supervision*: Anna Helmholtz; see letter 2296, n. 6.
6. *Mrs Gaskell's books*: Elizabeth Cleghorn Gaskell (1810–65), a prominent English novelist (*ODNB*). Her novels included *Mary Barton* (1848), *Ruth* (1853), *North and South* (1855), *Sylvia's Lovers* (1863), and *Wives and Lovers* (1866).

From Félix Foucou 31 May 1866 2441

Paris—31 mai 1866

Mon cher Monsieur Tyndall,

Mon petit garçon étant tombé malade la semaine dernière, je ne partirai que le 8 Juin pour l'Amérique. J'espère ainsi être en Angleterre à l'époque du meeting de la "British Association" et avoir le plaisir de vous y trouver.

En me transmettant votre acceptation conditionnelle, M. Hetzel m'a écrit:

"Un de mes amis croit comprendre que M. Tyndall se réserve l'avenir. Ce sera donc le droit de porter son livre ailleurs. Si cela était, je vous avoue que je ne me soucierais pas de servir d'engrais aux éditions futures. Une première édition qui réussit n'est en effet que cela. M. Tyndall est dans l'erreur s'il croit que sa réserve soit de nature à être acceptée par qui que ce soit ayant l'envie de faire le succès du livre et de s'y attacher sérieusement. On ne plante pas des arbres dans les maisons où l'on n'est que locataire."

J'ai cru, Mon cher Monsieur Tyndall, devoir vous transcrire cette déclaration, afin de vous mettre à même de savoir sur quel pied marcher. Les informations que j'ai prises, ces derniers jours, me font craindre en effet, qu'il ne soit pas possible de trouver à Paris un éditeur pour la 1ère édition seule. Mais je comprends la justesse de l'objection que vous soulevez, et je souscris d'avance à toute combinaison qui vous paraîtra de nature à sauvegarder vos intérêts.

Je n'ai pas besoin d'ajouter que si le sacrifice de ma traduction était nécessaire pour assurer ce résultat, je le ferais, malgré le prix que j'attache à voir mon nom se présenter au public sous les auspices du votre.

Dans mon manuscrit j'avais joint déjà l'accident survenu dans l'Engadine en 1864.

Je pars vendredi de la semaine prochaine. D'ici là, soyez assez bon pour m'écrire ce que je dois faire du manuscrit et des clichés.

Mille compliments affectueux | Félix Foucou

Paris—31 May 1866

My dear Mr Tyndall,

My little boy fell ill last week, so I will not be able to leave for America[1] before June 8th. I hope to be in England then around the time of the meeting for the "British Association",[2] and to have the pleasure of finding you there.

While sending me your conditional acceptance, Mr. Hetzel[3] wrote to me:

"One of my friends understands that Mr. Tyndall plans for the future. Thus, he will have the right to bring his book[4] elsewhere. If that is the case, I confess that I would not care to be used as a fertilizer for future editions. A first edition that succeeds is precisely that. Mr. Tyndall is mistaken if he thinks that his plan is of the kind to be accepted by anyone who has the desire to make a book successful and to work on it seriously. We do not plant trees in homes where we are only tenants."

I thought, my dear Mr. Tyndall, that I should transcribe this statement to let you know which steps to take. The information that I have received in the last few days made me worry that it would not be possible to find a publisher in Paris only for the first edition. But I understand the soundness of the objection that you raise, and I agree in advance to any combination that you will see as safeguarding your interests.

I do not need to add that if the sacrifice of my translation was necessary to ensure this result, I would do it, despite the price that I attach to my name appearing in public under the auspices of your own.

In my manuscript, I already added the accident that occurred in Engadin in 1864.

I am leaving on Friday next week. Until then, please write to me what I should do with the manuscript and the proofs.

Many affectionate compliments. | Félix Foucou

RI MS JT/1/F/39

---

1. *leave for America*: see letter 2431, n. 3.
2. *meeting for the "British Association"*: Tyndall attended the thirty-sixth annual meeting of the BAAS, held in Nottingham from 22–30 August 1866 (*Brit. Assoc. Rep. 1866*).
3. *Mr. Hetzel*: Pierre-Jules Hetzel.
4. *his book*: for a two-volume French edition of *Glaciers of the Alps* which, as this letter indicates, fell through.

To the Editors of the *Phil. Mag.*[1] [May 1866][2] 2442

LXXXIII. *On Aqueous Vapour and Terrestrial Radiation.* | *By* M. G. NEUMAYER. | *To the Editors of the Philosophical Magazine and Journal.*

GENTLEMEN,

I TRANSMIT to you a communication from M. Neumayer,[3] which I trust will have the effect of stimulating other meteorologists to enter upon and pursue the line of inquiry which he has so ably and zealously begun. There is, in my estimation, no result of physical science more certain than that the aqueous vapour of our air exerts a powerful influence upon the radiation from our earth; and there is none, I imagine, more likely to guide the really scientific meteorologist to results of permanent value.[4]

JOHN TYNDALL.

Phil. Mag., 31 (1866), p. 510

1. *Editors of the Phil. Mag.*: see letter 2245, n. 1.
2. *[May 1866]*: the month and year are given based on Neumayer's letter to Tyndall, dated 4 May 1866, and the month it was published in the *Phil. Mag.* (in the supplement following the June 1866 issue).
3. *communication from M. Neumayer*: Georg Balthasar von Neumayer had sent Tyndall a letter on 4 May 1866 (letter 2419), and Tyndall forwarded it to the *Phil. Mag.*
4. *of permanent value*: meteorology was a subject of great interest to Tyndall. See S. Dry, *Waters of the World: The Story of the Scientists Who Unraveled the Mysteries of our Oceans, Atmosphere, and Ice Sheets and Made the Planet Whole* (Chicago: University of Chicago Press, 2019), chapters 2 and 3; R. Jackson, 'Eunice Foote, John Tyndall and a Question of Priority', *Notes and Records*, 74 (2020), pp. 105–18; and I. M. McCabe, 'Second best as a researcher, second none as a populariser? The atmospheric science of John Tyndall FRS (1820–1893)' (PhD Dissertation, University College London, 2012). Tyndall's mountaineering exploits unsurprisingly led to his interest in the workings of the atmosphere. In an 1862 letter to Michael Faraday, following the successful climb of the Galenstock in the Switzerland, Tyndall wrote: 'We did not hope to see any thing from the summit, nevertheless we attained it and were rewarded:—not by the clearness of the prospect, but by the changes of the atmosphere, which were quite marvelous—sometimes shrouding all sometimes melting as if by magic and revealing the mountains. The atmosphere is a wondrous factory; the grand origin of all its power being overhead, lifting the snows and driving the clouds by his individual might' (see letter 1915, *Tyndall Correspondence*, vol. 8).

To Charles Lyell 1 June 1866 2443

1st June 1866.

My dear Sir Charles,

Water in all its forms is a powerful radiator—It is a great radiator as snow, it is a great radiator as ice, it is a great radiator in the liquid condition.

In virtue of this power of radiation were the pole surrounded by snow it would be intensely chilled by a long winter. Were it surrounded by water, the water would soon become ice. Were it surrounded by ice that ice like the snow would become intensely chilled—in fact about to the same extent.

I can however imagine equatorial currents of sufficient intensity to resist the formation of ice—Unless we know the rate of transfer & the temperature of the water nothing can be said upon this point.

Taking the difference of weight into account I do not think the radiation from a sheet of ice or of snow or of water, would be materially different from that of an expanse of dry land.

All the compounds which constitute dry land have more or less of water as a constituent, and this water though in the state of chemical union still exerts a powerful action as a radiator.

I do not quite know the drift of your questions. But I would not consider the difference in radiative power between water & land is likely to furnish the solution of the glacial epoch.[1]

Yours very truly | John Tyndall

Excuse the little delay—I have been awfully busy & I at first intended to go more fully into the matter—The materials however sind nicht vorhanden[2]

RI MS JT/1/T/1053
RI MS JT/1/TYP/3/839–40

1. *glacial epoch*: in letter 2434, Lyell had asked Tyndall about variations in the earth's orbit affecting climate. See also letter 2243, esp. n. 2.
2. *sind nicht vorhanden*: are not available (German).

To William Stanley Jevons[1] 2 June 1866 2444

2nd June 1866

Dear Sir

The accompanying paragraph bears upon your subject.[2] It is copied from an article "On the Constitution of the Universe," published in the Fortnightly Review for December 1st 1865 p. 143.[3]

I would supplement the statements of the paragraph by adding that I see no prospect of any substitute being found for coal as a source of motive power. We have, it is true, our winds & streams and tides; and we have the beams of the sun as a source of the power derivable from trees and animals. But these are common to all the world. We cannot make head against a nation which, in addition to those sources of power, possesses the power of coal. We may enjoy a multiple of the energy both physical & intellectual, and still be unable to hold our own against a people which possesses abundance of coal; and we should have, in my opinion, no chance whatever in a race with a nation whose energy & intelligence are approximately equal to our own.

It is no new thing for me to affirm in my public lectures that the destiny of this nation is not in the hands of its statesmen but in those of its coal owners; and that while the orators of St Stephens[4] are in blessed unconsciousness of the fact the very lifeblood of this country is flowing away.

You may add to this note such extracts from my preceding one as you may think it desirable to publish. My opinion is very much at your service.

Yours very sincerely | John Tyndall
W. Jevons Esquire
I agree with you in thinking the talk about electricity as a source of practical power fit to replace coal as so much nonsense.

John Rylands Library, University of Manchester, Jevons Archive

---

1. *William Stanley Jevons*: see letter 2435, n. 1.
2. *The accompanying paragraph bears upon your subject*: see letter 2435 from Tyndall to Jevons. A reply from Jevons, to which this letter is a reply, is missing. The paragraph from 'The Constitution of the Universe' (see n. 3) that Tyndall was referring to, which Tyndall presumably provided to Jevon as an enclosure and is missing, is probably: 'Wherever two atoms capable of uniting together by their mutual attractions exist separately, they form a store of potential energy. Thus our woods, forests, and coal-fields on the one hand, and our atmospheric oxygen on the other, constitute a vast store of energy of this kind—vast, but far from infinite. We have, besides our coal fields, bodies in the metallic condition more or less sparsely distributed in the earth's crust. These bodies can be oxydised, and hence are, so far as they go, stores of potential energy. But the attractions of the great mass of the earth's crust are already satisfied, and from them no further energy can possibly be obtained. Ages ago the elementary constituents of our rocks clashed together and produced the motion of heat, which was taken up by the ether and carried away through stellar space. It is lost for ever as far as we are concerned. In those ages the hot conflict of carbon, oxygen, and calcium produced the chalk and limestone hills which are now cold; and from this carbon, oxygen, and calcium no further energy can be derived. And so it is with almost all the other constituents of the earth's crust. They took their present form in obedience to molecular force; they turned their potential energy into dynamic, and gave it to the universe ages before man appeared upon this planet. For him a residue of potential energy remains, vast truly in relation to the life and wants of an individual, but exceedingly minute in comparison with the earth's primitive store' (p. 143).
3. *Fortnightly Review for December 1st 1865 p. 143*: J. Tyndall, 'The Constitution of the Universe', *Fortnightly Review*, 3 (1865), pp. 129–44.
4. *orators of St Stephens*: St. Stephen's Chapel was the original sitting place of the House of Commons, so the reference is to Members of Parliament.

From Joseph Plateau 8 June 1866 2445

Gand, 8 Juin 1866

Mon cher Monsieur Tyndall,

J'ai l'honneur de vous adresser trois exemplaires de mon dernier mémoire, veuillez accepter l'un d'eux, offrir, de ma part, le second à la Société Royale, et le troisième à l'Institution Royale. Encore une série Sur les figures d'équilibre d'une masse liquide sans pesanteur! direz-vous peut-être. C'est que ce titre n'est pas tout à fait le plus convenable, il s'adaptait parfaitement à mes premières séries, mais alors je ne me rendais pas bien compte du but général de l'ouvrage. Quand cet ouvrage sera complet, ce qui exigera encore quatre séries, il constituera la Statique expérimentale et théorique des liquides soumis aux seules forces moléculaires, et pourra, je l'espère, être considéré comme fermant une tranche nouvelle de la physique.

Vous trouverez, dans la série actuelle, les résultats définitifs de mon étude du liquide glycérique, et l'indication nette des meilleurs procédés à suivre pour la préparation de ce liquide.

Recevez, mon cher Monsieur Tyndall, mes remerciements pour l'obligeance que vous avez eue de remettre à la Société Royale l'exemplaire du mémoire de mon fils, et agréez l'assurance de tous mes sentiments d'affection.

J. Plateau.

Ghent, 8 June 1866

My dear Mr Tyndall,

I have the honour of sending you three copies of my latest memoir.[1] Please accept one of those, and offer the second one on my behalf to the Royal Society, and the third one to the Royal Institution. Yet another series Sur les figures d'équilibre d'une masse liquide sans pesanteur![2] you might say. But this title is not quite fitting; it fitted perfectly my earlier series, but I had not taken into account the general purpose of the book. When this work is done, which would still require another four series, it will be the Statique expérimentale et théorique des liquides soumis aux seules forces moléculaires,[3] and I hope that it could be considered as closing a new phase of physics.

In the current series, you will find the final results of my study of the glyceric liquid and a clear indication of the best methods to follow for its preparation.

Please accept, my dear Mr Tyndall, my gratitude for the kindness that you have shown by giving my son's[4] memoir to the Royal Society;[5] and accept my feelings of affection.

J. Plateau.

RI MS JT/1/P/98

1. *my latest memoir*: see letter 2332, n. 3.
2. *Sur les figures d'équilibre d'une masse liquide sans pesanteur*: see letter 2332, n. 3.
3. *Statique expérimentale et théorique des liquides soumis aux seules forces moléculaires*: see letter 2332, n. 3.
4. *my son's*: Félix Plateau (see letter 2332, n. 1).
5. *memoir to the Royal Society*: see letter 2332, n. 2.

From Johann Conrad Bohn 13 June 1866 2446

Giessen d. 13 Juni 1866

Hochgeehrter Herr Tyndall!

Am 10ten April forderten Sie mich auf Ihnen das Manuscript meiner Uebersetzung Ihres Buches zur Ansicht zu schicken und versprachen: „I will not delay it longer than necessary". Ich schickte Ihnen hierauf am 12n April einen Theil meines M.S.; am 26n April und in noch förmlicherer Weise am 7n Mai entzogen Sie mir das Recht zur Herausgabe der Uebersetzung Ihres Buches. Da Sie demnach mein M.S. nicht mehr nöthig haben, ersuchte ich Sie am 25tn Mai nun die am 10n April versprochene gefällige Rücksendung desselben. Eine solche ist bis heute nicht erfolgt und ich bin daher genöthigt wiederholt jenes Ersuchen an Sie zu stellen.

Ihr ergebener Diener | Dr Bohn

Giessen, 13 June 1866

Highly esteemed Mr. Tyndall!

On April 10th you requested that I send you the manuscript of my translation of your book[1] for review and promised: "I will not delay it longer than necessary". Consequently, I sent you on April 12th a part of my M.S.;[2] on April 26th and in a more formal way on May 7th you withdrew from me the right to publish the translation of your book.[3] As you therefore no longer need my M.S., on May 25th I requested the kind return of the same,[4] as promised on April 10th.[5] Until today this has not occurred and I am therefore forced to repeat this request to you.

Your humble servant | Dr Bohn

RI MS JT/1/B/116

1. *my translation of your book*: the German translation of *Heat* (1865); see letter 2296, n. 6.
2. *I sent you on April 12th a part of my M.S.*: see letter 2407.
3. *on April 26th and in a more formal way on May 7th . . . translation of your book*: letters missing.
4. *on May 25th I requested the kind return of the same*: see letter 2437.
5. *as promised on April 10th*: letter missing.

From William Francis 20 June 1866 2447

Printing Office, | Red Lion Court, | Fleet Street, E.C. | June 20th /66

My dear Tyndall,

I should prefer Hull and Selby Purchase at £105 to North Eastern 5 % at par The first is a guaranteed 4 1/2 per cent Stock, with a good prospect of 5 or 6 %; the other is merely a preference Stock.

London and North Western are also worth buying at present prices and so also are Lancashire and Yorkshire at £120.

I shall be very glad to have the abstract of your lecture for the Magazine[1] if you will send me a prepared copy. I also take this opportunity of reminding you that I have not yet received your promised notice of some paper of Helmholtz and that the month is fast passing away.[2]

Yours sincerely | William Francis | Prof. Tyndall, F.R.S.

RI MS JT/1/TYP/12/3984
LT Typescript Only

1. *abstract of your lecture for the Magazine*: J. Tyndall, 'Experiments on the Vibrations of Strings', *Phil. Mag.*, 32 (1866), pp. 68–76. Tyndall delivered a Friday Evening Discourse at the RI, 'Experiments of the Vibrations of Strings', on 15 June 1866 (*Roy. Inst. Proc.*, 4 (1862–6), pp. 685–94).
2. *your promised notice of some paper of Helmholtz*: H. Helmholtz, 'On the Regelation of Ice', *Phil. Mag.*, 32 (1866), pp. 22–3. See letter 2440, n. 1.

From Emily Tennyson[1] 21 June 1866 2448

June 21st 1866

My dear Professor,

Do you remember a certain arched cliff not a hundred miles from this?[2] Well notwithstanding—such is my faith in you I would trust him[3] with you and believe that great would be the good and pleasure of his going.[4]

He has a feeling, I think, that he might be a hindrance to you because of his eyes.

Then will your times suit? Or rather can he make his bend to yours? May we have a time to tell us when it is? tho' I fear that he is not likely to go.

With many thanks and much kind regard from us

believe me | most truly yours | Emily Tennyson

We have been talking of Switzerland or rather of Aix & Chamberry[5] while the boys[6] are with us. We should leave home about the 3rd of July if we did take them & return (I at least, and the boys) about the middle of August because of school. This even is not settled.

RI MS JT/1/T/5

---

1. *Emily Tennyson*: Emily Sarah Tennyson (née Sellwood, 1813–96), Lady Tennyson from 1883, was the wife, manager, and secretary of Alfred Tennyson. She married Tennyson in 1850 and together they had two sons, Hallam and Lionel, in 1852 and 1854. Emily has been recognized by her contemporaries and by historians for the important role she played in Tennyson's work. She answered his correspondence, edited and set his poetry to music, and read French, German, and Italian (*ODNB*).
2. *a certain arched cliff*: probably a reference to Tyndall's trip to the Isle of Wight in April 1866 (see letter 2409, n. 7), when he 'Saw Tennyson, roamed the downs, climbed down and up the cliffs of Scratchels bay, and returned much benefitted' (Journal, RI MS JT/2/13c/1330). Additionally, Tyndall wrote to Annette Roche in November 1864: 'A fortnight ago I flew down for a day to the Isle of Wight. I spent an evening with Tennyson, and on the following day he accompanied me in a climbing excursion up the chalk cliffs' (see letter 2171, *Tyndall Correspondence*, vol. 8).
3. *him*: Alfred Tennyson.
4. *his going*: this refers to eventually abortive plans that Tennyson should accompany Tyndall on his annual excursion in the Alps.
5. *Aix & Chamberry*: towns in eastern France, Aix-les-Bains on the eastern shore of the Lac du Bourget, and Chambéry just south of Aix-les-Bains.
6. *boys*: Hallam Tennyson (1852–1928), second Baron Tennyson, the eldest son of Alfred and Emily Tennyson. He was educated at home by his mother (as was his brother) before attending Marlborough College (1866–72), followed by Trinity College, Cambridge. Following the death of his father Hallam inherited the barony in 1892, and subsequently wrote his father's biography, *Tennyson: A Memoir* (London: Macmillan and Co., 1897). He served as governor of South Australia (1899–1902) and governor-general of Australia (1903–4) (*ODNB*). Lionel Tennyson (1852–86), the second son of Alfred and Emily Tennyson. He was educated at home by his mother (as was his brother) before attending Eton College, followed by Trinity College, Cambridge. In 1878 he married Eleanor Locker with whom he had two children, Alfred (b. 1878) and Charles (b. 1879). He worked for the India Office and while travelling in India in 1885 he developed a fever and died in 1886 (Anon, *Lionel Tennyson* (London: Macmillan and Co., 1891)).

To Juliet Pollock 22 June [1866][1] 2449

Royal Institution of Great Britain | 22nd June

My dear Mrs. Pollock

On Tuesday next at 3 P.M. I expect a small audience here to see the experiments on vibrating strings.[2] The doors shall be open at that hour for you and your friend.[3]

Yours ever | John Tyndall

RI MS JT/1/T/1168
RI MS JT/1/TYP/6/2081

---

1. *[1866]*: the year is given by reference to Tyndall's Friday Evening Discourse at the RI on 15 June 1866 (see n. 2).
2. *the experiments on vibrating strings*: on 15 June 1866, Tyndall delivered a Friday Evening Discourse at the RI, "Experiments on the Vibrations of Strings' (see letter 2447, n. 1). He repeated this discourse at the RI at 3 PM on 26 June 1866 for a distinguished small audience that included the Duke of Argyll (George Campbell) and his Duchess and daughter; Lord Belper (Edward Strutt) and his wife Lady Belper; Ellen Frances Lubbock (née Hordern, 1834–79), the daughter of an Anglican minister who married John Lubbock in 1856; and Ellen Busk (see letter 2343, n. 1); as well as Tyndall's friend Thomas Hirst, who noted his attendance and some other guests in his journal entry for 26 June 1866 (RI MS JT/2/32d; the version of this entry on p. 1787 in Brock and MacLeod, *Hirst Journals* is incomplete).
3. *your friend*: not identified.

To Edward Strutt 22 June [1866][1] 2450

22nd June

Dear Lord Belper

On Tuesday next, the 26th, at 3 P.M. I shall feel great pleasure in showing you the experiments mentioned on Wednesday evening.[2]

Of course all doors shall be open for Lady Belper[3] also if she should care to come.

Yours very faithfully | John Tyndall

Nottinghamshire Archives DD/BK/7/4

---

1. *[1866]*: the year is given by reference to Tyndall's Friday Evening Discourse at the RI on 15 June 1866 (see letter 2449, n. 2).
2. *the experiments mentioned on Wednesday evening*: see letter 2449, n. 2.
3. *Lady Belper*: Amelia Harriet Strutt (née Otter, 1817–90), Lady Belper, the wife of Edward Strutt. They were married in 1837 and had six children (*ODNB*).

To Emily Tennyson[1] 22 June [1866][2] 2451

Royal Institution of Great Britain | 22nd June

My dear Mrs Tennyson

I know that grand arched cliff[3] perfectly, but no danger shall pour upon your husband[4] if you entrust him to me.

I shall be ready to start at the time you mention. I would propose going from Paris to Bale,[5] and from Bale to Thun;[6] we are there in the presence of the Alps.

Why he showed me all the bays and inlets of the back of the Island[7] from the top of his own house—why therefore does he complain of his eyes?

We could readily find a quiet "Pension" for you in the neighborhood of Thun or Interlaken or Meyringen.[8] It would be very jolly to have the boys with us. Hallam[9] could use his legs no doubt, but Lionel[10] would I fear need a pony sometimes.

I hope he will make up his mind to come. Let him do so & I will tell him what to pack in his knapsack

Yours very truly | John Tyndall

I intended before the death of his wife[11] to persuade Mr Carlyle to come, but that is now out of the question. Besides I fear his sleep might not be always good

Tennyson Research Center Letters 6242

---

1. *Emily Tennyson*: see letter 2448, n. 1.
2. *[1866]*: the year is given by relation to letter 2448 from Emily Tennyson, dated 21 June 1866.
3. *cliff*: on the Isle of Wight (see letter 2448, n. 2).
4. *your husband*: Alfred Tennyson.
5. *Bale*: see letter 2309, n. 3.
6. *Thun*: a town in the Swiss canton of Bern.
7. *Island*: the Isle of Wight where Tyndall often visited the Tennyson family at their island

home Farringford House in the village of Freshwater Bay. The family occupied this home since 1853 (and it was kept by the family after Tennyson's death in 1892), but from 1869 the Tennyson's main residence was Aldworth House near Haslemere.

8. *Meyringen*: Meiringen, a town in the Swiss canton of Bern.
9. *Hallam*: see letter 2448, n. 6.
10. *Lionel*: see letter 2448, n. 6.
11. *his wife*: Jane Welsh Carlyle had died suddenly on 21 April 1866 (*ODNB*). See letter 2414, n. 4.

From François-Napoléon-Marie Moigno — 23 June 1866 — 2452

Mon cher Mr Tyndall

Je suis désolé de ne vous avoir pas vu, plus désolé encore de n'avoir pas pu forcer Mr Giraud à rendre ses comptes avant mon départ de Paris. Vous êtes pour moi un remords, et je ne récupérerai la paix qu'après vous avoir payé loyalement ce qui vous [revient] de la publication de vos admirables leçons sur la chaleur.

Je vais visiter le Great Eastern, je serai de retour je l'espère vers deux heures, faites que je vous retrouve à Jermyn Street, il me serait trop pénible de quitter Londres ce jour vous avoir serré la main. Mr Wheatstone m'a appris une bonne nouvelle, vous passerez une semaine à Paris, j'en suis tout heureux.

Votre admirateur et votre ami | F. Moigno | 23 juin 1866

Acceptez le bon souvenir de mon [affection] respectueuse et dévouée.

My dear Mr Tyndall

I am sorry that I did not see you, and even sorrier that I could not compel Mr Giraud[1] to complete his accounts before I left Paris. I feel remorse, and I will not feel at peace until I properly pay you your earnings, which come from the publication of your admirable lessons on heat.

I go to visit the Great Eastern,[2] I will return I hope at around 2 o'clock, let me meet you at Jermyn Street,[3] it would be too painful for me to leave London that day without having shaken hands with you. Mr Wheatstone[4] told me some good news, that you will spend a week in Paris, and I am very glad.

Your admirer and your friend | F. Moigno | 23 June 1866

Please accept the [remembrance] of my respectful affection and devotion.

RI MS JT/1/M/109

1. *Mr Giraud*: Étienne Giraud.
2. *to visit the Great Eastern*: the SS *Great Eastern* (also called the Leviathan) was at the time the largest ship ever built. Designed by Isambard Kingdom Brunel (1806–59), its launch on 31 January 1857 on the River Thames was a public spectacle. At this time the ship was being used to lay down underwater telegraph cables.
3. *meet you at Jermyn Street*: at the Royal School of Mines where Tyndall lectured (see letter 2259, n. 17).
4. *Mr Wheatstone*: Charles Wheatstone (1802–75), an English experimental philosopher and inventor. He had been a musical instrument maker before becoming professor of experimental philosophy at King's College, London (1834–75). There he developed the electric telegraph with William Fothergill Cooke as well as a number of optical instruments, including the stereoscope which led him into controversy with David Brewster (*ODNB*).

From Joseph Plateau 23 June 1866 2453

Gand 23 Juin 1866.

Mon cher Monsieur Tyndall

Votre lettre m'a vivement contrarié, je ne sais à quoi attribuer le retard dans la réception de mes exemplaires, qui ont été mis à la poste le 8 ou le 9 de ce mois. J'ai pris des informations à la poste, où l'on m'a assuré que tout était parti, et que, si les brochures étaient égarées, cela devait provenir de l'administration anglaise, et non de l'administration belge. Maintenant, je vous serais extrêmement obligé si vous vouliez bien m'informer, par un mot de lettre, si mes exemplaires vous sont enfin parvenus ou non. En cas de négative, je m'empresserai de vous en adresser d'autres. J'en avais envoyé également un à M. Grove; si vous avez occasion de voir le savant sous peu, j'espère que vous aurez la bonté de lui demander si la chose lui est arrivée, et de me le dire en même temps dans votre lettre.

Mille pardons, mon cher Monsieur Tyndall, pour les petits embarras que je viens vous causer; j'attends votre réponse, et vous prie de croire à tous mes sentiments d'affection.

J. Plateau.

P.S. Veuillez me donner votre adresse exacte, je serai plus sûr des résultat de mes envois.

Ghent 23 June 1866.

My dear Mr. Tyndall

Your letter[1] has deeply upset me; I do not know what could be the reason for the delay in the delivery of my copies, which were sent to the post office on the 8th or the 9th of this month. I made inquiries at the post office, where I was assured that everything was sent, and that if the brochures were lost, it would be due to the English administration and not the Belgian administration. Now, I would be much obliged to you if you would inform me by letter whether my copies have finally reached you or not. If not, I will hasten to send you new ones. I have also sent one to Mr. Grove.[2] If you have the opportunity to see the savant soon, I hope that you will kindly ask him if it reached him, and inform me about it in that same letter of yours.

I deeply apologize, my dear Mr. Tyndall, for the small trouble that I have caused you; I am awaiting your reply and I ask you to accept all of my affections.

J. Plateau.

P.S. Please give me your exact address, so that I will be more certain about the results of my shipments.

RI MS JT/1/P/99

1. *Your letter*: letter missing, but presumably a letter in which Tyndall informed Plateau that the copies of his memoir have not arrived (see letter 2445).
2. *Mr. Grove*: William Robert Grove.

From George Campbell[1] 26 June 1866 2454

PRIVY SEAL OFFICE | June 26/66

My Dear Dr. Tyndall

I am sorry to say that a cabinet has been called for 3.30 today[2]—But I shall be at the Institution at 3 & see as much as I can.[3]

Y[ou]rs t[rul]y | Argyll

RI MS JT/1/A/96
RI MS JT/1/TYP/1/67

1. *George Campbell*: George Douglas Campbell (1823–1900), Duke of Argyll, a Scottish politician and scientist. In the peerage of Scotland he was eighth Duke of Argyll and was first Duke of Argyll in the peerage of the United Kingdom. Campbell served over ten years as Lord Privy Seal and from December 1868 to February 1874 was Secretary of State for India. He was a proponent of educational reform and led the Argyll Commission (1865–8) into the Scottish school system. As an amateur scientist he wrote on ornithology, geology, and evolution. In his work on evolution he defended the special creation of humans apart from other species, publishing *The Reign of Law* in 1868 (*ODNB*).
2. *a cabinet has been called for 3.30 today*: Campbell was Lord Privy Seal in Lord Palmerston's liberal government and continued in that role in Lord John Russell's cabinet following Palmerston's death in 1865.

3. *I shall be at the Institution*: to see Tyndall's experiments on the vibrations of strings at the RI (see letter 2449, n. 2).

To Jane Barnard 27 June 1866 2455

27th June 1866

My dear Miss Barnard

Your dear uncle[1] in the days of his health was kind enough to take charge of a document[2] for me. He told me that he would seal it up and write upon it "for Dr. Tyndall" or some words to that effect, so that in case of his failing it might delivered up to me. If you, any time that it is convenient to you, would so far imitate his goodness as to secure this document for me I should be truly obliged to you.

It seems I missed Dr. Bence Jones by only a short interval on Sunday. I was exceedingly pleased to learn that he had gone out to see your uncle and I have told him so.

Yours ever sincerely | John Tyndall

RI MS JT/1/T/107
RI MS JT/1/TYP/12/4192

1. *Your dear uncle*: Michael Faraday.
2. *document*: not identified.

To the Editors of the *Phil. Mag.*[1] [June 1866] 2456

GENTLEMEN,

PROFESSOR HELMHOLTZ has forwarded to me the following Note,[2] which forms an appendix to the translation of his excellent lecture upon glaciers,[3] recently published in the *Revue des Cours Scientifiques*. You will, I doubt not, willingly give it a place in the Philosophical Magazine.

JOHN TYNDALL.

H. Helmholtz, 'On the Regelation of Ice', *Phil. Mag.*, 32 (1866), pp. 22–3, on p. 22

1. *Editors of the Phil. Mag.*: see letter 2245, n. 1.
2. *forwarded to me the following Note*: see letter 2440.
3. *his excellent lecture upon glaciers*: see letter 2440, n. 1.

To Thomas Archer Hirst [8][1] July 1866 2457

J. Tyndall | Sunday July 1866.

My dear Tom,

It is some consolation to me to find you punished for forsaking me;[2] and there is still time if you will convert your theoretic into a practical repentance. That document from South Kensington makes me exactly four & twenty pounds richer than ordinary.[3] Fifteen pounds of it you know belongs to yourself. Hence added 15 to 24 you have 39 and this will be quite sufficient for a moderate man in Switzerland.

But it is only too likely that you will permit your penitence to remain in theoretic stage: so I must reconcile myself to my circumstances. I started on Thursday[4] at the Fenchchurch[5] Station for Blackwall.[6] There got on board the Trinity Yacht:[7] purposing to land at Havre and inspect the lighthouse there.[8] We steamed down the river[9] but the wind arose; and the sea when we reached it leaped into billows and the angry crests were taken hold of by the storm and scattered far & wide. On rounding the North Foreland[10] we found the Downs[11] crowded with vessels which had run in there for shelter: so we followed their example and cast anchor too.

The wind abated; but my trip to Havre was ended. We steamed off to Dover[12] and there at 4 o'clock in the morning I was set ashore. I started by the 9 35 boat for Paris and arrived there without accident.

It pleases me to hear of those feelings of devotion to our friend,[13] because it satisfies me that <u>my</u> devotion towards her is not the result of subjective illusion. I have met many women as you know, and talked to many, and been more or less influenced by many but she rises, even in presence of my cool intellectual judgment, above them all.

Goodbye | Yours affect[ionate]ly | John.

RI MS JT/1/T/909
RI MS JT/1/HTYP/535

---

1. *[8]*: the day is given based on when Tyndall left for the Alps, on Thursday, 5 July 1866. The Sunday following was 8 July.
2. *forsaking me*: Hirst did not join Tyndall in the Alps this summer; he left London on 17 July for Ireland for just over a month, and departed from Dublin on 21 August to travel to Nottingham for the thirty-sixth annual meeting of the BAAS (from 22–30 August; *Brit. Assoc. Rep. 1866*), where, as he wrote in his journal later in October, he was to 'play my new part in the proceedings of the British Association' (Brock and MacLeod, *Hirst Journals*, p. 1790) as he was appointed General Secretary of the BAAS. While in Ireland he stayed with and visited various friends, and on 6 August 'went to Leighlinbridge to see Mrs and

Miss Tyndall The old lady was very feeble but calm and resigned to what appeared to me to a very dreary life. Miss Tyndall I found better than I expected. I also called to see Mrs Stewart [Steuart] of Stewart [Steuart] Lodge' (p. 1790).

3. *That document from South Kensington*: this probably refers to a payment for Tyndall's role as an Examiner for Science for the Department of Science and Art (sometimes Science and Art Department) at the South Kensington Museum. See letter 2230, n. 2.
4. *I started on Thursday*: Tyndall's journey to the Alps began on Thursday, 5 July 1866, and he returned to London on 20 August 1866 (see letter 2466). For his activities during this time, see letters 2460, 2461, and 2464. He noted that it was a particularly poor summer for mountaineering.
5. *Fenchchurch station*: Tyndall misspelt 'Fenchurch'; a railway station in London, still in operation.
6. *Blackwall*: a railway station in London until 1926.
7. *Trinity Yacht*: the *Princess Alexandra*, Trinity House's steam yacht.
8. *land at Havre and inspect the lighthouse there*: the Cape de La Heve lighthouse is just northwest of the French coastal city of Le Havre.
9. *river*: the Thames.
10. *North Foreland*: a headland on the coast of southeast England.
11. *Downs*: the name of an area of sheltered sea off the east coast of Kent between Deal and Sandwich, which provided a safe anchorage for ships negotiating the dangerous Goodwin Sands while sailing through the Straits of Dover (*OED*).
12. *Dover*: Dover, England to Calais, France is a common passage across the English Channel.
13. *our friend*: not identified.

To the Privy Council on Education 9 July 1866 2458

To the Right Honourable the Lords of the Committee of Her Majesty's Most Honourable Privy Council on Education.[1]

My Lords,

IN accordance with the request contained in the letter of the Lord President dated the 17th February 1866, and with your Lordships' Minute of the 10th March we, the commissioners thereby appointed, have carefully considered the subject of the new College of Science referred to us, and have now the honour to submit the following Report.

2. From the general Meeting on Scientific Institutions and Instruction in Dublin—that of the 21st September 1865—it appears that your Lordships consider that, for the various reasons therein given, "the Museum of Irish Industry"[2] now existing in Dublin "should," on its reorganization as contemplated by the Minute, "have a wider scope given to it than that of a School of Mines: that it should become a College for affording a complete and thorough course of instruction in those branches of science which are

more immediately connected with and applied to all descriptions of industry, including Agriculture, Mining and Manufactures; that it should in this way supplement the elementary scientific instruction already provided for by the science schools of the Department; and that it should assist in the training of teachers for these school."

3. At the same time the Minute of 10th March 1866, states that "as the sphere of action sketched out in this (the preceding) Minute will be somewhat new and beyond the limits hitherto placed on the action of the Science and Art Department[3] in respect of the encouragement of science, My Lords have appointed a Commission to advise them on the subject;" and it proceeds "My Lords consider that it is desirable that the College should, on its establishment, commence with a clear and defined object, a well considered course of study, and a proper staff of professors. They therefore request the Commission to consider these subjects and report generally on the scope of the instruction to be given, the examinations for testing it, and the certificates, &c., to be awarded to successful students."

4. Her Majesty's Government having thus decided generally on the necessity for the enlargement of the sphere of action of the existing Government institution for scientific instruction in Dublin, it would appear that the matters on which Your Lordships desired that we should advise may be most conveniently taken under the following heads.

I. The precise sphere of action of the College and the object at which it should aim.
II. The scope and subjects of instruction.
III. The staff of professors necessary.
IV. The course of instruction and its duration.
V. The examinations and granting of certificates.

I.—*The precise sphere of action of the College and the object at which is should aim.*

5. We think the object of the College should be to supply, as far as practicable, a complete course of instruction in Science applicable to the Industrial Arts, especially those which may be classed broadly under Mining, Agriculture, Engineering, and Manufactures, and to aid in the instruction of teachers for the local Schools of Science.

6. We do not consider that the practical applications of Science to Industry, or the Arts themselves, should be undertaken by the New College of Science as a special part of its teaching. Its aim should rather be to impart a sound and through knowledge of those branches of Science which may be so applied, leaving it to the student subsequently to specialize his knowledge and turn his attention in the direction he may find most suitable; but practical subjects when capable of being rendered illustrative of scientific principles

should in all cases be introduced in the course of instruction.

7. Under existing circumstances, however, due to the division and redistribution of duties between the Royal Dublin Society[4] and the College of Science, and in consequence of the representation made by the Secretary, we think it may be advisable to attach a chair of Agricultural Science to the College. This should be looked upon as experimental and its continuance be understood to be dependent on its success.

II.—*The scope and subjects of instruction.*

8. The subjects of instruction should be the following:-

1. Applied Mathematics.
2. Descriptive Geometry, and Mechanical Drawing.
3. Mechanism.
4. Physics.
5. Chemistry.
6. Botany.
7. Zoology.
8. Geology.
9. Mineralogy.
10. Agriculture
11. Mining.
12. Metallurgy.
13. Machinery.
14. Surveying.

9. We propose that under Applied Mathematics should be taken the application of Mathematics to those Sciences which are generally included under the head of Mechanics, viz., Statics, Dynamics, Hydrostatics, and Hydrodynamics, as well as to some other branches of Physics.

10. Under Mechanism should be treated only the relations of motion, or the study of machines merely as contrivances for changing one kind of motion into another, apart from any considerations of force,-a science which has been termed Kinematics.

11. Under Machinery then would be treated the application of Mechanics and Mechanism to machines used in the industrial arts.

12. As probably many students will enter at first without a sufficient knowledge of pure mathematics, the professor of Applied Mathematics should give such preliminary instruction as may be necessary to students entering for the Associateship (see par. 18). He should not be called upon to give any such instruction to the occasional students, who can obtain it easily for themselves, and who should not be allowed to join the Applied Mathematics class

without being thoroughly prepared in the elements of Pure Mathematics.

13. Chemistry should include both Lectures and Laboratory practice.

14. The requisite appliances and assistance for instructing the students practically in this and other branches of science should be provided.

III.—*The Staff of Professors.*

15. To teach the foregoing subjects professors will be required in-

1. Applied Mathematics and Mechanism.
2. Physics.
3. General Chemistry.
4. Applied Chemistry.
5. Botany.
6. Zoology.
7. Geology.
8. Mineralogy and Mining.
9. Agriculture.
10. Descriptive Geometry, Mechanical Drawing, Machinery and Surveying.

Professorships of Physics, General Chemistry, Applied Chemistry, Botany, Zoology, Geology, and Agriculture already exist in connexion with the Science and Art Department. This proposal therefore contemplates the addition of three new professorships.

16. It will be necessary for the professor of Applied Mathematics and Mechanism, and the professor of Descriptive Geometry, Machinery, and Surveying, to devote their whole time to the College; whilst from the other professors a minimum of three lectures a-week during the session will be required. As the school developes, assistants or tutors to teach under the professors may be needed in many of the subjects.

IV.—*The Course of Instruction.*

17. The course of instruction should extend over three years. There should be two terms in each year. In the first two years the instruction should be general. In the last year it should be specialized under the heads of Agriculture, Mining, Engineering, and Manufactures. The following is the scheme proposed.

FIRST YEAR.

1st Term.	2nd Term.
Applied Mathematics.	Applied Mathematics.
Physics.	Physics.
Descriptive Geometry.	Botany.
	Descriptive Geometry.

SECOND YEAR

Applied Mathematics	Chemistry.
Chemistry.	Laboratory practice.
Laboratory practice.	Zoology.
Mechanical Drawing.	Mechanical Drawing.

THIRD YEAR

Division A. Mining.

Geology with demonstration in Palaeontology, Mineralogy, Mining and Assaying.

Division B. Agriculture.

Geology, Agriculture Science, and Land Surveying.

Division C. Engineering.

Mechanism, Machinery, Mechanical Drawing, and Surveying.

Division D

Mechanism, Applied Physics, and Applied Chemistry.

V.—Examinations and Certificates.

18. A diploma of Associateship of the College should be given to students who pass in all the subjects of the first two years, and take a first class in all the subjects in one division of the third year.

19. Persons should also be permitted to enter for the separate courses and to receive certificates after examination.

20. The examinations should be conducted by officers appointed by the Science and Art Department conjointly with the professors of the College.

21. In conclusion we would beg leave strongly to urge that it is important and advisable to have instruction given in Political Economy. Probably the best way of promoting a knowledge of this important science would be to establish evening lectures to be attended by the students but open to all other persons. Any objection that might be taken to a government department dealing with this subject might, we would suggest, be met by requesting one or other of the Universities to appoint the lecturer for the time being.

We have the honor to be, | My Lords, | Your Lordships' obedient Servants,

(Signed) ROSSE.[5]

TALBOT DE MALAHIDE.[6]

W. B. CARPENTER.[7]

B. M. COWIE.[8]

JOHN FOWLER.[9]

E. FRANKLAND.

W. H. GREGORY[10] (with the exception of paragraph 21 to which I object.)

H. D. HARNESS,[11] COL, R.E.

A. W. HOFMANN.[12]

THOMAS H. HUXLEY.

J. BEETE JUKES[13]

ROBERT KANE.[14]

MYLES O'REILLY[15] (with the exception of paragraph 21 to which I object.)

LYON PLAYFAIR.

E. SABINE,[16] LIEUT.-GENERAL, R.A.

WARINGTON W. SMYTH.[17]

WILLIAM K. SULLIVAN.[18]*

JOHN TYNDALL.

J. F. D. DONNELLY,[19] CAPT. R. K. AND SECRETARY OF COMMISSION.

South Kensington, 9th July 1866.

* Professor Sullivan signs with the folllowing reservation, with the exception of paragraph 13, which does not express the decision of the Commission; the first part of paragraph 16, so far as relates to the words 'devote their whole time;' and the last sentence of paragraph 21, to which I strongly object."

---

1. *Her Majesty's Most Honourable Privy Council on Education*: in 1865 the Education Committee of the Privy Council established a committee chaired by the Earl of Rosse, former President of the RS, to consider the question of establishing a College of Science in Ireland. Following their report, the Royal College of Science was duly established in 1867. See C. Cullen, '"A pure school of science": the Royal College of Science for Ireland and scientific education in Victorian Ireland', in J. Adelman and É. Agnew (eds), *Science and Technology in Nineteenth-Century Ireland* (Dublin: Four Courts Press, 2011), pp. 136–49.
2. *"the Museum of Irish Industry"*: a Dublin museum founded in 1854 for displaying material from the Irish mining and manufacturing industries.
3. *Science and Art Department*: the Department of Science and Art at the South Kensington Museum (see letter 2230, n. 2).
4. *Royal Dublin Society*: founded in 1731 by the Dublin Philosophical Society to promote economic activity, the society was originally named the 'Dublin Society for improving Husbandry, Manufactures and other Useful Arts and Sciences'.
5. *ROSSE*: William Parsons (1800–67), Lord Rosse, Third Earl of Rosse, an Anglo-Irish astronomer noted for his 'Leviathan of Parsontown' at Birr Castle in Ireland completed in 1845. He was President of the RS from 1848–54 (*ODNB*).
6. *TALBOT DE MALAHIDE*: James Talbot (1805–83), fourth Baron Talbot of Malahide in the peerage of Ireland, and first Baron Talbot de Malahide in the peerage of the United Kingdom, an Anglo-Irish politician and archaeologist (*ODNB*).
7. *W. B. CARPENTER*: William Benjamin Carpenter (1813–85), a prominent English

physiologist and devout Unitarian who dedicated much of his intellectual life to advocating for the compatibility of science and religious belief (*ODNB*).

8. *B. M. COWIE*: Benjamin Morgan Cowie (1816–1900), Vicar of St Lawrence Jewry and later dean of Manchester and then dean of Exeter (*ODNB*).
9. *JOHN FOWLER*: John Fowler (1817–98), first baronet, a civil engineer (*ODNB*).
10. *W. H. GREGORY*: William Henry Gregory (1816–92), an Anglo-Irish politician and later colonial governor (*ODNB*).
11. *H. D. HARNESS*: Henry Drury Harness (1804–83), a British army officer (*ODNB*).
12. *A. W. HOFMANN*: August Wilhelm von Hofmann.
13. *J. BEETE JUKES*: Joseph Beete Jukes (see letter 2307, n. 7).
14. *ROBERT KANE*: Robert John Kane (see letter 2245, n. 1).
15. *MYLES O'REILLY*: Myles William Patrick O'Reilly (1825–80), an Irish politician who served as a Member of Parliament from 1862–79. He was appointed the Assistant Commissioner of Intermediate Education (*ODNB*).
16. *E. SABINE*: Edward Sabine.
17. *WARINGTON W. SMYTH*: Warington Wilkinson Smyth (1817–90), an English mining engineer and geologist (*ODNB*).
18. *WILLIAM K. SULLIVAN*: William Kirby Sullivan (1822–90), an Irish chemist, historian, and educationalist. He was Professor of Theoretical Chemistry at the Catholic University of Ireland until taking on the same position at the new Royal College of Science in 1867 (T. S. Wheeler, 'Life and Work of William K. Sullivan', *Studies: An Irish Quarterly Review*, 34 (1945), pp. 21–36).
19. *J. F. D. DONNELLY*: John Fretcheville Dykes Donnelly (1834–1902), an Army officer and promoter of scientific education. In 1859 he became Director of Science in the Department of Science and Art at the South Kensington Museum (*ODNB*).

From Joseph Plateau 9 July 1866 2459

Gand, 9 Juillet 1866.

Mon Cher Monsieur Tyndall,

J'ai eu l'honneur de vous écrire, il y a une quinzaine de jours, pour vous exprimer combien j'étais contrarié de ce que vous n'ayez pas reçu mes exemplaires; je vous priais instamment de vouloir bien m'informer si ces exemplaires vous étaient enfin parvenus; j'ajoutais que si vous avez l'occasion de voir M. Grove, vous m'obligeriez de lui demander s'il avait reçu l'exemplaire que je lui avais adressé; enfin je vous priais de me donner votre adresse exacte, afin de donner plus de certitude à la réussite d'un nouvel envoi. Votre silence m'étonne; je crains que tous les exemplaires que j'ai mis à la poste à destination de l'Angleterre n'aient été égarés; je viens donc vous supplier de ne plus différer votre réponse; si elle est négative, je m'empresserai de vous envoyer d'autres exemplaires.

Agréez, Mon cher Monsieur Tyndall, l'assurance de tous mes sentiments d'affection.

J. Plateau | professeur à l'Université | place du Casino, 22.

Ghent, 9 July 1866.

My dear Mr Tyndall,

I had the honour of writing to you two weeks ago, to tell you how upset I was that you did not receive my copies.[1] I urged you to kindly inform me if those copies have finally reached you. I added that, should you have the opportunity to see Mr. Grove,[2] I would be obliged if you ask him if he received the copy that I sent him. Finally, I asked you to give me your exact address so that I will be more certain in the success of a new shipment. Your silence surprises me,[3] and I fear that all of the copies that I mailed to England were lost. So I am begging you to not delay your response. If you have not received them, I will hasten to send you new copies.

Please accept, my dear Mr. Tyndall, all of my affections.

J. Plateau | Professor at the University | 22 Place du Casino

RI MS JT/1/P/100

1. *how upset I was that you did not receive my copies*: see letter 2453.
2. *Mr. Grove*: William Robert Grove.
3. *Your silence surprises me*: any letter from Tyndall to Plateau on this is missing.

To Thomas Archer Hirst 31 July 1866 2460

Bel Alp | 31st July 1866

My dear Tom

I have thought more than once of writing to you since I came to the mountains:[1] but one gets lazy or out of gear for writing & so the thing is neglected. The weather on the whole has been barbarous. I just went to Engsteln[2] & climbed the Titlis.[3] Thence to Rosenlaui,[4] purposing to climb the Wetterhorn.[5] Thence driven by bad weather to the Faulhorn.[6] There snowed up, quitting it calves deep in cold wintry snow. Thence by the Scheinigen Platte[7] to Lauterbrunnen[8] & the Schmadribach.[9] Thence over the Petersgrat[10] to Platten.[11] There housed & entertained by the curé.[12] The house was dirty but I escaped the inflictions described by other travellers—I was not bitten into the condition of smallpox by fleas. From Platten I went over the Lötschen-Lucke[13] down the Lötschen glacier to the Aletsch[14] of which it is a branch. Then down the Trunk glacier[15] to our old quarters of the Aeggischhorn.[16] The weather as I have said has been barbarous, and at the Aeggischhorn I have remained a week; for the air is good, the food is good & the beds are good. During the dull drifting days I wrote a little about miracles, as I suppose I must make some subject of this kind in its scientific aspects the staple of my lecture at Edinburgh.[17] If one had done something good might be said upon the subject: but with me to say any thing good requires[18]

I came over here yesterday. I met at the AEggischhorn a fine tall fellow, an Oxford man named Girdlestone[19] with whom I had some scientific correspondences.[20] He was exceedingly desirous that we should try the Aletschhorn[21] ourselves here without guides or porters. I agreed. We attacked it this morning and got on gallantly for some hours. But the heavens blackened utterly. The summit which first rose clear above us became enveloped: we worked up to a bunch of stones where we proposed having a bit of food. But before we had finished a gust of storm so violent struck us that we instantly backed up and commenced the descent. We paused again—exceedingly unwilling to yield as long as there was any hope. But hope was more and more shut out by the utter blackness of the air. The dirtiest Manchester smoke was nothing to it. And now the snow began to fall and my companion who had quite set his heart upon the ascent saw that all possibility of effecting it was cut off. Down therefore we went. The snow thickened, and the flakes increased in size and finally took the form of inverted umbrellas as they fell. That is they were built so as to form hollow cones with their apexes turned downwards, thus

I have often seen conical hailstones. These snowflakes were the rudiments of such hail. My health and strength have much increased since I came to Switzerland—I wish you were here—but wherever you are I hope you also are strong. I really am so without a plan,—so much at the mercy of the weather—that I hardly know where you could write to me. Try Hotel de la Poste | Visp Canton de Valais.

Thine aff[ectionate]ly | John Tyndall

BL Add MS 63902, ff. 33–4
RI MS JT/1/HTYP/536

1. *since I came to the mountains*: see letter 2457, n. 4.
2. *Engsteln*: the Engstelnalp is an alpine peak near Meiringen, a town in the Swiss canton of Bern.
3. *Titlis*: an alpine peak in the Uri Alps, on the border between the Swiss cantons of Obwalden and Bern.
4. *Rosenlaui*: a small town in the Rosenlaui Valley, near the Rosenlaui glacier,

5. *Wetterhorn*: an alpine peak above the Swiss town of Grindelwald (in the Canton of Bern).
6. *Faulhorn*: a peak in the Bernese Alps, between Lake Brienz and Grindelwald.
7. *Scheinigen Platte*: Schynige Platte is a small mountain ridge and viewpoint in the Bernese Highlands.
8. *Lauterbrunnen*: a village in the Swiss canton of Bern.
9. *Schmadribach*: a waterfall above Lauterbrunnen.
10. *Petersgrat*: a glaciated mountain ridge, used as a pass, located on the border of the Swiss cantons of Bern and Valais.
11. *Platten*: now Blatten, a village in the Lötschental, the largest valley on the northern side of the Rhône valley in the Swiss canton of Valais.
12. *curé*: priest.
13. *Lötschen-Lucke*: a high glacier col (or pass) between the Lötschental and Concordiaplatz, a confluence of glaciers on the main Aletsch glacier.
14. *Lötschen glacier to the Aletsch*: the Aletsch is the longest glacier in the Alps, which flows past the Eggishorn.
15. *Trunk glacier*: meaning the main glacier.
16. *Aeggischhorn*: Eggishorn, a peak of the Bernese Alps in the Swiss canton of Valais.
17. *my lecture at Edinburgh*: Tyndall was working on a lecture about miracles for the Edinburgh Philosophical Institution planned for later that year. As he later wrote in his journal, in June 1867 (the same month he published on miracles in the *Fortnightly Review*), he had changed it to an experimental topic: 'They had pressed me to lecture at Edinburgh, but I had long refused, and in the end thought of giving them the substance of the article on miracles and special providences, which has since appeared in the Fortnightly Review. They were at first very courageous, but retreated when the[y] became really acquainted with what I proposed to say: gave them therefore an experimental lecture instead. It was an awful trouble to me to get there and back' (RI MS JT/2/13c/1333). In the end, his lecture in Edinburgh on 14 December 1866 was on the topic of 'Sonorous Vibrations', and 'Professor Tyndall, during the lecture and exhibition, was frequently applauded' ('Philosophical Institution—Professor Tyndall on Sonorous Vibrations', *Caledonian Mercury*, 15 December 1866, p. 3). The referred to article in *Fortnightly Review*, published in June 1867, is: J. Tyndall, 'Miracles and Special Providences', *Fortnightly Review*, 7 (1867), pp. 645–60.
18. A page appears to be missing at this point in both the original handwritten letter and LT's typescript.
19. *Oxford man named Girdlestone*: Arthur G. Girdlestone (1842–1908), an Anglican clergyman and mountaineer, later author of *The High Alps without Guides: Being a Narrative of Adventures in Switzerland* (London: Longmans, Green, and Co., 1870), in which he wrote of his encounter with Tyndall (pp. 68–71) (*Times*, 15 December 1908, p. 1).
20. *some scientific correspondences*: there are no letters to or from Girdlestone in the extant Tyndall correspondence.
21. *Aletschhorn*: the second highest mountain of the Bernese Alps.

To Juliet Pollock 31 July 1866 2461

31st July 1866

My dear Mrs. Pollock.

The day of the morning when I shook hands with you through the door[1] I started from Blackwall[2] in the Trinity yacht,[3] purposing to steam to Havre;[4] but the weather became strong and we had to anchor in the Dover.[5] Next morning at 4. I was discharged in the streets of Dover & took my passage to Calais in the ordinary boat. I halted some days at Paris to buy instruments & see friends. Thence to the mountains, meeting a guide at Thun.[6] Over the Thuner See,[7] over the Brienzer See,[8] up to Meyringen[9] where I spent a night. Thence to Engstelen,[10] through green valleys with clear streams gambling through them, and here & there a noble cataract twisting from the mountain side. The weather at first was glorious, and when warm I took like a fish to the cold clear streams, tracing my muscles by numerous plunge baths. From Engstelen I climbed the snowy mass of the Tiltis[11] whence I had a glorious view. Returned, halting at Rosenlaui[12] purposing to climb the Wetterhorn,[13] but here alas the weather changed. The break was magnificent; thunder deafening & lightning dazzling. I took shelter under a pine and watched during a long portion of a dark night the elemental war. From Rosenlaui to the Faulhorn[14] where I was snowed up & quitted it plodding through deep new snow, with a blast which would not have done discredit to an English 1st of January in my tack. Dropped down to Lauterbrunnen[15] and ascended the valley to its end. From here crossed a series of snow fields and glaciers obtaining views of inexpressible magnificence and finally dropped down upon a dingy village in which there was no view. So we went to the Cure's.[16] I cannot speak highly of his fare—nor of his cleanliness. It was home but for a night: Next morning over along glaciers and across a beautiful snow Col; which landed us on one of the arms of the great Aletsch glacier[17]—a splendid snow corridor which is almost too lovely to strike you with grandeur. Thence to the AEggischhorn,[18] where the weary traveller found food rest & a welcome.

The weather has been barbarous, and I have been able to do but little in the way of mountaineering. Today in company with a friend from Oxford[19] I tried a great mountain[20] we took no guides through the mountain has a reputation of being very difficult. For a time we advanced gallantly. The keen morning air bracing the body & clarifying the mind. But black clouds gathered as we approached the heights, and we were struck by a storm gust so forcible that it drove us down in *[2–3 words illeg]*. Thick snow fell afterwards; and wetted well for our paces we regained our hotel after 6 hours capital exertion.

My health & strength have much improved since I came here. but the weather is abominable & defeats all plans. I would ask you to write to me if

I could name a post town. But I know not what a day may bring forth. You might if you had any friendship in your heart try

Hotel de la Poste | Visp | Canton de Valais | Switzerland.

Thine ever John Tyndall

BL Add MS 63902, ff. 374–5

---

1. *The day of the morning when I shook hands with you through the door*: 5 July 1866.
2. *Blackwall*: see letter 2457, n. 6.
3. *Trinity yacht*: see letter 2457, n. 7.
4. *Havre*: see letter 2457, n. 8.
5. *Dover*: see letter 2457, n. 12.
6. *Thun*: a town in the Swiss Canton of Bern, on the western side of the Thuner See.
7. *Thuner See*: a lake in the Swiss Canton of Bern, north of the Alps and just west of Interlaken.
8. *Brienzer See*: a lake in the Swiss canton of Bern, north of the Alps and just east of Interlaken.
9. *Meyringen*: Meiringen, a town in the Swiss canton of Bern.
10. *Engstelen*: see letter 2460, n. 2.
11. *Tiltis*: see letter 2460, n. 3.
12. *Rosenlaui*: see letter 2460, n. 4.
13. *Wetterhorn*: see letter 2460, n. 5.
14. *Faulhorn*: see letter 2460, n. 6.
15. *Lauterbrunnen*: see letter 2460, n. 8.
16. *Cure's*: priest.
17. *Aletsch glacier*: see letter 2460, n. 14.
18. *AEggischhorn*: see letter 2460, n. 16.
19. *a friend from Oxford*: Arthur Girdlestone (see letter 2460, n. 19).
20. *great mountain*: the Aletschhorn, the second highest mountain of the Bernese Alps.

To Thomas Archer Hirst [July 1866][1] 2462

Thursday

My dear Tom,

Do not go to Ireland—I have told them[2] that they are to write to you if any thing important occurred, and that if necessary you would go to Ireland. I had a letter[3] written by her[4] which seems to indicate that she is not quite so bad as I imagined.[5]

Yours affe[ctionate]ly. | John Tyndall

RI MS JT/1/T/908
RI MS JT/1/HTYP/535/2

1. *[July 1866]*: the month and year are written on the letter below 'Thursday', but appear to have been added later.
2. *them*: probably Tyndall's mother Sarah Tyndall and sister Emily Tyndall. Since Hirst took a trip to Ireland while Tyndall went to the Alps, Tyndall likely asked Hirst if he could look in on his mother and sister in Leighlinbridge. In his journal entry begun on 23 October 1866, Hirst recounted his time in Ireland months earlier, commenting that he 'went to Leighlin Bridge to see Mrs and Miss Tyndall The old lady was very feeble but calm and resigned to what appeared to me to a very dreary life. Miss Tyndall I found better than I expected. I also called to see Mrs Stewart [Steuart] of Stewart [Steuart] Lodge' (Brock and MacLeod, *Hirst Journals*, p. 1790).
3. *letter*: letter missing.
4. *her*: probably Sarah Tyndall.
5. *not quite so bad as I imagined*: this is possibly a reference to Tyndall's mother, who died the following year.

From Henry Bence Jones 7 August 1866 2463

My dear Friend

I was very glad to hear some news of you for except a line to Mrs Faraday[1] no one knew anything of you since your departure. All has been going on much as usual here excepting the riots which for a time created a sensation & have shewn the people their power to destroy park railings, but of which we shall get a broader Park Lane and new & better rails.[2]

Mr Faraday has been rather better the account I have just got this morning is the best I have had and on your return he will see you I have no doubt; though he says very little yet that little is very nice. I think of going to him tomorrow if possible, though there is little to be done yet my time here is now so short that I like to do all I can. I was in the Institution[3] yesterday & saw all the old furnaces down from that shelf in your laboratory & the temptation to change was too strong not to direct them to be put away elsewhere. So you will have a shelf more I hope for room

I wish you had finer weather and been braced up with mountain work.[4] It has been just as bad and cold here and now we are to have cholera to clean us up & Mr Glaisher says he can see it as a blue mist and scarlet fever as a yellow mist!!![5]

I have had a long letter for du Bois[6] amazed but rejoicing in Prussian successes. Louis Nap must make a three weeks campaign against England to compensate French glory for the shine it has had taken out of it by Prussian ditto.[7] Prussia will be glad to help him in any little project of this kind. Thus I interpret du Bois expressions that Prussia is beginning to hate England for not liking Prussian thieving & lying. If you come back by Paris & see Dumas[8] try & get him to fulfill his promise of lecturing for us.[9] I am trying to get Prevost Paradol who the Emperor has just shut up, but I have no great hope of success.[10] I wish I had you here to read my speech to the chemical section[11] & to get a few bright thoughts from you. I am very tired & wishing for the 1st of September when I hope to get to quiet at Folkestone[12] for two months & glad shall I be if many Sundays you will run away from London & come for a walk on the cliff at Folkestone where I hope we shall have no cholera to disturb us & you shall have the best brandy I can get in Calais[13] and the best mutton from the downs.

Goodbye dear friend. I hope you will come back happy and well and that all your doings will prosper is the warm wish

of yours most truly | H. Bence Jones | Aug 7. 1866

I have no news for you.

RI MS JT/1/J/56
RI MS JT/1/T/TYP/2/687–8

---

1. *Mrs Faraday*: Sarah Faraday.
2. *riots which for . . . new & better rails*: thousands of supporters of the Reform League had been involved in riots in London's Hyde Park 23 July 1866, causing injuries to protestors and the death of one police officer, as well as damage to the parks railings (J. Carlisle, *Picturing Reform in Victorian Britain* (Cambridge: Cambridge University Press, 2012), p. 122).
3. *Institution*: the RI.
4. *finer weather and been braced up with mountain work*: at this time Tyndall was in the Alps. His journey began on Thursday, 5 July 1866, and he returned to London on 20 August 1866 (see letter 2466). For his happenings during this time, see letters 2460, 2461, and 2464. He noted that it was a particularly poor summer for mountaineering.
5. *cholera to clean us up & Mr Glaisher says he can see it as a blue mist and scarlet fever as a yellow mist!!!*: a cholera outbreak in England in June 1866 claimed 14,378 lives, with 5,596 in London (E. A. Underwood, 'The History of Cholera in Great Britain', *Proceedings of the Royal Society of Medicine*, 41 (1948), pp. 165–73). James Glaisher had speculated on the connection between cholera and a blue mist in the sky of England (J. L. Hunt, 'James Glaisher FRS (1809–1903) Astronomer, Meteorologist and Pioneer of Weather Forecasting: "A Venturesome Victorian"', *Quarterly Journal of the Royal Astronomical Society*, 37 (1996), pp. 315–47, on pp. 325–6). In a letter to the *Times* of 7 August 1866 ('The Cholera Mist Again', p. 12), Glaisher described first seeing the blue mist during the cholera epidemic of 1854, and not again until 30 July 1866 when he observed it while at the Royal Observatory, Greenwich: 'This blue mist is apparent on all sides; it extends fully to the top of the trees, though it is not then so easy to distinguish. It is most easily discernible through as much atmosphere as possible, viewed from under a tree, looking under other trees. Thus seen, the boundary walls of Greenwich Park and all objects near them are coloured blue;

or through gaps in trees, if there are others at a sufficient distance to form a background, when it resembles the smoke from a wood fire. The intensity of the blue is increased when viewed through a telescope with a low power. It is of great importance to know whether it is general over the country'. Further, in the discussion following the reading of a paper at the BAAS annual meeting in Nottingham—'On Electro-negative Fog, and more particularly the Dry Mist which occurred over London and its neighbourhood in June, 1861, and August, 1866'—Glaisher 'stated that when engaged in the investigation of the state of the atmosphere during the epidemic of 1854, when the blue mist was last recorded, he had found that where ozone existed freely in the air, and the test-papers were changed brown there were healthy districts, and where the test papers were left unchanged in those districts the mortality was greatest. Although during the past two years, in consequence of the anticipations of the arrival of the epidemic, he had constantly watched for this peculiar mist, he had not seen it until very recently. He did not assign to it any connection with the actual development of cholera, indeed, where the "blue mist" was most intense appeared to be coincident with the most healthy, but there was a very remarkable coincidence in its appearance at the time of the late outbreak of disease, and certainly, as a mere physical phenomenon, it was well worthy of investigation' ('The British Association', *The Gentlemen's Magazine and Historical Review*, 221 (October 1866), pp. 471–91, on p. 474). The same statement appeared in *The Illustrated London News*, 1 September 1866, p. 201. Glaisher's comments regarding this blue mist also appeared in *The Lancet*, 3 November 1866, p. 497; and S. Haughton, 'A Scientific Inquiry into Some of the Causes Alleged to Produce Asiatic Cholera', *The Medical Times and Gazette*, 9 March 1867, pp. 247–8, which included the poem: 'Let Physicians and Sanitarians beware! | The blue fog brings cholera morbus! | The yellow fog brings scarlet fever! | The green fog is the sure precursor of black death!' (p. 247).

6. *duBois*: Emil Dubois-Reymond (1818–96), a German anatomist and physiologist from Berlin who focused on the emerging field of electrophysiology (*CDSB*).
7. *Louis Nap must . . . by Prussian ditto*: Napoleon III, Charles-Louis Napoléon Bonaparte (1808–73), the French emperor from 1852–70. The reference is to the Austro-Prussian War (14 June–23 July 1866) which had resulted in Prussian gains at the expense of the Austro-Hungarian Empire.
8. *Dumas*: Jean Baptiste André Dumas (1800–84), a French chemist and politician (*CDSB*).
9. *promise of lecturing for us*: while it does not appear that Dumas gave a lecture or series of lectures organized by the RI, he would lecture at the RI a few years later, on 17 June 1869 at 8pm, delivering the first of the Faraday Memorial Lectures for the British Chemical Society (Faraday died in August 1867). From the society's journal: 'M. Dumas delivered the "Inaugural Faraday Lecture: after which a vote of thanks to the Lecturer was proposed by Dr. Tyndall, and seconded by Dr. Odling. The lecture was delivered in the Theatre of the Royal Institution' (*Journal of the Chemical Society*, 22 (1869), p. 269). The lecture, lauding Faraday and philosophizing on the questions of science, was published in *Chemical News*, 20 (1869), pp. 1–7.

10. *trying to get Prevost Paradol*: Lucien Anatole Prévost Paradol (1829–70), a French journalist and writer. He was a liberal opponent of the imperial government (T. Chevalier (ed), *Encyclopedia of the Essay* (London: Fitzroy Dearborn Publishers, 1997), p. 300). It does not appear that Paradol lectured for the RI.
11. *my speech to the chemical section*: Bence Jones delivered an address to the Chemical Section of the BAAS at its thirty-sixth annual meeting in Nottingham from 22–30 August 1866 (*Brit. Assoc. Rep. 1866*, pp. 28–33).
12. *Folkestone*: Folkestone is a port town on the English Channel in Kent, where Bence Jones had a home at 5 Albion Villas since 1851.
13. *Calais*: Calais, France, a coastal town from which travelers would cross the English Channel to or from Dover, England.

To Rudolf Clausius 11 August [1866][1] 2464

Brigg Valais[2] 11th Aug.

My dear Clausius.

When I came to Switzerland[3] I hoped that I should not leave it without seeing you: but this atrocious weather has defeated all my plans. I came over about the middle of July; went to Engsteln,[4] a beautiful spot, climbed the Titlis[5]—Went to Rosenlaui;[6] with a view of climbing the Wetterhorn;[7] was driven away by thunder & lightning and terrible rain. Found myself snowed up on the Faulhorn.[8] Went over the Petersgrat[9] to Platten;[10] and here I had one of the very few glorious days with which I have been favoured this year. The view from the Petersgrat is indescribably grand. From Platten I crossed the Lötschsattel;[11] and went down the Aletsch glacier[12] to the Aeggischhorn[13] where I remained, sometimes rambling as far as the Bel Alp[14] for several days hoping that I might be able to pass up the Aletschhorn.[15] One day which promised well a friend[16] & myself tried the mountain without guides but we were driven back by a snowstorm. Defeated thus I was tempted to accompany a friend to Italy.[17] We went first to Baveno,[18] thence over the Lago Maggiore[19] to Luino;[20] thence by carriage to Lugano.[21] Thence to Palezza by water[22]—again by carriage to Menaggio,[23] whence we took a small boat to Belaggio.[24] Even here the bad weather followed us and on Sunday I was caught upon the lake by a thunderstorm. From Bellagio we took a sailing boat to Como,[25] thence by rail to Milan, and back again by Orta, Varallo, Pontegrande, and Macugnaga,[26] and over the Moro[27] again to Switzerland. And here I find the weather worse than ever. I should have much liked to see you Mrs. Clausius[28] & the children.[29] But I fear now that it will be out of my power to do so.

I want you to buy a good strong watch for me.[30] Several years ago Mr. Heinrich Fick,[31] brother of the Professor,[32] bought a watch for me in Zürich which proved a very good one. Inasmuch as the watch will often be in danger from magnets I should not like to go beyond 225 or 250 francs in price. Would you kindly write me a line to the Hotel des Trois Rois Basel[33] letting me know whether you could manage this for me—If Geneva were better than Zürich Soret,[34] I know, would execute the little commission for me.*

Mr. Girdlestone is here with me and desires his kindest remembrances.—Goodbye dear Clausius.

Always yours | John Tyndall

*Ein Secundenzeiger würde nöthig sein.[35]

Am 18n morgens werde ich von Basel abreisen.[36]

I shall be in Thun[37] on the 17th & shall call at the post office for letters.

RI MS JT/1/T/237

1. *[1866]*: the year is given by relation to letters 2460 and 2461.
2. *Brigg Valais*: now Brig-Glis, a town in the Swiss canton of Valais.
3. *came to Switzerland*: see letter 2457, n. 4.
4. *Engsteln*: see letter 2460, n. 2.
5. *Titlis*: see letter 2460, n. 3.
6. *Rosenlaui*: see letter 2460, n. 4.
7. *Wetterhorn*: see letter 2460, n. 5.
8. *Faulhorn*: see letter 2460, n. 6.
9. *Petersgrat*: see letter 2460, n. 10.
10. *Platten*: see letter 2460, n. 11.
11. *Lötschsattel*: a peak of the Bernese Alps in the Swiss canton of Valais.
12. *Aletsch glacier*: see letter 2460, n. 14.
13. *Aeggischhorn*: see letter 2460, n. 16.
14. *Bel Alp*: now Belalp, an alpine village where in 1877 Tyndall and his wife Louisa would build a home on Alp Lusgen, and spend their summers there.
15. *Aletschhorn*: see letter 2460, n. 20.
16. *a friend*: Arthur Girdlestone (see letter 2460, n. 19).
17. *accompany a friend to Italy*: also Arthur Girdlestone, who later wrote that after their failed attempt on the Aletschhorn, 'for three weeks we remained together in hopes of renewing our attack, but each day the weather was against us, and at length we quitted the neighbourhood finally, the Professor for England, and I for the sole peak which the weather of that summer allowed me to ascend—though not unaccompanied with guides—that of the Schreckhorn' (A. Girdlestone, *The High Alps without Guides: Being a Narrative of Adventures in Switzerland* (London: Longmans, Green, and Co., 1870), p. 71). Tyndall wrote in *Hours of Exercise in the Alps* that following the attempt on the Aletschhorn the 'perfectly unmanageble weather of July 1866 caused me to cross with Mr. Girdlestone into Italy, in the hope that a respite of ten or twelcve days might improve the temper of the mountains' (p. 261). When they returned, 'snow fell so heavily as to cover the pastures for 2,000

feet below the hotel' (p. 264). While the Aletschhorn was not attainable, they 'climbed the Sparrenhorn, and found its heavy-laded Kamm almost as hard as that of Monte Rosa' (p. 264).

18. *Baveno*: an Italian town on the west shore of Lago Maggiore.
19. *Lago Maggiore*: Italy's second largest lake, south of the Alps.
20. *Luino*: a small Italian town on the eastern shore of Lake Maggiore, near the border with Switzerland.
21. *Lugano*: a town in Switzerland near the border with Italy.
22. *Palezza by water*: across Lago Lugano to Porlezza, an Italian town on the lake's northeastern end.
23. *Menaggio*: east of Porlezza, an Italian town on the western shore of Lago Como.
24. *Belaggio*: an Italian town across Lago Como to the southeast of Menaggio.
25. *Como*: an Italian city to the southwest of Belaggio at the southwestern end of Lago Como.
26. *Orta, Varallo, Pontegrande, and Macugnaga*: heading northwest from Milan is Lago Orta, with the towns of Varallo, Pontegrande, and Macugnaga falling generally in a northwestern direction toward Zermatt, Switzerland.
27. *Moro*: Monte Moro, the key pass between Macugnaga and Saas (Switzerland), below the mountain of the same name.
28. *M^{rs}. Clausius*: Adelheid Clausius.
29. *the children*: see letter 2270, n. 5, and letter 2387, n. 15.
30. *buy a good strong watch for me*: see also letter 2494, and two letters from spring 1867 (letters 2581 and 2585, *Tyndall Correspondence*, vol. 10).
31. *M^{r}. Heinrich Fick*: brother of Adolf Eugen Fick.
32. *Professor*: Adolf Eugen Fick (1829–1901), a German physician and Professor of Physiology at the University of Zurich (*CDSB*).
33. *Basel*: see letter 2309, n. 3.
34. *Soret*: Jacques Louis Soret.
35. *Ein Secundenzeiger würde nöthig sein*: a seconds hand would be necessary (German).
36. *Am 18^{n} morgens werde ich von Basel abreisen*: I depart from Basel on the morning of the 18th (German).
37. *Thun*: see letter 2461, n. 6.

From Kathleen Ginty[1] 18 August 1866 2465

London | August 18th /66

My dear Dr Tyndall,

Miss Lucy Wright[2] came with me this afternoon to call on the Baron De Mauã,[3] having just received the dreadful news of my father's death.[4] My darling mother[5] wished that I should see you while here and I went to the Royal Institution and was disappointed at not doing so this evening. We return to Manchester by the first train to-morrow morning.

Dear Dr. Tyndall you were once fond of my mother will you write a few words to help and comfort her? Most of our friends are in Rio and I think we shall have to return there by the next English Packet. I enclose you our address in Manchester.

Believe me your's sincerely | Kathleen Ginty
Euston Square Hotel | Saturday evening
Mrs. Ginty | 20 Peru Street | Higher Broughton | Manchester

RI MS JT/1/TYP/11/3647
LT Typescript Only

1. *Kathleen Ginty*: a daughter of William and Margaret Ginty.
2. *Miss Lucy Wright*: possibly a daughter of Tyndall's friend Richard Pears Wright (see letter 2279, n. 9).
3. *the Baron De Mauã*: Irineu Evangelista de Sousa (1813–89), a Brazilian industrialist who received the title of Baron (1854) and later Viscount (1874) of Mauá. Kathleen's father William Ginty worked for Baron Mauá after the family relocated to Rio de Janeiro in 1854. See A. Marchant, *Viscount Mauá and the empire of Brazil: a biography of Irineu Evangelista de Sousa, 1813–1889* (Berkeley: University of California Press, 1965).
4. *the dreadful news of my father's death*: William Ginty, a good friend of Tyndall's, had died suddenly on 13 July 1866 in Brazil, where he had emigrated with his family. His wife and children were in England when they received news of his death.
5. *My darling mother*: Margaret Ginty.

To Margaret Ginty 20 August 1866 2466

20th. Aug. 1866.

My dear Mrs. Ginty,

On my return form the Alps this day[1] I received the account of your terrible loss.[2] I can but imperfectly tell you how it grieves me. It brings the past so vividly before me, when he[3] and I were boys running races together, he depending upon his wonderful swiftness and I on my power of holding out. No friend that I have had of late has stirred up so many recollections, and the letter of your daughter[4] brings the particular one to mind that I was once fond of you. I wish I could give you comfort in your affliction. Heavy duties will now devolve upon you but the care of your children[5] will in due time bring healing to your heart. I have just looked again at your daughter's note and find that you are in Manchester. I thought you were in Rio. Will you write me a line addressed to the Reception Room of the British Association, Nottingham. I am going there to-morrow to fulfil an engagement made some months ago.[6] Write to me pray and say whether you think I could do any thing to help you in your affliction and to lessen the weight of this terrible stroke.

Yours ever truly | John Tyndall.

Thinking you were in Rio, I began this letter on Foreign paper.

RI MS JT/1/TYP/11/3648
LT Typescript Only

1. *My return from the Alps this day*: Tyndall's journey to the Alps began on Thursday, 5 July 1866, and he returned to London on 20 August 1866, as this letter indicates. For his happenings during this time, see letters 2460, 2461, and 2464. He noted that it was a particularly poor summer for mountaineering.
2. *account of your terrible loss*: letter 2465.
3. *he*: William Ginty, a good friend of Tyndall's, had died suddenly on 13 July 1866 in Brazil, where he had emigrated with his family. His wife and children were in Britain when they received news of his death.
4. *your daughter*: Kathleen Ginty.
5. *your children*: William and Margaret Ginty had five children: Kathleen, William Roberts, Fanny, George, and Fred.
6. *an engagement made some months ago*: Tyndall attended the thirty-sixth annual meeting of the BAAS, held in Nottingham from 22–30 August 1866 (*Brit. Assoc. Rep. 1866*).

To Jane Barnard — 21 August [1866][1] — 2467

Royal Institution of Great Britain | 21st Aug

My dear Miss Barnard.

Just arrived from the Alps[2]—I have received the packet[3] though probably I shall never open it. Your uncle's[4] superscription[5] affected me very much.

Yours ever | John Tyndall.

Tomorrow I run down to Nottingham[6] afterwards I hope he will be able to see me

RI MS JT/1/T/123
RI MS JT/1/TYP/12/3/4192

1. *[1866]*: the year is given by reference to going to Nottingham (see n. 6).
2. *from the Alps*: see letter 2457, n. 4.
3. *packet*: contents of the packet are not identified, and were presumably sent by Michael Faraday (Tyndall referred to Faraday's 'superscription').
4. *your uncle's*: Michael Faraday.

5. *superscription*: a piece of text written or printed at the head or beginning of a document; or, a piece of text written, printed, or inscribed on or above something (*OED*).
6. *to Nottingham*: the BAAS held its thirty-sixth annual meeting at Nottingham from 22–30 August 1866 (*Brit. Assoc. Rep. 1866*).

To Thomas Archer Hirst [21][1] August 1866 2468

J. Tyndall | Tuesday Aug[u]st. 1866

Dear Tom

Arrived! all right. I start at 3 P. M. tomorrow for Nottingham.[2] Do in charity let me dine with you & yours somewhere tomorrow. They dine at 4 at Lord Belper's[3] & this is too early so I do not join them until after the address.[4]

Thine aff[ectiona]tely | John

I have had the terrible news of Ginty's death.[5]

RI MS JT/1/T/916
RI MS JT/1/HTYP/536

---

1. *[21]*: the day is given by relation to the dates of the 1866 meeting of the BAAS (see n. 2).
2. *I start at 3 P. M. tomorrow for Nottingham*: Tyndall attended the thirty-sixth annual meeting of the BAAS, held in Nottingham from 22–30 August 1866 (*Brit. Assoc. Rep. 1866*).
3. *Lord Belper's*: Edward Strutt.
4. *after the address*: presumably the Presidential Address of the BAAS annual meeting, delivered that year on 22 August 1866 by William Robert Grove. Tyndall might have been aware that Grove would use his address to express his support for the transmutation of species, but he did not distinguish between different theories of transmutation, only between transmutation and creationist accounts of species. Grove had written to Joseph Dalton Hooker asking for support of Darwin's theory of evolution by natural selection for his speech, and Darwin later wrote to Grove (in a missing letter) that he was displeased Grove in the end did not go into the specifics of his theory (F. Burkhardt, et al. (eds), *The Correspondence of Charles Darwin*, 24 vols (New York: Cambridge University Press, 2004), vol. 14, pp. xxi–xxii). For Grove's address, see *Brit. Assoc. Rep. 1866*, pp. lxxi–lxxix.
5. *terrible news of Ginty's death*: the death of Tyndall's friend William Ginty (see letter 2465, n. 4).

To Margaret Ginty 26 August 1866 2469

Derby | August 26th. 1866.

My dear Mrs. Ginty,

I have just received your dividend,[1] and forward you without delay a cheque for the amount.

My head is weary and I am going to run into the country for a few days. Be good enough to send me a line acknowledging the receipt of the cheque.[2]

Yours faithfully | John Tyndall.

RI MS JT/1/TYP/11/3649
LT Typescript Only

1. *dividend*: likely from the Gold Loan Stock, as an 1882 letter to Tyndall from William Ginty's daughter Fanny referred to dividends from the Gold Loan Stock in regard to her late father's will (RI MS JT/1/TYP/11/3682).
2. *the receipt of the cheque*: Tyndall was assisting with the Gintys' finances in the aftermath of William Ginty's sudden death. See letters 2465 and 2466.

To Margaret Ginty [26][1] August 1866 2470

Kingston, | Derby. | Sunday Aug. 66.

My dear Mrs. Ginty,

The letters containing the account of your great affliction have reached me here, and I now return them.[2] The stroke was one of which strong men die: men of brains and energies too powerful to last long. And among my acquaintances I hardly ever knew a man of energy and general intellectual power greater than those of your husband.[3] Your duty is now plain—You must as far as in you lies shake off the stunning effect of this shock. It will never do to lie prostrate under it. That might be permitted you if others were not depending on you. Your object must be to place your son[4] in some position which will enable him to acquire an independence and to lend a helping hand to you and his sisters.[5] From all I have heard of your eldest boy he is likely to do credit to the memory of his father and to you, and all your efforts must now be directed to getting him a start in life. No doubt the country in which his father won such a high reputation will open many avenues to the son of that father—So I hope that with prompt energy you will gird yourself up for the task before you. The future of your whole family depends on the manner in which you may now act.

You are certainly to be congratulated in the possession of such a friend as Mr. Holman.[6] I read his exceedingly intelligent and sympathising letter with the greatest interest. In him you will have a source of strength; but it will infinitely augment that strength to see you also prompt and energetic.

Any thing that I can do to assist you shall be done with pleasure; but remember the best service that mortal man could do you at present is to arouse you to a sense of the duties of your new position. Hitherto you have had a strong man to lean upon: by and bye you will have a similar prop in your sons:[7] but to secure this you must exert yourself <u>now</u>.

Good bye and believe me | ever yours sincerely | John Tyndall

RI MS JT/1/TYP/11/3650
LT Typescript Only

1. *[26]*: the day is given as this Sunday in August 1866 followed the letter from Kathleen Ginty dated 18 August 1866 (see letter 2465). Sunday, 19 August 1866 would have been too soon.
2. *letters containing the account of your great affliction have reached me here, and I now return them*: letters missing, but see letters 2465, 2466, and 2469.
3. *your husband*: William Ginty.
4. *son*: the eldest son was William Roberts Ginty.
5. *sisters*: the daughters were Kathleen and Fanny (who later married John Owen Unwin).
6. *Mr. Holman*: not identified, but presumably he later married Kathleen Ginty.
7. *your sons*: the two younger sons were George and Fred.

To Augusta Stanley[1] 26 August [1866][2] 2471

Kingston, | Derby. | 26th. Aug.

Dear Lady Augusta[3]

I should have joined your party had I not on that day been knee-deep in the snow of the Titlis.[4] Your card[5] only reached me on my <u>return</u>.

Yours most truly | <u>John Tyndall</u>

Private Collection

1. *Augusta Stanley*: Augusta Elizabeth Frederica Stanley (née Bruce, 1822–76), a courtier who served as lady in waiting to Queen Victoria before marrying Arthur Stanley (see letter 2372, n. 1) in 1863 (*ODNB*). She died on 1 March 1876, the day after Stanley officiated at Tyndall's wedding to Louisa Hamilton.

2. *[1866]*: the year is given by reference to climbing the Titlis (see n. 4).
3. *Augusta*: on the handwritten letter, '[Stanley]' has been written in after 'Augusta', but not in Tyndall's hand.
4. *in the snow of the Titlis*: in July, Tyndall had climbed the Titlis in the Uri Alps, on the border between the Swiss cantons of Obwalden and Bern (see letters 2460 and 2464).
5. *Your card*: letter missing.

From Gustav Wiedemann 2 September 1866 2472

Engelberg, 2ten Sept 1866

Mein lieber Tyndall!

So eben erhalte ich von Vieweg die Nachricht, daß die Angelegenheiten in Betreff der Uebersetzung Ihres Buches mit Bohn völlig geendet sind, und wir jetzt nach dem definitiven Rücktritt desselben freie Hand haben. Es können somit unsere Damen, Frau Helmholtz & meine Frau an die Arbeit gehen und wir wollen hoffen, daß die Uebersetzung ganz nach Wunsch ausfällt. Selbstverständlich arbeiten Helmholtz *[und]* ich dieselbe vor dem Druck *[1 word illeg]* durch & geben Sie auch unter unseren Namen heraus. Ueber die Thätigkeit der Damen bitten wird selbstere stündlich um *[Verschwiegegenheit]*. Leider schreibt nun Clausius, daß Sie die Schweiz schon verlassen haben. Sonst hätte ich Sie gebeten, auf ein Paar Tage hierher zu kommen. Helmholtz und Bunsen sind <auch hier> und Clausius hat sich angemeldet. Indeß werden wir alle unseren jetzigen Wohnsitz in den nächsten Tagen wieder verlassen. Wollen Sie mich mit einigen Worten erfreuen, bitte, so adressiren Sie nach:

Baden-Baden No 158A, Villa Koch, neue | Bendig Strasse,

wo ich auch bis Ende September mit meiner Familie die frische Herbstluft genießen will.

Viele Grüße von Clausius, der Sie fragen läßt, ob Sie seinen post restant nach Thun gesendeten Brief erhalten haben. Ebenso freundliche Empfehlungen von Helmholtz & Bunsen.—

In treuer freundschaftlicher Ergebenheit.

Ihr | G. Wiedemann

Engelberg, 2nd Sept 1866

My dear Tyndall!

I have just received a message from Vieweg,[1] that the affairs with Bohn[2] regarding the translation of your book[3] have been terminated completely, and that now, after the latter's withdrawal, we have free hand. Thus, our ladies, Mrs. Helmholtz & my wife,[4] can begin their work and we will hope that their translation will turn out as desired. Naturally, Helmholtz and I will work through the same *[1 word illeg]* before printing & will also publish it under our names. About the ladies' work we hourly ask the latter for confidentiality. Unfortunately, Clausius now writes that you have already left Switzerland. Otherwise, I would have asked you to come here for a few days. Helmholtz and Bunsen are here as well and Clausius has announced himself. However, we will all leave our present residence again in the next days. If you would like to delight me with a few words, please address them to:

Baden-Baden No 158A, Villa Koch, neue | Bendig Strasse

where I also will enjoy the fresh autumn air with my family until the end of September.

Many greetings from Clausius, who asks whether you have received his letter that was sent poste restante to Thun.[5] Friendly greetings likewise from Helmholtz & Bunsen.—

In loyal, amicable devotion.

Your | G. Wiedemann

RI MS JT/1/W/44

1. *Vieweg*: Eduard Vieweg.
2. *Bohn*: Johann Conrad Bohn.
3. *translation of your book*: the German translation of *Heat* (1865); see letter 2296, n. 6.
4. *Mrs. Helmholtz & my wife*: while Helmholtz and Wiedemann are credited with the translation, it was in fact their wives Anna and Clara, respectively, who carried out the work of translating the book.
5. *his letter that was sent poste restante to Thun*: letter missing. Thun is a town in the Swiss Canton of Bern, on the western side of the Thuner See. In his letter to Clausius from 11 August [1866], Tyndall wrote that he 'shall be in Thun on the 17th & shall call at the post office for letters' (letter 2464). In letter 2494, Clausius wrote: 'We regret not having seen you here this summer. I therefore sent my letter to Thun, instead of Basel, because I thought that in Thun you could still change the plans of your journey and travel to Zürich instead of Basel'. Poste restante means 'general delivery' (French).

To Margaret Ginty 3 September 1866 2473

The Athenaeum[1] | 3rd. Sep. 1866.

My dear Mrs. Ginty,

I can well imagine what your feelings are in your great bereavement.[2] But is your guide only "cold duty?" Is it not something to think that you are acting in the manner in which the man[3] who loved you would approve of? The best and indeed only return you can now make for all that love is to do your duty bravely towards his children and yours.[4] Depend upon it the health of your soul will be improved by every effort you make in this direction. Show yourself then the worthy wife of a brave and loving man. This is not to be done by repining. You must nerve your heart and show a good example to those around you. The problem of your life at present is to get your children properly established. This is a problem worthy of the wife of William Ginty, and the successful solution of which will bring peace and tranquility to your heart. If you could let me know when you are to arrive in London I will call upon you. The Baron[5] is very desirous that you should return to Rio[6] as soon as possible. You have now many firm and sympathising friends—and you have only to exert yourself to get by their aid an honourable career for your sons.

Good bye for the present and believe me | Yours ever sincerely | John Tyndall.

RI MS JT/1/TYP/11/3651
LT Typescript Only

---

1. *The Athenaeum*: see letter 2253, n. 8.
2. *your great bereavement*: Margaret Ginty's husband William Ginty had died suddenly on 13 July 1866 in Brazil, where he had emigrated with his family. His wife and children were in England when they received news of his death (see letters 2465, 2466, 2469, and 2470).
3. *man*: William Ginty.
4. *his children and yours*: see letter 2466, n. 5.
5. *The Baron*: see letter 2465, n. 3.
6. *Rio*: Rio de Janeiro, Brazil.

From Sarah Faraday — 4 September [1866][1] — 2474

The Green, | Hampton Court,[2] | 4th Sept.

Dear Dr. Tyndall,

My goodman[3] is pleased at the thought of seeing you, and I write at once in hopes that you will be able to come next Thursday or Friday, to take a meat tea with us. I have thought much of what you said about finding something for Mr. Faraday to do, which should employ his fingers; it is our great endeavour, for he is much happier when employed, but it is so difficult. Perhaps in your kindness you could suggest something, or take the initiative, so as to give the necessary interest, but it must require so little thought, or he gets quite confused and troubled. At any rate we will talk about it when we meet.

Believe me very sincerely yours, | S. Faraday

Faraday Correspondence, 6:4589
RI MS JT/TYP/12/4169

1. *[1866]*: the year is given by the editor of the Faraday correspondence, probably because Faraday was in decline (see n. 3).
2. *The Green, | Hampton Court*: the Faradays' house in London since 1858.
3. *My goodman*: Michael Faraday. As this letter makes clear, Faraday was by now increasingly infirm. He died the following year, on 25 August 1867 (Faraday's death and how this affected Tyndall both personally and professionally comprise a large part of *Tyndall Correspondence*, vol. 10).

From Émile Goubert[1] 4 September [1866][2] 2475

Paris 4 7bre

Monsieur et très vénéré maître,

Je n'ai pas l'honneur d'être connu de vous, mais je suis un admirateur passionné de vos beaux travaux.

À ce titre, veuillez me permettre de vous adresser, par la poste, un exemplaire d'une thèse que je viens de présenter à la Faculté de médicine. Elle contient certainement bien des erreurs et c'est avec une grande reconnaissance que je recevrais de vous l'indication de ces fautes échappées à mon ignorance. Je pense en effet faire publier cette thèse dans une dizaine de jours, mais j'attendrai, très vénéré maître, que vous ayez bien voulu me communiquer vos précieuses observations ou corrections ou additions, résolu d'ailleurs à prévenir le lecteur de l'important concours que je me permets de réclamer de votre bienveillante science.

Il est possible que vous ayez sur l'achromatopsie, vous ou vos collègues d'Angleterre, des observations ou des idées dont je n'ai pas eu connaissance. Je vous serais alors infiniment obligé de me les communiquer tout au long.

Veuillez agréer, Monsieur et bien vénéré maître, l'expression de ma profonde gratitude et de ma haute considération.

Dr Emile Goubert | Professeur de chimie et physique au collège Rollin | 6 rue Blainville près de Panthéon

Paris 4 September

Sir and much revered teacher,

I do not have the honour of being known to you, but I am an avid admirer of your beautiful work.

As such, please allow me to send you, by mail, a copy of a thesis that I have just presented to the Faculty of Medicine.[3] It certainly contains many errors and it is with great appreciation that I would receive from you any indication of the mistakes that I missed due to my ignorance. I am thinking of publishing this thesis in ten days, but I will wait, much venerated teacher, that you will kindly send me your valuable comments or corrections or additions. I am also determined to inform the reader of the significant support that I allow myself to claim for your benevolent science.

It is possible that on the subject of achromatopsy,[4] you or your colleagues in England would have observations or ideas about which I did not know. Therefore, I would be endlessly grateful if you would communicate them to me all throughout.

Please accept, Sir and revered teacher, the expression of my gratitude and my highest consideration.

Doctor Emile Goubert | Professor of chemistry and physics at the college of Rollin[5] | 6 Blainville Street near the Pantheon.

RI MS JT/1/G/9

1. *Émile Goubert*: Émile M. J. M. P. Goubert (d. 1867), a French physician, botanist, and geologist (Darwin Correspondence Project).
2. *[1866]*: the year is given by relation to letter 2479, dated 18 September 1866.
3. *copy of a thesis*: probably E. Goubert, *De la Perceptivité Normale et Surtout Anormale: de l'Oeil pour les Couleurs, Spécialement de l'Achromatopsie ou Cécité des Couleurs* (Paris: Adrien Delahaye, 1867).
4. *achromatopsy*: colour blindness (*OED*).
5. *college of Rollin*: the Collège Rollin was a private educational institution first established as the Collège Sainte-Barbe in 1820 and renamed in 1830.

To Thomas Archer Hirst [8 September 1866][1] 2476

Saturday

My dear Tom

I have this day seen Mrs Ginty.[2] She starts tomorrow to Rio, with three of her children—her eldest boy[3] and her two daughters.[4] Her eldest daughter is engaged to a young fellow in Rio, whose letters show considerable intelligence.

Two of her boys remain at school in England.[5]

Thus far the only documents which have reached me are the certificates of his death[6]—I do think that much is to be done; but if the work becomes onerous I will do as you say—hand it over to Wright.[7]

He has been making a great deal of enquiries about this place in Westminster.[8] The gentleman however requests £400 from him. I hardly know what to do. If the risk be not too great—I am willing to lend him £100: but I cannot afford to lose it. I wish you were here so that we might have a talk over the matter.

I have been devoting far more time than it is worth to an essay on Mozley's work on Miracles[9]—intending to make it the subject of my Edinburgh Lecture.[10] But on this head I must consult you when you return.

I saw Carlyle on Wednesday night:—he looked extremely well. He was furious against things in general, though his cheek was clear & his eye bright & happy looking. He came good half of the way home with Pollock[11] & myself. He had been invited by Lady Ashburton[12] to spend the winter with her at Mentone.[13] But he did not seem to entertain the thing. I offered to take him there at the end of December & he said that quite altered the complexion of the question. I shall probably go with him.

Faraday I have not yet seen; but I have had a very good account of him today. I also had a long note from Mrs Steuart.[14]

My Glaciers of the Alps[15] have brought me a dividend of between 7 and 8 guineas.

I have had a letter[16] from Wiedemann stating that Vieweg[17] has finally & completely arranged with Bohn.[18] The translation of the book will now proceed with rigour.

We have the senior United service people[19] as our guests at the Athenaeum.[20] We were theirs last year.

The weather is vile.

This I think is all the news.

Yours affectionately | John Tyndall

RI MS JT/1/T/664

---

1. *[8 September 1866]*: the date was written on the letter in another hand, probably Hirst's, and this fits with Tyndall having first heard in August 1866 of William Ginty's death.
2. *Mrs Ginty*: Margaret Ginty, whose husband William Ginty, a friend of Tyndall's, had died suddenly on 13 July 1866 in Brazil, where he had emigrated with his family. His wife and children were in England when they received news of his death (see letters 2465 and 2466). As this letter indicates, Tyndall was overseeing her financial affairs. See also letters 2469, 2470, and 2473.
3. *eldest boy*: William Roberts Ginty.
4. *two daughters*: Kathleen and Fanny Ginty.
5. *Two of her boys remain at school in England*: George and Fred Ginty. A letter from Margaret Ginty to Tyndall from February 1867 shows that the boys were at Linton House School, Colchester, Essex (letter 2570, *Tyndall Correspondence*, vol. 10).
6. *his death*: of William Ginty (see n. 2).
7. *hand it over to Wright*: possibly Tyndall's friend Richard Pears Wright (see letter 2279, n. 10). In between teaching at Queenwood (1851–63) and at University College School, London (1870–92), Wright opened his own school for boys at 6 South Parade, Southsea, Hampshire (D. T. Bottomley, *Science, Education and Social Vision of Five Nineteenth Century Headmasters* (PhD diss., Curtin University, 2018), p. 260). In 1863, when Queenwood

headmaster George Edmondson died, Tyndall thought that William Ginty may wish to take his sons out of Queenwood, and he suggested Wright's new school in Southsea (see letter 2051, *Tyndall Correspondence*, vol. 8).

8. *this place in Westminster*: not identified.
9. *an essay on Mozley's work on Miracles*: J. Tyndall, 'Miracles and Special Providences', *Fortnightly Review*, 7 (1867), pp. 645–60. This essay was a response to Oxford's regius professor of divinity James Bowling Mozley's (1813–78) *Eight Lectures on Miracles preached before the University of Oxford in the Year M.DCCC.LXV* (London: Rivingtons, 1865). See letter 2460, n. 17.
10. *my Edinburgh Lecture*: see letter 2460, n. 17.
11. *Pollock*: William Frederick Pollock.
12. *Lady Ashburton*: Louisa Baring.
13. *winter with her at Mentone*: Tyndall and Carlyle went together to visit Louisa Baring in Mentone—a Mediterranean coastal town in southeastern France, at the border with Italy—where she had a home, Villa Madonna (V. Surtees, *The Ludovisi Goddess: The Life of Louisa Lady Ashburton* (Michael Russell, 1984), p. 126). They left London on 21 December 1866, spent the evening of 22 December in Paris, and arrived in Mentone very late on 23 December. Tyndall returned to London on 29 December 1866 (see letters 2528 and 2531, and *Ascent of John Tyndall*, p. 202).
14. *a long note from M[rs] Steuart*: letter from Elizabeth Dawson Steuart is missing.
15. *Glaciers of the Alps*: *Glaciers of the Alps*.
16. *letter*: letter missing.
17. *Vieweg*: Eduard Vieweg.
18. *completely arranged with Bohn*: the German translation of *Heat* (1865) by Johann Conrad Bohn (see letter 2296, n. 6).
19. *United service people*: the United Services Club was a London gentleman's club founded in 1815 for members of the British Army and Royal Navy above a certain rank.
20. *Athenaeum*: see letter 2253, n. 8.

From William Frederick Pollock — 9 September 1866 — 2477

Chaldon | Bournemouth[1] | 9th Sept. 1866

My dear Tyndall

Decidedly this is a place for you to come to—a fine bay—the most lovely colours of heath & furze blossoms on land—fascinating pine woods—& a pleasing arrangement of human dwellings. The town—if town it can be called—may be described as a resolvable nebula or a cluster of villas, each in its own garden—& there is not a street, a terrace, a row, a crescent, a square, or d---d[2] parade in the whole place. We have a good house in a good position—& a good room in it for a good friend, if he will come. We have wells which we digged not & vines which we planted not, a nice little well working household of servants, left to us with the house—two cats (one of them in an interesting position—which makes the precise amount of feline population, uncertain almost from hour to hour), & a Pomeranian dog called Billy, who walks out with us. So come and see all this if you can. The time of the journey I fear is not much shorter than to Freshwater[3]—but there are more, and more convenient trains—especially one late in the afternoon which w[oul][d] get you to London at night after getting most of the day here—Aubrey de Vere[4] is here, but goes on Monday. Henry Taylor[5] however is en permanence[6]—& you know his merits: and ought to know his wife[7] & very charming girls.[8]

We have not heard yet anything of Fred[9] & his companion in Switzerland, but I suppose all is going well with them. Only one may hope that they have better weather than prevails here.

Juliet[10] thanks you for your letter.[11] The boys[12] send their loves.

Yours ever | W. F. Pollock

RI MS JT/1/P/236
RI MS JT/1/TYP/6/2075

1. *Chaldon | Bournemouth*: Chaldon was a home the Pollocks rented in the English seaside town of Bournemouth. On 6 September 1866, Pollock wrote in his journal: 'To Bournemouth, where we occupied a house for some weeks, rejoicing much in the close neighbourhood of Sir Henry and Lady Taylor and their family' (*Remembrances*, vol. 2, p. 153).
2. *d---d*: damned.
3. *Freshwater*: see letter 2451, n. 7.
4. *Aubrey de Vere*: Aubrey Thomas de Vere (1814–1902), an Irish poet and critic (*ODNB*). He was the cousin of Henry Taylor's wife Theodosia.
5. *Henry Taylor*: Henry Taylor (1800–86), a British poet, playwright, and diplomat who spent much of his career working in the Colonial Office (*ODNB*). Beginning in 1861, Taylor and his family spent summers at The Roost, his home in Bournemouth (U. Taylor, *Guests and Memories: Annals of a Seaside Villa* (London: H. Milford, Oxford University Press, 1924), p. 14).
6. *en permanence*: permanently (French; *OED*).
7. *wife*: Theodosia Alice Taylor (née Spring-Rice, 1818–91), daughter of Thomas Spring-Rice (1790–1866), first Baron Monteagle of Brandon in Kerry. Theodosia married Henry Taylor on 17 October 1839.
8. *very charming girls*: the Taylors had five children: Aubrey Ashworth Taylor (1845–70), who went to Balliol College, Oxford and died young; Eleanor Ashworth Towle (née Taylor, 1847–1911), a novelist; Ida Ashworth Taylor (1850–1929), a novelist and biographer (*ODNB*); Harry Ashworth Taylor (1855–1907), a captain in the British Army; and Una Ashworth Taylor (1857–1922), a novelist and author of *Guests and Memories: Annals of a Seaside Villa*, her memoir about her family's summer home and the guests they entertained there (H. Taylor, *Autobiography of Henry Taylor, 1844 to 1875* (London: Harrison and Sons, 1877); U. Taylor, *Guests and Memories: Annals of a Seaside Villa*).

9. *Fred*: Frederick Pollock.
10. *Juliet*: Juliet Pollock.
11. *your letter*: letter missing. The previous letter from Tyndall to Juliet was in July (2461).
12. *boys*: Walter Pollock and Maurice Pollock.

From Charles Gros-Renaud[1] 12 September 1866 2478

Déville 12 Sepbre 1866.

Tyndall à Londres.

Je serais très désireux de me remettre en rapport avec Mr Williamson pour des renseignements qui me seraient très précieux, mais ne sachant pas où adresser ma lettre pour qu'elle arrive à ce chimiste distingué; j'ai recours, cher Monsieur, à votre obligeance pour vous demander si vous ne pourriez pas par quelques mots me fournir l'adresse de Mr Williamson; je vous en serai extrêmement reconnaissant.

C'est la grande popularité de votre nom qui m'a inspiré de toute la liberté que je me permets à votre égard, et dans l'espoir, Monsieur, que vous veuillez bien réserver un bon accueil à ma lettre, je vous prie de vouloir recevoir en même temps que mes sincères remerciements, l'expression des meilleurs sentiments de votre tout dévoué

Gros-Renaud

adresse. Monsieur Chles Gros. Renaud | chimiste | de la maison Cordier | à Déville-lès-Rouen. | Seine Inférieure France

Deville 12 September 1866.

Tyndall in London.

I would be very eager to get in touch with Mr Williamson[2] to get some information that would be very precious to me, but not knowing where to send my letter so it reaches this distinguished chemist, I appeal to your kindness, dear Sir, to ask you if you could in a few words give me the address of Mr Williamson; I would be extremely grateful.

The great popularity of your name has inspired me all this freedom that I allow myself towards you, and in the hope, Sir, that you would welcome my letter, I beg you to receive at the same time my most sincere thanks, and the expression of the best feelings from your very devoted

Gros-Renaud

Address. Mr. Charles Gros Renaud | chemist | of Cordier house | in Déville-lès-Rouen. | Seine Inferieure France[3]

RI MS JT/1/R/13

1. *Charles Gros-Renaud*: Charles Gros-Renaud (dates unknown), a French chemist.
2. *Mr Williamson*: probably chemist Alexander Williamson (see letter 2287, n. 7).
3. *Déville-lès-Rouen*. | *Seine Inferieure France*: Déville-lès-Rouen is a town in the Seine-Inférieure (now Seine-Maritime) department of the Normandy region of northern France.

From Émile Goubert[1] 18 September 1866 2479

Paris, 18 7bre 1866

Monsieur et très vénéré maître,

Je suis sensible à l'honneur que vous m'avez fait en m'écrivant une lettre aussi élogieuse. Je serais heureux de profiter, en faisant le tirage à part de ma thèse, de l'intéressante observation que vous voulez bien me communiquer. Malheureusement vous ne m'avez pas dit dans quel Recueil vous avez publié ce travail. Peut être avez-vous eu l'intention de me faire honneur de cette brochure, mais je ne l'ai pas encore reçue.

Je crois que cette observation se rapporte à l'achromatopsie fausse, telle que je l'ai décrite notamment page 116

Veuillez agréer, bien vénéré maître, l'expression de ma haute considération

E. Goubert | professeur | Paris—6 rue Blainville

Paris, 18 September 1866

Sir and much venerated teacher,

I deeply appreciate the honour that you bestowed upon me by writing such a laudatory letter[2] to me. I would be happy to, while printing my thesis[3] aside, take advantage of the interesting observation that you are kind enough to inform me of. Unfortunately you have not told me in which Collection you published that work.[4] Perhaps you had the intention of honouring me with that brochure, but I have not received it yet.

I believe that this observation relates to the false achromatopsy, as I described it in particular on page 116.

Please accept, much revered teacher, the assurance of my highest consideration.

E. Goubert | Professor | Paris—6 Blainville st.

RI MS JT/1/G/10

1. *Émile Goubert*: see letter 2475, n. 1.
2. *laudatory letter*: Tyndall had presumably responded to letter 2475, but the letter is missing.
3. *my thesis*: see letter 2475, n. 3.
4. *that work*: possibly J. Tyndall, 'On a peculiar case of Colour Blindness', *Phil. Mag.*, 11 (1856), pp. 329–33. This is cited by Goubert in his thesis on p. 126.

To the Editor of the *PMG*[1] 20 September 1866 2480

THE PRUSSIAN SOLDIERY. |
To the EDITOR *of the* PALL MALL GAZETTE.

SIR,—Your remarks on the morale of the Prussian soldiery[2] induce me to send you an extract from a journal written in Dresden on the 30th of May, 1849,[3] a few days after the insurrection to quell which the King of Saxony had invoked the aid of Prussia.

Dresden stands on an elbow of the Elbe, the old town resting on the convex, and the new town on the concave side of the river. The former was the scene of strife. With the exception of the castle, which covers the bridge, the town has been entirely in the hands of the insurgents. The traces of fierce conflict are everywhere visible. Corner houses seem to be particularly unfortunate; they are naturally chosen for a stand, and suffer accordingly. The "Thurmhaus" is an edifice of this description; the Stadt Rome hotel in the New Market Place is another. Both are torn by cannon balls, now presenting the appearance of the riddled corner house of the Faubourg St. Antoine last year. The Post-office was in possession of the military, and opposite to it is a spectacle manufactory held by the insurgents. This is almost wholly destroyed, mottled with musketry and shattered by cannon. The people set fire to the old opera house to prevent its being taken by the military. At each end of the Zwinger stood a temple; that adjacent to the opera house is wholly destroyed. The statues, blackened by smoke, standing on the grim gaunt walls, seem pictures of desolation. On the New Market Place stands a fine church—the Frauenkirche. This was taken possession of by the troops, and from it they poured destruction upon the surrounding houses. On Thursday, the 3rd of May, the fight began; on Sunday the Prussians arrived to aid the Saxon military. The fight continued until the following Wednesday, when towels and tablecloths were held out by the insurgents, and the firing ceased.

In the afternoon I had a row on the river, returned on foot. Hearing shouting I followed the noise, and finally plunged into a crown of Prussian soldiers who were amusing themselves in a field near the caserne. About three thousand men were assembled. The amusements were various. For footracers, who started ten at a time, a space was enclosed by ropes; close to this was a swing with its patrons. On a table stood a soldier reciting poetry, amid the bravos of his hearers. Farther on was a party singing with taste and feeling

"Ein Deutsche hat ein edles Herz."[4]

On a platform was a band of musicians, and close to them a party of thirty soldiers waltzing. A long string of about a hundred men beat time to the music by throwing themselves into all imaginable attitudes. The officers mingled freely with the men. A number of girls, some of them very handsome, were walking about in the midst of the throng. They were hardly taken notice of, and neither in word nor gesture was the slightest indecency shown by any individual of this crown of men.

JOHN TYNDALL. | *Athenaeum Club*,[5] *Sept.* 20, 1866.

PMG, 21 September 1866, p. 3

---

1. *Editor of the PMG*: see letter 2314, n. 1.
2. *Your remarks on the morale of the Prussian soldiery*: the *PMG* had published many accounts of the Austro-Prussian War, as had other British newspapers at this time. Tyndall was probably responding to: 'Modern Chivalry', *PMG*, 19 September 1866, p. 1.
3. *extract from a journal written in Dresden on the 30th of May, 1849*: RI MS JT/2/13a/63–4.
4. *"Ein Deutsche hat ein edles Herz"*: A German has a noble heart (German).
5. *Athenaeum Club*: see letter 2253, n. 8.

To Hamilton Hume[1] 20 September [1866][2] 2481

ATHENAEUM[3] | 20th. Sept[embe]r

Sir

Please accept two guineas for the Eyre Defence Fund[4] from—
Y[ou]r. obed[ien]t Servant | John Tyndall

Wellcome Library MS. 7777/59

---

1. *Hamilton Hume*: Hamilton Hume (1797–1873), an Australian explorer and fellow of the Royal Geographical Society who helped to organize the Eyre Defence and Aid Fund Committee (see n. 4), serving as its secretary (*ODNB*).
2. *[1866]*: the year is given by mention of the Eyre Defence and Aid Fund Committee, which was established in 1866 (see n. 4).
3. *ATHENAEUM*: see letter 2253, n. 8.
4. *Eyre Defence Fund*: on 11 October 1865, an insurrection had broken out among black peasants protesting the British colony of Morant Bay in Jamaica. After several of the Jamaican protestors were shot by police and volunteer militia, they burned the court house and nearby buildings and killed a number of police and militia. In retaliation, the Governor of Jamaica Edward Eyre declared martial law, allowing government troops to kill indiscriminately the black population as well as executing the mixed race member of the Assembly, George William Gordon, who had no involvement in the insurrection. The 'Jamaica case', or 'Eyre controversy', engaged the British public for many years. See W. F. Finlason, *The History of the Jamaica Case: Being an Account, Founded Upon Official Documents, of the Rebellion of the Negroes in Jamaica* (London: Chapman and Hall, 1869); E. B. Underhill, *The Tragedy of Morant Bay: A Narrative of the Disturbances in the Island of Jamaica in 1865* (London: Alexander & Shepheard, 1895); and *Ascent of John Tyndall*, pp. 199–202.

The Governor Eyre Defence and Aid Committee was established and chaired by Thomas Carlyle to support Edward Eyre against accusations of mass murder for his direction of the execution of the Jamaican rebels at Morant Bay. Other supporters of Eyre included Charles Dickens, Alfred Tennyson, and John Ruskin. The Jamaica Committee (active 1866–9) was formed in 1866 in opposition to the violent suppression of the uprising of black peasants. With John Stuart Mill as chairman, the committee advocated prosecuting Eyre, but he would be unsuccessfully tried for private criminal charges. Other notable members of the Jamaica Committee included Thomas Huxley, Charles Lyell, Herbert Spencer, and Charles Darwin (*ODNB*). Tyndall's support for Eyre's violent suppression of the Morant Bay rebellion is discussed in the introduction to this volume.

From Matthew Fontaine Maury — 20 September 1866 — 2482

41 Clarendon Terrace | N.W. | 20 Sept. '66

My dear Tyndall,

A friend[1] inquires for "Quelques documents intéressants relatif aux glaciers"[2] Will you be so kind as to let me know when your own[3] are to be had,:

I would be thankful also if you would have the kindness to refer me to other works & papers upon the same subject and oblige

yours truly | M. F. Maury

Prof. John Tyndall | F.R.S. &c &c &c | 28 Jermyn St.[4]

RI MS JT/1/M/67

1. *friend*: not identified.
2. *"Quelques documents intéressants relatif aux glaciers"*: Some interesting documents relating to glaciers (French).
3. *your own*: see letter 2309, n. 5.
4. *28 Jermyn St.*: location of the Royal School of Mines, where Tyndall had held the Chair of Natural Philosophy since 1859 (see letter 2259, n. 17).

From Matthew Fontaine Maury — 24 September 1866 — 2483

41 Clarendon Ter[race]. | 24 Sept—'66

My dear Tyndall

Pray allow me to express hearty thanks for your kindness & the glacier information.[1]

y[ou]rs truly | M. F. Maury

RI MS JT/1/M/6

1. *the glacier information*: see letter 2482.

From Frederick Pollock 25 September [1866][1] 2484

Most High-worshipful Herr Professor,

Where you are I know not,[2] neither when this letter will be finished, still less where I shall post it, at this present writing we are at Visp,[3] the road is flooded between this and <*1–2 words missing*>, and carriage traffic cut off, and this Tuesday afternoon Sep. 25th we have been prisoners in Visp with absolutely nothing to do: the hotel possessing no books but an old Bradshaw.[4] Tomorrow we shall have to walk to Turtmann,[5] and if no further grief come to the road we shall get to Sion[6] one day later than we intended. My friend[7] and I have both caught the glacier fever, and come to the conclusion that 8000 feet above the sea is the least altitude at which you can be thoroughly jolly, and if I come here another year I think I must try some of the high passes. This time, being novices on the ice, we have kept within very modest bounds. Our best day was last Thursday, when we came over to the Riffel[8] by the S. Theodule[9] and went up the Breithorn[10] on the way. It was glorious weather, and the snow in wonderful good order. Our guide was a Val Tournanche[11] man named Solomon Meyart,[12] a very good fellow. He took part in an attack on the Matterhorn[13] this summer with M. Giordano[14] who it seems is the President of the Italian A.C.[15] They got up to the point called by your name[16] and were then baffled by a storm.

The Monte Rosa[17] range strikes me as grander than anything to be seen at Chamouni[18] or in the Oberland.[19] I would give something to go up Monte Rosa with you. The first ascent this year was made, I grieve to say, by a New York snob[20] who proclaimed in more than one visitor's book the dangerous nature of his exposition to <u>Mount</u> Rosa as he calls it.

The final arete[21] certainly looks queer from a distance, but I suppose when you come to it patience is the chief thing required.

On Saturday we strolled up to the Gornergrat[22]—and soon after our return we looked out and saw the clouds marching up over the Theodule like an army in line of battle, throwing out skirmishers of curling vapour right and left. Gradually they enveloped the Matterhorn, rose and blackened steadily—it was a grand sight but not encouraging, and without delay we packed our knapsacks and fled down to Zermatt,[23] not an hour too soon. The rain kept us there all Sunday. If only it had cleared for a few days—visions flitted before us of a dignified retreat by the Col de le Valpelline[24] and the great S. Bernard[25]—the former by all accounts must be thoroughly worth doing.

But the rain it raineth every day, and the alpenstock[26] must go into winter quarters.

Geneva Sep. 27th. Yesterday we made our way from Visp to Turtmann along the hills with some nasty scrambling now and then, and got on to Sion by diligence[27] in the evening after waiting at Turtmann some hours for the Simplon[28] mail to arrive. At Sion we found a great party of German students making merry in the inn—having by great good luck put up in a house not frequented by our countrymen.

And to-day we have returned to civilization, and felt quite ashamed of sitting down at a magnificent table d'hote[29] in our shooting-coats. Dinner is a good thing, but better is the nameless meal taken in the midst of snow-fields. We leave to-morrow afternoon for Paris.

Yours truly | F. Pollock.

RI MS JT/1/TYP/6/2051–2
LT Typescript Only

---

1. *[1866]*: the year is given since this letter would have been written after Tyndall's ascent to the Pic Tyndall in 1862 but probably before his actual ascent of the Matterhorn in 1868. Also, William Frederick Pollock's letter to Tyndall dated 9 September 1866 refers to his son 'Fred & his companion in Switzerland' (letter 2477).
2. *Where you are I know not*: by this time, Tyndall had already returned to London from his annual excursion to the Alps (see letter 2457, n. 4).
3. *Visp*: a town in the Rhône valley in the Swiss canton of Valais, at the junction of the valley to Zermatt.
4. *Bradshaw*: colloquial designation of 'Bradshaw's Railway Guide', a time-table of all railway trains running in Great Britain (*OED*).
5. *Turtmann*: an alpine town in the Swiss canton of Valais between Visp to its east and Sion to its west.
6. *Sion*: a town in the Rhône valley, and the capital of the Swiss canton of Valais.
7. *My friend*: not identified.
8. *Riffel*: the Riffelhaus on the Riffelberg, underneath the Riffelhorn, a mountain in the Pennine Alps, located south of Zermatt in the Swiss canton of Valais. The hotel was conveniently higher up the mountainside towards Gornergrat for early morning ascents of surrounding peaks, and is where Tyndall started for his ascents of Monte Rosa. Opened in 1853, the hotel was managed by a brother of Joseph Seiler of Zermatt who managed many alpine hotels, including the Hôtel du Glacier du Rhône.
9. *S. Theodule*: the Col de St. Theodule, the summit of the Théodule Pass, connecting Zermatt and Breuil and running between the Matterhorn and the Breithorn.
10. *Breithorn*: a peak in the Pennine Alps on the border between Switzerland and Italy.
11. *Val Tournanche*: or Valtournanche, a town in the Aosta Valley region of northwestern Italy.
12. *Solomon Meyart*: not identified.

13. *Matterhorn*: Tyndall first attempted the Matterhorn, one of the high peaks of the Alps located on the border between Switzerland and Italy, in August 1860 (*Hours of Exercise in the Alps*, pp. 27–52), and again, unsuccessfully, in August 1862 (*Hours of Exercise in the Alps*, pp. 153–65).
14. *M. Giordano*: Felix Giordano (1825–92), an Italian engineer and mountaineer who worked in the Mining Corps of Italy, becoming Inspector General in 1870 (*Dizionario Biografico Degli Italiani*, 94 vols (Rome: Istituto della Enciclopedia Italiana, 1960–2019), vol. 55, pp. 264–6).
15. *Italian A.C.*: the Club Alpino Italiano, founded in 1863.
16. *point called by your name*: Pic Tyndall, a lower summit on the western shoulder of the Matterhorn first reached by Tyndall and Johann Joseph Bennen in 1862 (see *Hours of Exercise in the Alps*, pp. 153–65).
17. *Monte Rosa*: the second highest mountain in the Alps located between Switzerland and Italy near the town of Zermatt.
18. *Chamouni*: Chamonix, a town in the valley north of Mont Blanc in south-eastern France.
19. *Oberland*: the geographically higher region of the Swiss canton of Bern.
20. *New York snob*: not identified.
21. *arete*: *arête*, a sharp ascending ridge or 'edge' of a mountain; the local name in French Switzerland, whence it has become a technical term with mountain-climbers (*OED*).
22. *Gornergrat*: a rocky ridge, located southeast of Zermatt, Switzerland, and which overlooks the Gorner glacier.
23. *Zermatt*: an alpine town in Switzerland surrounded by many of the range's high peaks, including Monte Rosa, the Weisshorn, and the Matterhorn.
24. *Col de le Valpelline*: the pass between the Valpelline and Zermatt.
25. *S. Bernard*: Great St Bernard, an alpine pass in the Swiss canton of Valais near the border with Italy, the lowest pass within the fifty miles between the two highest mountains of the Alps, Mont Blanc and Monte Rosa. The pass connects the Swiss canton of Martigny to Aosta in Italy.
26. *alpenstock*: a long iron-tipped staff used in hill and mountain climbing (*OED*).
27. *diligence*: a public stage-coach (*OED*).
28. *Simplon*: a town in the Swiss canton of Valais.
29. *table d'hote*: a shared table for diners at a hotel, restaurant, tavern, etc., at which a set menu is served at a stated time (*OED*).

From Henry Bence Jones 30 September [1866][1] 2485

Sunday | Sep 30

My dear friend

Thanks many for your kind letter.[2] I should long since have written to you but I had no good news to tell you & though I have many times been near writing bad news, yet I did not like to add this to your troubles.

I am still in doubt what I ought to write but there can be no harm in your hearing the present state of things. I have been here[3] a month, & I have been ill all the time with my breath partly the result of a bad cold partly the result of heart disease. At first for a fortnight after strict confinement to my bed room blisters mercury &c I thought I was getting much better & a very fine day I walked on a level a mile; after this (they say, that it was all wrong) at any rate in the last fortnight I have gone wrong again & am going thro' my first cure again!! This is all very depressing & the probability of relapse after relapse & no way out of my troubles is some nights very vivid & I think in the morning I will write to you & tell you that I shall have to give up one of my greatest pleasures trying to do good for the Institution.[4] I expect Sir H. Holland[5] back this week & you & he must take counsel together. There is no hurry the courses of lectures are almost settled & I have a list of those to be asked to give Friday evenings but who my successor sh[oul][d] be I know not. Busk[6] would not take the lecture arrangements. If he w[oul][d] he w[oul][d] be a famous man. Gladstone[7] I dare say w[oul][d] do every thing & if I am not to get better you must think of him. You are the person most concerned if I have to go[8] & it wants someone with whom you can speak plainly when you become the head of the house. The present interregnum[9] has troubles in store for the future rule & someone who can work thoroughly with you hereafter is most necessary. So please let this be in your thoughts & speak plainly to Sir H. Holland. I shall not leave you if I can help it, but sometimes I think I am hastening to a wreck & that a few months or less will see me waterlogged & going down. May I be all wrong.

I hope you will have no return of your illness. I am glad your book on Sound[10] gets on. I saw a nice review yesterday in the Sat[urda][y] by you.[11]

I have not written to any one because I did not want to talk about my illness to anyone until I saw my way more clearly to the future. At present it is all dark

Good bye dear friend | Ever yours most sincerely | H Bence Jones

RI MS JT/1/J/86
RI MS JT/1/TYP/2/696–7

1. *[1866]*: the year is given by relation to letter 2463, dated 7 August 1866.
2. *your kind letter*: letter missing. This would presumably be Tyndall's reply to Bence Jones's letter dated 7 August 1866 (letter 2463).
3. *here*: probably Folkestone (see letter 2463, n. 12).
4. *Institution*: the RI, where Bence Jones served as Secretary from 1860–72.
5. *Sir H. Holland*: Henry Holland.
6. *Busk*: George Busk.
7. *Gladstone*: John Hall Gladstone (see letter 2228, n. 3).

8. *if I have to go*: Bence Jones had a number of years left in him, as he did not die until 20 April 1873.
9. *interregnum*: a cessation or suspension of the usual ruling power; a period of freedom from some authority (*OED*).
10. *your book on Sound*: *Sound* (1867).
11. *nice review yesterday in the Sat[urda]y by you*: 'The Oberland and Its Glaciers (review)', *Saturday Review of Politics, Literature, Science and Art*, 22 (29 September 1866), pp. 399–400. While no author was given, it must have been clear to Bence Jones that Tyndall wrote this book review, considering the first lines: 'In crossing from Folkestone to Boulogne, or from Dover to Calais, every traveller may notice the greenish colour of the sea, which is more decided about the prow and paddles, where the water is partially churned up into a finely-bubbled foam. Whence does this colour of the sea-water arise? From the fact that it has the power of abstrating from the solar light its red constituent, the mixture of remaining colours imparting to the water its peculiar hue. But what becomes of the red light thus absorbed? It is converted into heat—applied, in fact, to warm the water and to promote its evaporation'. This scientific lesson continued long before Tyndall even mentioned the book being reviewed.

From Juliet Pollock 3 October [1866][1] 2486

Chaldon | Bournemouth[2] | Oct. 3.[d]

My dear Mr. Tyndall

Your handwriting seemed an echo to my thoughts in which you have occupied a considerable place while I have watched the gentle or the turbulent waves, or the glowing gorse and heather, or while I have been enjoying the society of the H. Taylors.[3]—in which I should like much to see you. Such fine noble simple ingenious girls! the daughters the Mother so lovely and entertaining, and the father one of the most interesting men among the distinguished whom I have had the happiness of knowing. That you & he would be friends speedily I am certain.

The weight of his thought is associated with a certain slowness of manner which I believe some people think tedious, but you would not, because his intellect would have its due value with you & you would recognize too the tenderness, benevolence, and noble candour of his nature—for this reason & for many others I have felt more than regret that you have not been able to come to us this holiday.

I am deeply concerned to hear of your indisposition we shall be in town on the 16th of Oct.

Mrs Cameron[4] (dear thing!) has just come to the close of a weeks visit to us which has been truly pleasant to all parties. There were those, Mr. Tyndall, (who shall be nameless) who prepared themselves to see jealousies between two friends who both admired the Poet so much;—but they have been obliged to change their attitude, & to confess that women can be occasionally both zealous & liberal. When you add to this the generosity of the wife's behaviour you will reach a sum of feminine amiability perhaps unknown to you before.

You will have seen my husband[5] probably in London yesterday we expect him home with Fred[6]—to night.

I am pleased by Fred's enthusiasm.

Walter[7] is looking well & Moss[8] is also improved in looks,—and now I am wanted for a walk so good bye. Such a sun to day such an air to invite me out it is quite irresistible

Your ever attached friend | Juliet Pollock

RI MS JT/1/P/200
RI MS JT/1/TYP/6/2076

1. *[1866]*: the year is written in on the letter (possibly in a different hand), and LT included it in her typescript.
2. *Chaldon | Bournemouth*: see letter 2477, n. 1.
3. *the H. Taylors*: Henry Taylor and his family (see letter 2477, n. 5, n. 7, and n. 8).
4. *Mrs Cameron*: Julia Margaret Cameron (1815–79), a British photographer who was well known in the nineteenth century for her style of portraits. Notable subjects were Alfred Tennyson, Thomas Carlyle, Charles Darwin, John Herschel, Henry Taylor, and Ellen Terry (*ODNB*). She was a frequent visitor to the Taylor family's summer home in Bournemouth, The Roost. Henry Taylor's daughter Una published a memoir about The Roost and the guests they entertained there—*Guests and Memories: Annals of a Seaside Villa* (London: H. Milford, Oxford University Press, 1924)—which included Cameron's portraits of Taylor, James Spedding (see letter 0094, n. 13), and Benjamin Jowett (1817–93), a theologian who held the Regius Professorship of Greek at Oxford (*ODNB*).
5. *my husband*: William Frederick Pollock.
6. *Fred*: Frederick Pollock.
7. *Walter*: Walter Pollock.
8. *Moss*: Maurice Pollock.

To Mary Anne Coxe 4 October 1866 2487

4th. Oct. 1866.

I do like them: they bring old old days to mind when the poems of Scott[1] and Burns[2] were the delight of my life. When Straths Bens, lochs and Corries[3]—purple heather & misty glen—constituted my ideal scenery. No note in Byron[4] stirred me more than that:

"And wild & high the Cameron's gathering rose
The war-note of Lochiel which Albyn's hills
Have heard & heard too have her highland foes
When in the noon of night that pibroch thrills."[5]

I quote from memory and may trip.

I should like to pass away in some such manner. I hate the thought of creeping through old age. This to my fancy is the way "a mountaineer" would wish to die.

I wish to heaven I could comfort you.[6] | J. T.

RI MS JT/1/T/251
RI MS JT/1/TYP/1/289

---

1. *Scott*: Walter Scott (1771–1832), a Scottish novelist, poet, and historian (*ODNB*).
2. *Burns*: Robert Burns (1759–96), a Scottish poet and lyricist (*ODNB*).
3. *Straths Bens, lochs and Corries*: Scottish words for valleys (*straths*), mountains (*bens*), lakes (*lochs*), and circular hollows on mountain sides (*corries*).
4. *Byron*: George Gordon Byron (1788–1824), Lord Byron, an English poet and politician (*ODNB*).
5. *"And wild & high . . . that pibroch thrills"*: from Byron's long narrative poem 'Childe Harold's Pilgrimage', Canto III, stanza 26 (L. Byron, *Childe Harold's Pilgrimage. A Romaunt* (London: John Murray, 1860), p. 79).
6. *comfort you*: Coxe's need to be comforted is not identified for certain, but possibly because of her nephew Alick Cumming's death in early 1865 while serving in India (see letter 2252).

From Matthew Fontaine Maury — 4 October 1866 — 2488

7 Upper Gloucester Str | Dorset Sq[ua]r[e] | 4—10—'66

My dear Tyndall,

The glaciers are elegant.[1] We read them o'nights, and they read as good as a 'winter evening's tale'. Thank you my friend for so much instructive reading and so many thought provoking remarks.

y[ou]rs truly | Maury

RI MS JT/1/M/69

---

1. *the glaciers are elegant*: see letters 2482 and 2483.

To Jane Barnard [8 October 1866][1] 2489

Monday

My dear Miss Barnard

Our friend's illness is not a light one but I have strong hopes that he will overcome it.[2] His heart is not as it ought to be and one of his lungs is also out of order. But he looked bright, and I trust he is not in danger.

Should his present unsatisfactory condition continue he will remove to town so as to be near good advice. If he improves he intends to remain some time at Folkestone.[3]

I do not well know what we should do without him. And I strongly hope on all our accounts & for the sake of the Institution[4] that he will overcome his enemy and be soon himself again.

I am rejoiced to hear of the good condition of your uncle[5]—give my kindest regards both to him & your aunt.[6]

Believe me | Yours most truly | John Tyndall

RI MS JT/1/T/129
RI MS JT/1/TYP/1/4193

1. *[8 October 1866]*: the date is not on the handwritten letter, but LT included it in her typescript.
2. *Our friend's illness . . . will overcome it*: in letter 2485, dated 30 September 1866, Henry Bence Jones had shared with Tyndall about his declining health.
3. *Folkestone*: see letter 2463, n. 12.
4. *Institution*: the RI.
5. *your uncle*: Michael Faraday.
6. *your aunt*: Sarah Faraday.

To Rudolf Clausius 12 October 1866 2490

12th. Oct. 1866.

My dear Clausius

Many times during the last five or six weeks I have been on the point of writing to you. Hirst, from time to time, has conveyed to me your kindly greeting, and this morning he showed me a note from you relative to your election to the Royal Irish Academy.[1] It gives me great delight my dear Clausius to find you so highly and so generally appreciated in this country. With regard to the delay it is of no importance. Probably the Secretary[2] has been taking his vacation & has forgotten to send you the diploma. Hirst however has written to Salmon[3] to ascertain all about the matter.

I went to Switzerland in July.[4] I dare say you think me an unsocial fellow for not calling to see you more frequently; but the fact is that when I leave London I am usually so worn down, & so out of humour with myself and the world that the mountain wilds are the only places fit for me. I had exceedingly bad weather. I went first of all to the Engsteln Alps,[5] climbed the Titlis[6] and went on to Rosenlaui,[7] intending to climb all the mountains in the neighbourhood. But here the weather broke, and ever afterwards it remained bad. On the 20th of July I was snowed up at the summit of the Faulhorn.[8] The only two consecutive fine days that I had afterwards were in crossing the Petersgrat[9] to Platten,[10] and in crossing from Platten over the Lötschsattel[11] to the Aeggischhorn.[12] These were glorious days & I do not know that any summit in the whole Alps commands a finer view than the Petersgrat. I oscillated for some time between the Aeggischhorn & the Bell Alp,[13] intending to climb the Aletschhorn,[14] and some of the other higher peaks without guides; but I was defeated in the only attempt which I was able to make by a heavy snowstorm. In company with a friend—you know him perhaps, Girdlestone of Oxford[15]—I got high up the Aletschhorn, but we were shaken down by the snowstorm. I went to Italy for 8 or 10 days thinking matters might mend, but I found them worse than ever when I returned. At the Bell Alp we had a fall of snow so heavy that avalanches shot down across the customary paths. The cattle all suffered very much.

I should have gone to Zurich on my return, but in the first place I was not sure that you were there, and secondly I had made an engagement in Paris which did not permit me to lose a day. I occupy at present the post formerly occupied by Faraday with regard to Lighthouses,[16] and my engagement in Paris had reference to this question. I concluded from the time-bills that I should have an hour's delay in Thun,[17] and therefore told you that I should call at the Post office. But when I reached Thun from the Giessbach[18] I found that I had not a minute's delay; hence your letter[19] which Hirst informs me you wrote to Thun never reached me.

Some time ago I had a letter from Wiedemann[20] informing me that Vieweg[21] had made the final arrangements with Bohn,[22] and that the translation of the book would now go on rapidly.[23]

No doubt the newspapers have made you acquainted with the news about the Atlantic cable.[24] Those engaged in the work have been honoured, some with Knighthood, others with the hereditary title of Baronet. William Thomson who held some post in connexion with the electrical department, is now Sir Wm. Thomson—having been knighted He well deserves it, and I have no doubt he likes it.[25]

I am engaged publishing a series of six lectures on Sound:[26] the proof sheets are now coming in, and I may ask you, if you have the time, to glance at them for me. Correct fundamental notions on all branches of physics are very much needed in England. There is also a large class of cultivated persons whom it is desirable to interest in science—These are my reasons for publishing the little work.

Give my love to the children[27] and kind regards to your wife.[28] How is Johnny[29] getting
on?
Yours ever John Tyndall.

RI MS JT/1/T/195

---

1. *your election to the Royal Irish Academy*: Clausius was elected an Honorary Member of the Royal Irish Academy on 16 March 1866 (*List of the Council and Officers and Members of the Royal Irish Academy; Dublin, 31st of July, 1876* (Dublin: Academy House, 1876), p. 23).
2. *Secretary*: the Secretary of the Royal Irish Academy at this time was William Reeves (1815–92), an Irish antiquarian and Bishop in the Church of Ireland (*ODNB*; *List of the Council and Officers and Members of the Royal Irish Academy*, p. 19).
3. *Salmon*: possibly George Salmon (1819–1904), an Irish mathematician and theologian of Trinity College in Dublin who was elected a member of the Royal Irish Academy in 1843 and at this time was a Vice-President of the academy (*ODNB*; *List of the Council and Officers and Members of the Royal Irish Academy*, p. 20).
4. *I went to Swizerland in July*: see letter 2457, n. 4.
5. *Engsteln Alps*: see letter 2460, n. 2.
6. *Titlis*: see letter 2460, n. 3.
7. *Rosenlaui*: see letter 2460, n. 4.
8. *Faulhorn*: see letter 2460, n. 6.
9. *Petersgrat*: see letter 2460, n. 10.
10. *Platten*: see letter 2460, n. 11.
11. *Lötschsattel*: see letter 2464, n. 11.
12. *Aeggischhorn*: see letter 2460, n. 16.
13. *Bell Alp*: Belalp (see letter 2464, n. 14).
14. *Aletschhorn*: see letter 2460, n. 21.
15. *Girdlestone of Oxford*: Arthur Girdlestone (see letter 2460, n. 19).
16. *regard to Lighthouses*: see letter 2423, n. 3.
17. *Thun*: see letter 2461, n. 6.
18. *Giessbach*: probably Giessbach Falls, located near the Brienzersee, Switzerland to the east of Thun.
19. *your letter*: letter missing.

20. *letter from Wiedemann*: Gustav Wiedemann; see letter 2472.
21. *Vieweg*: Eduard Vieweg.
22. *Bohn*: Johann Conrad Bohn.
23. *translation of the book would now go on rapidly*: the German translation of *Heat* (1865); see letter 2296, n. 6.
24. *news about the Atlantic cable*: work on re-laying the Atlantic telegraph cable after the failed attempt of 1858 was completed on 27 July 1866.
25. *I have no doubt he likes it*: Tyndall was pointedly underplaying Thomson's involvement with the Atlantic telegraph cable.
26. *publishing a series of six lectures on Sound*: *Sound* (1867).
27. *children*: see letter 2270, n. 5, and letter 2387, n. 15.
28. *your wife*: Adelheid Clausius.
29. *Johnny*: Rudolf John Clausius.

From the Jamaica Committee[1] 12 October 1866 2491

Jamaica Committee | 65 Fleet St. | 12 October 1866.

The Jamaica Committee present their compliments to Professor Tyndall and beg to request his attention to the accompanying pamphlet[2] relating to the atrocities committed by the authorities in Jamaica not in suppression of the so-called rebellion as Prof. Tyndall probably supposed when he gave his name to the Eyre Committee,[3] but for days and weeks after (as Eyre[4] and his officials show) all resistance had ceased, and when any offense ought to have come under the cognisance of regular legal tribunals.

The Jamaica Committee cannot believe that Professor Tyndall can approve of the violent and illegal assumption of power by <u>military</u> authorities.[5] Such authorities may be quite right in their mode of employing force to put down insurrection, but offences alleged to have been committed <u>before</u> the rebellion cannot with any legal propriety or with any safety to the public be tried by courts martial.

Gordon's[6] execution did not take place until Mr Eyre had our notes of the so-called trial in his possession, and endorsed the finding of the court-martial although the notes proved that <u>no facts</u> against Gordon had been proved. Professor Tyndall F.R.S.

RI MS JT/1/TYP/8/2565
LT Typescript Only

---

1. *Jamaica Committee*: see letter 2481, n. 4
2. *pamphlet*: not identified.

3. *Eyre Committee*: see letter 2481, n. 4.
4. *Eyre*: Edward Eyre.
5. *that Professor Tyndall . . . by military authorities*: Tyndall's support for Eyre's violent suppression of the Morant Bay rebellion is discussed in the introduction to this volume.
6. *Gordon's*: George William Gordon.

To Edward Livingston Youmans [12 October 1866][1] 2492

I quite expected the little hailstorm, and was astonished to find what you said (for the view was very strong for such a place) so heartily appreciated. Believe me, when I say from my very heart, the paper surprised me as much as it delighted me.

J. Fiske, *Edward Livingston Youmans: Interpreter of Science for the People* (New York: D. Appleton and Company, 1894), p. 226

---

1. *[12 October 1866]*: this letter fragment from Tyndall to Youmans was quoted in a letter of Youman's dated 13 October 1866 in which he described Tyndall's attendance at his lecture on 'The Scientific Study of Human Nature' on 10 October 1866, and Tyndall's subsequent defence of his position against his detractors (the 'hailstorm'). According to Youmans, he wrote to Tyndall the day after the lecture (11 October), and received the reply from Tyndall (probably 12 October). See J. Fiske, *Edward Livingston Youmans: Interpreter of Science for the People* (New York: D. Appleton and Company, 1894), pp. 224–6. Youmans's lecture was published as E. L. Youmans, *Observations on the Scientific Study of Human Nature: A Lecture Delivered Before the London College of Preceptors* (New York: D. Appleton and Company, 1867).

From Charles Brooke[1] 13 October 1866 2493

L. *On the Dynamical Theory of Electricity.* | *By* C. BROOKE, *F.R.S.* | *To Professor Tyndall, F.R.S.* | 16 Fitzroy Square, W. | October 13, 1866.

MY DEAR SIR,

I HAVE lately been led, in relation to a new edition of my 'Elements of Natural Philosophy'[2] now in the press, to reconsider the whole question of electricity, and I have become strongly impressed with the reality of the dynamical theory. I have hence naturally been induced to look about for confirmatory facts; and a strong, and (to the best of my knowledge and belief) a new fact has presented itself to me. It has long been known that if a bar of antimony and one of bismuth be connected, and a current be transmitted from antimony to bismuth, heat will be developed at the point of junction; and, on the contrary, if the current pass from bismuth to antimony, cold will be produced.[3] Now it struck me that if in the former, as in all other cases in which heat is developed in the passage of a current, a portion of *electric potential* is (as I believe) converted into *thermic potential*, or heat, there ought in the latter case to be an inductive conversion of thermic into electric potential; and if so, there should be a *loss* of current in the first instance, and a *gain* in the second; and such appears to be the fact. On duly balancing the thermo-element above mentioned in a Wheatstone's[4] bridge, the deflection of the needle followed the direction of the current, and the anticipated loss or gain of current was fully realized. If you can find room for this notice in the Philosophical Magazine,[5] you will greatly oblige,

Yours faithfully, | CHARLES BROOKE.

Phil. Mag., 32 (1866), p. 378

---

1. *Charles Brooke*: Charles Brooke (1804–79), an English surgeon and inventor of measuring instruments (*ODNB*).
2. *new edition of my 'Elements of Natural Philosophy'*: C. Brooke, *The Elements of Natural Philosophy; or an Introduction to the Study of the Physical Sciences*, 6th edn (London: John Churchill and Sons, 1867). This sixth edition was Brooke's third, as the first three were authored by Golding Bird (1814–54), an English physician (*ODNB*).
3. *cold will be produced*: the thermoelectric effect had originally been identified by Thomas Seebeck (1770–1831) in 1821.
4. *Wheatstone's*: Charles Wheatstone (see letter 2452, n. 4).
5. *in the Philosophical Magazine*: Tyndall had been one of the editors of the *Phil. Mag.* since 1854.

From Rudolf Clausius 18 October 1866 2494

ZÜRICH, 18/10/66.

Lieber Tyndall,

Wir haben sehr bedauert, Dich diesen Sommer nicht hier zu sehen. Ich adressirte deshalb meinen Brief nach Thun und nicht nach Basel, weil ich dachte, in Thun könntest Du Deinen Reiseplan noch ändern und nach Zürich anstatt nach Basel reisen; aber wenn Du einmal in Basel wärest, so würdest Du nicht wieder zurückreisen.

Ich muss Dir nun aus dem nach Thun geschickten Briefe den Theil wiederholen, der sich auf die Uhr bezieht.

Der Uhrmacher, von dem Deine frühere Uhr gekauft war, ist gestorben; ich habe aber bei drei anderen Uhrmachern nachgefragt, und habe übereinstimmend folgenden Bescheid erhalten.

Ankeruhren mit Secundenzeiger kann man erhalten zu Preisen von 160 frs bis 280 frs, welche sich vorzugsweise in der Goldstärke unterscheiden. Die zwischen 200 frs und 250 frs seien darunter am meisten zu empfehlen.

Der Uhrmacher, welcher hier als der intelligenteste gilt, sagte mir noch, es seien jetzt eine Art von Uhren sehr beliebt, welche ganz besonders enpfehlenswerth seien, nämlich auch Ankeruhren mit Secundenzeiger, aber mit Compensation und in Kälte und Hitze (d.h. in Kasten, die aussen mit Eis oder glühenden Kohlen umgeben sind) regulirt. Eine solche Uhr, die Halbchronometer genannt werde, koste bei genügender Goldstärke 300 frs.

Wenn Du es wünschest dass ich Dir eine Uhr besorge, so bin ich gern dazu bereit, und Du brauchst nur zu schreiben, von welcher Sorte Du sie haben wünschest.

Die Correcturbogen deines Buches über den Schall werde ich mit vielem Vergnügen durchlesen. Ich würde das Buch, wenn es erschienen wäre, doch lesen, und es verursacht mir daher keinen Zeitverlust, wenn ich schon in den Correcturbogen lese.

Bei uns geht alles gut. Im Sommer hatten sämmtliche Kinder den Keuchhusten, aber jetzt ist er beinahe vorüber.

Johnny entwickelt sich sehr niedlich und weiss sich als Junge unter den Mädchen schon gehörig geltend zu machen.

Mit herzlichen Grüssen von meiner Frau und mir,

Dein | Clausius

ZURICH, 18/10/66.

Dear Tyndall,

We regret not having seen you here this summer. I therefore sent my letter to Thun,[1] instead of Basel,[2] because I thought that in Thun you could still change the plans of your journey and travel to Zurich instead of Basel; but if you were already in Basel, you would not travel back.

I now have to repeat to you the part of the letter sent to Thun that dealt with the watch.[3]

The watchmaker from whom you bought your previous watch has died; but I have enquired with three other watchmakers and found the following concurrent information.

Lever-escapement watches with a second hand can be obtained for prices from 160 frs to 280 frs, the difference being mainly in the quality of the gold. The ones between 200 frs and 250 frs can be recommended the most.

The watchmaker who locally counts as the most intelligent told me also that there now is a type of watch that is very popular and which is particularly to be recommended. These are also lever-escapement watches with a second hand, but with compensation and adjusted to cold and heat (that is, in cases that are surrounded by ice or glowing coals). Such a watch, called a half-chronometer, at sufficient gold quality would cost 300 frs.

If you wish that I purchase a watch for you, I will gladly do it and you only have to write which type you would like to have.

The proof sheets of your book on sound I will read with great pleasure.[4] I would have read it anyway once the book was published and it therefore is no waste of time for me if I read the proofs.

Everything here is well. All the children[5] had whopping cough in the summer, but now it is almost over.

Johnny is turning into a cute boy and has learned to fiercely assert himself as a boy among the girls.

With best wishes from my wife[6] and myself

Your | Clausius

RI MS JT/1/TYP/7/2296–7
LT Typescript Only

---

1. *sent my letter to Thun*: letter missing. Thun is a town in the Swiss Canton of Bern, on the western side of the Thuner See. In letter 2472, Wiedemann wrote to Tyndall that Clausius 'asks whether you have received his letter that was sent poste restante to Thun'.
2. *Basel*: see letter 2309, n. 3.
3. *the watch*: see letter 2464, including n. 30.
4. *proof sheets of your book on sound*: *Sound* (1867).
5. *children*: see letter 2270, n. 5, and letter 2387, n. 15.
6. *my wife*: Adelheid Clausius.

To the Editor of *The Athenaeum*[1] 13 October 1866 2495

SCIENCE | ICE: DOES IT EXPAND OR CONTRACT BY COLD?

Royal Institution, Oct. 13, 1866.

In the experiment referred to by your distinguished Correspondent, Dr. Rae,[2] water, not ice, was inclosed in a cast-iron bottle, and surrounded by a freezing mixture. The bottle was burst by the expansive force exerted by the water in passing to the solid state. Dr. Rae is right in believing that ice, once formed, contracts by cold; it has, indeed, a very high co-efficient of contraction.

To illustrate the changes of form of which ice is capable, I have sometimes squeezed it into iron moulds. These experiments may possible have been confounded with that in which the iron bottle was riven.

Dr. Rae's observations on the effect of contraction and expansion on a large scale are exceedingly interesting. I fear, however, that those who have made the glaciers of the Alps their study will feel some difficulty in accepting his views regarding glacier motion in Greenland.

JOHN TYNDALL.

The Athenaeum, 20 October 1866, p. 502

1. *Editor of* The Athenaeum: see letter 2366, n. 1.
2. *experiment referred to by your distinguished Correspondent, Dr. Rae*: Tyndall was referring to a letter to *The Athenaeum* by John Rae (1813–93), an Arctic explorer (*ODNB*), published in the 13 October 1866 issue on pp. 470–1.

To Harriet Jane Moore[1] [October 1866][2] 2496

Wednesday

My dear Miss Moore

In your dear mother[3] I have lost one of my dearest friends. She was to me a wonder & a delight. But how beautiful her release. It is typical of what humanity might long to see the common lot. I can well appreciate your sorrow for I share it in my own proper heart.

Yours truly | John Tyndall

RI MS JT/1/T/1075
RI MS JT/1/TYP/3/868/3

1. *Harriet Jane Moore*: Harriet Jane Moore (1801–84), a British watercolor artist who painted scenes of Michael Faraday's apartment, study, and laboratory at the RI during the 1850s. Moore was the eldest child of surgeon James Carrick-Moore and Harriet Moore (née Henderson). She was elected a member of the RI in 1852. See F. A. J. L. James, 'Harriet Jane Moore, Michael Faraday, and Moore's Mid-Nineteenth-Century Watercolours of the Interior of the Royal Institution', in J. Hamilton (ed), *Fields of Influence: Conjunctions of Artists and Scientists, 1815–1860* (University of Birmingham Press, 2001), pp. 111–28.
2. *[October 1866]*: the year and month are given as LT included these on her typescript.
3. *your dear mother*: Harriet Moore (née Henderson, 1779–1866), was a daughter of the actor John Henderson (*c.* 1747–85) and Jane Figgins (*c.* 1752–1819). Harriet married surgeon and vaccine promoter James Carrick-Moore (1762–1860). Together they had five children: Harriet, Louisa (1802–53), Julia (1803–1904), John Carrick (1805–98), and

Graham Francis (1806–83). The wealthy Moores had close connections with the RI and frequently entertained Tyndall at their Wimbledon home.

From Juliet Pollock 2 November 1866 2497

59 Montagu Square W[1] | Nov[embe]r. 2d 1866.

My dear Mr. Tyndall

Because you are capable of forgetting your friends (at their dinner hour—or tea table—not otherwise whatever your angry Valentine's say) I write to remind you that you dine with us on Wednesday next at 7 o'clock and beg that you will stick up this note as a frontlet.[2]

I am ever (you must have observed it, I am sure) | Most truly yours | Juliet Pollock

RI MS JT/1/P/202
RI MS JT/1/TYP/6/2079

1. *59 Montagu Square W*: the Pollocks' residence in Marylebone, London.
2. *frontlet*: something worn on the forehead (*OED*); Pollock was probably asking Tyndall to place the note on his forehead so that if he looked in a mirror he would be reminded of this engagement.

To the Jamaica Committee[1] 7 November 1866 2498

PROFESSOR TYNDALL'S REPLY to the JAMAICA COMMITTEE

Gentlemen,—You have done me the honour of sending me certain papers "relating to the atrocities of the authorities in Jamaica," during the suppression of the late "alleged rebellion."[2]

I am far from being insensible to the opinion which your courtesy in sending me these papers implies: namely, that I am incapable of defending any line of conduct to which the term "atrocious" could with fairness be applied.

Amid the many destracting lights which fall upon this question, the notions which I have been able to form regarding Jamaica, and the interpretation which I have been able to put upon occurrences you so earnestly denounce, are these:

That through the absence of necessary guidance and governance, the negroes of the colony had fallen into a very chaotic and bad condition, presenting a deplorable contrast with their brethren in other colonies, who, favoured with less leisure, have become far better men. The Jamaica negroes are, I am informed, as a general rule, lazy and profligate, while many of the authorities I have consulted, and who have lived among these people, add a third and weightier adjective, which I refrain from writing down.

It is not, I believe, contended, even by their friends, that the Jamaica negroes are not lazy: it is only urged in extenuation that white men under the same circumstances would be lazy too. Be it black or white, the man who illustrates this doctrine, and who lives as the negroes of Jamaica are reported to live, soon ceases to be a civilised being.

But he may be harmless and docile notwithstanding. Though absolutely useless he may not be dangerous. Is this the character of the West Indian negro, or is it not?

History, written without reference to the Jamaica insurrection—written long years before the committal of the deeds you denounce—gives the following answer to this question:

> "This vast conspiracy (that of the negroes in St. Domingo[3]), productive in the end of calamities in the island, unparalleled even in the long catalogue of European atrocity, had for its object the total extirpation of the whites, and the establishment of an independent black government over the whole colony. So inviolable was the secrecy, so general the dissimulation of the slaves, that the awful catastrophe was in no way apprehended by the European proprietors, and a conspiracy which embraced nearly the whole negro population of the island, was revealed only by the obscure hints of a few faithful domestics, who, without betraying their comrades, warned their masters of the approach of an unknown and terrible danger. The explosion was sudden and dreadful, beyond anything ever before seen among mankind. At once the beautiful plains in the north of the island were covered with fires, the labours of a century were devoured in a night, while the negroes, like unchained tigers, precipitated themselves on their masters, seized their arms, massacred them without pity, and threw them into the flames . . . The cruelties exercised on the unhappy captives on both sides in this disastrous contest exceeded anything recorded in history. The negroes marched with spiked infants on their spears instead of colours, they sawed asunder the male prisoners, and violated the females on the dead bodies of their husbands."—Alison's[4] *History of Europe*, Vol, VI., p. 104.[5]

It was people of this race and temper, who, called together by the sound of horns and shells, assembled in hundreds round the Court House of Morant Bay on the 11th of October, 1865. It was such people who then and there fell upon our countrymen, and, after driving them by fire from the Court House, slew, hacked, and mutilated them; and who, after having done this, proceeded to spread the flame of insurrection elsewhere.

At the time here referred to discontent and disaffection prevailed throughout the whole island of Jamaica. This I gather from your own documents. The colony then resembled the dry grass of a prairie, ready to be set on fire from beginning to end by a spark of successful insurrection.

But lest what I gather from your documents should be tinctured by any bias of mine, I offer to your notice the first of a series of letters bearing upon this subject, which comes to my hand.

On the 20th of July, 1865, Mr. Hudson,[6] of Jamaica, writes thus to Mr. Salmon,[7] of the same island:

Spring Vale, July 20, 1865.

DEAR SIR,—I think it incumbent upon me to inform you of certain reports of a serious nature which are in circulation in my neighbourhood.

For the past few weeks there have been rumours of a rebellion among the peasantry, but I did not attach any importance to them, but conceived they originated from want of understanding the causes of the recent meetings held in this and the adjacent parishes until lately. I have been led by sufficient information to believe they are more than passing reports, and which require some notice.

It appears that there is an organised plot among the people in this vicinity for a rebellion about the early part of August ensuing. My reasons for believing there is truth in the matter are these:

A house girl of mine has stated several times to my family's hearing that the black people have agreed upon rising in arms at the above period for the purpose of murdering the upper classes, and destroying and afterwards seizing their properties. She does not deny that she obtained this information from her friends. For the present she will not say anything more on the subject, but only state, "We must wait and see." Other persons have likewise given us intelligence of a similar nature. Dr. M'Catty[8] mentioned something to me of a like import which tends greatly to strengthen them. He stated he had already acquainted you with the purport of his information.

I do not think it would be wise to treat these rumours merely as "idle tales." My district is hot to join with other parts of the island in a move of the kind, if made.

My delicate state of health prevents my calling to see you, but I shall be glad to give you whatever further information in my power, and to receive any directions from you on the subject.

It would not be prudent to allow these strong reports to continue, without preparing, as far as we are able, against an attempt of disturbance, should it turn out to be correct.

I have, &c., | (Signed) JOHN HUDSON. | The Hon. John Salmon, Custos.

"A house girl of mine":—You, gentlemen, are far better acquainted than I am with what an illustrious member of your body calls "the law of inseparable association,"[9] and no doubt you will recognise an illustration of this law in the promptitude with which the warnings preceding the troubles in St. Domingo suggest themselves on reading this letter.

On the 11th of October, as above stated, the much dreaded action began. The governor of the island had then thrown upon him the responsibility of preserving the property, lives, and honour of 14,000 British men and women, who were scattered in isolated and unprotected positions among a negro population of 350,000.

And it is to be remarked, that the leaders of these people were well aware of the advantages of their position. That they were not held back from revolt and rapine by the fear of subsequent retribution. In the neighbouring island of Hayti they had the example of a people which had successfully defied the power of France. In reply to an intimation of the swift vengeance of England in case of insurrection, Mr. G. W. Gordon[10] was able to point out that "all the power of the great Napoleon could not put down the rising in Hayti, and that was successful, for the troops died of disease before they could meet the people in the mountains."[11] And when the power displayed by England in crushing the Indian mutiny was urged upon his attention, Mr. Gordon was able to respond, "India is not at all a case in point, for India is a flat country, and the English troops would overrun and conquer it; but this is a mountainous country, and before the British troops could reach the mountains they would die of disease here."[12]

Taking all these things into account I endeavoured to clearly realise the position of the Governor who had this problem suddenly thrust upon his attention by the butchery at Morant Bay. How, with the force at his disposal, to preserve the lives of 7,000 British men, and the honour of 7,000 British women from the murder and the lust of black savages capable of the deeds which history shows to be theirs in St. Domingo; and a re-enactment of which had, to all appearance, begun under the said Governor's eyes.

I will not dwell upon the measures taken then and there by the Governor to solve this problem. For his deeds in the first instance have won for him the praise of his censors, and have challenged without hostile response, the judgment of yourselves. But you denounce him for extending martial law beyond the necessary period, and for inflicting the punishment due to rebellion "for days and weeks after all resistance had ceased."

I am not prepared to question the truth of these allegations; I am not prepared to deny that the period of punishment was too long, or that its character was too severe. Now that the smoke of battle has cleared away, you perhaps see more clearly the character of the field. But I would invite you to transport yourselves to that field while the smoke still hung upon it; to remember that a former rebellion in Jamaica which every body supposed to be quelled in May broke out "with redoubled fury in June;" to think of Governor Eyre with the blood of his slaughtered countrymen before his eyes; with the memories of St. Domingo in his mind; with the consciousness that the whole island round him was near its point of combustion, and with no possible means of estimating *how* near. In what way, gentlemen, some of your number would have acted under these circumstances, God only knows. In all probability they would have acted kindly and calamitously, for philanthropists can unconsciously become shedders of blood. How others of your number would have acted I believe I *do* know. They might have acted with more tact, and thus given less offence, but their measure of mercy to the murderers would not, I am persuaded, have exceeded that of Governor Eyre.

It appears to me that you practically forget the circumstances which I am here endeavouring to bring before your minds. There seems a tendency on your part to tone down the crimes of the negro and to bring his punishments into relief. You speak for example, of the massacre at Morant Bay as "a local riot." But looking at the historic antecedents what right had Governor Eyre to come to this conclusion? Had he come to it and acted on it, he might have earned for himself the execration of the English people, if not of mankind. To his knowledge the whole island was disaffected, and history had informed him what a disaffected negro population can do. He knew how they could plot in secret, and spring like tigers on their victims the moment their plans were matured. Only think of his position, knowing that such things had occurred, and with a horrible practical intimation before him that they were about to occur again. The height and span of a bridge, gentlemen, must be regulated with reference to the larger floods, and the action of Governor Eyre had to be determined, not by the hypothesis of a local riot, but by the contemplation of the calamities which were certain to overwhelm the whole island if the insurrection were permitted to expand. No narrower assumption than this could, under the circumstances, be entertained. Had the local riot hypothesis been demonstrable *at the time*—could it have been proved that the insurrection was confined to a single parish, and that the culprits at large would not spread the fire of rebellion throughout the colony, I would join you in condemning Governor Eyre. But no such thing was demonstrable. Had he a had force at his command sufficient to cope with insurrection, even though the whole island were engaged in it, I would let him answer for himself. But he had no such force. The power opposed to him, if permitted to organise itself, was sufficient to swallow him and those for whose lives he was accountable. He was, therefore, obliged to augment his relative strength, by damping everywhere the spirit of rebellion, and this could only be done by making the name, power, and determination of England terrible throughout the island. That errors should be committed, that cruelties should be perpetrated, that wrong should be done, is, I fear, under such circumstances, unavoidable; but the error, the cruelty, and the wrong, so far as they were committed, were, for the most part, beyond the control of Governor Eyre.

You have directed my attention to the case of Mr. Gordon, and I have looked into it again. I pass without comment the evidence of Lieut. Edenborough,[13] the extreme gravity of which cannot be overlooked, because, through the miscarriage of a letter, his evidence, though in the blue book, was not taken before the commissioners. From the speeches of Mr. Gordon in the House of Assembly; for his relationship to Mr. Paul Bogle,[14] one of the principal murderers of Morant Bay; from his letters to Mr. Lawrence,[15] his manager; from his conversations with Messrs. Harrison,[16] Beckwith,[17] Ford,[18] and others; from his published address to the negroes,[19] and from his other addresses, the conclusion seems indubitable, that he was the taproot from which the insurrection drew its main sustenance, and that Governor Eyre was justified in concluding that no man concerned in the murder of our countrymen was more guilty than he. When, therefore, the sentence on Gordon, after having been approved by his officers, was submitted to him, he did not stay judgment, but ratified the decision that others had pronounced. Whether, in the details of this proceeding, he committed or did not commit a legal blunder, I am not able to say. If he did, he has surely paid dearly for it by the ruin of his career. I notice throughout his entire conduct the clear conviction on his part that he was doing his duty, the undoubting trust in the justice of his countrymen which caused him to neglect the personal safeguards which a more adroit person would have employed. To associate the term "murder" with his name is, in my opinion, without justification. There is no propriety in the term unless it is backed by assumptions which no man has a right to make. And if in the prosecution, which, to the detriment of this country, is now invoked against Governor Eyre, the execution of Gordon, from a lawyer's point of view, be deemed indefensible, I trust the common sense of Englishmen will perceive that it was caused by those whose own bloody atrocities had, for the time, stifled mercy and abrogated law.

Such, gentlemen, are the notions and interpretations which your courtesy in writing to me induces me to lay before you. And such being my notions, when asked to join the Eyre Defence Committee,[20] I felt that cowardice, or a dislike to the worry involved in the act, was the only reason which opposed my doing so. Concluding from what I had observed in society, that Governor Eyre was more likely to suffer from timidity than from conviction—that the dread of you, gentlemen, and of those whom you represent, was very widely spread—I overcame my strong reluctance to mingle in any matter of the kind. I pledge you my word of honour, gentlemen, that during this enquiry I have had laid before me the testimony of men of approved valour, coolness of judgment, and capacity, who have occupied positions of the most critical danger, who have extricated themselves and others from such positions in a manner which has secured for them the unanimous applause of the people of this country, and who from data gathered on the spot conclude that in the month of October, 1865, Governor Eyre saved the whole white population of Jamaica from massacre. I have had the solemn assurance of such men, that England has few sons as noble, few men so equal to a great emergency, few who have done the state such service as Governor Eyre. And I now call upon you, with the most solemn emphasis that man can address to man, not to permit personal pride, the love of victory, or the desire to substantiate that to which you have committed yourselves, because you are so committed, to influence your further action in this matter. I call upon you in the name of all that is wise and dignified in human nature,—in the name of all that is just and manly in the English character,—not to permit the folly of Governor Eyre's admirers, if such folly should, in your estimation exist, to colour your judgment in this grave question. And whether you will hear, or whether you will forbear, I call upon the men of England who share the views here set forth, to throw themselves without delay between you and Governor Eyre, and to prevent you from adding to the harm which he has already experienced at your hands.

I have the honour to remain, Gentlemen, | with profound esteem for many of you, | and with brotherly affection for some, | Your obedient Servant, | JOHN TYNDALL

Athenaeum Club,[21] | November 7, 1866.

RI MS JT/1/TYP/8/2566–7

---

1. *Jamaica Committee*: see letter 2491, n. 4 (for which this letter is a reply). Tyndall's letter was published in *The Morning Post*, 8 November 1866, p. 2; in H. Hume, *The Life of Edward John Eyre, Late Governor of Jamaica* (London: Richard Bentley, 1867), Appendix B; and in W. F. Finlason, *The History of the Jamaica Case*, 2nd edn (London: Chapman and Hall, 1869), pp. 368*vv*–368*xx*.
2. *suppression of the late "alleged rebellion"*: see letter 2491, n. 4.
3. *St. Domingo*: capital city of the Dominican Republic.
4. *Alison's*: Archibald Alison (1792–1867), a British historian and lawyer (*ODNB*).
5. *History of Europe, Vol, VI., p. 104*: A. Alison, *History of Europe from the Commencement of the French Revolution in MDCCLXXXIX to the Restoration of the Bourbons in MDCCXV*, 14 vols (London: John Blackwood and Sons, 1849–50), vol. 6 (1849), pp. 103–4.
6. *Mr. Hudson*: John Hudson was a justice of the peace.
7. *Mr. Salmon*: John Salmon was *custos rotulorum*—an official serving as the principal Justice of the Peace, chief magistrate, and representative of the Governor-General in a parish (*OED*)—and president of the Legislative Council.
8. *Dr. M'Catty*: not identified.
9. *"the law of inseparable association"*: Tyndall was referring to John Stuart Mill (1806–73), a philosopher, economist, and women's rights advocate (*ODNB*), who was the chairman of the Jamaica Committee from 1866 (see letter 2491, n. 4). On the law of inseparable association, see J. S. Mill, *An Examination of Sir William Hamilton's Philosophy* (London: Longman, Green, Longman, Roberts & Green, 1865), pp. 251–70.
10. *Mr. G. W. Gordon*: George William Gordon.
11. *"all the power . . . in the mountains"*: see F. Finlason, *The History of the Jamaica Case*, p. 80.
12. *"India is not . . . of disease here"*: see F. Finlason, *The History of the Jamaica Case*, p. 80

13. *Lieut. Edenborough*: Edenborough, serving in the Confederate Army of the United States, was reportedly contacted by Gordon when his vessel was called at Jamaica some months before the rebellion, asking him to sell his vessel and armaments (F. Finlason, *The History of the Jamaica Case*, p. 164).
14. *Mr. Paul Bogle*: Paul Bogle (1820–65), a Jamaican Baptist deacon and friend of George William Gordon who led the protestors in the Morant Bay rebellion. He was tried, convicted, and hanged on 24 October 1865 (E. Bute and H. J. P. Harmer (eds), *The Black Handbook: The People, History and Politics of Africa and the African Diaspora* (London: Bloomsbury, 2016), p. 10).
15. *Mr. Lawrence*: Henry James Lawrence (d. 1865), Gordon's manager who was executed for alleged involvement in the rebellion (J. Gorrie, *Illustrations of Martial Law in Jamaica* (London: Jamaica Committee, 1867), pp. 70–72; T. C. Holt, *The Problem of Freedom: Race, Labor, and Politics in Jamaica and Britain, 1832—1938* (Baltimore: John Hopkins University Press, 1992), p. 459).
16. *Harrison*: James Harrison was an estate manager and attorney in Jamaica.
17. *Beckwith*: Arthur Beckwith, a minister in Kingston, Jamaica (F. Finlason, *The History of the Jamaica Case*, p. 79).
18. *Ford*: James Ford of the Bank of Jamaica (H. Hume, *The Life of Edward John Eyre, Late Governor of Jamaica* (London: Richard Bentley, 1867), p. 262).
19. *published address to the negroes*: Gordon's address, titled 'State of the Island', was seen as inciting Jamaicans to insurrection (M. Sheller, 'Hidden Textures of Race and Historical Memory: The Rediscovery of Photographs Relating to Jamaica's Morant Bay Rebellion of 1865', *The Princeton University Library Chronicle*, 72 (2011), pp. 533–67, on pp. 565–6).
20. *Eyre Defence Committee*: see letter 2481, n. 4.
21. *Athenaeum Club*: see letter 2253, n. 8.

To Thomas Archer Hirst 8 November 1866 2499

Nov. 8. 1866

Dear Tom

I send you a few separate copies[1] to be distributed as you wish.

Y[ou]rs hurriedly | J. T

RI MS JT/1/T/665

---

1. *few separate copies*: the words 'Letter to Jamaica Committee.' are written across the top of the letter. For Tyndall's letter to the committee, see letter 2498.

To Thomas Henry Huxley [8 November 1866][1] 2500

Thursday.

My dear Huxley,

I will not say that your letter to the Pall Mall Gazette [2] caused me to write what I now send you.[3] It may merely have hastened that which under the circumstances I should have written. It is right that you and I feeling so differently should express our respective opinions regarding the conduct of Governor Eyre.[4] But that we should feel so differently on any subject whatever is to me a source of pain more keen than these words are likely to make you aware of.

Yours affectionately | John Tyndall.

IC HP 8.49
RI MS JT/1/TYP/9/2906
Typed Transcript Only

---

1. *[8 November 1866]*: the date is given by relation to Tyndall's letter to the Jamaica Committee, dated 7 November 1866 (a Wednesday).
2. *your letter to the Pall Mall Gazette*: 'Mr. Huxley and Governor Eyre', *PMG*, 31 October 1866, p. 3. Thomas Huxley was a member of the Jamaica Committee which favoured the prosecution of Edward Eyre (see letter 2481, n. 4). Huxley's letter to the *PMG* was also published in: L. Huxley (ed), *Life and Letters of Thomas Henry Huxley*, 2 vols (London: Macmillan and Co., 1900), vol. 1, pp. 279–81. Unlike Tyndall, who insisted on interpreting the Morant Bay rebellion in terms of race, Huxley focused on the legal aspects in his response.
3. *what I now send you*: Tyndall, a member of the committee defending Edward Eyre, was referring to his own letter to the Jamaica Committee (see letter 2498).
4. *Governor Eyre*: Edward Eyre.

From Clémentine Soret 9 November 1866 2501

Cher Monsieur

Mon mari me dit qu'il vous écrit et je ne puis résister au désir de glisser quelques lignes dans sa lettre pour vous dire toute ma reconnaissance pour les beaux livres que vous m'avez envoyés, et tout l'intérêt avec lequel je les lis.—J'ai exprouvé une si grande joie de pouvoir recommencer à suivre par la pensée le célèbre "Professeur" sur les hautes cimes, que j'ai dévoré Mountaineering dans la matinée qui a suivi son arrivée—heureuse plus que je ne puis dire de repasser par les vives impressions que m'avait fait éprouver la lecture et la traduction de ses ascensions au Mt Rose—

Ainsi donc merci et encore merci pour l'aimable pensée et pour la peine que vous avez bien voulu prendre en vue de me procurer un tel Plaisir! Merci! Avec la prière de ne pas oublier vos amis de Genève qui espèrent et croient en votre promesse faite aux brineaux en Août 65 de revenir les trouver "le plus tôt possible?" Merci et au revoir à bientôt nous l'espérons vivement tous deux—

Votre bien dévoué et reconnaissante—| Clémentine Soret

Genève le 9 Novembre 66.

Dear Sir

My husband[1] tells me that he is writing to you, and I cannot resist the temptation to slip a few lines into his letter to let you know my gratitude for the beautiful books that you sent me[2] and how carefully I read them.—I was so overjoyed to once again be able to follow the thought of the famous "Professor" on the high peaks that I devoured Mountaineering the morning after its arrival. No words can express my happiness in experiencing the same vivid impressions that I took rom reading and translation his ascensions of Monte Rosa.[3]—

So thank you and thank you again for the kind thought and the trouble you took to procure such a pleasure for me! Thank you! I pray that you will not forget your friends in Geneva who hope and believe in your promise made to the [Brineaus][4] in August '65,[5] to come back and visit them "as soon as possible?" Thank you and goodbye, we both hope to see you soon.

Your devoted and grateful—| Clémentine Soret

Geneva, 9 November '66.

RI MS JT/1/S/81

1. *My husband*: Jacques Louis Soret.
2. *beautiful books that you sent me*: presumably Tyndall had sent her copies of *Glaciers of the Alps* and *Mountaineering in 1861*. In a letter to Tyndall earlier that year (2378), Louis Soret wrote that he was 'a little embarrassed to tell you that Mrs Soret did not receive the books' and suggested that perhaps the bookseller made in error in shipment.
3. *M^t^. Rosa*: Monte Rosa, the second highest mountain in the Alps located between Switzerland and Italy near the town of Zermatt.
4. *[Brineaus]*: this is a conjectured reading of this word. It seems reasonable that Soret was here referring to a family name; if indeed 'Brineaus', they are not identified.
5. *promise made to [Brineaus] in August 65*: this may refer to Soret and Tyndall having met each other during the meeting of the Société Helvetique des Sciences Naturelles (Swiss Society of Natural Sciences, now Swiss Academy of Natural Sciences, founded in 1815) from 21–3 August 1865, and Tyndall having promised to visit again.

From Emelia Gurney[1] 12 November 1866 2502

2 Palace Gardens, W. | Nov. 12th

Dear Professor Tyndall

I thank you very much for remembering me.[2] I do not think I have thanked you for your lecture on radiation[3] which you so kindly sent me, and now has come your letter to the Jamaica Committee.[4] You can well imagine how intensely interesting to me the Jamaica subject is, and how painful many of the loud words far away from righteous judgement have been on both sides.

Your letter is a noble exception to the utterances chiefly "from the teeth outwards". We are both heartily thankful for its spirit and its aim, and to me it seems so completely the just view of the whole case.

When we went out to Jamaica it seemed to us as if there were a really earnest desire that the truth might be brought to light now it seems as if all moral feeling was so mixed up with party faction that the noble desire for justice and truth is all fretted away.

I wish you could ever come to us. Perhaps we may be able some time this winter to fix upon a day when you are not too engaged or too busy.

Very truly yours | Emelia Gurney.

RI MS JT/1/TYP/8/2570
LT Typescript Only

1. *Emelia Gurney*: Emilia Gurney (1823–96), an English social reformer and campaigner for women's rights (*ODNB*).
2. *remembering me*: letter missing.
3. *your lecture on radiation*: *On Radiation*.
4. *your letter to the Jamaica Committee*: letter 2498.

To James Colvile[1] [12 November 1866][2] 2503

Royal Institution of Great Britain | <u>Monday</u>

Dear Sir James

It will give me great pleasure to join you on Friday.

I regret that it was not possible for me to answer your note[3] earlier.

Yours very truly | <u>John Tyndall</u>

[Envelope] Sir James Colvile | 3 Rutland Gate | Hyde Park

[Postmark]: LONDON SW | NO 12 | 66

BL Add MS 60634, ff. 119–20

1. *James Colvile*: James William Colvile (1810–80), a lawyer and judge who served in the courts of Bengal in the 1840s and 1850s, and was later appointed to the Judicial Committee of Privy Council (*ODNB*).
2. *[12 November 1866]*: the date is given based on the postmark on the envelope. '1866' is also written on the letter below 'Monday', in a different hand.
3. *your note*: letter missing.

To Margaret Ginty 13 November 1866 2504

13th. Nov. 1866.

My dear Mrs. Ginty,

The very sad but very beautiful letter which you wrote to me on the 8th. of October has reached my hands.[1]

The business matters therein referred to shall be at once attended to.[2]

I am most glad to hear that you are actively engaged. It is said that light regarding those higher things which you mention in letter can only be gained, by performing to the utmost that which we know to be our duty. Continue in this line and comfort will come in the end.

I will write to you again the moment I see my <*1 or more words missing*>

Yours ever | John Tyndall.

RI MS JT/1/TYP/11/3652
LT Typescript Only

1. *very sad but very beautiful letter*: letter missing.
2. *business matters therein referred to shall be at once attended to*: Tyndall was overseeing Ginty's financial affairs following the sudden death of her husband William Ginty (see letters 2465, 2466, 2469, 2470, and 2473).

From Joseph Dalton Hooker 13 November 1866 2505

Kew, Nov. 13th 1866.

Dear Tyndall

Thanks for your communication, far from scolding you, I quite take your view of the matter.[1] One must not judge such cases on abstract principles. Your illustration of the bridge that must be built to stand the heaviest flood is excellent. It would be a mistake to justify all and every act committed in a great and terrible emergency, but it is worse, it is a crime to judge of the errors committed by the light of after events. There are many other considerations that lean me to your side and that could find no place in a public contention: one is that the public applauds the making Governors of Colonies out of Geog. . . ." 9 days' wonders"[2] for such after all was poor Eyre,[3] what education or experience had he of politics, law or governance in peace? & what that could fit him for extremities? if such men do make mistakes is it for our public, who egg the Govt to give such men such ill-chosen rewards as Governorships, to visit their mistakes with cruel persecution.

I am as averse as anyone to bloodshed, I hate the name of war, I go much further, I hold that the taking the life of the lower animals <u>for sport</u>, is wrong, and I only tolerate sport on the grounds that bear no investigation; for the same reason I regard the negro's life as sacred as a pheasant's, or yours, or mine. I know no distinction but in degree, but when you or I, or the negro, or pheasant, become dangerous to law,—order, and the general weal they all are subject to the same nemesis; and that the Negro in Jamaica and even in the free towns of W. Africa is pestilential I have no hesitation in declaring, nor that he is a most dangerous savage at the best.

I have not read Huxley's letter[4] but indeed I loathe the whole subject. The sight of the heaps of dead bodies in Demerara[5] after the massacres shocked me inexpressibly. One can view calmly the results of the battle in hospital, but not the blood shed by massacre whether of black or white. In the Jamaica affair it was massacre in both cases, and I therefore cannot bring myself to take any public side in the question of impeachment or defence.

Ever yours affectionately | J. D. Hooker.

RI MS JT/1/TYP/8/2571
LT Typescript Only

---

1. *I quite take your view of the matter*: letter missing. Presumably Tyndall had sent Hooker a copy of his letter to the Jamaica Committee in support of Governor Eyre (see letter 2498).
2. *"9 days' wonders"*: 'nine-days wonder' is an idiom referring to something that generates interest for only a short amount of time, originating with Chaucer.
3. *Eyre*: Edward Eyre.
4. *I have not read Huxley's letter*: see letter 2500, n. 2.
5. *the heaps of dead bodies in Demerara*: Hooker was comparing the Morant Bay rebellion (see letter 2481, n. 4) to the Demerara rebellion of 1823, an uprising involving about 10,000 slaves in the colony of Demerara-Essequibo, now British Guinea.

To Joseph Dalton Hooker [14 November 1866][1] 2506

Athenaeum.[2] Wednesday.

My dear Hooker

I knew that your extended knowledge and wise head would prevent you from becoming a partisan; but I am rejoiced to find that you are not against me.[3] I read of those blood-sheddings with a horror equal to your own, but I am persuaded the Jamaica Committee[4] has not taken the right way to end them.

Had they not been so violent, a healthy and calm public opinion would have been declared against all proved atrocities, in an unmistakeable manner; and Governor Eyre[5] would not have found himself backed by the determinate English feeling which now supports him.

Yours ever | John Tyndall

I am going down to Harrow tomorrow to talk to the boys:[6]—this is work more to my taste.[7]

RI MS JT/1/TYP/8/2592
LT Typescript Only

---

1. *[14 November 1866]*: the date is given by relation to letter 2505 (dated 13 November 1866, a Tuesday), and that Tyndall went to Harrow in November 1866 (see n. 6).
2. *Athenaeum*: see letter 2253, n. 8.
3. *you are not against me*: see letter 2505.
4. *Jamaica Committee*: see letter 2481, n. 4.
5. *Governor Eyre*: Edward Eyre.
6. *down to Harrow tomorrow to talk to the boys*: Tyndall presumably went to Harrow in November 1866 to lecture to the students there (see letter 2517 from Frederic Farrar).
7. *work more to my taste*: Harrow is a well-known English public school, originally established in 1572.

From James Clark 16 November 1866 2507

Bagshot Park, Surrey. | November 16. 1866.

Dear Mr Tyndall

Ever since I read your admirable defence[1] of Mr Eyre,[2] I have thought of writing to express how grateful I felt as one of Mr Eyre's friends to you for your most judicious and well timed paper, coming after that of Mr Huxley,[3] which gave me as much pain—on Mr Huxley's own account as well as on Mr Eyre's—as yours gave me pleasure. Mr Eyre's position has anxieties & responsibilities with the safety of a colony hanging upon his decisions, are quite ignored most unjustly. That any man in his position may not have committed errors in degree does not touch the question. He saved the colony by his judgement and decision, and those who are now ready to condemn him for what they are pleased to consider over severity, would have been equally loud in their condemnation had he by a little more leniency lost the colony & the English been served as the French were in Hayti.

There is an unfeeling, thoughtless, unreasoning outcry against Eyre which is surprising to a thinking rational being.

But I merely meant to thank you in a few words. Pray excuse my long note, and

Believe me yours sincerely | James Clark.

I told Sir Henry Holland of my intention to write you & he immediately begged me to do so & he would be the bearer of my note.[4]

RI MS JT/1/TYP/8/2574
LT Typescript Only

1. *your admirable defence*: see letter 2498.
2. *Mr. Eyre*: Edward Eyre.
3. *that of Mr Huxley*: see letter 2500, n. 2.
4. *bearer of my note*: see letter 2508. This last sentence is written by hand on LT's typescript.

From Henry Holland 16 November 1866 2508

My dear Tyndall,

I bring you this letter from Sir James Clarke;[1] and I desire to express to you my own thorough adhesion to his opinion of your excellent letter.

Yours ever | H. H.

RI MS JT/1/TYP/8/2573
LT Typescript Only

1. *this letter from Sir James Clarke*: Clark, not Clarke; see letter 2507.

To Clémentine Soret 17 November 1866 2509

17th Nov. 1866

Dear Madame Soret,

I should think all the books that I possess well bestowed if the bestowal of each of them caused me half the pleasure that your letter[1] has given me. The clay of this earth must lie very light upon the soul that inspired that letter. I consider myself more fortunate, more highly honoured, and certainly more highly gratified in receiving such a letter than in receiving the diploma of the Academy of Sciences.[2] You have made me the debtor by thus writing to me.

Believe me | Most truly yours | John Tyndall

Would you thank M. Soret for me for his letter an extract of which I have sent to the Philosophical Magazine.[3,4]

RI MS JT/5/4
Transcript Only

---

1. *your letter*: letter 2501.
2. *receiving the diploma of the Academy of Sciences*: in 1859, Tyndall had been made an honorary member of the Société Helvetique des Sciences Naturelles (Swiss Society of Natural Sciences, now Swiss Academy of Natural Sciences, founded in 1815) (A. S. Eve and C. H. Creasey, *Life and Work of John Tyndall* (London: Macmillan & Co., 1945), p. 394).
3. *an extract of which I have sent to the Philosophical Magazine*: J.-L. Soret, 'On Aqueous Vapour and Solar Radiation', *Phil. Mag.*, 22 (1866), p. 478.
4. The following words of Soret's are written, presumably by Tyndall, on the upper right of the letter: '"Le congrès pour le 50ème anniversaire de la fondation de la Société Helvétique des Sciences Naturelles en Suisse, avant lieu au mois d'Aout de 1865 à Genève c'est là que j'ai fait la anniversaire de l'ami de mon mari le Profr. Tyndall." on letter dated 28th Sept. 1865'. This appears to be quoted from a missing 28 September 1865 letter from Clementine Soret in which she shared that she celebrated Tyndall's birthday (2 August) at the meeting of the Swiss Society of Natural Sciences (now Swiss Academy of Natural Sciences, founded in 1815) from 21–3 August 1865, where they likely first met (see letter 2501, n. 5).

From Helen Tyndall[1] 22 November [1866][2] 2510

Weston Super Mare,[3] Nov. 22d

Lieber Herr[4] Professor,

I am not going to inflict a long epistle upon your overtasked eyes, but I must thank you for your letter to that would-be-philanthropic committee,[5] who would so willingly forget that though the negro may be "a man and a brother"[6] he is lower than the brute when his nature is left to itself! If Governor Eyre's[7] supporters had borrowed your firm, cool and clear arguments he would have had less reason to exclaim "Save me from my friends" whose very injudicious support has been more injurious than opposition. Though possibly England expects every man to do his duty, it cannot be said that she rewards those who do it best!

In looking over the names of the opposing committee, those of any note are individuals who may be very scientific and learned, but who have had no personal experience of colonial life or negro character. It would be difficult to me to feel sympathy for a cause supported by John Bright[8] and Tom Hughes.[9]

Alas for my pleasant picture! Will you please make a mental promise to write us word when you have a spare moment, and feel inclination to bestow it on friends far off in the quiet West, who would value every hour as a great boon and do their best to give you rest and quiet.

Nella[10] will shortly be in town for a few days but I fear will not have a chance of seeing you.[11] We return to Perridge[12] on Saturday. All three join in kindest regards with

Yours very sincerely | Helen Tyndall

RI MS JT/1/TYP/8/2577
LT Typescript Only

---

1. *Helen Tyndall*: a relative of Tyndall's, the wife of John William Warre Tyndall (1811–97), an Oxford-educated barrister and author, notably of *The Island of Sardinia* (1849) (E. Walford, *The County Families of the United Kingdom or Royal Manual of the Titled and Untitled Aristocracy of England, Wales, Scotland, and Ireland* (London: Chatto & Windus, 1893), p. 1022). In letter 2515, she wrote in regard to Tyndall visiting her daughter Nella in London: 'How kind and cousinly of you ever to think of bestowing some of your valuable moments on a visit'.
2. *[1866]*: the year is given by reference to Tyndall's letter to the Jamaica Committee, dated 7 November 1866 (see n. 5).
3. *Weston Super Mare*: Weston-super-Mare, a seaside town in Somerset, England.
4. *Lieber Herr*: Dear Sir (German).

5. *your letter to that would-be-philanthropic committee*: Tyndall's letter to the Jamaica Committee in defense of Edward Eyre is letter 2498.
6. *negro may be "a man and a brother"*: a reference to the famous phrase "Am I not a man and a brother?" coined by Josiah Wedgwood (1730–95), founder of the highly successful Wedgwood pottery company, abolitionist, and maternal grandfather to naturalist Charles Darwin (*ODNB*). In 1787 Wedgewood had designed an anti-slavery medallion depicting a male slave in chains with this phrase, the image of which was further used in print for abolitionist causes.
7. *Governor Eyre's*: Edward Eyre.
8. *John Bright*: John Bright (1811–89), a British Liberal politician and supporter of free trade (*ODNB*). He was a member of the Jamaica Committee (B. Cash, *John Bright: Statesman, Orator, Agitator* (London: I. B. Tauris, 2012), pp. 194–5). The following passage appears in the brief discussion of Bright's involvement in the Eyre controversy in G. B. Trevelyan, *The Life of John Bright* (London: Constable and Company, 1914): 'Except Tyndall, the men of finest scientific mind—Darwin, Huxley, Mill, Leslie Stephen, Sir Charles Lyell—ranged themselves on the side of law and humanity, while those whose cut it sometimes was to complain of the hardness of the scientific attitude to life—Carlyle, Ruskin, Kingsley, Tennyson—showed by their own conduct how prone sentimentalists are to inconsiderate worship of brute force and the "strong man"' (p. 346–7).
9. *Tom Hughes*: Thomas Hughes (1822–96), an English lawyer, politician, and author (*ODNB*), most notably of the novel *Tom Brown's School Days* (1857). Hughes was a member of the Jamaica Committee, and had previously supported the abolition of slavery and the Union cause in the American Civil War (T. Hughes, 'Thomas Hughes and the American Civil War', *The Journal of Negro History*, 18 (July 1933), pp. 322–9).
10. *Nella*: daughter of Helen Tyndall.
11. *seeing you*: letter 2515 shows that Tyndall indeed planned to visit her.
12. *Perridge*: Perridge House in Shepton Mallet, Somerset, England, the home of John William Warre Tyndall, the husband of Helen Tyndall (W. G. Searle, *The History of the Queens' College of St Margaret and St Bernard in the University of Cambridge. 1446–1560* (Cambridge: Deighton, Bell & Co., 1867), p. 368).

From Hermann Helmholtz 24 November 1866 2511

Heidelberg. 24 Novb. 1866.

Verehrter Freund

die Verbesserungen einzelner Stellen aus „On Heat" fielen auf Wiedemann's Theil, und habe ich sie diesem zugesendet. Meine Frau hat die ersten sieben Capitel zu übersetzen, und ein grosser Theil des Manuscripts ist schon an Vieweg gesendet; wir erwarten in diesen Tagen die ersten Correcturbogen zu empfangen. Von Ihrer Abhandlung On Calorescence habe ich ein Exemplar, und sollten Sie keines für Wiedemann übrig haben, oder dieser keines besitzen, so werde ich ihm das meinige senden.

Ich glaube nicht, dass ich mit meiner Frau in der nächsten Zeit nach England kommen werde; wir haben erst noch andere Reisen vor. Ich nehme Sie deshalb beim Worte, dass Sie uns hier in Heidelberg Ihren Besuch machen wollen; ein Zimmer in unserem Hause ist jederzeit bereit, Sie aufzunehmen, und wenn wir auch in unserer Nachbarschaft keine Matterhörner haben, so bieten unsere waldigen Berge doch immerhin schöne erfrischende Wege genug, auf denen ein Londoner seine ermüdeten Nerven gut erfrischen und ausruhen kann.

Ich habe schon im Philosophical Magazine gelesen, wie Sie die Zusammensetzung der Klänge den Zuhörern in der Royal Institution mit electrischem Lichte illustrirt haben, und bin Ihnen dankbar, dass Sie die Grundlage der Tonempfindungen auf diese Weise den Leuten verständlich machen. Ich bin jetzt beinahe fertig mit dem Drucke meiner Physiologischen Optik, und hoffe sie Ihnen in den nächsten Wochen übersenden zu können. Es ist das eine lange Arbeit gewesen, und ich bin froh, damit endlich zu Ende zu sein.

Beste Grüsse. Dem Bunsen geht es gut; Kirchhoff aber ist seit dem März lahm in Folge einer Quetschung des Fussgelenks, und hat ein halbes Jahr gar nicht gehen können. Er kann auch jetzt noch nur langsam und vorsichtig gehen.

Ihr treuer | H. Helmholtz

Heidelberg. 24 November 1866.

Dear friend

The corrections of some passages from "On Heat"[1] fall to Wiedemann's lot and I have sent them to him. My wife[2] has to translate the first seven chapters and a large part of the manuscript has already gone to Vieweg;[3] we expect the first proofs any day. I have a copy of your essay On Calorescence[4] and if you have no copy left for Wiedemann, or if he does not have one, I will send him my copy.

I do not think that I will come to England with my wife any time soon; we have planned other trips first. I therefore take your word that you want to visit us here in Heidelberg; we always have a room in our house ready to accommodate you and, even if we do not have anything like the Matterhorn in our surroundings, our forested mountains always offer nice, refreshing paths enough, on which a gentleman from London can rest and refresh his exhausted nerves.

I already read in the Philosophical Magazine that you have illustrated the composition of sounds for your listeners at the Royal Institution with electric light[5] and I am grateful to you for making the basis of sound impressions comprehensible to people in this way. I am now nearly ready with the printing of my physiological optics[6] and hope to send it to you in the coming weeks. It has been a long work and I am glad to have it finished at last.

Best regards. Bunsen is well but Kirchhoff[7] has been lame since March because he crushed his ankle and he was unable to walk at all for half a year. Even now, he still can only walk slowly and carefully.

Yours truthfully | H. Helmholtz

RI MS JT/H/44

---

1. *"On Heat"*: the German translation of Tyndall's *Heat* (1865).
2. *my wife*: Anna Helmholtz. While Hermann Helmholtz and Gustav Wiedemann are credited with the translation, it was in fact their wives Anna and Clara, respectively, who carried out the work of translating the book.
3. *Vieweg*: Eduard Vieweg.
4. *On Calorescence*: see letter 2333, n. 2.
5. *illustrated the composition of sounds for your listeners at the Royal Institution with electric light*: see letter 2447, n. 1.
6. *my physiological optics*: see letter 2290, n. 9.
7. *Kirchhoff*: Gustav Kirchhoff.

From David Masson[1] 24 November 1866 2512

3, Rosebery Crescent, | Edinburgh: | Nov 24, 1866

My dear Tyndall,

Thanks for your kind note.[2] I should like very much to give a Friday evening lecture some time at the Royal Institution, but cannot undertake one this season. I have had to decline all such things, right & left, for some time to come.

I see that you are to be here soon to lecture at the Philosophical Institution.[3] If you have not preengaged yourself elsewhere, it would give M[rs] Masson[4] & me much pleasure if you would take up your lodging with us while you are here. At any rate we do hope you will be able to dine with us one of your days. As I calculate (but it is only a calculation, & I have not the prospectus before me) your lecture will be on Friday Dec. 14. If so, and you could give us your company on Saturday Dec. 15, that would suit well. Do, if you can, send me a time <u>by return</u> to say "yes", as I should like to ask a few friends to meet you, and here in Edinburgh they book you a century in advance.

Yours very truly | David Masson

RI CG4/2/55

---

1. *David Masson*: David Mather Masson (1822–1907), a Scottish biographer, literary scholar,

and editor. In 1865 he became Regius Professor of Rhetoric and English Literature at the University of Edinburgh (*ODNB*).

2. *your kind note*: letter missing.
3. *lecture at the Philosophical Institution*: see letter 2460, n. 17.
4. *M*[rs] *Masson*: Emily Rosaline Masson (née Orme, 1835–1915), a leader of the Edinburgh National Society for Women's Suffrage (E. Crawford, *The Women's Suffrage Movement: A Reference Guide, 1866–1928* (New York: Routledge, 2001), p. 390).

From Henry Bence Jones [before 25 November 1866][1] 2513

Dear Friend,

A second letter today from Sir Henry Holland tells me of his loss;[2] so will you as he will not be at the meeting tell the managers[3] that M[r] Hind[4] declines to lecture & ask someone (DelaRue?[5]) to direct the enclosed to M[r] Huggins:[6] failing him I thought of asking Max Muller[7] until your note[8] regarding M[r] Conway[9] came.

Now regarding M[r] Conway I would rather he gave us a Friday evening this next season[10] and then, if he talks well, his course;[11] than that we should engage him, not knowing how he speaks, for you know an untried man is very much more likely to fail than to succeed. If however you know him as a speaker & that he does well then if Huggins refuses we will try if you & the managers please this man & I will get the other two lectures taken by some one else. If you do not know anything of M[r] Conway's talking power then ask him to give a Friday evening.

Will you tell the managers that to give me more time I propose that D[r] Frankland should give six lectures on Chemistry[12] on Thursdays & Saturdays March 7 9. 14. 16. 21 & 23 & that Mr Huggins or some one else should take the six Tuesdays after Easter[13] April 30 to June 4.

This will make the courses of lectures before Easter all full & give more time.

I am going on well but not so fast as I want

Ever yours | H. B. J.

RI MS JT/1/J/115
RI MS JT/1/TYP/3/691

---

1. *[before 25 November 1866]*: the date is given by relation to letter 2514, dated 25 November [1866].
2. *tells me of his loss*: Holland's second wife Saba Holland (née Smith, 1802–66) had recently died on 2 November 1866 (*ODNB*).

3. *managers*: the following were elected Managers of the RI on 1 May 1866: Warren de la Rue (1815–89), George Everest (1790–1866), John Hall Gladstone (1827–1902), William Robert Grove (1811–96), Caesar H. Hawkins (1798–1884), John Lubbock (1834–1913), John Carrick Moore (1805–98), William Pole (1814–1900), William Frederick Pollock (1815–88), Lewis Powell (*c.* 1797–1867), Edward Sabine (1788–1883), Philip Henry Stanhope (1805–75), Joseph William Thrupp (1799–1873), Charles Wheatstone (1802–75), and Philip James Yorke (1799–1874). Henry Holland (1788–1873) was elected President, William Spottiswoode (1823–83) was elected Treasurer, and Henry Bence Jones (1813–73) was elected Secretary ('Annual Meeting', *Roy. Inst. Proc.*, 4 (1862–6), p. 615; *ODNB*).
4. *Mr Hind*: Probably John Russell Hind (1823–95), an English astronomer known for his discovery of ten asteroids and his superintendence of the Nautical Almanac from 1853–91 (*ODNB*).
5. *DelaRue?*: Warren de la Rue (1815–89), a British astronomer and chemist, and partner in the printing firm De La Rue founded by his father (*ODNB*).
6. *Mr Huggins*: Williams Huggins (1824–1910), an English astronomer known for his work in spectroscopy. He made observations from a private observatory at his home and developed methods and instruments for observing celestial spectra (*ODNB*). Huggins did not deliver any lectures at the RI in 1867. He had already delivered a Friday Evening Discourse, 'On the Physical and Chemical Constitution of the Fixed Stars and Nebulae', on 19 May 1865 (*Roy. Inst. Proc.*, 4 (1862–6), pp. 443–9), and he would give another Friday Evening Discourse a few years later on 5 March 1869, 'On some further Results of Spectrum Analysis as applied to the Heavenly Bodies' (*Roy. Inst. Proc.*, 5 (1866–9), pp. 475–9).
7. *Max Muller*: Friedrich Max Müller (see letter 0238, n. 6). Müller did not deliver any lectures at the RI in 1867. Previously, he delivered a Friday Evening Discourse on 29 May 1863, 'On the Vedas, or the Sacred Books of the Hindus' (*Roy. Inst. Proc.*, 4 (1862–6), pp. 135–40), and also delivered a series of lectures on language in spring 1861, published as *Lectures on the Science of Language delivered at the Royal Institution of Great Britain in April, May, and June, 1861* (London: Longman, Green, Longman, and Roberts, 1861), and again in spring 1863, published as *Lectures on the Science of Language at the Royal Institution of Great Britain delivered in February, March, April and May 1863. Second Series* (London: Longman, Green, Longman, Roberts & Green, 1864).
8. *your note*: letter missing.
9. *Mr Conway*: Moncure Daniel Conway (1832–1907), an American minister, social reformer, and author. While his father was a Virginia slave owner, Conway was an ardent abolitionist. In 1863 left the United States for England where he spent twenty-two years (*ODNB*; *ANB*).
10. *Friday evening this next season*: Conway delivered a Friday Evening Discourse at the RI, 'On New England', on 22 February 1867 (*Roy. Inst. Proc.*, 5 (1866–9), pp. 59–62).
11. *his course*: it does not appear that Conway delivered a course of lectures at the RI in 1867.
12. *six lectures on Chemistry*: Frankland delivered six lectures, 'On Coal Gas', on Thursdays and Saturdays from 7 to 23 March 1867 (*Roy. Inst. Proc.*, 5 (1866–9), p. 4).
13. *six Tuesdays after Easter*: it appears that William Allen Miller (1817–70), an English

chemist and MD who held the chair of Chemistry at King's College London and was treasurer of the RS from 1861–70 (*ODNB*), delivered a course of not six but four lectures, 'On Spectrum Analysis with some of the Results of Recent Astronomical Observations', on Tuesdays from 14 May to 4 June 1867 (*The Athenaeum*, 12 January 1867, p. 33).

From Henry Bence Jones 25 November [1866][1] 2514

Dear friend kind friend good f[rien]d

I am better

I have written to Blackie[2] to ask if he will give two extra afternoon lectures on Plato on 3^d & 10 April[3] & a Friday on any other subject on the 5th. This may suit us both. I am so sorry to trouble you about these Friday evenings[4] but till today I could not write a line.

I had hoped I had got Max Muller[5] but he is off again or rather not fixed I have written again to him.

I trust to get home this week & then a few minutes talk will settle all

Ever ever yours | H B J | Nov 25

Present scheme

Jan 18	Tyndall[6]
25	Max Muller Cornwall language
Feb 1	Odling[7] Grahams diffusion
8	Rev W. Farrar[8] instruction in science
15	(Atlantic cable?)[9]
22	M. D. Conway[10] New England
March 1	Abel[11] Small arms
8	(Wolds of Yorkshire tombs?)[12]
15	E B Tylor[13] on low brains
22	Dr Pettigrew[14] on flying machines
29	Balfour Stewart[15] on Sun as variable star
April 5	Prof. Blackie[16]
12	Frankland[17]

RI MS JT/1/J/114
RI MS JT/1/TYP/3/692

1. *[1866]*: the year is given by content (their discussion of lecturers for Friday Evening Discourses at the RI in 1867).
2. *Blackie*: John Stuart Blackie (1809–95), a Scottish classical scholar and Professor of Greek at Edinburgh University (*ODNB*).

3. *two extra afternoon lectures on Plato on 3[d] & 10 April*: Blackie delivered lectures 'On Plato' at the RI on 30 April and 7 May 1867 (*Roy. Inst. Proc.*, 5 (1866–9), p. 24).
4. *trouble you about these Friday evenings*: see letter 2513. For other Friday Evening Discourses delivered in 1867 not referred to here, see *Roy. Inst. Proc.*, 5 (1866–9), pp. iii–iv.
5. *Max Muller*: Friedrich Max Müller (see letter 2415, n. 6) did not end up giving a Friday Evening Discourse in 1867 (see letter 2513, n. 6).
6. *Tyndall*: on 18 January 1867, Tyndall delivered a Friday Evening Discourse at the RI, 'On Sounding and Sensitive Flames' (*Roy. Inst. Proc.*, 5 (1866–9), pp. 6–12).
7. *Odling*: on 25 January 1867, the English chemist William Odling (1829–1921) delivered a Friday Evening Discourse at the RI, 'On Mr. Graham's Recent Discoveries on the Diffusion of Gases (*Roy. Inst. Proc.*, 5 (1866–9), pp. 12–18). Odling was originally slated for 1 February, but this date was taken by the Scottish engineer J. Scott Russell (1808–82), 'On the Crystal Palace Fire' (*Roy. Inst. Proc.*, 5 (1866–9), pp. 18–24). Odling worked on classifying the elements and refining Dmitri Mendeleev's periodic table of elements. Before becoming Fullerian Professor of Chemistry at the RI in 1868, he was a lecturer in chemistry at St Bartholomew's Hospital Medical School, and from 1872–1912 he was a professor of chemistry at Worcester College, Oxford. Odling was elected FRS in 1859 (*ODNB*).
8. *Rev W. Farrar*: on 8 February 1867, the Anglican clergyman William Frederic Farrar (1831–1903) delivered a Friday Evening Discourse at the RI, 'On Some Defects in Public School Education' (*Roy. Inst. Proc.*, 5 (1866–9), pp. 26–44). From 1855–70, Farrar was assistant master at Harrow, during which time he wrote three novels and a number of books on philology. Farrar was a supporter of Darwin and adapted his ideas to the development of language during the 1860s, and it was Darwin who nominated Farrar FRS in 1866. On Darwin's death Farrar, by then a canon of Westminster, arranged for his burial at Westminster Abbey. From 1895 until his death Farrar was Dean of Canterbury (*ODNB*).
9. *(Atlantic cable?)*: on 15 February 1867, the English telegraph engineer Cromwell Fleetwood Varley (1828–83) delivered a Friday Evening Discourse at the RI, 'On the Atlantic Telegraph' (*Roy. Inst. Proc.*, 5 (1866–9), pp. 45–59). Varley served as chief engineer on the laying of the 1865 Atlantic telegraph cable. Varley was a believer in spiritualism and conducted investigations of his own (*ODNB*).
10. *M. D. Conway*: Moncure D. Conway (see letter 2513, n. 9 and n. 10).
11. *Abel*: the 10 March 1867 Friday Evening Discourse at the RI was not delivered by the chemist and explosives expert Frederick Augustus Abel (1827–1902). Instead, the British engineer Vivian Dering Majendie (1836–98) lectured on 'On Military Breech-Loading Small Arms' (*Roy. Inst. Proc.*, 5 (1866–9), pp. 62–75).
12. *(Wolds of Yorkshire tombs?)*: on 8 March 1867, the English clergyman and archaeologist William Greenwell (1820–1918) delivered a Friday Evening Discourse at the RI, 'On the Yorkshire Wold Tumuli' (*Roy. Inst. Proc.*, 5 (1866–9), pp. 78–83).
13. *E B Tylor*: on 15 March 1867, the English anthropologist Edward Burnett Tylor (1832–1917) delivered a Friday Evening Discourse at the RI, 'On the Traces of the Early Mental Condition of Man' (*Roy. Inst. Proc.*, 5 (1866–9), pp. 83–93). Tylor had previously

published *Researches into the Early History of Mankind and the Development of Civilization* (London: John Murray, 1865), and later published *Primitive Culture*, 2 vols (London: John Murray, 1871).

14. *Dr Pettigrew*: on 22 March 1867, the Scottish anatomist, naturalist, and aviation pioneer James Bell Pettigrew (1832–1908) delivered a Friday Evening Discourse at the RI, 'On the various modes of Flight in relation to Aeronautics' (*Roy. Inst. Proc.*, 5 (1866–9), pp. 94–107). Pettigrew invented a flying machine based on his observations of bird flight.
15. *Balfour Stewart*: on 12 April 1867, not 29 March, Balfour Stewart (see letter 2381, n. 6), delivered a Friday Evening Discourse at the RI, 'On the Sun as a Variable Star' (*Roy. Inst. Proc.*, 5 (1866–9), pp. 138–44).
16. *Prof. Blackie*: on 3 May 1867, not 5 April, John Stuart Blackie delivered a Friday Evening Discourse at the RI, 'On the Music of Speech in the Greek and Latin Languages' (*Roy. Inst. Proc.*, 5 (1866–9), pp. 145–55).
17. *Frankland*: on 29 March 1867, not 12 April, Edward Frankland delivered a Friday Evening Discourse at the RI, 'On the Water Supply of the Metropolis' (*Roy. Inst. Proc.*, 5 (1866–9), pp. 109–26).

From Helen Tyndall[1] 30 November [1866][2] 2515

Perridge, Shepton Mallet.[3] | Nov. 30th.

My dear Mr Tyndall

How kind and cousinly of you ever to think of bestowing some of your valuable moments on a visit to our "one chick"[4] and how pleased and glad she will be to see you!

She is till Dec 12th the guest of our friends the Gwyn Jeffreys[5] at 25 Devonshire Place and she is enjoying the pleasure of seeing old friends, though still retaining her love of a country home life.

I wish you could rest your burnt-out eyes with a view of such a sunset as I am now looking at! but which warns me that our primitively early post hour is at hand.

My dear husband,[6] whose bronzed face speaks volumes for country air, desires his kindest regards and a gentle reminder (not meant as a worry) that your room is ready at any season in an hour.

Yours ever sincerely | Helen Tyndall

RI MS JT/1/TYP/8/2578
LT Typescript Only

1. *Helen Tyndall*: see letter 2510, n. 1.

2. *[1866]*: the year is given by relation to letter 2510, dated 22 November [1866].
3. *Perridge, Shepton Mallet*: see letter 2510, n. 12.
4. *our "one chick"*: daughter Nella (see letter 2510).
5. *our friends the Gwyn Jeffreys*: John Gwyn Jeffreys (1809–85), a prominent lawyer in south Wales, conchologist, and a leading member of the Royal Institution of South Wales (*ODNB*).
6. *My dear husband*: presumably John Warre Tyndall (see letter 2510, n. 1).

From Ellen Busk[1] [November 1866][2] 2516

Dear Mr Tyndall

Thank you so much for sending me your beautiful letter[3] with the whole of which I cordially agree. I could have wished you had smitten the exact ground on which they stand, viz., whether the execution of Gordon was legal or not, and whether such a proceeding be made into a precedent.[4] Now I conceive the answer to their question to be given by the universal admission of the thinking public in England that in the strict sense of the word that execution was illegal. "The act of indemnity" passed in Jamaica presupposed illegal acts and justifies them, does it not?[5] Wherever circumstances justify a future act of indemnity, let the execution of Gordon form a precedent and welcome. Meantime in our tolerably well organized society it is absurd to speak of fear of such precedents.

Excuse haste. I am much harassed just now by constant attendance on my brother's[6] illness, but I wanted to thank you.

Ever yours truly, | Ellen Busk.

RI MS JT/1/TYP/8/2576
LT Typescript Only

1. *Ellen Busk*: see letter 2343, n. 1.
2. *[November 1866]*: the month and year are given by reference to Tyndall's letter to the Jamaica Committee, dated 7 November 1866 (see n. 3).
3. *your beautiful letter*: see letter 2498, in which Tyndall wrote to the Jamaica Committee in support of Edward Eyre.
4. *a precedent*: George William Gordon.
5. *does it not*: a 'Bill of Indemnity' is a statute passed to protect people who have committed some illegal act which would otherwise cause them to be subjected to legal penalties.
6. *my brother's*: not identified.

From Frederic W. Farrar[1] [November 1866][2] 2517

Harrow.[3]

My dear Sir,

We intended to meet in the library; but as that is not lighted with gas, and cannot be darkened, we will meet in one of the other rooms which I hope will be convenient in these respects for the experiments which I am so glad to hear you hope to make.[4] As we are most anxious to save you all further trouble, I will call at the Royal Institution (or any other place if that is inconvenient) on Tuesday afternoon between 3 and 4, and then can hear from you any arrangement you may like to have made for Thursday.

We anticipate the greatest advantage from your visit, and are fully sensible of the honour you do us by coming.

Many thanks for your very powerful letter about the Eyre Defense Committee.[5] I entirely agree with the sentiments you express. Unless all travellers and the great majority of Ethnologists have wilfully lied, the negro is naturally wild, animal, lazy, sanguinary: he does belong distinctly to an inferior race, and a race which cannot be treated as on an equality with civilised white races.[6] The cause of truth and liberality cannot possibly gain by ignoring this fact.

I am, dear sir, | Very sincerely yours | F. W. Farrar.

RI MS JT/1/TYP/8/2593
LT Typescript Only

---

1. *Frederic W. Farrar*: Frederic William Farrar (see letter 2514, n. 7).
2. *[November 1866]*: the month and year are given by reference to Tyndall's letter to the Jamaica Committee, dated 7 November 1866 (see letter 2498).
3. *Harrow*: Farrar was assistant master at Harrow from 1855–70.
4. *hope to make*: presumably to have Tyndall lecture for the students at Harrow.
5. *very powerful letter about the Eyre Defense Committee*: for Tyndall's letter to the Jamaica Committee in support of Edward Eyre, see letter 2498. See also letter 2481, n. 4.
6. *civilised white races*: on the racism of most Victorian anthropology, see, e.g., G. Stocking, Jr., *Victorian Anthropology* (New York: Free Press, 1987); D. Livingstone, *Adam's Ancestors: Race, Religion, and the Politics of Human Origins* (Baltimore: Johns Hopkins University Press, 2008); and E. Sera-Shriar, *The Making of British Anthropology, 1813–1871* (London: Pickering and Chatto, 2013).

From Ellen Lubbock[1] [November 1866][2] 2518

Dear Mr Tyndall

Thanks for the letter.[3] It is excellent. I feel most strongly on the subject and am delighted to find that you do also. If that brave and unfortunate man[4] is brought to trial like a criminal I declare I should like to go and stand in the dock by his side.

Believe me | Ever yours truly | Ellen Lubbock.

RI MS JT/1/TYP/8/2575
LT Typescript Only

1. *Ellen Lubbock*: Ellen Frances Lubbock (née Hordern, 1834–79), the daughter of an Anglican minister. She married John Lubbock in 1856, and they had three sons and three daughters.
2. *[November 1866]*: the month and year are given by reference to Tyndall's letter to the Jamaica Committee, dated 7 November 1866 (see letter 2498).
3. *letter*: see n. 2.
4. *man*: Edward Eyre. See letter 2481, n. 4.

From François-Napoléon-Marie Moigno [November 1866][1] 2519

Mon cher M[r]. Tyndall,

J'ai fait venir de Londres par l'intermédiaire de M[r]. Francis les clichés (blocs) de votre petit mémoire sur la plasticité de la glace, mais il m'a été impossible de retrouver le texte anglais qui donne l'explication de ces figures; et j'ose vous prier instamment de m'envoyer par la poste un nouvel exemplaire de cet extrait des Proceding's.

Mon procès avec M[r]. Giraud avance, et se terminera je l'espère heureusement. Il persiste toujours avec une effronterie sans *[1 word illeg]* à prétendre qu'il ne nous doit rien à vous et à moi pour la Chaleur. L'affaire serait bientôt réglée si vous aviez conservé mes lettres relatives à cette opération et si vous me les adressiez. J'ai bien votre lettre de juin ou juillet et elle suffira au besoin, mais les lettres antérieures auraient un très grand poids dans la balance.

Je commencerai dans la prochaine livraison la publication du mémoire sur la calorescence, avec toutes les figures, et j'en ferai faire un joli tirage à part.

Quand me sera-t-il donné de traduire les leçons sur le son.

Je me prépare à faire une grande campagne pendant l'exposition. J'aurai au Champ de Mars une belle salle de conférence, et je vous inviterai à y déployer votre talent tant admiré.

A vous de cœur | l'abbé F. Moigno

My dear Mr Tyndall,

I brought from London, via Mr. Francis,[2] the pictures (blocks) of your small dissertation on the plasticity of ice.[3] However, it was impossible for me to find the English text that explains those figures; and I dare to ask of you insistently to send me a new copy of that part of the Proceedings.[4]

My trial with Mr. Giraud[5] progresses, and I hope it will end happily. He always persists with an effrontery without *[1 word illeg]* to claim that he does not owe us anything, to you and to me, for the Heat.[6] The matter would be soon resolved if you would have kept my letters about this transaction and if you would send them to me. I do have your letter from June or July,[7] and it will be sufficient for this, but earlier letters would greatly tip the scales.

In the next delivery, I will begin the publication of the dissertation on the caloresence,[8] with all the figures, and I will have a nice printing made separately.

When will I be given to do the translation of the lessons on sound?[9]

I am preparing to do a large campaign during the exposition.[10] I will have a nice conference room at the Champ de Mars,[11] and I will invite you to display your admirable talent there.

Yours sincerely | Father F. Moigno

RI MS JT/1/M/139

1. *[November 1866]*: the year and month are given by reference to articles in Moigno's journal *Les Mondes* in November and December 1866.
2. *Mr. Francis*: William Francis.
3. *dissertation on the plasticity of ice*: probably a reference to J. Tyndall, 'Professor Helmholtz on Ice and Glaciers,' *Phil. Mag.*, 30 (1865), pp. 393–407. A translation of this essay was published in the 27 December 1866 issue of Moigno's *Les Mondes*: J. Tyndall, 'Plasticité de la glace', *Les Mondes*, 12 (1866), pp. 682–6.
4. *Proceedings*: Moigno perhaps thought that Tyndall's paper had been published in the *Roy. Inst. Proc.* rather than the *Phil. Mag.*
5. *Mr. Giraud*: Étienne Giraud, a Parisian publisher.
6. *the Heat*: J. Tyndall, *La chaleur: considérée come un mode de mouvement*, trans. F. Moigno (Paris: Étienne Giraud, 1864).
7. *your letter from June or July*: letter missing, but probably related to Moigno's letter of 23 June 1866 (letter 2452).

8. *dissertation on the calorescence*: a translation of Tyndall's *On Radiation* appeared in Moigno's *Les Mondes* starting in the 29 November 1866 issue (J. Tyndall, 'Radiation. Sur la calorescence', *Les Mondes*, 12 (1866), pp. 523–31).
9. *translation of the lessons on sound*: Moigno's French translation of Tyndall's *Sound* (1867) appeared in 1869 (J. Tyndall, *Le son*, trans. F. Moigno (Paris: Gauthier-Villars, 1869).
10. *exposition*: the International Exposition (or Exposition Universelle) of 1867, held in Paris from 1 April to 3 November.
11. *Champ de Mars*: the Palais (Palace) du Champ de Mars at the International Exposition. The Champ du Mars was a large public space in Paris where the exposition was held.

From J. E. Plowden[1] 3 December 1866 2520

Athenaeum[2] | 3/12/66.

Dear Sir,

My friend, General Le Grand Jacob[3] favoured me on Saturday with a sight of your very valuable and conclusive letter about poor Governor Eyre[4] whom my cousin by marriage has succeeded as Governor of Jamaica, Sir J. P. Grant.[5]

I believe many friends of mine reciprocate your noble sentiments respecting the late Governor and would like no doubt to see your letter, General Sir J. C. ....[6] and others if you can do me the favour to spare a few copies.

I beg leave to enclose my card and address.

In haste, | Yours truly | J. Plowden

RI MS JT/1/TYP/8/2579
LT Typescript Only

1. *J. E. Plowden*: not identified, but as shown in letter 2521, he was a Colonel.
2. *Athenaeum*: see letter 2253, n. 8.
3. *General Le Grand Jacob*: George Le Grand Jacob (1805–1881), a British army officer in the East India Company until his retirement in 1861. He played a significant role in the suppression of the Indian Mutiny in 1857 (*ODNB*).
4. *your very valuable and conclusive letter about poor Governor Eyre*: for Tyndall's letter to the Jamaica Committee in support of Edward Eyre, see letter 2498.
5. *Sir J. P. Grant*: John Peter Grant (1807–93), who had also played a role in supressing the Indian Mutiny in 1857, succeeded Eyre as Governor of Jamaica. He had previously been Lieutenant Governor of Bengal (*ODNB*).
6. *General Sir J. C. ....*: not identified.

From J. E. Plowden[1] 5 December 1866 2521

Dudley Lodge, Kingston | Surrey, | 5th Dec. 1866.

Colonel Plowden feels much indebted to Professor Tyndall for so kindly complying with his request for copies of his very masterly letter.[2] Colonel Plowden will use all his endeavour to subserve the late Governor Eyre's[3] cause. He deserves ennoblement rather than persecution.

J. E. Plowden.

RI MS JT/1/TYP/8/2580
LT Typescript Only

1. *J. E. Plowden*: see letter 2520, n. 1.
2. *his request for copies of his very masterly letter*: see letter 2520.
3. *Governor Eyre's*: Edward Eyre. See letter 2481, n. 4.

From William Erle[1] 10 December 1866 2522

12 Princes Gardens, | 10 December 1866.

My dear Professor Tyndall

I have received a copy of your letter to the Jamaica Committee.[2] I hope I am indebted to you for it, at all events I am indebted to you for some of the wholesome vigour which flows into every man's mind who reads and understands that letter.

Yours sincerely | W. Erle.

No answer.

RI MS JT/1/TYP/8/2581
LT Typescript Only

1. *William Erle*: while LT's typescript of this letter shows the signer to be 'V. Eyre', which she may have thought to be a relative of Edward Eyre, it appears this is a mis-transcription of "W. Erle', for William Erle (1793–1880), an English lawyer and judge who, like Tyndall, was a member of the Governor Eyre Defence and Aid Committee (W. F. Finlason, *The History of the Jamaica Case*, 2nd edn (London: Chapman and Hall, 1869), pp. 562*m*–562*o*; this book was dedicated to Erle). Also, Erle resided at 12 Princes Gardens, South Kensington, the address given in this letter (*List of Members, with Date of Election* (Royal Society of Arts, 1878), p. 17).
2. *your letter to the Jamaica Committee*: for Tyndall's letter to the Jamaica Committee in support of Edward Eyre, see letter 2498.

To William Erle[1] [10 December 1866][2] 2523

Royal Institution of Great Britain | Monday

My dear Sir William

You will not prevent a man from speaking out of the abundance of his heart, and therefore you will not prevent me from saying how grateful I am to M[rs]. Russel Gurney[3] for sending you that letter.[4] I should hardly have ventured to do so myself, lest instead of your approval I should have invoked your censure. It gratifies me more than I can say to find that you like it.

Believe me dear Sir William | Yours most sincerely | John Tyndall

Oxford MSS.Don c.72 f.76

---

1. *William Erle*: see letter 2523, n. 1.
2. *[10 December 1866]*: the date is given by relation to letter 2523. Erle wrote to Tyndall on 10 December 1866, a Monday, and this letter being a Monday suggests a reply on the same day.
3. *M[rs]. Russel Gurney*: Emelia Gurney (née Batten, 1823–96), the wife of Russell Gurney (1804–78), an English lawyer and politician (*ODNB*), who in January 1866 was appointed a Commissioner of the Jamaica Rebellion inquiry (E. M. Gurney (ed), *Letters of Emelia Russel Gurney* (London: James Nisbet & Co., 1902), p. 27).
4. *that letter*: Tyndall's letter in support of Edward Eyre to the Jamaica Committee (see letter 2498).

From George Biddell Airy 13 December 1866 2524

Royal Observatory, Greenwich, | London, S.E. | 1866 December 13.

My dear Sir

Some weeks ago I received, from persons (partly known to me and partly not in my acquaintance) connected with the "South Australian Institute" at Adelaide,[1] letters requesting assistance in selecting apparatus for Experimental Lectures for the various Scientific Institutions rising in that country, which are affiliated to the South Australian Institute, and to which the apparatus would travel in succession. These gentlemen wrote to me, in consequence of my being personally known to one of them, but at the same time they referred—and rightly—to you, as better able to assist them than any other person.

A succession of troublesome business (in the run of chances) has prevented me from looking at the papers which accompanied their letters till now. And I find this to be the state of things. They do possess some apparatus, of which they send a list. And they send other lists, in great detail and great extent, but in very good order, referring to a marked copy of Elliott's Catalogue[2] (which they wish to have returned). And, giving clear instructions in an orderly form, they wish that the instruments marked be purchased for them, or others substituted, at the discretion of the English friend who will consent to aid them.

Now it is quite impossible for me to help them materially in this matter. There is the fundamental and personal difficulty that I am thirty years out of practice of lecturing and out of knowledge of apparatus &c. And there is the accidental and official difficulty that I have not the time.

In this emergency, will you as requested by them give your assistance? I should be gratified by your doing so, for I honour the spirit of these colonists and should be glad to see their wishes carried out. If you will consent to entertain this matter, I will send for your information all the papers.

I am, my dear Sir, | Yours very truly | G B Airy
Professor Tyndall

RI MS JT/1/A/40
RI MS JT/1/TYP/1/30

1. *"South Australian Institute" at Adelaide*: the South Australian Institute had been established by Act of Parliament in 1855 'to comprise a Public Library and Museum, and, by means of public lectures, classes, and otherwise, to promote the general study and cultivation of all or any of the various branches or departments of art, science, literature and philosophy', and acted as a kind of umbrella body for a number of scholarly societies (*Report of Commission appointed to Inquire into the Whole Question of the New Institute and Museum* (Adelaide: W. C. Cox, Government Printer, 1874), p. iv).
2. *Elliott's Catalogue*: Elliot Brothers was a British scientific instrument maker, founded by William Elliot (1780/1–1853) and run by his two sons from 1857 forward (*ODNB*). The company served many prominent academics such as Maxwell, Wheatstone, and Lord Rayleigh. Most of their products included galvanometers and batteries, though their later products were for avionics. See H. R. Bristow, 'Elliott, instrument makers of London: Products, customers and development in the 19th century', *Bulletin of the Scientific Instrument Society*, 36 (1993), pp. 8–11.

From Thomas Carlyle 13 December 1866 2525

CHELSEA,[1] 13 December 1866.

Dear Tyndall,

To.day I have again a little note from the Lady A.[2] If there be still a possibility (that is if the Dumfriesshire coil[3] be got rid out) before the 21st,—send me your Edinburgh Address,[4] and I will warn you at once. Till the 17th (or 18th, is it?) a message hits you, I think, in Edinburgh.

Gluck auf dem Wege[5] to you.

Yours always | T. Carlyle.

RI MS JT/1/TYP/1/199
LT Typescript Only

1. *Chelsea*: Carlyle lived at No 5 Cheyne Row, Chelsea since 1834.
2. *Lady A.*: Louisa Baring (Lady Ashburton).
3. *Dumfriesshire coil*: not identified.
4. *send me your Edinburgh Address*: Tyndall was in Edinburgh to deliver a lecture to the Edinburgh Philosophical Institution on 14 December (see letter 2460, n. 17).
5. *Gluck auf dem Wege*: wishing you the best for your travels (German).

From James Coxe[1] 14 December 1866 2526

Edinburgh | Dec. 14th 1866.

My dear Dr Tyndall

I was unlucky at not finding you at home,[2] and so must acquit myself by pen of the messages with which I was charged.

1. Will you kindly use the carriage tomorrow, close or open, from your rising up to your lying down, or any lesser portion of the time, in any way you think fit?

2. Will you dine with us in the very quietest way on Sunday. Campbell of Islay[3] whom I dare say you know has written to say he will dine at Kinellan[4] on that day (7 o'clock) to see Cumming,[5] who is better—but quite unable to support himself on his legs.

What a misfortune it is that we are in so lamentable a condition when you are here.

I hope to see you this evening | Very truly Yours | J. Coxe.

RI MS JT/1/TYP/1/291
LT Typescript Only

1. *James Coxe*: see letter 2409, n. 5.
2. *at home*: Tyndall was in Edinburgh to deliver a lecture to the Edinburgh Philosophical Institution on the very day of this letter (see letter 2460, n. 17), so 'at home' refers to wherever he was staying.
3. *Campbell of Islay*: John Francis Campbell (1821–85), a Scottish folklorist and author of *Popular Tales of the West Highlands*, 4 vols (Edinburgh: Edmonston and Douglas,

1860–62). He was known as Campbell of Islay even though his family had been obliged to sell the island in 1847 to pay off their debts (*ODNB*).

4. *Kinellan*: the Edinburgh estate of James Coxe and Mary Anne Coxe.
5. *Cumming*: probably William Fullerton Cumming (see letter 2409, n. 6).

From George Busk 17 December 1866 2527

15 Harley St. | Dec. 17th—1866.

Divine Phebus,[1]

I am desolated in being compelled to deprive myself of the pleasure and profit of accompanying you in your intended circuit seeing that I am at present pretty completely tied up. But I do hope when you pervade the Pass of Llanberis[2] you will remember me with gratitude as its discoverer I don't doubt that is and must be the direction to which your feet will not to say nose will be directed—but if not it ought to be, especially as it will be snowed up.

With the utmost reverence | your worshipper | G. B.

P.S. It is a shame however to leave the Phil. Club[3] in the lurch—what shall I do for my precious X, if you all go away to Wales.[4]

P.S. I hope there will be somebody in my place to take care of you.

P.S. When at Beddgelert[5] count the suspicious ones. Six I should say.

P.S. Your best route if Welsh will be the rail to Machynlleth[6] and round by Tan y Bulch[7] and Beddgelert to Carnarvon[8] only then you will miss the Pass.[9]

RI MS JT/TYP/1/179
LT Typescript Only

1. *Divine Phebus*: in Greek mythology, Phebus (Phoebus in Latin) was the god governing the sun. This could be Busk's way of acknowledging either Tyndall's work on the absorption of radiant heat or the polarization of light and the blue colour of the sky. Tyndall had previously referenced this god in a published poem entitled 'The Sky. Apostrophe to Friendship', which began: 'The sun has fallen beneath the western hill, | And o'er me spreads in slumber deep and still | The wondrous sky, without a single star | To gem its azure bosom—from afar | Phoebus, though hidden, flings a feeble spray, | Quenching their feebler fires in dubious day, | O, holy sky! How beautiful art thou!' (*The Carlow Sentinel*, 3 August 1844, p. 3); also published in R. Jackson, N. Jackson and D. Brown (eds), *The Poetry of John Tyndall* (London: UCL Press, 2020), pp. 147–9.
2. *Pass of Llanberis*: the Llanberis Pass in Snowdonia travels between the mountain ranges of the Glyderau and the Snowdon massif.
3. *Phil. Club*: Busk was referring to a December 1866 meeting of the Philosophical Club of the RS, founded in 1847 by William Robert Grove in the aftermath of the reform of the RS's governance. Tyndall was elected a member on 10 May 1855 (Journal, RI MS

JT/2/13c/745). See also T. G. Bonney, *Annals of the Philosophical Club of the Royal Society, Written from Its Minute Books* (London: Macmillan and Co., 1919).

4. *my precious X, if you all go away to Wales*: LT's typescript shows '14' which we think is a mistranscription of 'X'. Busk was referring to a short trip by the members of the X Club, a monthly dining club founded in 1864 for the discussion and promotion of science in British society, its members being Tyndall, George Busk, Edward Frankland, Thomas Hirst, Joseph Hooker, Thomas Huxley, John Lubbock, Herbert Spencer, and William Spottiswoode (see *The X Club*).
5. *Beddgelert*: a village in the Snowdonia area of Gwynedd, Wales. Busk's reference to six 'suspicious ones' is not identified.
6. *rail to Machynlleth*: the Newtown and Machynlleth Railway, built by David Davies, had been completed a few years earlier in 1862.
7. *Tan y Bulch*: the correct spelling is Tan y Bwlch.
8. *Carnarvon*: now Caernarfon, a port town in Gwynedd, Wales.
9. *Pass*: see n. 2.

To Roderick Impey Murchison [before 21 December 1866][1] 2528

ROYAL SCHOOL OF MINES | JERMYN STREET[2]

Dear Sir Roderick.

I go with M^r^. Carlyle on Friday morning to Mentone.[3]

Carlyle smokes clay pipes—He wishes to take a score of them with him.

He also smokes his own tobacco—he wishes to take two or three pounds with him.

He also drinks his own special brandy & wishes to take a quart or so with him.

He has not asked me to make any stir in the matter—but I want to take these things for the grand old man's comfort. I want therefore to be secure as to the French Douaniers[4]—Can you do any thing for me? A note from the Ambassador[5] would set all right.

I have already found his Excellency's note[6] regarding instruments of the greatest service to me.

I am persuaded that if I write to the Emperor[7] himself he would grant me the request to be allowed to pass.

Yours ever | John Tyndall

Please Remember I start out early on Friday night so that I ought to have your note, if at all, on Thursday

Science Museum, Wroughton, Ms 2108/79b

1. *[before 21 December 1866]*: the date is given based on Tyndall and Carlyle having left London for Mentone on 21 December 1866, placing this letter within the week before that date.
2. *ROYAL SCHOOL OF MINES | JERMYN STREET*: Tyndall had held the Chair of Natural Philosophy at the Royal School of Mines since 1859 (see letter 2259, n. 17).
3. *to Mentone*: see letter 2476, n. 13.
4. *Douaniers*: Customs officers (French).
5. *Ambassador*: not identified.
6. *his Excellency's note*: not identified; if this is a letter to Tyndall, it is missing.
7. *Emperor*: Napoleon III, Charles-Louis Napoléon Bonaparte (1808–73), the French emperor from 1852–70.

To the Editor of the *PMG*[1] 21 December 1866 2529

CORRESPONDENCE. | COLLIERY EXPLOSIONS. | *To the* EDITOR *of the* PALL MALL GAZETTE.

SIR,—On the 28th of September, 1844, a fatal explosion occurred in the Haswell colliery, situated seven miles east of Durham. At the instance of Sir James Graham, then Secretary of State, Mr. Faraday and Sir Charles Lyell inspected the scene of the disaster, and embodied their views and recommendations in a report, which is published in the 26th volume of the *Philosophical Magazine* for 1845.[2]

The attention of these two eminent men was soon converged upon a single point, the ventilation of the "goaf."[3] As the coal is taken away the roof of the mine must be supported. This is first done by residual pillars of coal, then by wooden props, called "juds." After a time the juds are withdrawn and the roof falls in. The heap of fragments covering the floor of the mine in consequence of this falling in of the roof is called a "goaf." The goaf, according to Mr. Faraday and Sir Charles Lyell, is the chief seat of danger in a coal mine. The roof of the mine above the goaf constitutes a kind of flattened dome, underneath which the fire-damp collects, and from which it trickles into the workings. If any portion of it be there fired by the lamp of the miner the explosion is instantly propagated backwards to the whole mass contained in the space above the goaf. The product of the direct union of the fire-damp with the air is an irrespirable gas, carbonic acid. The quantity of this gas is increased by the combustion of the fine coal dust whirled up in the explosion. Thus the miner is burnt by the fire-damp and smothered by the choke-damp or carbonic acid.

To sweep out the fire-damp from the dome of the goaf with cold air, which on account of its greater weight, flows along the bottom of the mine, is the great difficulty. Mr. Faraday and Sir Charles Lyell recommended the carrying of a pipe from the goaf to the up-cast shaft of the mine. They dwelt, moreover, strongly upon the necessity of a better education for the miners, remarking that men of natural intelligence and power were sure to be found amongst them who, if properly instructed, would be able to examine the air of the mine at intervals, and thus foretell the approach of danger. As an instance of the recklessness of ignorance, Mr. Faraday once told me that he saw a miner (I believe on the occasion here referred to) endeavouring to light his pipe by forcibly sucking the flame of his safely lamp through the surrounding wire gauze. Hence the desirability of using a light which shall render the employment of a flame by the miner unnecessary. These underground regions may certainly be flooded.

Faraday made the great discovery that by causing a bobbin of wire to move near a magnet electric currents are obtained which exhibit all the effects of currents from the voltaic battery. By employing a large apparatus Mr. Holmes[4] so intensified these currents as to render them capable of producing the electric light, which, as your readers know, is of almost sunlike brilliancy. His machine has been employed successfully for some years at the lighthouse of Dungeness.[5] In Paris, in the atelier[6] of M. Berlioz,[7] I saw this year an improved magneto-electric machine, which is nightly employed in the lighthouses of La Hève, near Havre.[8] The performance of the machine was excellent. Such machines consist of a number of permanent magnets suitably arranged, and a number of bobbins of wire kept in rapid rotation by a steam engine. The steam engine, I may remark, is only a go-between by which the power of the coal and atmospheric oxygen is converted first into electricity, and afterwards into the electric light. The cost of a machine capable of producing this sunlike light would be between four and five hundred pounds. It would remain in good order for I know not how many years, and the expense of working it at the mouth of a coal pit would be next to nothing. The light would illuminate a large area. It could be caused to shine *in vacuo*, and the exhausted vessel containing it might be wholly immersed in cold water without prejudice to the light; all danger of explosive contact between the light and the air of the mine being thus avoided. The direct rays of this light could not of course illuminate tortuous galleries, or surfaces turned form the light. But by bits of looking-glass its rays might be turned in any direction, the reflected rays accomplishing what the direct rays fail to accomplish. In mines of large extent two or more of these lights might be employed. Possibly also, on account of the difference of refrangibility between it and air, the fire-damp might be actually rendered visible by means of the electric light, and its insidious advance detected. Mr. Wilde, of Manchester,[9] has recently devised a magneto-electric machine of far greater power than either that of Mr. Holmes or that of M. Berlioz. I have witnessed the splendid performance of Mr. Wilde's machine, which, if it stand the test of continuous working, will supersede all others.

Other methods have been proposed for illuminating coal workings. The machine to which I have referred, or, indeed, a far cheaper one, may be applied to realize those methods. Instead of concentrating the power of the machine on one splendid light, it may be caused to send luminous discharges through glass tubes containing rarefied gases, and these tubes may be employed at the points where the minor works. Their light would be superior to that of the safety-lamp, and their application absolutely without danger. In fact, the safe illumination of coal mines is wholly a question of expense. It is possible to carry into these underground regions the light of day, this light being the transmuted force of the very coal worked by the miner. The only question is, will it pay the coal-owner to have the darkness of the mine converted into daylight?

JOHN TYNDALL. | *Royal Institution*, *Dec.* 20, 1866.

PMG, 21 December 1866, p. 3

1. *Editor of the PMG*: see letter 2314, n. 1.
2. *Philosophical Magazine for 1845*: in the aftermath of the Haswell Colliery explosion, in which 95 people were killed, as Tyndall indicated, Faraday and Charles Lyell participated in the inquest and subsequently submitted a report on the disaster to the government of Sir Robert Peel, containing a number of recommendations. See 'Messrs. Faraday and Lyell's Report to the Rt. Hon. Sir James Graham, Bart., Secretary of State for the Home Department, on the subject of the Explosion at the Haswell Collieries, and on the means of preventing similar accidents', *Phil. Mag.*, 26 (1845), pp. 16–35. See also F. A. J. L. James & M. Ray, 'Science in the pits: Michael Faraday, Charles Lyell and the home office enquiry into the explosion at Haswell Colliery, country Durham, in 1844', *History and Technology*, 15 (1999), pp. 213–31.
3. *goaf*: an empty space from which coal has been extracted in a longwall mine, sometimes filled with waste material (*OED*).
4. *Mr. Holmes*: Frederick Hale Holmes (b. *c.* 1812), a British pioneer of electric lighting and professor of Chemistry at the Royal Panopticon of Science and Art (*Literary Gazette*, 23 July 1853, p. 722).
5. *lighthouse of Dungeness*: in 1862, Trinity House introduced commercial electricity to the Dungeness lighthouse in coastal Kent, England, built in 1790.
6. *atelier*: workshop (French; *OED*).
7. *M. Berlioz*: Auguste Prosper Berlioz (1800–88), of Auguste Berlioz and Company, a Paris firm specializing in magneto-electric machines.
8. *lighthouses of La Hève, near Havre*: see letter 2457, n. 8.
9. *Mr. Wilde, of Manchester*: Henry Wilde (1833–1919), a British electrical engineer and inventor (*ODNB*).

From James Clark 26 December 1866 2530

Bagshot Park | Surrey. | December 26. 1866.

My dear Professor Tyndall,

If you happened to be present at a course of lectures given in the Royal Institution on Insanity[1] by the late Dr Conelly,[2] I should feel obliged by your opinion of the lectures, how were they attended, and the effect they produced. I scarcely expect that you can reply to all these points, but I thought it just possible that you might, or, if not, that you might know some one who could. At any rate I hope you will kindly excuse the trouble that I am giving you.

I am distressed to find that difficulties are accumulating round our unfortunate friend Eyre[3]—Several of the black population are prosecuting him for unlawful imprisonment—this is too bad—Surely no Governor was ever treated with such persecuting cruelty. I should almost hope the Government would be excited by this new phase of the persecution, to take some steps to put a stop to the thing; and yet one does not well see what they can do.

These new sources of persecution are I expect the doings of the anti-slavery fanatics.

Yours sincerely | James Clark

RI MS JT/1/TYP/1/219
LT Typescript Only

1. *course of lectures given in the Royal Institution on Insanity*: Conolly twice delivered Friday Evening Discourses at the RI on the subject of insanity: 'On the Past and Present Condition of the Insane, and the Characters of Insanity' on 22 April 1853 (*Roy. Inst. Proc.*, 1 (1851–4), p. 303); and 'On the Characters of Insanity' on 17 February 1854 (*Roy. Inst. Proc.*, 1 (1851–4), pp. 375–81). See also A. Scull, 'A Victorian alienist: John Conolly, FRCP, DCL (1794–1866)', in W. F. Bynum, R. Porter, and M. Shepherd (eds), *The Anatomy of Madness: Essays on the History of Psychiatry*, 3 vols (London: Routledge, 2018), vol. 1, pp. 103–50.
2. *the late Dr Conelly*: John Conolly (1794–1866), an English psychiatrist and author of *The Treatment of the Insane without Mechanical Restraints* (London: Smith, Elder & Co., 1856) and other key texts, had died on 5 March 1866.
3. *our unfortunate friend Eyre*: Edward Eyre. See letter 2481, n. 4.

To Eliza Spottiswoode [29 December 1866][1] 2531

My dear Mrs. Spottiswoode,

Your friend must have seen me by second sight: as I did not reach London till this Saturday morning after a very rough and disagreeable passage across the Channel. The journey to Mentone was prosperous and I left Carlyle in a beautiful place and in the hands of a beautiful woman.[2]

Now look on this picture. I lecture at the Royal Institution on Friday the 18th,[3] and a terrible job it is to prepare for it. I lecture again on Tuesday the 22nd,[4] devoting Monday to the preparation of the Lecture. I lecture again on Thursday the 24th,[5] and from morning till night on the Wednesday my assistant[6] and myself will be engaged in the work of preparation. Under these circumstances you will see how impossible it is for me to grant your request. I would do so if it were within the compass of my power and if you can defer this "winding up" until I get my lectures over and my book off my hands[7] I will do all that I can to gratify you. I hope to be pretty free towards the end of March or the beginning of April.

Yours ever | John Tyndall

RI MS JT/1/TYP/3/1243
LT Typescript Only

---

1. *[29 December 1866]*: the date is given by reference to Tyndall's returning to London from Mentone. He arrived in Mentone on 23 December, a Sunday, making the Saturday of his return 29 December (see letter 2476, n. 13).
2. *in the hands of a beautiful woman*: Tyndall and Thomas Carlyle had visited Louisa Baring in Mentone (see letter 2476, n. 13).
3. *lecture at the Royal Institution on Friday the 18th*: on 18 January 1867, Tyndall delivered a Friday Evening Discourse at the RI, 'On Sounding and Sensitive Flames' (*Roy. Inst. Proc.*, 5 (1866–9), pp. 6–12).
4. *lecture again on Tuesday the 22nd*: Tyndall delivered twelve lectures 'On Vibratory Motion, with Special Reference to Sound' at the RI in 1867, on Tuesday and Thursday afternoons from 22 January to 28 February (RI Guard Book, vol. 3, p. 59; *Roy. Inst. Proc.*, 5 (1866–9), p. 4).
5. *lecture again on Thursday the 24th*: see n. 3.
6. *my assistant*: either John Cottrell, a laboratory assistant at the RI between 1866–1885, or William Fletcher Barrett (see letter 2245, n. 31). Tyndall's previous assistant Charles Anderson had died in January 1866 (see letter 2348, n. 1 and n. 2).
7. *my book off my hands*: Tyndall's forthcoming *Sound* (1867), published the following July.

From Charles Pritchard[1] 29 November 1866 2532

Nov. 29, 1866.

My dear Dr. Tyndall

Let me have the satisfaction of tendering you the hearty thanks of I hope one honest man to another for your Stand-up-for-the-right in the defence of Gov[ernor]. Eyre.[2]

Yours very sincerely | C. Pritchard.

RI MS JT/1/TYP/8/2577
LT Typescript Only

---

1. *Charles Pritchard*: the sender is probably Charles Pritchard (1808–93), a British astronomer, clergyman, and headmaster at Clapham. In 1870, Pritchard became Savilian professor of astronomy at the University of Oxford (*ODNB*). He was one of the founders of the Metaphysical Society, a club formed in 1869 for the scientific discussion of theological questions. On the Metaphysical Society, see C. Marshall, B. Lightman, and R. England (eds), *The Metaphysical Society (1869–1880): Intellectual Life in Mid-Victorian England* (Oxford: Oxford University Press, 2019).
2. *in the defence of Gov. Eyre*: referring to Tyndall's letter to the Jamaica Committee in support of Edward Eyre (see letter 2498).

From Amy Harriet Lubbock[1] [December 1866][2] 2533

Mein lieber Herr Tyndel.

Es thut mir sehr leid daß Sie uns nicht zu Weihnachten besuchen können. Da unsere liebe Mutter uns einen Weihnachtsbaum versprochen hat, hätten wir uns sehr gefreut wenn Sie hier gewesen wären, um ihn so wie vor zwei Jahren anzuzünden. Rolph kann jetzt schon ganz gut gehen, er ist ein sehr hübscher kleiner Junge und liebt mich über Alles. Ich gebe Norman Stunden und Conny gibt Gertrude Stunden, sie sind beide sehr gute Kinder und haben sehr Fortschritte gemacht. Ich habe die Deutsche Sprache sehr lieb und ich hoffe wenn Sie zurückkommen werden, recht viel Deutsch mit Ihnen sprechen zu können. Johnny und Conny senden viele <herzliche> Küße, und ich grüße und küße Sie auch bestens, und hoffe daß wenn Sie zurückkommen werden Sie mir viel von dem schönen Itallien erzählen werden. Meine liebe Mutter läßt Sie herzlich grüßen. Es thut ihr sehr leid Sie gestern Abend nicht gesehen zu haben.

Von Ihrer Sie liebenden | Amy Harriet Lubbock.

My dear Mr Tyndall.

I am very sorry that you cannot come to visit us for Christmas.[3] Since our dear mother[4] has promised us a Christmas tree, we would have been very happy if you were here to light it has you had done two years ago. Rolph[5] can already walk quite well, he is a very handsome little boy and adores me. I give Norman[6] lessons and Connie[7] gives Gertrude[8] lessons, they are both very good children and have made great progress. I am clearly quite fond of the German language and I hope that when you will come back, to be able to speak with you quite a lot of German. Johnny[9] and Conny both send warm kisses, and I too greet and kiss you warmly, and hope that when you will come back you will tell me a lot about beautiful Italy.[10] My dear mother sends kind regards. She is very sorry to not have seen you yesterday evening.

From your loving | Amy Harriet Lubbock.

RI MS JT/1/L/40

---

1. *Amy Harriet Lubbock*: Amy Harriet Lubbock (1857–1929), eldest child of John and Ellen Lubbock ('Obituary', *Times*, 21 May 1929, p. 16; R. Birkbeck, *Notes on the History and Genealogy of the Family of Lubbock* (London: Mitchell and Hughes, 1891), pp. 36–7).
2. *[December 1866]*: the month and year are given by reference to Tyndall's trip to Mentone (see n. 3). Also, Rolfe Lubbock was born on 19 Sep 1865, making him several months older than one year in December 1866, and certainly able to 'already walk quite well'.
3. *cannot come to visit us for Christmas*: Tyndall traveled to Mentone, France with Thomas Carlyle to visit Louisa Baring (see letter 2476, n. 13).
4. *dear mother*: Ellen Lubbock (see letter 2518, n. 1).
5. *Rolph*: Rolfe Arthur Lubbock (1865–1909), the third son and youngest child of John and Ellen Lubbock (née Horden) ('The Hon. R. A. Lubbock', *Times*, 21 May 1909, p. 11; R. Birkbeck, *Notes on the History and Genealogy of the Family of Lubbock*, pp. 36–7). Tyndall was his godfather.
6. *Norman*: Norman Lubbock (1861–1926), the second son of John and Ellen Lubbock (R. Birkbeck, *Notes on the History and Genealogy of the Family of Lubbock*, pp. 36–7).
7. *Connie*: Constance Mary Lubbock (1860–92), the second daughter of John and Ellen Lubbock (R. Birkbeck, *Notes on the History and Genealogy of the Family of Lubbock*, pp. 36–7).
8. *Gertrude*: Gertrude Lubbock (1863–1934), the third daughter of John and Ellen Lubbock (R. Birkbeck, *Notes on the History and Genealogy of the Family of Lubbock*, pp. 36–7).
9. *Johnny*: John Birkbeck Lubbock (1858–1929), the first son of John and Ellen Lubbock (R. Birkbeck, *Notes on the History and Genealogy of the Family of Lubbock*, pp. 36–7).
10. *about beautiful Italy*: Amy must have been mistaken in writing 'Italy'. Mentone is in France but at the border with Italy.

To Henry Sidgwick[1] [1866][2] 2534

Royal Institution Saturday

Dear Mr. Sidgwick

I wanted to rake up my memory before replying to your important question.[3]

Good teachers are the real want of the present day & their absence the peak difficulty. They are not there because they have not been much in demand. I had a reason to dwell upon this point at the late meeting of the British Association of Nottingham.[4]

The difficulty in your case is enormously enhanced by the range of subjects which you require your teacher to undertake. <u>I do not hesitate to say that the man who would agree to teach all these subjects must be himself deficiently instructed</u>.

Why not limit yourself for the present to such subjects as can be taught in a real & healthy manner? The backbone of modern Science as regards having an education is formed by Chemistry and Physics. I do not think you will get a man who know these two subjects well to teach the others with any profit—hardly with sincerity.

I know at the present moment but one man likely to suit you. He is profound & of acknowledged eminence as a chemist, and has a thorough knowledge of physics. He is a good mathematician. He is a Fellow of the Royal Society and an examiner of the University of London. He is a man of lofty character and of rare general intellectual power. He is a German, but has been teaching boys in England for the last 15 years. He is not very fluent as a lecturer, though he lectures at present four times a week. But he is first rate as a teacher, and is sure to gain the respect of those he teaches. His name & address are

Dr. Debus[5]
Queenwood College
Near Stockbridge | <u>Hants</u>

I recommended him some time ago to a professorship at Sandhurst,[6] but they preferred an Englishman. For the sake of himself and of Science I wish sincerely he could get into a position which would give due scope to his real powers.

I am | yours very truly | <u>John Tyndall</u>

Trinity College Cambridge, Houghton. Add.ms.c.95/173

---

1. *Henry Sidgwick*: Henry Sidgwick (1838–1900), an English philosopher and economist who from 1859–69 was a lecturer in classics at Trinity College, Cambridge (*ODNB*).

2. *[1866]*: the year is given based on the reference to the recent BAAS meeting at Nottingham (see n. 4).
3. *your important question*: the letter to which this is a response is missing. Sidgwick was presumably soliciting Tyndall's advice regarding a teaching position.
4. *late meeting of the British Association of Nottingham*: the BAAS held its thirty-sixth annual meeting at Nottingham from 22–30 August 1866 (*Brit. Assoc. Rep. 1866*).
5. *Dr. Debus*: Heinrich Debus.
6. *Sandhurst*: Royal Military College, Sandhurst.

BIOGRAPHICAL REGISTER

This register contains the names and biographical details of people who are mentioned three or more times in the letters included in this volume. Biographical information about people mentioned only once or twice is contained in the notes appended to the appropriate letter(s).

Conventions

The sources for the biographical entries are standard biographical dictionaries, such as the *Complete Dictionary of Scientific Biography*, the *Oxford Dictionary of National Biography*, Poggendorff's *Biographisch-literarisches Handwörterbuch der exacten Wissenschaften, Historisches Lexikon der Schweiz, Neue Deutsche Biographie*, and the *Dictionary of Nineteenth-Century British Scientists*. These have not been specified in the entries unless necessary. Additionally, the editors have found material from the Tyndall correspondence and from the journals of Tyndall and T. A. Hirst very useful. Where additional sources add important material, they are given in parentheses at the end of the entry.

Airy, George Biddell (1801–92), was educated at Cambridge University, graduating senior wrangler in 1823. In 1826 he was appointed Lucasian Professor of Mathematics, and in 1828 to the Plumian Chair of Astronomy and directorship of the Cambridge University observatory. In 1835 he was appointed astronomer royal, a position he held until 1881, and set about improving the instrumentation and research activities of the Royal Observatory, Greenwich. He vigorously pursued research on a wide range of topics, including optical diffraction, calculating the Earth's mean density, and analyzing the inequalities in the motions of the Earth and Venus. During the period covered by this volume Airy was interested in electricity and magnetism, as shown by his correspondence with Tyndall about experiments with induction coils.

Akin, C. K. (Károly) (1830–93), was a Hungarian physicist, writer on physics and chemistry, and corresponding member of the Hungarian Academy of Sciences (1868). He worked with the deputy professor of experimental philosophy at Oxford University on "calcescence," and in 1864 came into conflict with Tyndall over infrared experiments. Akin and Tyndall initially enjoyed cordial relations, but they later soured as Akin first privately and then publicly accused Tyndall of having plagiarized his own research. Akin, who was less well connected in the scientific community at this time, lost out to Tyndall, who was able to marshal his large professional network in his support.

Baring, Louisa Caroline (née Stewart-Mackenzie, 1827–1903), Lady Ashburton, was a philanthropist, patron of the arts, and second wife of statesman William Bingham Baring, 2nd Baron Ashburton (1799–1864). After her husband's death Baring amassed a significant collection of paintings and sketches, which filled her numerous residences where she entertained Britain's social and cultural elites, including Tyndall, the Carlyles, John Ruskin, George Campbell (the Duke of Argyll), the Walter Trevelyans, Edward Lear, W. W. Story, Robert Browning, Samuel Wilberforce (the Bishop of Oxford), Richard Monckton Milnes (Lord Houghton), John Everett Millais, and Charles Kingsley.

Baring, Mary "May" Florence (1860–1902), was the daughter of Louisa Baring, Lady Ashburton, and William Bingham Baring, 2nd Baron Ashburton. She married William Compton, 5th Marquess of Northampton in 1884, and held the title Marchioness of Northampton from 1897.

Barnard, Jane (1832–1911), was a niece of Michael and Sarah Faraday who lived at the RI and assisted the Faradays. She was the daughter of Sarah's brother, John Barnard, and Michael's younger sister Margaret. Like her aunt and uncle, she was a member of the Sandemanian Church.

Bohn, Johann Conrad (1831–97), was a German translator and experimental physicist. Privatdocent at Munich University and extraordinary professor at the University of Giessen in 1858, Bohn may have known Tyndall, since they both were students at Marburg. Bohn contacted Tyndall on 9 September 1865 about preparing a translation of Tyndall's *Heat* (1865). The translation was completed sometime early in 1866 and submitted to the German publishing house Vieweg Verlag (Eduard Vieweg). However, in March Bohn had annulled the contract, explaining to Tyndall that Vieweg accused him of submitting a flawed translation, but insisted that Vieweg's objections to his work were "unfounded." Tyndall consulted others, including Heinrich Debus and Rudolf Clausius, who agreed with Vieweg's judgment. Following their advice, Tyndall had Anna Helmholtz and Clara Wiedemann complete the translation, although it was their husbands, Hermann Helmholtz and Gustav Wiedemann, who received publication credit.

Brewster, David (1781–1868), was a Scottish mathematician, astronomer, editor, and educator, known for his work in optics and his role in founding the BAAS. He was a regular contributor to numerous periodicals, by which he mostly earned his living. He was one of three founding editors of the *Edinburgh Magazine* and editor of the *Edinburgh Encyclopedia* (1808–30). He was appointed principal of the United College of Saint Salvator and Saint Leonard at the University of Saint Andrews in 1838, and principal of the University of Edinburgh in 1859.

Bunsen, Robert (1811–99), was a German chemist who received his PhD from the University of Göttingen, where he then lectured (1833–36). Bunsen subsequently held positions at the Polytechnikum at Kassel, the University of Marburg (1839–51), the University of Breslau, and, from 1852, the University of Heidelberg. He made many contributions to chemistry, including the Bunsen burner and the discovery of cesium and rubidium, and developed a battery that bears his name. He made extensive use of both electrolysis and spectrum analysis to analyze chemical substances. A renowned researcher, he was also an extremely popular teacher, mentoring both Tyndall and Edward Frankland at Marburg. He received the RS's Copley Medal in 1860, and its Davy Medal, with Gustav Kirchhoff, in 1877.

Busk, George (1807–86), was a naval surgeon who retired in 1855 and studied zoology and paleontology. He was one of the founders of the Microscopical Society in 1839 and edited the *Quarterly Journal of Microscopical Science* (1853–68). He served as treasurer of the RI (1873–86), and secretary of the zoological section of the Linnean Society (1857–68). In 1864 Busk joined Tyndall, Edward Frankland, Thomas Hirst, Joseph Hooker, Thomas Huxley, and three others as members of the X Club, a monthly dining club in which they promoted science in British society.

Carlyle, Jane Welsh (1801–66), was the wife of Thomas Carlyle and a literary figure in her own right. She died suddenly on 21 April 1866, as covered by letters in this volume.

Carlyle, Thomas (1795–1881), was a Scottish essayist, novelist, and historian, and noted intellectual figure in the Victorian era. He was a major literary influence on Tyndall, who avidly read his publications. After reading *Past and Present* (1843), Tyndall commented that Carlyle "must be a true hero. My feelings towards him are those of worship 'transcendental wonder' as he defines it" (Journal, 18 July 1847, RI MS JT/2/13a/231). Tyndall met Carlyle and his wife Jane Carlyle in the 1850s and wrote to Hirst about their meetings.

During the 1860s Tyndall became a close friend of Carlyle's, supporting him in the Eyre controversy and accompanying him to Louisa Baring's home in Mentone, France, in December 1866.

Clark, James (1788–1870), was a British physician and physician-in-ordinary to Queen Victoria between 1837 and 1860.

Clausius, Adelheid (née Rimpham, d. 1875), was the wife of Rudolf Clausius. They married in Zurich in 1859 and had six children. Both she and her husband corresponded with Tyndall. They named their son Rudolf John, after Tyndall, and he became his godfather.

Clausius, Rudolf John "Johnny" (b. 1864), was the son of Adelheid and Rudolf Clausius. He was named after Tyndall, who was also his godfather. He later completed a doctorate of law in 1890 at Friedrich Wilhelms University Bonn (now University of Bonn), and became a lawyer in Cologne.

Clausius, Rudolf Julius Emanuel (1822–88), was a renowned German mathematician and physicist and one of the founders of modern thermodynamics. His famous paper on the theory of heat, "Ueber die bewegende Kraft der Warme," was published in the *Annalen der Physik* in 1850. He taught physics at the Royal Artillery and Engineering School in Berlin in 1850 before moving to a professorship in mathematical physics at the Polytechnikum in Zurich in 1855. He then moved to the University of Würzburg in 1867, and finally to the University of Bonn in 1869. During the 1850s Tyndall arranged publication of English translations (by Hirst and himself) of Clausius's works in the *Phil. Mag.* In 1858 Clausius nominated Tyndall to the Naturforschende Gesellschaft in Zurich for his work on Swiss glaciers. Clausius and his wife Adelheid named their son Rudolf John, after Tyndall, who was his godfather.

Cookson, Henry Wilkinson (1810–76), was elected vice-chancellor of Cambridge University on five occasions (1848, 1863, 1864, 1872, and 1873). A staunch Protestant, in 1870 he defended the maintenance of religious tests in the university. On 30 May 1865 Cookson and the university Council of the Senate awarded Tyndall an honorary LLD.

Coxe, Mary Anne (née Cumming, 1811–ca. 1875), was the daughter of Lesley Baillie and Robert Cumming of Logie, County Moray. Her mother was the subject of Robert Burns's "Bonnie Lesley," as noted in the novelist Charles Gibbon's dedication of *In Honour Bound*, vol. 1 (London: Richard Bentley and Son, 1874) to Lady Coxe. In 1841 she married the Scottish physician James Coxe (1811–78), and became Lady Coxe after he was knighted in 1863. Tyndall claimed to have toasted her on the summit of Mont Blanc in 1857.

de la Rive, Arthur-Auguste (1801–73), was a Swiss physicist and one of the founders of the electrochemical theory of batteries. He served as professor of natural philosophy at the Academy of Geneva and worked principally on electricity, although he also investigated the specific heats of gases and calculated the temperature of the Earth's crust. His principal book was *Traité d'électricité théorique et appliquée*, 3 vols. (Paris: J.-B. Baillière et H. Baillière, 1854–58).

Debus, Heinrich (1824–1915), was a German chemist who studied at Marburg (1845–48), where he met Tyndall, Edward Frankland, and Thomas Hirst. Debus moved to England in 1851 and taught chemistry at Queenwood College until 1867, when he began teaching at Clifton College, Bristol. Later Debus lectured at Guy's Hospital Medical School (1870–88) and became professor of chemistry at the Royal Naval College, Greenwich (1873–88).

Doyle, Abbey/Abby, was a hired caretaker for Tyndall's mother, Sarah, and sister, Emily. Tyndall was often not pleased with Doyle, but received assurances from Elizabeth Dawson Steuart that she was attentive to his mother and sister. Letters in *Tyndall Correspondence*, vol. 10, show that Tyndall's view of Doyle soured as she requested money from him.

Draper, John William (1811–82), was an English-born American chemist known for contributions to early photography, photochemistry, and intellectual history. Born at Saint Helens, near Liverpool, England, Draper was the son of a scientifically inclined Methodist minister. He went to study chemistry at the University of London, which had recently been inaugurated in 1829. Draper immigrated to America in 1832 after his father's death, and thereafter studied medicine at the University of Pennsylvania. In 1836 he was appointed professor of chemistry, physiology, and natural philosophy at Hampden-Sydney College, followed by a call to the University of the City of New York (now New York University) in 1839 as professor of chemistry and natural philosophy. There he founded the university's School of Medicine in 1840. In 1850 he became the school's president, which he held until 1873. Draper also spoke at the BAAS meeting at Oxford in 1860, which ignited the infamous exchange between Bishop Samuel Wilberforce and Thomas Huxley. In 1869 Draper also became founding president of the American Union Academy of Literature, Science and Art, and in 1876 founding president of the American

Chemical Society. In 1875 he was awarded the Rumford Medal by the American Academy of Arts and Sciences for his research on radiant heat—an honor Tyndall himself held in 1864 for his work on the absorption and radiation of heat by gases and vapors. Draper is known for his massive *A History of the Intellectual Development of Europe* (1863), and probably best known for his *History of the Conflict between Religion and Science* (1874), which many historians of science claim is an exemplar of the "conflict thesis," the idea that science and religion are fundamentally and irrevocably in conflict.

Escher, Arnold (von der Linth, 1807–72), was a Swiss geologist. A student of Horace-Bénédict Saussure in Geneva and Alexander von Humboldt in Berlin, he became professor of geology at the École Polytechnique in Zurich in 1856. He produced, in collaboration with Bernhard Studer, a detailed geological map of Switzerland in 1853. For his father's work diverting the Linth River the family was given the surname von der Linth.

Eyre, Edward John (1815–1901), was an English colonial governor, controversially known for his violent repression of the 1865 Morant Bay rebellion, while he served as governor of Jamaica (1862–65). In the years following the rebellion, London society was divided over whether Eyre had acted justly or criminally, resulting in the founding of the Jamaica Committee, which sought to prosecute Eyre, and the Eyre Defence and Aid Committee, which sought to defend Eyre's actions (which Tyndall supported).

Faraday, Michael (1791–1867), was a natural philosopher, Fullerian Professor of Chemistry at the RI, and a member of the Sandemanian Church. His extensive research focused on electricity and encompassed electrochemistry, electromagnetism, diamagnetism, and magneto-optics (the Faraday effect). Faraday also held the positions of director of the laboratory (1825–67) and superintendent of the house (1852–67) at the RI. He was Tyndall's patron and helped him obtain the position of professor of natural philosophy at the RI in 1853. Despite significant differences in their views on religion and on the nature of magnetism, they held the highest mutual regard for each other. Tyndall wrote to Hirst that "[i]t falls to the lot of few to have such a noble affectionate scientific father as your friend John possesses in Faraday. Indeed I know of no material advantage which I could set for an instant in competition with the friendship of that unspoiled and beautiful soul." Following Faraday's death in August 1867, Tyndall's *Faraday as a Discoverer* (London: Longmans, Green, 1868) was his tribute to his mentor. In 1867 Tyndall succeeded Faraday as superintendent of the house and director of the laboratory at the RI.

Fischer, Theodore, was a German publisher based in Cassel (now Kassel).

Folie, François-Jacques-Philippe (1833–1905), was a Belgian astronomer and mathematician (Lucien Godeaux, "Notice sur François-Jacques-Philippe Folie," *Annuaire de l'Académie royale des sciences, des lettres et des beaux-arts de belgique* 108 [1942]: 1–33).

Forbes, James David (1809–68), was a Scottish physicist and glaciologist who held the chair of natural philosophy at the University of Edinburgh from 1833, until he became principal of the college of Saint Salvator and Saint Leonard (the United College), at Saint Andrews University in 1860. While much of his research focused on heat, he was famous for his studies of glaciers. An avid mountaineer, Forbes published *Travels through the Alps of Savoy* (Edinburgh: A. and C. Black) in 1843. His theory that a glacier is a viscous body brought him into conflict with Tyndall and Huxley in the late 1850s.

Foucou, Félix (1831–70), was a French civil engineer, inventor, and geologist, with a particular interest in petroleum deposits (which led him, in 1870, to the United States, where he died). Foucou also wrote for and edited scientific periodicals, and was the author of *Histoire du travail: La nature et l'homme* (Paris: J. Hetzel, 1868) (Pierre Larousse, *Grand dictionnaire universel du XIXe siècle*, vol. 8 [Paris: Larousse, 1872], 652).

Francis, William (1817–1904), was a chemist, translator, and editor. He was the illegitimate son of the printer and publisher Richard Taylor. Francis learned the printing trade in Germany in the 1830s before training as a chemist with Justus von Liebig and Gustav Rose in Giessen, where he obtained his doctorate in 1842. He made many translations for his father's *Scientific Memoirs* and the *Phil. Mag.* and became the editor of the latter in 1851. In 1852 he entered a partnership with his father, establishing the Taylor & Francis publishing firm. He first met Tyndall in 1850 and became one of Tyndall's close confidants (W. H. Brock and A. J. Meadows, *The Lamp of Learning: Two Centuries of Publishing at Taylor & Francis*, 2nd ed. [London: Taylor & Francis, 1998]).

Frankland, Edward (1825–99), was an English chemist and one of Tyndall's closest friends. He worked as an assistant to the chemist Lyon Playfair in 1845 before teaching at Queenwood College in 1847, where he met Tyndall. He and Tyndall studied together at the University of Marburg in 1848. After completing his degree in 1849 Frankland worked briefly in Liebig's laboratory at Giessen and taught at the Putney College of Engineering, near London, before assuming the chair of chemistry at Owens College, Manchester. In 1851, just before moving to Manchester, he married Sophie Fick (1821–74), whom he had met in Marburg. In 1857 Frankland returned to London as a lecturer at Saint Bartholomew's Hospital, and was awarded the RS's Royal

Medal after Tyndall nominated him. In 1864, Frankland joined Tyndall, Thomas Hirst, Joseph Hooker, Thomas Huxley, and four others as members of the X Club, a monthly dining club in which they promoted science in British society.

Ginty, Margaret (née Roberts, b. ca. 1826), was the wife and widow of Tyndall's friend William Ginty. She married William in Ireland in 1846, after which they relocated to Manchester. As of the 1851 census they had three children: Kathleen, William Roberts, and Fanny; George and Fred were born later. In 1854 the family moved to Rio de Janeiro, Brazil, where William Ginty died in 1866. Margaret would return to England within a few years of her husband's death.

Ginty, William Gilbert (ca. 1820–66), was a close friend of Tyndall from Irish Ordnance Survey days and an ally in the campaign for better treatment of the civil assistants employed in the English Ordnance Survey (see Geoffrey Cantor and Gowan Dawson, eds., *The Correspondence of John Tyndall, Vol. 1, The Correspondence, May 1840–August 1843* [Pittsburgh: University of Pittsburgh Press, 2016]; and Melinda Baldwin and Janet Browne, eds., *The Correspondence of John Tyndall, Vol. 2, The Correspondence, September 1843–December 1849* [Pittsburgh: University of Pittsburgh Press, 2016]). Ginty became a prominent Manchester businessman, and was involved in various civil engineering projects, including the construction of the Liverpool–Bolton railway. He served briefly as honorary secretary of the Manchester Athenaeum (elected late 1853) before moving to Rio de Janeiro in April 1854 to manage the Imperial Gas Works Company established by Baron de Maua. He was awarded a Brazilian knighthood—Knight of the Imperial Order of the Rose—in 1857 for his contributions to major construction projects. Ginty enjoyed a lifelong correspondence with Tyndall, and died suddenly from unspecified causes in mid-1866.

Giraud, Étienne (fl. 1850–60s), was a French publisher and bookseller whose firm was located at 20 rue Saint Sulpice in Paris. He published Abbé Moigno's journal *Les mondes* (1863), as well as the first French edition of Tyndall's *Heat* (1863) which appeared as *La chaleur* (1864).

Glaisher, James (1809–1903), was an English astronomer and meteorologist, who worked at the Royal Observatory Greenwich (1835–74), managing the meteorological department there from 1841 until his retirement. Glaisher was perhaps best known for his pioneering balloon ascents during the 1860s with Henry Tracey Coxwell (1819–1900). The observations they made during these ascents were published in *Brit. Assoc. Rep. 1863–66.*

Gordon, George William (1820–65), was a wealthy mixed-race Jamaican businessman and politician, who was arrested, tried, and executed, at the direction of Edward Eyre, on suspicion of his having planned the 1865 Morant Bay rebellion. He was an advocate for the suffering black poor and a critic of colonial government. Gordon maintained an extensive correspondence with English evangelical critics of colonial policy. He also established the Native Baptist Church, where Paul Bogle (1820–65) served as deacon and activist.

Griffith, George (ca. 1834–1902), was an English science educator. He taught at Oxford from 1857 to 1867, but left after the professorship of experimental philosophy was given to Robert Clifton in 1865. He was afterward the first natural science master at Harrow (1867–93). Griffith also took an active role in the BAAS, serving as its assistant general secretary for a total of thirty years (1862–78, 1881, and 1890–1903) (B. P. L., "George Griffith," *Nature* 66 [1902]: 64).

Grove, William Robert (1811–96), was a Welsh judge and electrochemist. He was called to the bar in 1835 but also took an interest in science. He held a professorship of experimental philosophy at the London Institution (1841–47) and developed a gas battery (the first fuel cell), for which he was awarded the RS's Royal Medal in 1847. In the late 1840s financial constraints turned his attention back to practicing law. He became a queen's counsel in 1853, was appointed a judge of the court of common pleas in 1871, and then the queen's bench in 1880. He remained a highly respected member of the scientific community and was elected president of the BAAS in 1866.

Helmholtz, Anna (née Mohl, 1834–99), was a Berlin salonnière and translator of scientific works. She married Hermann Helmholtz in 1861. With Clara Wiedemann, she translated Tyndall's *Heat* (1863) and *Sound* (1867), which appeared as *Die Wärme, betrachtet als eine Art der Bewegung* (Braunschweig: Friedrich Vieweg, 1867) and *Der Schall: acht Vorlesungen gehalten in der Royal Institution von Grossbritannien* (Braunschweig: Friedrich Vieweg, 1869). While Anna Helmholtz and Clara Wiedemann translated the works, only their husbands were credited.

Helmholtz, Hermann Ludwig Ferdinand von (1821–94), was a German physicist and physician. Educated in military medicine at the Friedrich-Wilhelms Institute in Berlin, he became the first to hold a regular post as teacher of anatomy at the Berlin Academy of Arts in 1848. His interest in physics was already developed, as in 1847, Helmholtz published his paper "Über die Erhaltung der Kraft," which discussed the mathematical principles of the conservation of energy. This work was based on his study of muscle metabolism, and he argued that there was no such thing as a vital "force." Rather, factors such as mechanics, heat, and electricity were all manifestations of a single "force"; no energy was expended during muscle movement. He became professor of physiology at the University of Königsberg, and became professor for anatomy and physiology at the University of Bonn in 1855. It was at Bonn that he increasingly turned his attention to physics, which continued during his time at Heidelberg from 1858 onward. Called "an intellectual giant" by James Clerk Maxwell, Helmholtz was involved in hydrodynamics and electrodynamics.

Herschel, John Frederick William (1792–1871), was an English mathematician and astronomer. He was the son of the famous astronomer William Herschel (1738–1822) and himself became one of the most eminent scientists in Victorian Britain. He graduated from Cambridge University in 1813 as senior wrangler. He was awarded the RS's Copley Medal in 1821, and a second Copley Medal in 1847. His *Preliminary Discourse on the Study of Natural Philosophy* (London: Longmans, Rees, Orme, Brown & Green, 1830) was widely read and greatly admired. He was knighted in 1831, followed by being made a baronet in 1838, on his return from mapping the southern skies in Cape Town, South Africa. In 1849 he published *Outlines of Astronomy* (London: Longman, Brown, Green, and Longmans), a popular presentation of astronomy that went through eleven editions by 1871. He made many contributions to the science of photography and also edited the admiralty's *Manual of Scientific Enquiry* (London: John Murray, 1849), a science field guide for the use of the Royal Navy. From 1850 to 1855 he served as master of the mint, and upon his death was buried next to Isaac Newton in Westminster Abbey.

Hetzel, Pierre-Jules (1814–86), was a French editor and publisher known for publishing the works of Jules Verne. He also published Michael Faraday's *Histoire d'une chandelle*, translated into French by W. Hughes (Paris, J. Hetzel, 1865), and would publish Félix Foucou's *Histoire du travail: La nature et l'homme* in 1868. His firm was located in Paris at 18 Jacob Street.

Hirst, Thomas Archer (1830–92), was an English mathematician, educator, and Tyndall's closest friend and correspondent. His friendship with Tyndall began in Halifax, Yorkshire, in 1845, when they were both working for the surveyor Richard Carter. Encouraged by Tyndall—who acted as a sort of mentor to him—Hirst studied at the Universities of Marburg, Göttingen, and Berlin between 1849 and 1852. In 1853 he returned to England with a PhD and replaced Tyndall as mathematics teacher at Queenwood College. Like Tyndall, he did not get on with the headmaster, George Edmondson. He resigned in June 1856 to look after his sick wife, Anna, who died in Paris in 1857. During the 1860s Hirst established himself as an active and respected member of the London scientific community. He taught mathematics at University College London (1860–64), and subsequently held professorships of physics and mathematics (1865, 1867–70). He joined the RS Council (1864); was a founder of the X Club, a monthly dining club in which nine members (Tyndall included) promoted science in British society; and served as general secretary of the BAAS (1866–70).

Hofmann, August Wilhelm (1818–92), was a German chemist whom Prince Albert called on and promoted as extraordinarius (associate professor) and director of the Royal College of Chemistry in 1845, which in 1853 became part of the Government School of Mines.

Holland, Henry (1788–1873), first baronet, was an English physician and travel writer educated at the University of Edinburgh. He began a practice in London in 1816 serving members of London's wealthy and fashionable society. He became physician-extraordinary to Queen Victoria (1837), and physician-in-ordinary to Prince Albert (1840). From 1865 to 1873 he served as president of the RI.

Hooker, Joseph Dalton (1817–1911), was a British botanist who specialized in taxonomy and geographical distribution of plants, and ultimately became the most famous British botanist of the Victorian era. He was the younger son of William Jackson Hooker (1785–1865), the director of the Royal Botanic Gardens, Kew. After studying medicine at the University of Glasgow he became a naval surgeon and spent the years 1839–43 on James Clark Ross's geomagnetic survey of the southern oceans, where he collected plants in addition to serving as surgeon. Subsequently he explored and collected plants in India (1847–51), achieving considerable mountaineering success, before joining his father as deputy director of Kew Gardens in 1855. Hooker was an intimate friend of Charles Darwin and Thomas Huxley. He was first introduced to Tyndall by Huxley in November 1855 and the three became close friends. Together with Huxley, in August 1856 Hooker helped Tyndall with his first studies of alpine glaciers. With Charles Lyell, Hooker was instrumental in bringing Darwin's theory of natural selection to public attention. In 1864 Hooker joined Huxley, Tyndall, and others in founding the X Club, a monthly dining club in which they promoted science in British society. He succeeded his father as director of Kew in 1865. He served as president of the BAAS (1868), and was knighted several times and received many scientific awards, including the RS's Royal Medal (1854) and Copley Medal (1887).

Huxley, Henrietta Anne (née Heathorn) (1825–1915), was born in the West Indies to Henry Heathorn and Sarah Henrietta Richardson (née Harris), and attended school in Neuwied, Germany, for two years. In 1843 she, her mother, and her half-sister moved to Australia to join her father. She was keeping house for her sister Oriana and brother-in-law William Fanning in Sydney, Australia, when she met Thomas Huxley in 1847, who was then a surgeon and zoologist on HMS *Rattlesnake*. After an eight-year engagement, she arrived in London in 1855 and married Huxley on 21 July. She assisted her husband in his work by translating German and drawing diagrams for his lectures. She wrote poetry and at the age of eighty-six published a volume of her collected poems, *Poems of Henrietta A. Huxley with Three of Thomas Henry Huxley* (London: Duckworth, 1913) ("Death of Mrs. Huxley," *Times*, 6 April 1915, 12.)

Huxley, Thomas Henry (1825–95), was a British zoologist, educator, science popularizer, and one of Tyndall's closest friends. Following medical training at Charing Cross Hospital in London, he was a ship's surgeon aboard the HMS *Rattlesnake* expedition to Australia (1846–50) under Captain Owen Stanley. After eking a living by writing for Chapman's *Westminster Review*, in 1854 he became professor of natural history at the Royal School of Mines, London. He held this position until 1872, and continued with the School of Mines when it then moved to South Kensington until 1885. While in Australia he met Henrietta Heathorn, whom he married in July 1855 after an eight-year engagement. He joined Tyndall at the RI as Fullerian Professor of Natural History (1855–58, 1866–69). In the summer of 1856, and again in 1857, he and Tyndall studied the movement of glaciers and climbed in the Alps together (they published their research in January 1857 as "On the Structure and Motion of Glaciers," *Phil. Trans.* 147 [1857]: 327–46). In 1864 Huxley joined Tyndall, Edward Frankland, Thomas Hirst, Joseph Hooker, and four others as members of the X Club, a monthly dining club in which they promoted science in British society.

Jones, Henry Bence (1813–73), was a wealthy English physician and medical chemist. He was educated at Harrow School and Trinity College, Cambridge, before studying medicine at Saint George's Hospital, London. Attracted by the potential applications of chemistry to clinical medicine, he studied organic chemistry with Justus Liebig at Giessen (1841–43) before assuming a career as a successful physician in private practice and at Saint George's Hospital. In May 1842 he married a cousin, Lady Millicent Acheson, daughter of the 2nd Earl of Gosford, with whom he had seven children. He published important research on protein chemistry and made an abstract in English of Emil du Bois-Reymond's work on electrophysiology (1852) that annoyed

Carlo Matteucci, the Italian pioneer of electrophysiology. In 1853 he became a manager (trustee) of the RI and was elected secretary in 1860 (a position he served in until 1873). He frequently entertained Tyndall at both his London home and Folkestone, where he kept a summer house.

Kirchhoff, Gustav Robert (1824–87), was a German physicist. In Königsberg, Kirchhoff studied physics under Franz Neumann and mathematics under Friedrich Julius Richelot. His formulation of the laws of electrical circuits was part of his doctoral dissertation, which he finished at Königsberg in 1847. During an associate professorship at the University of Breslau in 1850, he became acquainted with Robert Bunsen, who made sure that Kirchhoff followed his move to Heidelberg in 1854. Together with Bunsen, he worked on thermal radiation and spectrum analysis. Regarding his law of thermal radiation, he would fight a long struggle for recognition of priority with the Scottish physicist Balfour Stewart (1828–87). Kirchhoff and Tyndall were close friends and colleagues. Tyndall had recommended him for the RS's Rumford Medal, which he received in 1862; Kirchhoff was unable to attend the ceremony and Tyndall accepted it in his place.

Liebig, Justus (1803–73), was a noted German chemist and educator. He opened the field of organic chemistry and made substantial contributions to agricultural chemistry, plant and animal physiology, and the chemistry of foodstuffs, in addition to introducing the laboratory-based teaching method.

Lubbock, John (1834–1913), was an English banker, politician, and evolutionary anthropologist. An early supporter of Darwin's *On the Origin of Species* when it was published in 1859, Lubbock became associated with Thomas Huxley, Joseph Hooker, and other friends and supporters of Darwin. He was also one of the founders of prehistoric archaeology—in 1865 he published the highly influential *Pre-historic Times: As Illustrated by Ancient Remains*, in which he coined the terms *paleolithic* and *neolithic*—and an author of books on insects for the popular reader. In 1856 he married Ellen Frances Horden, and the couple were among Tyndall's closest friends. He was the youngest member of the X Club and joined Tyndall on his climbing adventures in the Alps on a number of occasions. When his father died in 1865, he became Sir John Lubbock, 4th baronet. Starting in 1870, he served numerous times in parliament and was made 1st Baron Avebury in 1900.

Lyell, Charles (1797–1875), was a Scottish geologist and lawyer, whose three-volume *Principles of Geology* (London: John Murray, 1830–33) advocated a "uniformitarian" view of the processes of geological change. Educated at Exeter College, Oxford, he studied law at Lincoln's Inn, London. He was briefly professor of geology at King's College, London (1831–33), after which he lived on private means. He married Mary Elizabeth Horner in July 1832; she assisted her husband in his scientific work. A prominent figure in the Geological Society of London, Lyell was twice elected as its president (1835–37; 1849–51). He was knighted in 1848. Lyell and Joseph Hooker were instrumental in bringing Darwin's theory of natural selection to public attention. The discovery in 1859 that humans had lived alongside the mammoths of the last ice age led to Lyell's last major publication, *Geological Evidences of the Antiquity of Man* (London: John Murray, 1863).

Magnus, Heinrich Gustav (1802–70), was a German chemist and physicist. A pupil of the German chemist Eilhard Mitscherlich in Berlin and of J. J. Berzelius in Stockholm, he was professor of technology and physics at the University of Berlin from 1833 until his death. In Berlin, he developed a superbly equipped physics laboratory at his home, to which he invited young German physicists to work, as well as Tyndall in 1851. Although his early research was primarily in chemistry, by the 1850s he had begun to investigate a range of physical problems, as his letters to Tyndall demonstrate.

Maury, Matthew Fontaine (1806–73), was an American naval officer, astronomer, oceanographer, and meteorologist. His study of the sea began while he was a midshipman on board the USS *Brandywine*; however, a leg injury ceased a career at sea. Maury turned his attention to laborious study of navigation, meteorology, winds, and currents. In 1842 he became naval superintendent of the Depot of Charts and Instruments, which was later renamed the United States Naval Observatory and Hydrographical Office, where he instituted a system to organize the logbooks and records in which shipmasters wrote down their observations of sea conditions. Using data from old logbooks, Maury received international recognition for his *Wind and Current Chart of the North Atlantic* (1847), and further published *Sailing Directions* (1851), *Physical Geography of the Sea* (1855), and *Physical Geography of the Seas and Its Meteorology* (1861). When the American Civil War began in 1861, the Virginia-born Maury left the US Navy—and thus his position at the Naval Observatory and Hydrographical Office—and joined the Confederacy as Chief of Sea Coast, River and Harbor Defenses. He later helped to found the Virginia Agricultural and Mechanical College (now Virginia Tech) and in 1868 he became professor of meteorology at the Virginia Military Institute until his death. On Maury's legacy with regard to slavery, see P. K. Hardy and H. M. Rozwadowski, "Maury for Modern Times: Navigating a Racist Legacy in Ocean Science," *Oceanography* 33, no. 3 (2020): 10–15.

Mayer, Julius Robert von (1814–78), was a German physicist, chemist, and physician. In 1842 he published a short text that contained one of the first statements on the conservation of energy and the mechanical equivalent of heat. Mayer was a scientific outsider; his arguments in this paper were based on philosophical principles and thought experiments, and his papers on heat and the conservation of force received little recognition in the 1840s. In the late 1850s and early 1860s, Tyndall sought to revive Mayer's work and promote him as codiscoverer of energy conservation, along with James Joule. The ensuing priority dispute between Mayer and Joule, which was largely fought out between their proxies—Tyndall on the one hand, and William Thomson and Peter Guthrie Tait, on the other—was carried on in increasingly acrimonious terms both in private correspondence and then in the public arena.

Milnes, Richard Monckton (1809–85), 1st Baron Houghton, was a traveler, poet, author, and politician. He was a member of the famous Apostles Club, an intense group of undergraduates at Trinity College, Cambridge, who attempted to make sense of the religious changes that were taking place in mid-Victorian Britain in a spirit of freedom and openness. Members—which then included Alfred Tennyson, Arthur Hallam, Richard Chenevix Trench, Joseph Williams Blakesley, and many others—worried over natural science, natural theology, and Samuel Taylor Coleridge's distrust of natural theology.

Moigno, François-Napoléon-Marie (1804–84), was a French Catholic priest, physicist, writer, and editor. Educated at a Jesuit seminary, he studied science alongside theology and was given the mathematical chair at Jesuit House in Paris in 1836. After leaving the order in 1843 he became the science editor of *L'epoque*, and later became editor of the scientific journal *Cosmos* (1852–62). He founded the journal *Les mondes* in 1863 and undertook the translation and publication of many of Tyndall's works in French, including *Radiation* (1867), *Le son* (1869), *La chaleur* (1864), and *Faraday inventeur* (1868) ("L'abbé Moigno," *Nature*, 24 July 1884, 291–92).

Murchison, Roderick Impey (1792–1871), was a Scottish geologist and geographer who published over 180 scientific papers as well as several books. He was elected fellow of the Geological Society in 1825, and was twice its president. He was one of the founders of Royal Geographical Society of London (1830) and from 1843 served as its president for sixteen years. He used his role in the RGS in particular to promote imperial exploration. He was also one of the founders of the BAAS (1831) and was president of the geographical and geological sections ten times in total.

Neumayer, Georg Balthasar von (1826–1909), was a German meteorologist, magnetician, hydrographer, and oceanographer who studied at Munich University, earning a PhD in 1849, and was for a time chair in physics at Hamburg. Having a great interest in terrestrial magnetism, Neumayer worked on a magnetic survey of Bavaria, carried out research at the Hobart Magnetic Observatory in Van Diemen's Land (now Tasmania), and then established the Flagstaff Observatory in Melbourne with support from the RS, BAAS, and Alexander Humboldt (R. A. Swan, "Neumayer, Georg Balthasar von (1826–1909)," in *Australian Dictionary of Biography*, ed. G. Serle and R. B. Ward, vol. 5 [Melbourne: Melbourne University Press, 1974], 55–64).

Plateau, Joseph Antoine Ferdinand (1801–83), was a Belgian physicist. Born in Brussels, Plateau studied at the University of Liège, where he graduated in 1829 with a doctorate in mathematics and physics. In 1832 Plateau invented a device that could give the illusion of a moving image, which he called the phenakistoscope. In 1835 he was appointed professor of experimental physics at the University of Ghent. Plateau is best known for his experimental work in which he investigated what is now known as the Plateau problem. From 1840 he conducted experiments to demonstrate that a minimal surface could be formed from a soap film, made of a solution of soapy water and glycerine, which was stretched on a wire frame.

Playfair, Lyon (1818–98), was a Scottish chemist who worked in Justus Liebig's laboratory in Giessen (1839–41). In 1842 he became professor of chemistry at the Royal Manchester Institution. In 1845 he became chemist to the Geological Survey of Great Britain and professor of chemistry at the Royal School of Mines, London, and in 1848 was elected a fellow of the RS. In 1853 Playfair was appointed secretary for science in the Department of Science and Art. In this position, he attempted to establish a national system of science education. Playfair maintained a close relationship with the royal family and especially Albert, the prince consort, and for a time educated the queen's sons Albert and Alfred. In 1858 he became professor of chemistry at Edinburgh University.

Pollock, Frederick (1845–1937), was a 3rd baronet, jurist, and the eldest son of Juliet and William Frederick Pollock. He attended Eton College (1858–63) before entering Trinity College, Cambridge, where he obtained his BA (1867) and MA (1870). He was admitted to Lincoln's Inn in 1868, was called to the bar in 1871, and married Georgina Harriet Deffell in 1873. He was a distinguished jurist, author of many legal treatises, and became professor of jurisprudence at Oxford University (1883–1903). Pollock was an avid alpine mountaineer.

Pollock, Juliet (née Creed, 1819–99), was a daughter of Catherine Herries and the Reverend Henry Creed, vicar of Corse, Gloucestershire. She married William Frederick Pollock, the barrister and author, in 1844, with whom she had three sons. She and her husband were loyal supporters of the RI, attended Tyndall's lectures, and frequently invited Tyndall to dinner, and were among Tyndall's closest friends. Juliet Pollock was the author of several books, including *Julian and his Playfellows* (London: Grant and Griffith, 1852), *New Friends* (London: John W. Parker and Son, 1858), and *Macready as I Knew Him* (London: Remington, 1885).

Pollock, Maurice Emilius (1857–1932), was the third son of Juliet and William Frederick Pollock. He became a sculptor and married Lydia Helen Roberts (1853–84) in 1880, and Mabel Mary McPherson (1860–1937) in 1889. "Moss" was his family's nickname for him (*Remembrances*, 2:75).

Pollock, Walter Herries (1850–1926), the second son of Juliet and William Frederick Pollock. He was educated at Eton College (1862–65) and was admitted to Trinity College, Cambridge, where he obtained a BA (1871) and MA (1875). He was admitted to the inner temple in 1870 and called to the bar in 1873. Pollock became a poet and journalist, and was editor of the *Saturday Review* (1884–94). He married Emma Jane Pipon in 1876.

Pollock, William Frederick (1815–88), was a 2nd baronet, lawyer, and author. The eldest son of Jonathan Frederick Pollock, 1st baronet (1783–1870), and Frances Rivers (d. 1827), he married Juliet Creed in 1844. He was admitted to Trinity College, Cambridge, in 1832, obtaining his BA (1836) and MA (1840). Admitted to the inner temple in 1833 and called to the bar in 1838, he then became a master of the court of exchequer in 1846 and was appointed queen's remembrancer in 1874. Pollock's two-volume *Remembrances* illustrates how the family circulated among London's elite and entertained at their Montagu Square home. Both William and Juliet were loyal supporters of the RI, and Tyndall was frequently among their guests.

Regnault, Henri Victor (1810–78), was a French chemist, physicist, and photographer. A pupil of Justus Liebig's at the University of Giessen, he was professor of physics at the Collège de France (1841–54), and director of the famous porcelain factory at Sèvres (1854–70).

Roche, Annette (1770–1853), was an active supporter of literary, philanthropic, and mercantile movements in Cork, Ireland. She was a good friend of Tyndall's and cared for his mother and sister by managing money, arranging accommodations, and choosing suitable clothing. In 1864 she had arranged for the erection of a tombstone over the grave of Tyndall's father in Ireland.

Sabine, Edward (1788–1883), was an Anglo-Irish astronomer, geophysicist, and military officer who led the British effort to establish magnetic observatories all over the globe. A graduate of the Royal Military Academy at Greenwich, he reached the rank of general in the 1850s. He accompanied J. C. Ross's Arctic expedition (1818) and W. E. Parry's expedition (1819). He received the RS's Copley Medal in 1821 for magnetic observations in the Arctic on those expeditions. Sabine was closely connected with the administration of the RS and served as its treasurer (1850–61) and president (1861–71).

Soret, Clémentine (née Odier, 1831–99), was the wife of Jacques-Louis Soret. They were married in 1853, and had a son, Charles Soret (1854–1904), who would himself become a Swiss physicist and chemist best known for his work on thermodiffusion also called the Soret effect.

Soret, Jacques-Louis (1827–90), was a Swiss chemist and professor of medicine at the University of Geneva. He is best known for determining the chemical composition of ozone.

Spencer, Herbert (1820–1903), was an English philosopher, social theorist, sociologist, biologist, and anthropologist. He was baptized in the Methodist tradition, despite misgivings of his father, who rejected institutional religion in favor of "rational" religion. Unsurprisingly, Spencer himself at an early age rejected all possibility of belief in an immanent or personal God, and later became a self-professed agnostic. Through family connections, at the age of twenty-eight he eventually secured a post as subeditor at the *Economist*, which put him at the heart of literary London. He published his first book, *Social Statics*, in 1850 and his second book, *Principles of Psychology*, in 1855. Spencer was a respected philosopher, but one of his most lasting contributions was his *Principles of Biology* (1864), in which he coined the phrase "survival of the fittest." He made evolution his guiding principle in his multivolume synthesis of all knowledge. In spite of his agnosticism, however, Spencer affirmed the existence of the "Unknowable," a shadowy deity that was behind the evolutionary process and, he believed, would ultimately reconcile the growing conflict between science and religion. Along with Tyndall and seven others, Spencer was a member of the X Club, a monthly dining club in which they promoted science in British society. Spencer was the only member who was not a fellow of the RS. In 1865 he joined the Jamaica Committee, which advocated the prosecution of Governor Edward Eyre for his violent repression of the Morant Bay rebellion.

Spottiswoode, William (1825–1883), was an English mathematician and physicist. He attended Balliol College, Oxford, where he graduated first class in mathematics. Although he became a partner in his father's printing firm of Eyre and Spottiswoode of New Street Square, he always reserved time for his scientific pursuits—he even gave scientific lectures in a school he set up for his employees. In 1861 he married Eliza Taylor Arbuthnot, eldest daughter of William Urguhart Arbuthnot, member of the council for India. Parties at the Spottiswoode home were considered the height of the London season, attracting men of science, politics, and society. In 1864 he was one of the founding members of the X Club, which met regularly at the Saint George's Hotel in Albemarle Street near Piccadilly. His financial acumen was put to good use in the cause of science. He was treasurer of the BAAS (1861–74), the RI (1865–73), and the RS (1871–78), and served as president of the BAAS (1878) and the RS (1878–83).

Steuart/Stewart, Elizabeth Dawson (née Duckett, 1802–93), was born into a very wealthy family with an estate of twelve thousand acres and a gothic mansion (Duckett's Grove, now a ruin) situated several miles northeast of Carlow. In 1820 she married William Richard Steuart (1792–1852), and they lived at Steuart's Lodge, Leighlinbridge. She took a keen interest in Tyndall's career and was probably his most consistent and long-lived patron. Steuart frequently helped Tyndall arrange financial support for his relatives and acquaintances in Ireland, and provided updates on the hired help for his mother, Sarah, and sister Emily.

Stokes, George Gabriel (1819–1903), was an English mathematician and physicist. He graduated from Cambridge University as senior wrangler and Smith's Prizeman in 1841 and served as Lucasian Professor of Mathematics at Cambridge University (1849–1903). He augmented his income as a lecturer at the Royal School of Mines (1849–56). In the 1850s Stokes turned his attention to optics, and in 1852 published on fluorescence, a term he coined, showing that the phenomenon was the result of changes in the wavelength of light. For this work he was awarded the RS's Rumford Medal (1852), and was later awarded the Copley Medal (1893) for his contributions to science. As secretary of the RS (1854–85) he was responsible for refereeing papers for the *Phil. Trans.*, a task for which he frequently sought Tyndall's assistance.

Strutt, Edward (1801–80), 1st Baron Belper, was a British liberal politician. He was raised to the peerage as Baron Belper, of Belper, in the County of Derby, in 1856.

Tennyson, Alfred (1809–92), 1st Baron Tennyson from 1883, was poet laureate from 1850 in succession to William Wordsworth. Born in Lincolnshire, he was educated at Trinity College, Cambridge. In 1850 he published *In Memoriam* (London: Edward Moxon). In 1853 he moved to Farringford, on the Isle of Wight, where he published *Maud* (London: Edward Moxon, 1855). Tyndall greatly admired Tennyson's poetry and met him in 1858. He subsequently developed a close friendship with the Tennysons and stayed with them at their home in Farringford. Tennyson declined honors or titles from Queen Victoria four times between 1865 and 1880, but accepted a barony in 1883.

Thomson, James (1822–92), was a mechanical engineer and older brother of William Thomson, Lord Kelvin. As a teenager, he attended Glasgow University, receiving his BA (1839) and his MA (1840) in mathematics and natural philosophy. He worked as an engineer at firms in Glasgow, London, and Belfast before he became professor of engineering at Queen's University Belfast in 1854. Thomson applied his studies of pressure in steam engines to James David Forbes's theory of glacier motion. In 1857 he argued that liquefaction by pressure accounted for the plasticity of ice observed in glaciers, bringing him into disagreement with Tyndall's theory of regelation (James Thomson, "On the Plasticity of Ice, as Manifested in Glaciers," *Roy. Soc. Proc.* 8 [1857]: 455–58; *Glaciers of the Alps*, 408).

Thomson, William (1824–1907), later Lord Kelvin, was a distinguished Irish-Scottish mathematician, physicist, and inventor, and younger brother of James Thomson. He was educated at the University of Glasgow and Peterhouse College, Cambridge University, where he was second wrangler and Smith's Prizeman in 1841. He was professor of natural philosophy in Glasgow from 1846, paying frequent visits to London. He first met Tyndall at the Edinburgh meeting of the BAAS in 1850, and from the start the two physicists had a strained relationship. For much of his career Tyndall regarded Thomson as a scientific rival.

Tyndall, Emily (Emma) (1817–1904), was John Tyndall's older sister, whose education he repeatedly encouraged. She does not appear to have married and later in life suffered from a mental illness, which manifested as an extreme form of religiosity.

Tyndall, Sarah (née Macassey [Macasey/McAssey], 1793–1867), was Tyndall's mother. She was descended from a prosperous farming family with lands in Ballybrommell, County Carlow, Ireland, that included in earlier generations Quakers by the name Malone. In 1813 she married John Tyndall senior

(1792–1847). After her husband's death in March 1847, she remained in the family cottage at Leighlinbridge, Ireland, supported financially by Tyndall. Tyndall later told Thomas Hirst of his deep affection for her rare qualities.

Vieweg, Hans Heinrich Eduard (1797–1869), was a German publisher and owner of the Vieweg publishing house. As the son of Johann Friedrich Vieweg (1761–1835), the publisher and founder in 1786 of Vieweg Verlag in Berlin, Eduard modernized his father's printing house by obtaining an iron press from Paris in 1822 and a Columbian press from England the following year. When he took over, Eduard primarily published scientific works from leading scholars in chemistry, physics, physiology, natural history, mathematics, and technology. In addition to such scientific works, Eduard was also a newspaper publisher, including the moderately liberal *Deutsche Nationalzeitung aus Braunschweig und Hannover* and *Zeitung für das deutsche Volk*. He suffered a stroke in 1866 and remained bedridden until his death in 1869. The publishing house passed to his only son, Hans Heinrich Rudolf Vieweg (1826–90).

Wiedemann, Clara Laura (née Mitscherlich, 1827–1914), the daughter of the chemist Eilhard Mitscherlich (1794–1863), was a German translator who married Gustav Wiedemann in 1851. With Anna Helmholtz, she translated Tyndall's *Heat* (1863) and *Sound* (1867), which appeared as *Die Wärme, betrachtet als eine Art der Bewegung* (Braunschweig: Friedrich Vieweg, 1867) and *Der Schall: acht Vorlesungen gehalten in der Royal Institution von Grossbritannien* (Braunschweig: Friedrich Vieweg, 1869). While Clara Wiedemann and Anna Helmholtz translated the work, only their husbands were credited (*HLS*; B. Meijer et. al., eds., *Nordisk familjebok*, 2nd ed. [Stockholm: Nordisk familjeboks förlags, 1904–26]).

Wiedemann, Gustav Heinrich (1826–1899), was a German physicist and editor. He was trained at the University of Berlin by Heinrich Gustav Magnus where he came to know Hermann Helmholtz, his dissertation advisor on organic chemistry. He held professorships of physics at Basel (1854–63), Brunswick (1854–66), and Karlsruhe (1866–71) before taking the chair of physical chemistry at Leipzig in 1871. After the death of Johann Poggendorff in 1877, Wiedemann took over as editor of the *Annalen der Physik und Chemie*, which subsequently took on the name *Wiedemann's Annalen*, and his wife, Clara Wiedemann, assisted by translating the scientific papers.

Wilson, James Maurice (1836–1931), was an Anglican clergyman, schoolteacher, and astronomer, who, along with Tyndall and Thomas Huxley, served on a committee of the BAAS to study the question of promoting science education in England. In addition to being a vocal advocate of science education in Britain, Wilson was also the mathematics and science master at Rugby (1859–79) and headmaster of Clifton College (1879–90).

Wynne, George (1804–90), was an Irish surveyor, engineer, and military officer. Born in Dublin, he enlisted in the Royal Engineers (ca. 1825) and was commissioned as lieutenant in 1826. Wynne met Tyndall while employed in the Irish Ordnance Survey (1835–40). When he and Tyndall met, Wynne was in charge of the division in which Tyndall worked. Wynne later served with distinction in Greece, China, and the West Indies, rising to the rank of general by the time he retired in 1877. Over the period of this volume he had the rank of captain, and the position of inspector of railways (1847–58) under the board of trade, before being transferred to China. He married Augusta Osborne in 1834. Tyndall and Wynne had a high regard for each other ("The Late General Wynne," *Times*, 12 July 1890, 11).

Youmans, Edward Livingston (1821–87), was an American scientific popularizer, editor, and lecturer. Born in Coeymans, New York, he was raised in a home where books and ideas were highly valued. An attack of ophthalmia in 1835 damaged his eyesight, which prompted him to move to New York City for treatment in 1839. There he formed a lasting professional partnership with leading American publisher William Henry Appleton (1814–99), son of Daniel Appleton (1785–1849). In New York Youmans also studied chemistry under John William Draper and completed a medical degree through the University of Vermont. Youmans produced a stream of popular science books, including *A Class-book of Chemistry* (1851), *Alcohol and the Constitution of Man* (1853), and *The Handbook of Household Science* (1857). He possessed a genuine talent for rendering vast amounts of scientific material into clear, understandable prose. In *The Correlation and Conservation of Forces* (1864) and *Culture Demanded by Modern Life* (1867), Youmans insisted that science was not materialistic but led to an ever higher plane of knowledge and experience. To this end, Youmans founded the International Scientific Series in 1871 and the *Popular Science Monthly* in 1872, which he "conducted" until his death, when his brother William Jay Youmans (1838–1901) took it over. Youmans also served as vice-president of the Free Religious Association, which consisted of a diverse assortment of Unitarians, Universalists, spirituals, transcendentalists, scientific theists, and other disaffected religious minorities, and advanced the principles of free thought and moral philosophy without any reference to institutional Christianity.

INDEX

The index refers to the letters using their serial numbers. The letter 'n' denotes a footnote reference and appears if an item is not otherwise mentioned in the letter to which the footnote pertains. A number in bold signifies a letter to or from a correspondent.